D1207800

Metabolic Pathways

Third Edition

VOLUME III

Amino Acids and Tetrapyrroles

CONTRIBUTORS TO THIS VOLUME

Bruce F. Burnham
L. A. Fahien
David M. Greenberg
Ernest Kun
Victor W. Rodwell
H. J. Sallach

Metabolic Pathways

THIRD EDITION

EDITED BY

David M. Greenberg

University of California
San Francisco Medical Center
San Francisco, California

QP 514
.C483
V, 3

VOLUME III

Amino Acids and Tetrapyrroles

ACADEMIC PRESS   New York and London   1969

INDIANA
UNIVERSITY
LIBRARY

NORTHWEST

COPYRIGHT © 1969, BY ACADEMIC PRESS, INC.
ALL RIGHTS RESERVED.
NO PART OF THIS BOOK MAY BE REPRODUCED IN ANY FORM,
BY PHOTOSTAT, MICROFILM, OR ANY OTHER MEANS, WITHOUT
WRITTEN PERMISSION FROM THE PUBLISHERS.

ACADEMIC PRESS, INC.
111 Fifth Avenue, New York, New York 10003

United Kingdom Edition published by
ACADEMIC PRESS, INC. (LONDON) LTD.
Berkeley Square House, London W.1

LIBRARY OF CONGRESS CATALOG CARD NUMBER: 67-23160

Second Printing, 1972

PRINTED IN THE UNITED STATES OF AMERICA

List of Contributors

Numbers in parentheses indicate the pages on which the authors' contributions begin.

BRUCE F. BURNHAM,* Departments of Medicine and Biochemistry, College of Medicine, University of Minnesota, Minneapolis, Minnesota (403)

L. A. FAHIEN, Department of Pharmacology, University of Wisconsin Medical School, Madison, Wisconsin (1)

DAVID M. GREENBERG, University of California, San Francisco Medical Center, San Francisco, California (95, 237)

ERNEST KUN, Departments of Pharmacology and Biochemistry, and Cardiovascular Research Institute, University of California School of Medicine, San Francisco, California (375)

VICTOR W. RODWELL, Department of Biochemistry, Purdue University, Lafayette, Indiana (191, 317)

H. J. SALLACH, Department of Physiological Chemistry, University of Wisconsin Medical School, Madison, Wisconsin (1)

---

* Present Address: Department of Chemistry, Utah State University, Logan, Utah

Preface

A noble structure of understanding of the chemical processes of living organisms has been erected by the discoveries of biochemists in the comparatively short period of a little over thirty years. The tremendous modern development of biochemistry started with the announcement of the citric acid cycle by H. A. Krebs and W. A. Johnson in 1937. This was followed shortly by the clarification of the unique function of ATP as the universal acceptor and donor of energy in biochemical reactions. In the ensuing years, very rapid progress was made in identifying the individual reaction steps in the formation and dissimilation of the numerous organic constituents of living cells and in isolating, purifying, and characterizing the enzymes that catalyze the many metabolic reaction sequences.

In this same period, biochemists and geneticists teamed up to provide an explanation of how the organism directs the synthesis of the many required enzyme proteins by the discoveries of how the DNA of the genetic material is synthesized and replicated. The genetic code of triplet bases of purines and pyrimidines which specifies the kind and sequence of individual amino acids that make up each of the enzymes has been solved. Knowledge has been gained of the manner in which the information contained in the composition of DNA is transmitted through formation of a series of ribonucleic acids to direct the synthesis of the individual enzymes through the biological machinery of the cell. Information is being obtained rapidly on the mechanisms by which the formation and levels of activity of the different enzymes are regulated and integrated to meet the changing needs of the living organism during its life history.

This volume is devoted to an exposition of the metabolic reactions of the amino acids and the tetrapyrrole coenzymes. The amino acids, in addition to their great importance as the constituents of proteins, are also the precursors of a large number of other highly important biological compounds. The tetrapyrrole coenzymes, the iron porphyrins, and magnesium porphyrins of chlorophyll are essential for trapping and utilizing the energy required by biological reactions. A third type of tetrapyrrole coenzymes, the cobamides, have varied functions in an increasing number of biochemical reactions.

With a few exceptions, the sequential steps in the biosynthesis and dissimilation of the amino acids that are found in proteins have been identified.

Many of the enzymes involved have been extensively purified. Knowledge of many other of these enzymes is still fragmentary. Much attention is now being devoted to the study of the nature of the catalytic function of the enzymes and to the processes that control their synthesis, levels of activity, and decay. A great deal is known about the tetrapyrrole compounds but knowledge of the metabolism of these substances is less well developed.

The common abbreviations and symbols utilized in this work are the same as those accepted by the *Journal of Biological Chemistry* (see Vol. **244**, No 1., pp. 2–8, 1969). Abbreviations and symbols other than those in the accepted list of this journal are defined in the chapter where employed.

We gratefully acknowledge the valuable suggestions of Dr. Martin A. Apple and his generous help in correcting the proofs of Chapters 15 and 16.

DAVID M. GREENBERG

San Francisco, California
February, 1969

Contents

Chapter 14
Nitrogen Metabolism of Amino Acids

H. J. SALLACH AND L. A. FAHIEN

Chapter 15 (Part 1)
Carbon Catabolism of Amino Acids

DAVID M. GREENBERG

Chapter 15 (Part II)
Carbon Catabolism of Amino Acids

VICTOR W. RODWELL

Chapter 16 (Part I)
Biosynthesis of Amino Acids and Related Compounds

DAVID M. GREENBERG

Chapter 16 (Part II)
Biosynthesis of Amino Acids and Related Compounds

VICTOR W. RODWELL

Chapter 17
Selected Aspects of Sulfur Metabolism

ERNEST KUN

Chapter 18

Metabolism of Porphyrins and Corrinoids

Bruce F. Burnham

Contents of Other Volumes

CHAPTER 14

Nitrogen Metabolism of Amino Acids

H. J. Sallach
and
L. A. Fahien

I. SCOPE

This chapter is concerned primarily with the enzymic systems involved in the transformation or transfer of the amino and amide nitrogen moieties of amino acids and amino acid amides. The metabolism of the respective carbon chains of these compounds is discussed in Chapters 15 and 16. Reactions involved in purine biosynthesis and degradation are discussed in Volume IV, Chapter 19. Complete coverage of the literature since the last edition of this chapter (*1*) has not been attempted. Of necessity there has been a selection of the literature references dictated for the most part by the interests of the authors. In general, preference has been given to studies in mammalian systems and to those which emphasize the enzymic aspects of chemical transformations. Reviews which cover this general area are those of Cohen and Brown (*2*), which deals with the comparative biochemical aspect, and Meister (*3*), which is a comprehensive review of the biochemistry of amino acids.

II. GENERAL SURVEY OF NITROGEN METABOLISM

The dynamic state of the amino group of amino acids was clearly demonstrated by Schoenheimer and associates (*4,5*) in their classical isotopic experiments. When $^{15}NH_3$ or a specific ^{15}N-labeled amino acid was given to rats, a rapid redistribution of the isotope into almost all of the amino acids was observed. No appreciable incorporation of isotope into threonine or lysine was noted. The highest isotope concentration, apart from that in the amino acid fed, was generally found in glutamic and aspartic acids, indicating the primary role of the dicarboxylic acids in the nitrogen metabolism of the amino acids. The essential dietary nature of certain amino acids was established by Rose and associates (see *6*) who, in addition, demonstrated that α-keto acid analogs could replace the corresponding essential amino acids in supporting growth of rats. These *in vivo* studies clearly established that the mammal possesses enzyme systems for the deamination and reamination of amino acids.

Early *in vitro* studies by Krebs (*7,8*) demonstrated that preparations of mammalian liver and kidney catalyze the oxidative deamination of both the D- and L-isomers of the amino acids. Although subsequent studies have established the presence of a very active D-amino acid oxidase in mammalian liver and kidney, the same is not true for the L-amino acid oxidase. The known general mammalian L-amino acid oxidase does not have the activity nor the breadth of substrate specificity to account for the catabolism of all of the L-amino acids as a group. Furthermore, except for L-glutamate dehydrogenase, there is no evidence for the existence of highly specific L-amino acid oxidative systems in mammals. On the other hand, there is abundant evidence for the existence of very active transaminases in mammalian systems. Almost all of the naturally occurring amino acids have been shown to participate in transamination reactions. On the basis of available data, it now appears that the mechanism for the oxidative deamination of L-amino acids involves the coupled action of L-amino acid-α-ketoglutarate transaminases and glutamate dehydrogenase, as first suggested by Braunstein and associates (see 9). In this process of transdeamination, the amino acids first participate in a transamination reaction with α-ketoglutarate to form glutamate and the corresponding α-keto acid. The glutamate formed is then oxidatively deaminated by glutamate dehydrogenase to form ammonia and regenerate α-ketoglutarate.

Experimental evidence supports the fact that transdeamination is of primary importance in the oxidative deamination of L-amino acids. A requirement for α-keto acids has been demonstrated for the oxidation of L-amino acids in cell-free preparations of liver and kidney from the rat, rabbit, and hen (10). Studies with liver and kidney preparations from vitamin B_6-deficient rats have shown that in such preparations the deamination as well as transamination of L-alanine and L-aspartate is reduced, although glutamate dehydrogenase activity is unaffected (11). Rowsell (12) has found with rat liver and kidney preparations that glutamate formation by transamination of α-ketoglutarate and all L-α-amino acids tested occurs at a rate faster than the rate of aerobic oxidation of these amino acids. In addition, the rates of transamination of L-α-amino acids with α-ketoglutarate were found to be comparable to the rates of ammonia or urea formation in tissue slices. In recent studies with rat liver mitochondria (13) it has been established that ammonia production from alanine and aspartate, like that from glutamate, is coupled with both oxygen consumption and the phosphorylation of ADP. Furthermore, it was demonstrated that added α-ketoglutarate is required for ammonia formation from either alanine or aspartate. These observations are consistent with the fact that transdeamination is the route for the oxidative deamination of these two amino acids. Since the transamination of certain amino acids with pyruvate in rat liver preparations occurred more rapidly than the aerobic oxidation of the amino acids, it has been suggested that the oxidative deamination of some L-amino acids may involve an initial transamination with pyruvate (14). The resulting alanine would then be transaminated with α-ketoglutarate by the glutamate-pyruvate transaminase.

In addition to its production via the transdeamination system, ammonia may be formed by the nonoxidative deamination of certain of the amino acids, e.g., serine, threonine, cysteine, or histidine, or by the hydrolysis of glutamine. As will be discussed below, in certain systems, ammonia is formed from D- and L-amino acids through the action of D- and L-amino acid oxidases.

The fate of the ammonia produced in the above reactions is dictated by physiological needs. Three major synthetic routes exist for the fixation of ammonia at all phylogenetic levels, including the mammal. The coupled transaminase-glutamate dehydrogenase system is of general importance not only in the degradation of amino acids but also in their biosynthesis. Fixation of ammonia into glutamate via the action of glutamate dehydrogenase and subsequent transamination of this compound provides a route for the synthesis of many amino acids from the corresponding keto acids and ammonia. A second route for the fixation of ammonia is its conversion to glutamine, a physiologically important storage form of ammonia. The third pathway utilizing ammonia is its conversion to carbamyl phosphate. This compound may be utilized for the formation of pyrimidines or for the production of urea.

Mammals, unlike plants and microorganisms, have no self-regulatory mechanisms permitting the utilization from the environment of amino acids only to the degree required for cellular needs and hence must have some means for the disposal of excess ammonia derived from the catabolism of amino acids. This is accomplished by urea production and excretion. Early studies on the formation of the latter compound by Krebs and Henseleit (15) established that urea formation occurred via the cyclic interconversion of ornithine, citrulline, and arginine. Subsequent studies have established that carbamyl phosphate is the ammonia donor required for the conversion of ornithine to citrulline and that the other nitrogen of the urea molecule is derived from aspartic acid.

The overall reactions involved in the nitrogen metabolism of the amino acids are summarized in Fig. 1 and are considered in greater detail in the sections below. The dicarboxylic amino acids are of primary importance in intermediary nitrogen metabolism, and their reversible formation from α-ketoglutarate and oxaloacetate establishes the close relationship between protein and carbohydrate metabolism. The formation or utilization of ammonia in the interconversion of α-ketoglutarate in the reaction catalyzed by glutamate dehydrogenase (II) is of central importance. Glutamate formation permits the biosynthesis by transamination (I) of many amino acids including aspartate (XII). The reverse sequence of reactions results in the oxidative deamination of the amino acids. Ammonia is also formed from certain amino acids by nonoxidative deamination (III) or by the hydrolysis of glutamine or asparagine (IV). The oxidative deamination of D-amino acids when required, or of L-amino acids in certain systems, by the general amino acid oxidases (V) is another possible source of ammonia. The latter

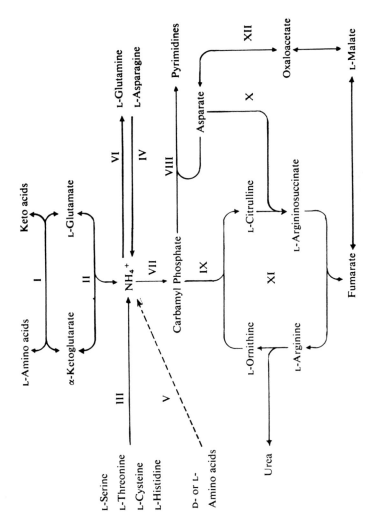

Fig. 1. Summary of the general reactions involved in the nitrogen metabolism of the amino acids.

compound may be stored in the form of glutamine (VI), a reaction discussed in Chapter 16, or be converted to carbamyl phosphate (VII). This compound and aspartate are important precursors of pyrimidines (VIII). Excess dietary ammonia resulting from the catabolism of amino acids is excreted as urea. Carbamyl phosphate (XI) and the amino group of aspartate (X) are utilized in urea biosynthesis (XI).

III. TRANSAMINATION

A. General Considerations

1. INTRODUCTION

Transamination reactions are those in which an amino group is transferred from an amino compound to a keto compound (or aldehyde) without the intermediate formation of ammonia. Although the usual reactants are mono- and dicarboxylic α-amino acids and α-keto acids, various other compounds including amino acids possessing β-, γ-, and δ-amino groups, amino acid ω-amides, and amines have been shown to participate in such reactions. Derivatives of vitamin B_6 function as cofactors in the transamination reaction. Recent reviews which cover this general topic include those of Braunstein (16), which deals with the role of pyridoxal-P, and of Meister (17), and of Snell (18), which are general surveys of amino group transfer reactions.

Enzymes which catalyze transamination reactions are termed transaminases or aminotransferases (19) and were first described by Braunstein and Kritzman (20). The transaminases are widely distributed in plants, microorganisms, and animal tissues (see 3,17,18). The overall transamination reaction may be formulated as shown in Eq. (1).

$$R_1—CO—COO^- + R_2—CH(NH_3{}^+)—COO^- \rightleftharpoons R_1—CH(NH_3{}^+)—COO^-$$
$$+ R_2—CO—COO^- \qquad (1)$$

2. SCOPE OF TRANSAMINATION REACTIONS

A large number of transamination reactions have been reported. However, the mere demonstration of such a reaction does not establish that separate and distinct enzymes are involved. For example, in extracts containing glutamate-oxaloacetate (L-aspartate : 2-oxyglutarate aminotransferase, EC 2.6.1.1) and glutamate-pyruvate (L-alanine : 2-oxoglutarate aminotransferase, EC 2.6.2.2) transaminases, a transamination reaction between aspartate and pyruvate can be shown in the presence of glutamate or α-ketoglutarate; however, no specific aspartate-pyruvate transaminase is known. In certain cases the existence of separate enzymes has been established by the physical separation and purification of the transaminases; in other cases it has been suggested on the basis of differences in response of the different activities to dietary variations, the *in vivo* and *in vitro* effects of hormones, inhibitors, and other factors (*21–24*). It is apparent from present evidence that a number of different transaminases exist.

Transaminases appear to be present in all biological systems studied. It is evident that each tissue has a characteristic distribution of trans-aminases (*12,14,25,26*). Furthermore, with the clear demonstration of separate mitochondrial and supernatant glutamate-oxaloacetate and glutamate-pyruvate transaminases (see Section III, C), it is apparent that different enzymes catalyzing the same reaction (isozymes) are found in different parts of the cell. At the present stage of our information it is not possible to define all of these enzymic activities in quantitative terms. In general it appears that the most active and widely distributed transaminase is the glutamate-oxaloacetate enzyme (*25,27–39*) (see Table I). It is also clear that α-ketoglutarate is the most active and widely

TABLE I

GLUTAMATE-OXALOACETATE TRANSAMINASE ACTIVITY OF ANIMAL
AND PLANT TISSUES AND MICROORGANISMS (*1*)

Source	μmoles transaminated/mg protein N/hr
Escherichia coli	397
Azotobacter vinelandii	703
Clostridium welchii	522
Oat seedlings (96 hr)	2522
Potato root	1464
Potato stem	1017
Potato leaf	290
Brain (rat)	1250
Liver (rat)	982
Kidney (rat)	781
Heart (rat)	1486

used keto acid substrate. A distribution of transminase activity involving α-ketoglutarate and a variety of amino acids is shown in Table II.

Earlier studies on transamination (see *17,18*) suggested that only L-α-amino acids and α-keto acids were active as substrates, and further that one of the substrates (either the amino acid or keto acid) had to be

TABLE II

RELATIVE TRANSAMINATION RATES WITH α-KETOGLUTARATE AND DIFFERENT AMINO ACIDS[a]

Amino acid	Pig Heart (25)	Pig Liver (25)	Pig Kidney (25)	Rat Kidney (12) (S)[b]	Rat Kidney (12) (P)[b]	Rat Liver (12) (S)	Rat Liver (12) (P)	Lupine (36)	N. crassa (35)	R. rubrum (37)	P. fluorescens (32)	E. coli (32)	L. arabinosus (38)	M. phlei (34)	B. subtilis (33)	M. tuberculosis (39)
Valine	100	100	100	100	100	100	100	100	100	100	100	100	100	100	100	100
Alanine	72	—	165	—	—	—	—	1310	66	71	0	80	0	56	17	4
α-Aminobutyrate	17	—	—	—	—	—	—	490	—	—	—	—	96	—	53	—
Arginine	12	201	390	—	—	—	—	678	369	128	100	—	0	7	0	—
Aspartate	190	404	1003	—	—	—	—	—	—	—	—	167	28	137	238	50
Cysteate	—	—	—	—	—	—	—	670	—	—	—	—	52	—	—	—
Cysteine	12	71	5	—	—	—	—	1177	—	—	—	—	0	0	0	—
Cystine	1	0	0	—	—	—	—	—	—	—	—	—	0	—	—	—
Glycine	3	88	0	—	—	—	—	109	—	—	40	67	0	10	0	—
Histidine	4	0	14	—	—	—	—	91	—	86	40	67	30	16	58	—
Isoleucine	88	72	73	100	—	100	—	172	100	86	40	—	104	105	102	109
Leucine	100	152	332	—	133	150	122	341	100	86	80	167	106	107	101	61
Lysine	6	6	32	50	100	—	—	141	—	14	40	67	0	14	0	—
Methionine	24	85	32	—	—	25	44	108	—	14	60	134	103	49	70	23
Norleucine	31	—	—	—	—	—	—	—	—	—	80	67	109	—	93	—
Norvaline	—	—	—	—	—	—	—	—	—	128	—	—	105	—	81	—
Ornithine	32	—	—	—	—	—	—	168	233	128	—	—	0	—	0	—
Phenylalanine	29	112	73	300	568	50	28	—	134	—	60	87	89	26	59	22
Serine	6	0	0	—	—	—	—	—	—	—	—	—	0	7	0	—
Threonine	1	0	8	—	—	—	—	—	—	—	32	67	0	3	0	—
Tryptophan	21	3	57	300	368	—	17	—	—	86	60	100	57	—	12	—
Tyrosine	35	221	181	900	333	—	17	—	—	128	48	80	68	—	15	—

[a] Valine is given an arbitrary value of 100 in each column; values in the different columns are not comparable.

[b] S: supernatant fraction; P: particulate fraction.

a dicarboxylic acid. Subsequent studies established that the scope of enzymic transamination is very broad and that practically all of the naturally occurring primary amino acids undergo this reaction (*25,32,40*). Present evidence clearly demonstrates that transamination occurs between monocarboxylic amino and keto acids (*14,21,26,41–45*), and that compounds with ω-amino groups, amines, and aldehydes are active in transamination as well (*44,46–74*). In addition it has been demonstrated that D-amino acid transaminases exist in microorganisms (*37,42, 75–78*). A selected list of transamination reactions illustrating the different types of compounds that are active as substrates is presented in Table III.

3. SUBSTRATE SPECIFICITY

Transaminases that have been highly purified from plant and animal sources have revealed a relatively high degree of specificity for the natural substrates, e.g., glutamate-oxaloacetate (*79–81*), glutamate-pyruvate (*79,81,82*) and glutamate-*p*-hydroxyphenylpyruvate (L-tyrosine: 2-oxoglutarate aminotransferase, EC 2.6.1.5) (*83,84*) transaminases. Activity with other substrates or analogs may be demonstrated in certain cases, but in general the rate of reaction is much less than that observed with the natural substrates (*17,18,85*). Certain exceptions may be noted. For example, transamination·between cysteinesulfinic acid and α-ketoglutarate (reaction 1 in Table III) is·catalyzed by purified preparations of glutamate-oxaloacetate transaminase at a faster rate than that observed with asparate and α-ketoglutarate (*86*). The branched-chain amino acid-α-ketoglutarate transaminase (reaction 2 in Table III) has been extensively purified, and it is evident that one enzyme functions with either leucine, valine, or isoleucine as substrates (*87–89*). On the other hand, partially purified preparations from *Escherichia coli* (*24,90*) and *Neurospora crassa* (*74,91*) appear to have broader substrate specificities. For example, with the enzyme that catalyzes the histidinol phosphate-α-ketoglutarate transamination (reaction 17 in Table III) it has been found that L-glutamate can be replaced by L-α-aminoadipate, L-arginine, or L-histidine and that α-ketoadipate and α-keto-δ-guanidino-valerate are active in place of histidinol phosphate (*74*). Hence a number of different transamination reactions may be demonstrated with the various combinations of the above substrates. Whether these differences reflect differences in the degree of purification or inherent differences in the properties of the *E. coli* and *N. crassa* enzymes has not been completely established; available evidence suggests that a single enzyme can catalyze a number of different transamination reactions.

TABLE III

Selected Examples of Transamination Reactions

Reaction	References
A. L-Amino acids	
1. L-Cysteinesulfinate + α-ketoglutarate ↔ β-sulfinylpyruvate + L-glutamate	(86)
2. L-Leucine + α-ketoglutarate ↔ α-ketoisocaproate + L-glutamate	(87–89)
3. L-Leucine + pyruvate ↔ α-ketoisocaproate + L-alanine	(12,14)
4. L-Serine + pyruvate ↔ β-hydroxypyruvate + L-alanine	(43,45)
B. ω-Amino acids	
5. Glycine + α-ketoglutarate ↔ glyoxylate + L-glutamate	(46)
6. β-Alanine + α-ketoglutarate ↔ malonate semialdehyde + L-glutamate	(44,47–52)
7. β-Alanine + pyruvate ↔ malonate semialdehyde + L-alanine	(44,52)
8. β-Aminoisobutyrate + α-ketoglutarate ↔ methylmalonate semialdehyde + L-glutamate	(51)
9. γ-Aminobutyrate + α-ketoglutarate ↔ succinate semialdehyde + L-glutamate	(47–49,53–57)
10. N-α-Acetyl-L-ornithine + α-ketoglutarate ↔ N-acetyl-L-glutamate semialdehyde + L-glutamate	(58,59)
11. L-Ornithine + α-ketoglutarate ↔ L-glutamate semialdehyde + L-glutamate	(50,60–65)
12. δ-Aminovalerate + α-ketoglutarate ↔ glutarate semialdehyde + L-glutamate	(49)
13. N-Succinyl-α, ε-diaminopimelate + α-ketoglutarate ↔ N-succinyl-α-amino-ε-ketopimelate + L-glutamate	(66)
C. L-α-Amino acid ω-amides	
14. L-Glutamine + β-oxaloacetylglycine ↔ α-ketoglutaramate + β-aspartylglycine	(67)
D. Amines	
15. Pyridoxamine + pyruvate ↔ pyridoxal + L-alanine	(68–70)
16. Putrescine + α-ketoglutarate ↔ γ-aminobutyraldehyde + L-glutamate	(71–73)
E. Other	
17. L-Histidinol phosphate + α-ketoglutarate ↔ imidazolacetol phosphate + L-glutamate	(74)
F. D-α-Amino acids	
18. D-Alanine + α-ketoglutarate ↔ pyruvate + D-glutamate	(78)

B. Mechanism of the Transamination Reaction

1. NONENZYME-CATALYZED REACTIONS

Nonenyzme-catalyzed reactions have been used as model systems in elucidating the mechanism of enzyme-catalyzed transamination reactions and of other pyridoxal phosphate-requiring enzyme systems. Extensive studies on the nonenzymic reactions of pyridoxal, pyridoxamine, and their phosphorylated derivatives with amino acids and keto acids in the presence or absence of metal ions have been carried out (for reviews see *3,18,16,92*). Snell and associates (*93–95*) demonstrated that a large number of amino acids undergo reversible nonenzymic transamination with pyridoxal in the presence of metal ions at elevated temperatures. Furthermore, it was found that in such systems an excess of the L-isomers of the amino acids was formed, suggesting that the reaction must proceed through an asymmetric intermediate which could be provided by a metal chelate (*96*). In addition to transamination, it was observed that pyridoxal plus metal ions catalyzed the dehydration of serine and threonine (*97*), the racemization of amino acids (*98*), the reversible cleavage of the hydroxyamino acids to glycine and the corresponding aldehyde (*99–101*) and other related reactions. Hence most of the reactions of amino acids which occur in living systems that are catalyzed by pyridoxal-P enzymes may occur nonenzymically in the presence of pyridoxal or more readily in the presence of pyridoxal plus a metal ion.

Studies with these model systems led to the independent formulation of a general mechanism for vitamin B_6-catalyzed reactions by Snell and associates (*94*; see *18*) and by Braunstein and associates (*102*; see *16*) (Fig. 2). The general mechanism involves the formation of an aldimine (Schiff base) between pyridoxal (or a pyridoxal-containing catalyst) and an amino acid as the initial activation step. The metal ion aids in the stabilization of the intermediate and the subsequent electronic transformations involved in the reactions. The structures shown in Fig. 2 possess a planar system of conjugated double bonds from the α-carbon of the amino acid to the nitrogen atom of the pyridoxal, thereby providing a mechanism for the shift of electrons from the bonds surrounding the α-carbon atom toward the strongly electrophilic heterocyclic nitrogen atom. This displacement of the electrons is evidently increased by the electronegative chelated metal ion which may be considered to function in a role analogous to that of an enzyme. By this process the bonds surrounding the α-carbon atom are weakened, and the amino acid is activated for any of a large number of subsequent reactions that have been demonstrated in both enzymic and nonenzymic systems. These reactions have been classified by Braunstein (16) and by Snell (103) according to the particular bond that is broken. For example, reactions resulting from the labilization of the α-hydrogen atom (a in Fig. 2) include transamination and racemization of amino acids in general and the dehydration of serine and threonine, all of which are considered in this chapter. The labilization of the carboxyl group (b in Fig. 2) results in decarboxylation, and that of the R group (c in Fig. 2) gives rise to the cleavage of the hydroxyamino acids in an aldolase type of reaction; these reactions are considered in Chapter 15.

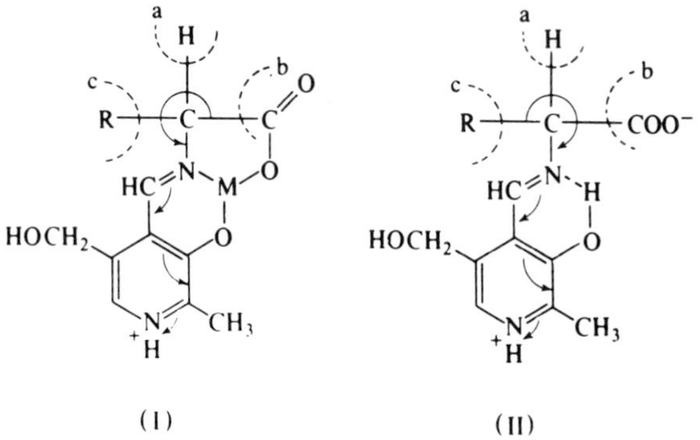

(I)   (II)

FIG. 2. Structures of aldimines formed between amino acid, pyridoxal, and metal ion (I) and between amino acid and pyridoxal (II) (132).

The mechanism proposed by Snell and co-workers (94; see 18) for the nonenzymic transamination reaction is shown in Fig. 3. Pyridoxal, amino acid, and metal ion react to form the chelated aldimine (I). Electron withdrawal from the bond to the α-hydrogen labilizes the proton and permits a shift in the position of the double bond to form the tautomeric Schiff base (II). In the reverse reaction the formation of a ketimine from the keto acid and pyridoxamine results in labilization of a proton of the methylene group at position 4 of pyridoxamine, giving rise to structure III. The three intermediates (I, II, and III) may be considered resonating forms of the same structure. The reactions shown represent one-half of a typical transamination reaction; the second half of the transamination reaction is effected by a reversal of the above sequence of reactions with a second keto acid as a substrate. While metal ions catalyze the reaction in aqueous systems, they are not an absolute requirement (104–106). Imidazole has been shown to be

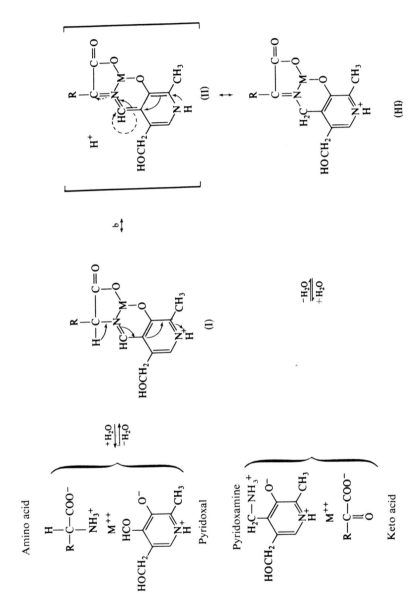

FIG. 3. Proposed mechanism for the role of pyridoxal-P and pyridoxamine phosphate in transamination (96).

an effective catalyst for certain nonenzymic transamination reactions in the absence of metal ions (107,108). On this basis it has been suggested that histidine residues and other polar groups of proteins may enhance the catalytic activity of pyridoxal-P in enzymic transaminations.

2. Enzyme-Catalyzed Transamination Systems

Pyridoxal-P and pyridoxamine-P have been shown to be the coenzymes for transaminases (17,20,109–111). The coenzymes function as amino group acceptors or donors in the overall transamination reaction as shown in Eqs. (2a) and (2b). The evidence in

R_1—CH(NH$_3$$^+$)—COO$^-$ + enzyme $\cdots$ pyridoxal-P $\rightleftharpoons$

$$R_1\text{—CO—COO}^- + \text{enzyme} \cdots \text{pyridoxamine-P} \qquad (2a)$$

R_2—CO—COO$^-$ + enzyme $\cdots$ pyridoxamine-P $\rightleftharpoons$

$$R_2\text{—CH(NH}_3{}^+)\text{—COO}^- + \text{enzyme} \cdots \text{pyridoxal-P} \qquad (2b)$$

support of the above role for the coenzymes in the transamination reaction includes the following:

1. Resolved transaminases (apotransaminases) are activated to an equivalent extent by either pyridoxal-P or pyridoxamine-P; activation by the latter compound, however, occurs at a slower rate (111).

2. Isotopic studies have demonstrated that the amino group of the initial substrate amino acid is the same as the amino group of the product amino acid and that ammonia is not an intermediate (112); the ^{15}NH$_2$-group of pyridoxamine appears in glutamate during trans-amination of α-ketoglutarate without the intermediate formation of ammonia (112); one proton is acquired from the media for each amino group transferred (113); the β-hydrogen of the amino acid is not involved in the transamination reaction (114–117).

3. Enzymic transamination occurs not only between a given amino acid and a structurally different α-keto acid [see Eq. (1)] but also between a given amino acid and its corresponding α-keto acid (118–121), and the α-hydrogen of the amino acid is liberated as a proton (118).

4. Pyridoxal-P and pyridoxamine-P forms of glutamate-oxaloacetate transaminase have been prepared (122–125). Incubation of substrate amounts of the pyridoxal-P form of the enzyme with either glutamate or aspartate leads to the formation of an equivalent amount of enzyme-bound pyridoxamine-P and the analogous α-keto acid, thereby provid-ing a direct demonstration of the first half of a typical transamination

reaction [see Eq. (2a)]. The reverse reaction in which the pyridoxamine-P form of the enzyme and α-keto acids were used as substrates has been demonstrated also [see Eq. (2b)].

5. The two half-reactions of transamination have been demonstrated in another manner by Wada and Snell (126). These investigators found that the apoenzyme of pig heart glutamate-oxaloacetate transaminase, but not the holoenzyme, catalyzes the reversible transamination of pyridoxamine with either α-ketoglutarate or oxaloacetate. According to these investigators, pyridoxamine and pyridoxal are apparently bound to the enzyme at the site(s) normally occupied by the phosphorylated coenzymes; however, the affinity for the nonphosphorylated derivatives is so low that these compounds behave as freely diffusible analogs of the coenzymes and hence act as cosubstrates rather than as coenzymes.

The above findings provide convincing proof of the amino donor-acceptor role of the coenzymes in the general mechanism of transamination as shown in Eqs. (2a, 2b).

It has been clearly established that the aldehyde group of pyridoxal-P does not exist in the free state when the coenzyme is bound to the transaminase. Spectrophotometric and kinetic studies (119–128) support the conclusion that the aldehyde group of the coenzyme is bound to an amino group of the protein by a Schiff base or aldimine linkage. A direct demonstration of such a linkage in glutamate-oxaloacetate transaminase has been provided by Hughes et al. (129). These investigators isolated a tetradecapeptide containing covalently bound pyridoxal-5′-P from a chymotryptic digest of the enzyme following reduction with sodium borohydride at pH 4.8. A partial structure of the peptide was determined: H-Ser-Thr-Glu-(Asp, Gly, Ala, Val, Ileu, ε-Pyridoxyllys, Lys)-Gly-Ser-Asp-Phe-OH. The linkage of pyridoxal-P to an ε-amino group of lysine has been established for other enzymes which require this coenzyme [see V below; also 18,92]. Since apotransaminases may be reactivated by pyridoxamine-P as well as by pyridoxal-P, functional groups other than the aldehyde group must be involved in the interaction of the coenzyme with the protein. Studies with coenzyme analogs and other investigations (see 16,18,103) indicate that ionic linkages of phosphate esters and pyridine nitrogen moieties are involved and an unsubstituted 3-hydroxyl group is required for activity. Although neither the hydroxymethyl group at position 5 nor the methyl group at position 2 is required for the nonenzymic reaction, inhibitory studies with coenzyme analogs have shown that every substituent on the pyridoxal-P affects the binding of the coenzyme to the apoenzyme (103). Individual apotransaminases are activated to different extents by a single coenzyme analog (130); different conditions are also required for the resolution of glutamate-oxaloacetate and glutamate-pyruvate transaminases from pig heart (126). These findings indicate a difference in the nature of binding of coenzymes to the individual apoenzymes.

A mechanism similar to that shown in Fig. 3 for nonenzymic transamination was proposed by Snell and associates (94,98) for enzymic transamination. In the latter case the apoenzyme is considered to play the role of the metal ion (M in Fig. 3). Although there have been some reports suggesting the requirement for a metal ion in enzymic transamination (see 3,18) available data on highly purified transaminases indicate that metal ions are not involved (69,131). A more detailed mechanism of transamination based on current views has been proposed recently by Snell (132) and is presented in Fig. 4. Since the pyridoxal-P exists, at least in part, in transaminases in the form of an aldimine (Schiff base) with the ϵ-amino group of a lysine residue, the first step in the reaction sequence with this form of the enzyme must involve a transaldimination (transimination) reaction with the substrate, i.e., a reaction of an amine with an imine resulting in the conversion of the "internal" Schiff base between pyridoxal-P and the enzyme into a Schiff base between pyridoxal-P and the substrate amino acid. A first appraisal would suggest that, if the aldehyde group of the coenzyme is bound in an aldimine linkage to the protein, it would not be free to form a similar derivative with the substrate amino acid. However, it has been pointed out that Schiff bases are still chemically equivalent to carbonyl compounds (133). Furthermore, studies with nonenzymic models (134–136) and with glutamate-oxaloacetate transaminase (137) have demonstrated that the Schiff base of pyridoxal-P is much more reactive with many carbonyl reagents than is free pyridoxal-P. Hence it appears that the enzyme substrate complex is formed much more rapidly because the aldehyde group of the coenzyme is bound in imine linkage to the protein. The next step in the mechanism is the tautomeric conversion of the aldimine into the ketimine. The tautomerization reaction may be facilitated by the proximal —NH_2 and —NH_3^+ groups on the protein (not necessarily restricted only to the ϵ-amino group of lysine that is released from the azomethine linkage by the formation of the enzyme-substrate complex) which serve as general acid-base catalysts. The hydrolysis of the double bond of the ketimine results in the liberation of an α-keto acid and the formation of enzyme-bound pyridoxamine-P, thereby completing the first half-reaction of a typical transamination; reversal of the above sequence of reactions with the α-keto acid substrate would complete the overall transamination reaction.

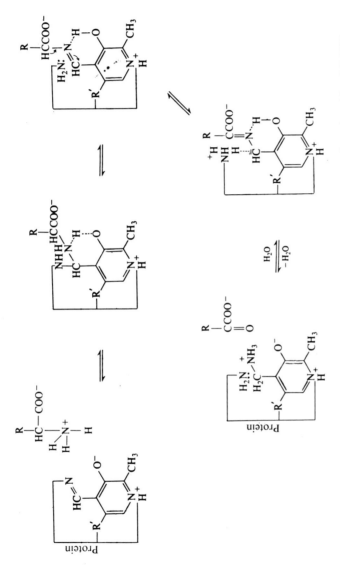

FIG. 4. Possible mechanism for the transamination reaction catalyzed by enzyme-bound pyridoxal-P (*132*).

3. Kinetic and Spectral Studies

In general the reaction of transaminases can be described by the mechanism

$$E + A \underset{k_2}{\overset{k_1}{\rightleftharpoons}} EX \underset{k_4}{\overset{k_3}{\rightleftharpoons}} E' + P \tag{3}$$

$$E' + B \underset{k_6}{\overset{k_5}{\rightleftharpoons}} EY \underset{k_8}{\overset{k_7}{\rightleftharpoons}} E + Q \tag{4}$$

where A and B are, respectively, substrate amino and keto acids; P and Q are, respectively, product keto and amino acids; and E and E' are, respectively, the phosphopyridoxal and phosphopyridoxamine forms of the enzyme [see Eqs. (2a) and 2b)]. The steady-state rate equation for mechanism (I) with respect to A, B, and Q, in the absence of added P, is

$$v = \cfrac{V_f}{1 + \cfrac{K_A}{(A)}\left[1 + \cfrac{(Q)k_8}{k_7}\left(1 + \cfrac{k_6}{(B)k_5}\right)\right] + \cfrac{K_B}{(B)}} \tag{I}$$

where v and V_f are, respectively, the initial and maximal velocities of the forward reaction; K_A and K_B are, respectively, the Michaelis constants of A and B; and k_7/k_8 and k_6/k_5 are, respectively, the dissociation constants of Q and B from the reactive complex EY.

Equation (I) predicts that, in the absence of added P and Q, a series of parallel lines will be obtained if double reciprocal plots are made of initial velocity versus the concentration of A at various fixed concentrations of B. If Q is added (so that its concentration approaches k_7/k_8), it will be a competitive inhibitor of A. Similar relationships can be developed for the product inhibition pattern of P. Thus the product amino acid will be a competitive inhibitor of the substrate amino acid and a noncompetitive inhibitor of the substrate keto acid. A nonreactive substrate analog I could be bound to the phosphopyridoxamine form of the enzyme,

$$E' + I \underset{k_{10}}{\overset{k_9}{\rightleftharpoons}} E'I \tag{5}$$

the phosphopyridoxal form of the enzyme,

$$E + I \underset{k_{12}}{\overset{k_{11}}{\rightleftharpoons}} EI \tag{6}$$

or to both forms of the enzyme. If it can be bound to both forms of the enzyme, it can be either a noncompetitive inhibitor of both A and B or a competitive inhibitor of one substrate if the concentration of the other substrate is sufficiently high with respect to its Michaelis constant. If the inhibitor is bound to only one form of the enzyme, for example, the phosphopyridoxal form, then it will behave as a competitive inhibitor of A and an uncompetitive inhibitor of B.

In addition to forming reactive complexes, such as EX and EY, substrates and products can form nonreactive complexes as shown in Eqs. (7–9). Complexes between the phosphopyridoxal form of pig

$$E' + A \underset{k_{14}}{\overset{k_{13}}{\rightleftharpoons}} E'A \tag{7}$$

$$E + B \underset{k_{16}}{\overset{k_{15}}{\rightleftharpoons}} EB \tag{8}$$

$$E' + Q \underset{k_{18}}{\overset{k_{17}}{\rightleftharpoons}} E'Q \tag{9}$$

heart glutamate-oxaloacetate transaminase and α-ketoglutarate (*125, 127,138*) and the phosphopyridoxamine form of glutamate-pyruvate transaminase and amino acids (*82*) are known to exist. In general, the dissociation constants of these complexes are much higher than Michaelis constants. Therefore substrates are inhibitory only at high concentrations.

The rate equation of the transaminase mechanism plus these additional complexes (again in the absence of added P) is

$$v = \cfrac{V_f}{\begin{aligned} &1 + \frac{K_A}{(A)}\left(1 + \frac{(B)k_{15}}{k_{16}}\right) + \frac{K_B}{(B)}\left(1 + \frac{(A)k_{13}}{k_{14}}\right) + \frac{(Q)k_8}{k_7} \\ &\times \left[\frac{K_A}{(A)}\left(1 + \frac{k_6}{(B)k_5} + \frac{(Q)k_6 k_{17}}{(B)k_5 k_{18}}\right) + \frac{1}{(B)}\left(\frac{k_6 k_{13} K_A}{k_5 k_{14}} + \frac{k_7 k_{17} K_B}{k_8 k_{18}}\right)\right] \\ &+ (I)\left(\frac{k_{11}K_A}{(A)k_{12}} + \frac{k_9 K_B}{(B)k_{10}} + \frac{(Q)k_6 k_8}{(A)(B)k_5 k_7}\right) \end{aligned}} \tag{II}$$

where the ratios k_{10}/k_9, k_{12}/k_{11}, etc., are the dissociation constants of the complexes shown in Eqs. (5–9).

There have been many experimental verifications of these relationships (*82,120,125,127,138*), and some representative values of kinetic constants are shown in Tables IV and V. As for most kinetic studies, different results have been obtained with different experimental conditions and enzyme preparations. As shown in Table VI, kinetic constants of anionic and cationic forms of the enzyme can be different (*139–141*).

Many substrate analogs seem to be bound to both the pyridoxal and pyridoxamine forms of transaminases with equal affinity (*127,142*).

TABLE IV

KINETIC CONSTANTS OF PIG HEART (EXTRAMITOCHONDRIAL)
GLUTAMATE-OXALOACETATE TRANSAMINASE $(138)^a$

Compound	Michaelis Constant (mM)	Dissociation constant of reactive complex (mM)
L-Aspartate	3.9	3.48
α-Ketoglutarate	0.43	0.71
Oxaloacetate	0.088	0.05
L-Glutamate	8.9	8.4

^{a} Experiments performed in 0.1 M sodium arsenate at pH 7.4 and 37°.

TABLE V

INHIBITION PATTERN OF GLUTAMATE-ALANINE TRANSAMINASE
BY SUBSTRATE ANALOGS (82)

Inhibitor	Variable substrate		k_{12}/k_{11} (mM)	k_{10}/k_9 (mM)
	Amino acid	Keto acid		
Glutarate	Noncompetitive	Noncompetitive	9.2	92
Maleate	Uncompetitive	Competitive	—	8.4
γ-Aminobutyrate	Noncompetitive	Noncompetitive	30	61 ·

TABLE VI

KINETIC CONSTANTS OF VARIOUS FORMS OF PIG HEART
GLUTAMATE-OXALOACETATE TRANSAMINASE (139)

Enzyme form	Michaelis constant	
	Aspartate (mM)	α-Ketoglutarate (mM)
Anionic 1	3.9	0·57
Anionic 2	4.9	0·61
Cationic (mitochondrial)	0.35	1·1

This would suggest that all four substrates utilize the same catalytic site on the enzyme. However, as shown in Table V, this type of inhibition pattern is not always found (82).

Transamination reactions are known to be even more complicated. For example, the complex EX (as described in the previous section) is known to consist of many intermediates. However, this would not alter the form of the rate equations (143). Kinetic studies of the mitochondrial glutamate-oxaloacetate transaminase of beef liver (140) and the rat liver tyrosine-α-ketoglutarate transaminase (144) are not consistent with the transamination mechanism and suggest that ternary complexes between holoenzyme and both substrates might exist.

Many spectral studies of the pig heart glutamate-oxaloacetate transaminase have been performed (120,122,125,127). Recent detailed spectral studies of this enzyme by Jenkins and his associates substantiate the interpretation of kinetic data (120,125). The theoretical development by these authors of this method should also be of great value to investigators of similar enzymes. Spectral and kinetic studies of this enzyme are consistent both with respect to mechanisms and to numerical values of constants. These experiments demonstrate further that the values of kinetic constants are the same at both high and low concentrations of enzyme. Values obtained from temperature jump experiments (145) are not consistent with kinetic and spectral data. The reason for this discrepancy is not clear.

While in theory the equilibrium constant of a transaminase reaction should be approximately equal to unity, values as high as 20 have been found (125,127,128,146,147).

C. Purified Amino Acid Transaminases

A large number of transamination reactions have been reported, and many of the individual enzymes have been studied. However, space does not permit an individual discussion of each of these enzymes nor, in fact an extensive consideration of any single enzyme. This section will deal with only a limited number of enzymes which have been brought to a highly purified state.

1. GLUTAMATE-OXALOACETATE TRANSAMINASE

Glutamate-oxaloacetate transaminase is the most active and widely distributed transaminase, a fact in accord with the fundamental roles of its four substrates in intermediary metabolism. This enzyme has been studied in far greater detail than any of the other transaminases, and much of our understanding of the mechanism of enzymic transamination has been obtained through work on this enzyme. A review chapter devoted exclusively to this one transaminase has appeared recently (*128*) and covers earlier studies on this enzyme.

Glutamate-oxaloacetate transaminase has been shown to exist in two molecular forms, one mitochondrial and the other extramitochondrial, in a variety of tissues including heart and liver of dog, rat, ox, and man and in kidney, skeletal muscle, and brain of rat (*131–139, 148–153*). The two isozymes differ in their physical, immunochemical, and kinetic properties (for the latter see Table VI). Upon electrophoresis at pH values ranging from 7.0 to 8.6 the mitochondrial isozyme is cationic and the supernatant isozyme is anionic (*140,141,148,149*). The mitochondrial isozyme is much more labile to heat treatment even in the presence of dicarboxylic acids, or other stabilizers, than is the supernatant isozyme (*139,154*). In view of this fact, earlier studies on this transaminase were apparently carried out with the supernatant form since the purification procedures usually employed included a heat-treatment step (*155*).

Glutamate-oxaloacetate transaminases of both mitochondrial and supernatant fractions of beef liver have been obtained in crystalline form (*140,154*). The former isozyme is cationic at pH 6.7 and has a molecular weight of approximately 100,000. The supernatant isozyme is anionic under the same conditions and has a reported molecular weight of 120,000. Both isozymes contain 2 moles of pyridoxal-P per mole of enzyme. The crystallization of the two isozymes from beef and pig heart has also been achieved (*154*). The properties of the isozymes from beef heart (electrophoretic mobilities, molecular weights, pyridoxal-P content, and absorption spectra) are essentially identical with those of the corresponding enzymes from liver. Pig heart isozymes show slight differences in electrophoretic mobilities from the corresponding beef enzymes, but are identical in other properties.

Immunological studies have been carried out (*141,154,156*). Rabbit antisera to crystalline supernatant and mitochondrial glutamate-oxaloacetate transaminases from beef liver inhibited 80% of the activity of the homologous enzymes from beef liver or heart, while the heterologous enzymes were unaffected (*156*). Similar results were obtained in studies with human heart enzymes (*141*). These studies have been extended to include a number of other species (*141,153,154,156*). These studies show that the mitochondrial and supernatant enzymes are completely distinct from each other immunochemically and that no distinction in the antigenic structure is noted between the corresponding enzymes in various tissues from any given animal; however, the antigenic structures of the enzymes from different species do vary *in part* from animal to animal.

Heart tissue was used by many investigators in earlier studies on glutamate-oxaloacetate transaminase (*50,79,157,158*). The procedures developed by Lis and her associates (*159,160*) and by Jenkins *et al.* (*155*) have been used with slight modifications in many recent studies. Either procedure results in an enzyme preparation that contains essentially the supernatant isozyme (*154*) and has 1 mole of pyridoxal-P per 55,000–57,000 gm of protein. Homogeneous preparations of the super-natant enzyme from pig heart (*161,162*) and from beef heart (*163,164*) have been prepared and used for structural studies. The reported molecular weight of the enzyme from pig heart, calculated from the coenzyme content (2 moles of pyridoxal-P per 116,000 gm of protein), is in good agreement with the value originally reported by Jenkins *et al.* (*155*) and that by Wada and Marino (*154*) (see above). The amino acid composition and sulfhydryl groups have been deter-mined; there are no disulfide bonds in the molecule which contains 14 residues of cysteine per mole. The quantitative determination of N-terminal amino acids demonstrated that there are 2 moles of N-terminal alanine per mole of protein. It was concluded that the enzyme (MW = 116,000) is composed of 2 polypeptide chains. The enzyme from beef heart is similar to that from pig heart in many respects (*163,164*). The reported molecular weight is 96,000 by sedimentation equilibrium studies (117,000 weight-average molecular weight by light scattering) with 1.5–1.7 moles of pyridoxal-P per 96,000 gm of protein (*163,164*). A minimum molecular weight of 48,000 is suggested. There are two N-terminal alanine and two C-terminal leucine residues per mole of enzyme; no disulfide bonds are present. Hence this enzyme is also com-posed of two polypeptide chains held together by noncovalent linkages. Additional studies by the same group (*165*) have suggested that the 2 polypeptide subunit chains have identical primary structures. The enzyme under slightly alkaline conditions dissociates into two com-ponents each of which has the same coefficient of sedimentation. Poly-anovsky and Ivanov (*166*) had previously shown that the supernatant isozyme from pig heart appears to dissociate upon dilution as judged by changes in the polarization of fluorescence. In a continuation of this work, Polyanovsky (*167*) has demonstrated that the succinylated form of the enzyme (MW = 117,000) dissociates into two subunits and that the monomeric form retains 65% of the activity of the native trans-aminase. The dissociation is pH-dependent, with alkalinity (pH 8.5–9.3) favoring monomer formation, whereas at neutrality the dimer exists. Other physical studies relating to the secondary and tertiary structure of this enzyme have been reported (*168,169*).

Present evidence indicates that the two isozymes of glutamate-oxaloacetate transaminase may exist in multiple forms. Martinez-Carrion *et al.* (*170*) report that the supernatant holotransaminase from pig heart, prepared by the procedure of Jenkins *et al.* (*155*) with the exception that succinate rather than maleate buffer was used in the heat stabilization step, may be separated into three components by chromatography on carboxymethyl-Sephadex. The components differ in specific activity, spectrophotometric properties, and mobility on starch gel electrophoresis. All three components, termed α-, β-, and γ-forms based on electrophoretic mobilities, are anionic at pH 8.0, have the same pyridoxal-P content, and identical amino acid compositions, and the latter are the same as that reported for a homogeneous preparation of the pig heart glutamate-oxaloacetate enzyme (*161*) from which they were separated. No heterogeneity was observed on immunochemical analysis by the Ouchterlony technique. Other investigators have observed multiple forms of the supernatant enzyme upon chromatography (*120,125,138,139*). A brief mention of the fact that crystalline supernatant glutamate-oxaloacetate transaminase from pig heart yields three anionic bands on electrophoresis has appeared (see footnote in *120*). It is not clear at present whether these different forms are biologically determined or reflect the isolation procedures employed. It has also been suggested that mitochondria may contain several cationic electrophoretic components (*170*). The transaminase from yeast, which has been obtained in a homogeneous form, has a reported molecular weight of 91,200 with a minimum of 4 moles of coenzyme per mole of enzyme (*171,172*).

2. GLUTAMATE-PYRUVATE TRANSAMINASE

The existence of mitochondrial and supernatant forms of glutamate-pyruvate transaminase has been established in a number of tissues including rat liver, heart, skeletal muscle, and kidney, pig heart, and rabbit liver (*173–182*). The mitochondrial isozyme from rat liver is extremely labile under most conditions and hence has resisted attempts at purification (*179,180,182*). The separation of the two isozymes has been achieved by chromatographic procedures (*178,181*), and the preparation of the two isozymes in a highly purified form has been reported (*178*). Extensive purification of the supernatant isozyme from beef

(*82*) and pig (*133*) heart has been achieved. Kinetic and spectral (*82,133, 183*) properties of these enzymes are similar to those of other transaminases.

3. D-ALANINE-α-KETOGLUTARATE TRANSAMINASE

This transaminase has been obtained from *Bacillus subtilis* in a form approximately 95% homogeneous, as estimated by the criteria of ultracentrifugation, free boundary electrophoresis, and starch gel electrophoresis (*78*). The molecular weight obtained by a sedimentation equilibrium method is 53,000 ± 4,000, and the holoenzyme is estimated to contain 1 mole of pyridoxal-P per protein molecule. In contrast to many other transaminases, this enzyme is insensitive to sulfhydryl reagents. D-Alanine, D-glutamate, D-aspartate, D-asparagine, and D-α-aminobutyrate were active as substrates; no activity was observed with the L-isomers. The final equilibrium constant at pH 8.3 was 1.7 in the direction of D-alanine formation. The reaction mechanism has been investigated (*78*) and, like L-amino acid transaminases, proceeds via the interconversion of pyridoxal and pyridoxamine forms of the enzyme.

4. OTHER TRANSAMINASES

In addition to the above enzymes a number of other transaminases have been brought to a high state of purity. They include the crystalline pyridoxamine-pyruvate transaminase of *Pseudomonas* (*70,126*), γ-aminobutyrate-α-ketoglutarate transaminase of mouse brain (*57*), and L-ornithine-α-ketoglutarate transaminase from rat liver (*63*). Many other transaminases have been partially purified. Earlier studies in this area were devoted largely to the demonstration of transamination reactions; at the present writing it would appear that the purification of the enzymes responsible for these reactions and a study of their properties is finally proceeding at a rate equal to or greater than that at which new transamination reactions are reported.

D. Amino Acid Amide Transaminases

The direct participation of glutamine and asparagine in transamination reactions without their prior deamination was demonstrated by Meister and associates (*50,67,184–191*). These investigators have established the existence of two separate transaminases which catalyze the reactions shown in Eqs. (10) and (11). Glutamine-α-keto acid transaminase (L-glutamine : 2-oxoacid aminotransferase, EC 2.6.1.15) (Eq. 10) is relatively specific for the amino donor, glutamine, but not for the α-keto acid. A large number of α-keto acids have been shown to act as substrates, including the analogs of the common amino acids (and asparagine) and, in addition, the keto acids, β-oxalacetylglycine and β-oxalacetylalanine. The latter are converted to the corresponding β-aspartyl peptides, aspartylglycine and aspartylalanine. Formation of α-peptides from the corresponding keto acids has not been observed.

$$
\begin{array}{c}
\text{COO}^- \\
| \\
\text{CO} \\
| \\
\text{R}
\end{array}
+
\begin{array}{c}
\text{COO}^- \\
\overset{+}{\text{H}_3\text{N}}-\text{C}-\text{H} \\
| \\
\text{CH}_2 \\
| \\
\text{CH}_2 \\
| \\
\text{CO}-\text{NH}_2
\end{array}
\longrightarrow
\begin{array}{c}
\text{COO}^- \\
\overset{+}{\text{H}_3\text{N}}-\text{C}-\text{H} \\
| \\
\text{R}
\end{array}
+
\begin{array}{c}
\text{COO}^- \\
| \\
\text{CO} \\
| \\
\text{CH}_2 \\
| \\
\text{CH}_2 \\
| \\
\text{CO}-\text{NH}_2
\end{array}
\xrightarrow{\text{H}_2\text{O}}
\begin{array}{c}
\text{COO}^- \\
| \\
\text{CO} \\
| \\
\text{CH}_2 \\
| \\
\text{CH}_2 \\
| \\
\text{COO}^-
\end{array}
+ \text{NH}_4{}^+ \quad (10)
$$

L-Glutamine   α-Ketoglutaramate   α-Ketoglutarate

$$
\begin{array}{c}
\text{COO}^- \\
| \\
\text{CO} \\
| \\
\text{R}
\end{array}
+
\begin{array}{c}
\text{COO}^- \\
\overset{+}{\text{H}_3\text{N}}-\text{C}-\text{H} \\
| \\
\text{CH}_2 \\
| \\
\text{CO}-\text{NH}_2
\end{array}
\longleftarrow
\begin{array}{c}
\text{COO}^- \\
\overset{+}{\text{H}_3\text{N}}-\text{C}-\text{H} \\
| \\
\text{R}
\end{array}
+
\begin{array}{c}
\text{COO}^- \\
| \\
\text{CO} \\
| \\
\text{CH}_2 \\
| \\
\text{CO}-\text{NH}_2
\end{array}
\xrightarrow{\text{H}_2\text{O}}
\begin{array}{c}
\text{CH}_2 + \text{NH}_4{}^+ \\
| \\
\text{CH}_2 \\
| \\
\text{COO}^-
\end{array}
\quad (11)
$$

L-Asparagine   α-Ketosuccinamate   Succinate

A partial list of keto acids which are active as substrates is shown in Table VII.

TABLE VII

EFFECT OF DIFFERENT α-KETO ACIDS ON GLUTAMINE AND GLUTAMIC ACID TRANSAMINATION WITH RAT LIVER PREPARATIONS (191)

	Glutamine		Glutamate
α-Keto acid	Glutamine disappearance (μmoles)	NH₃ formation (μmoles)	Glutamate disappearance (μmoles)
None	0	0	0
Pyruvic	10.2	10.6	10.9
Glyoxylic	8.1	8.7	8.0
α-Keto-ε-hydroxycaproic	9.7	10.2	0.1
α-Keto-ε-N-chloroacetylcaproic	2.6	2.8	0
α-Keto-γ-ethiolbutyric	7.8	8.2	0
α-Keto-β-cyclohexylpropionic	1.3	1.4	0
α-Keto-β-indolylpropionic	4.2	4.4	0

α-Ketoglutaramate, the intermediate in (Eq. 10), has been prepared by enzymic and chemical methods and has been shown to participate in the reaction as formulated. The glutamine-α-keto acid transamination reaction is irreversible owing to the ready deamidination of α-ketoglutaramate. The deamidination is catalyzed by an ω-amidase (185), which can be separated from the transaminase (186). The amidase hydrolyzes a number of amides including α-ketosuccinamate (Eq. 11), but does not deamidate glutamine or asparagine. The glutamine-α-keto acid transaminase is present in liver and kidney of mammals (184,192), muscle of the locust (193), plants (194), and N. crassa (195).

Asparagine-α-keto acid transaminase (L-asparagine : 2-oxoacid aminotransferase, EC 2.6.1.14) (Eq. 11) has been demonstrated to occur in liver and kidney, and there is evidence that this transaminase is different

TABLE VIII

FORMATION OF L-ASPARAGINE FROM α-KETOSUCCINAMATE BY
TRANSAMINATION (189)

Amino acid added	L-Asparagine formed (μmoles)	Amino acid added	L-Asparagine formed (μmoles)
None	0.71	L-Tyrosine	4.51
L-Alanine	7.46	L-α-Aminophenylacetic acid	0.72
L-α-Aminobutyric acid	8.62	L-Methionine	6.26
L-Norvaline	6.29	L-Ethionine	4.37
L-Valine	0.79	L-Glutamic acid	1.21
L-Norleucine	5.20	L-Glutamine	7.51
L-Leucine	4.59	L-Aspartic acid	1.01
L-Isoleucine	0.65	L-Histidine	5.54
L-Alloisoleucine	0.60	L-Tryptophan	1.72
L-t-Leucine	0.78	Glycine	0.87
L-Phenylalanine	4.15	L-Serine	0.22

from the glutamine transaminase (189). Asparagine acts with a large number of α-keto acids to form α-ketosuccinamic acid. The deamidation of the latter is catalyzed by the same ω-amidase which deamidates α-ketoglutaramic acid. However, the affinity of the deamidase for α-ketosuccinamic acid is low; this accounts for the reversibility of the transaminase system (185). A large number of amino acids and glutamine are active in forming asparagine from α-ketosuccinamate (see Table VIII).

IV. OXIDATIVE DEAMINATION

A. L-Glutamate Dehydrogenase

Glutamate dehydrogenase (L-glutamate-NAD[P] oxidoreductase, EC 1.4.1.3) catalyzes the reaction

$$
\begin{array}{l}
\text{COO}^- \\
|\\
\text{CH}_2 \\
|\\
\text{H}_2\text{O} + \text{CH}_2 \\
|\\
\text{CHNH}_3{}^+ \\
|\\
\text{COO}^-
\end{array}
\; + \text{NAD}^+ (\text{or NADP}^+) \rightleftarrows
\begin{array}{l}
\text{COO}^- \\
|\\
\text{CH}_2 \\
|\\
\text{CH}_2 \\
|\\
\text{C}{=}\text{O} \\
|\\
\text{COO}^-
\end{array}
\; + \text{NADH (or NADPH)} + \text{H}^+ + \text{NH}_4{}^+
$$

$$(12)$$

This important reaction participates in amino acid and carbohydrate metabolism and in the regulation of the utilization and production of ammonia.

In view of its physiological importance, it is not surprising that this mitochondrial enzyme has a widespread distribution (1,196–199), is regulated by many metabolites, and in many cases can react with either diphospho- or triphosphopyridine nucleotides (197). In animal tissue the activity of the enzyme is high in liver, kidney, and brain but is low in heart (196,197). The enzyme has been crystallized from bovine (200,201), chicken (202), frog (203), calf, pig (204), and human liver (205,206).

1. NATURE OF THE REACTION

Kinetic studies of both the bovine (207) and frog (203) liver enzymes reveal that the order of addition of substrates to the enzyme is sequential with the pyridine nucleotide adding first; in coenzyme oxidation, binding of $\text{NH}_4{}^+$ precedes α-ketoglutarate. As shown in Table IX, Michaelis constants tend to be higher and maximal velocities lower in the case of the frog enzyme.

The overall equilibrium constant (K) of the reaction has been measured kinetically and directly and ranges from 1×10^{-14} to 10×10^{-14} (molar)^2, the activity of water having been assumed to be 1 (199–203):

$$
K = \frac{[\text{NADPH}][\text{NH}_4{}^+][\alpha\text{-Kg}][\text{H}^+]}{[\text{NADP}^+][\text{Glutamate}]}
\tag{III}
$$

Using equilibrium dialysis, spectrophotometric and spectrofluorimetric techniques, Fisher has concluded that α-ketoglutarate and glutamate are not bound to the free enzyme. α-Ketoglutarate can be bound to the enzyme-NADPH complex but is not bound to the enzyme-NAD complex. Glutamate can be bound to the enzyme-NADH complex (*208,209*).

TABLE IX

Comparison of Kinetic Constants of Bovine and Frog
Liver Glutamate Dehydrogenase (*203,207*)[a]

Constant	Substrate or coenzyme	Bovine	Frog
Michaelis	NADPH	0.026 mM	0.2 mM
	Ammonia	3.2 mM	0.5 mM
	α-Ketoglutarate	0.7 mM	5 mM
	NADP	0.047 mM	0.5 mM
	Glutamate	1.8 mM	1.8 mM
Dissociation	NADPH	0.026 mM	0.2 mM
	NADP	0.23 mM	1.2 mM
Maximal velocity	NADPH oxidation	400 sec⁻¹	100 sec⁻¹
	NADP reduction	13 sec⁻¹	4 sec⁻¹

[a] Experiments were performed in 0.01 M Tris-acetate buffer containing 10 μM EDTA, pH 8.0, at 23°. Values are calculated on the basis of a molecular weight of 2.5×10^5 for the frog enzyme and 4×10^5 for the bovine enzyme.

Chen (*210*) has observed that the enzyme loses activity as well as protein fluorescence following ultraviolet irradiation. Fisher and Cross (*209*) have observed spectral changes in the 290 mμ region following the addition of glutamate to the enzyme-NADH complex. NADP markedly quenches the fluorescence of the frog liver enzyme (*203*). Since these spectral and fluorimetric changes are believed to be due mainly to tryptophan residues of the enzyme, it is suggested that this residue is part of the active center.

Binding of pyridine nucleotides, in the case of the beef liver enzyme, is apparently unrelated to thiol groups (*199*) and the NADP-specific yeast enzyme is apparently devoid of these groups (*211*). The hydrogen at the 4 position of the reduced nicotinamide is utilized in the oxidation of the coenzyme (*212*).

All glutamate dehydrogenases studied to date possess some alanine dehydrogenase activity (*197,213*).

2. MODIFIERS

The effects of a large number of modifiers of the initial rate of the glutamate dehydrogenase reaction have been studied (1,199). The theory of the effects of modifiers on multisubstrate enzyme-catalyzed reactions has not been developed. The glutamate dehydrogenase reaction is further complicated by the fact that double reciprocal plots of velocity versus diphosphopyridine nucleotide concentration are usually not linear. In spite of many limitations, conclusions have been made on the basis of kinetic data which seem to be reasonable and consistent with other experiments.

In general, animal enzymes possess, in addition to active sites, at least two types of allosteric sites (197). NAD and purine nucleotides are bound to one, and a second is specific for reduced pyridine nucleotides (214).

The nonanimal enzymes which have been studied do not seem to possess allosteric sites (197,215,216). The NAD-dependent enzyme of N. crassa is inhibited by purine nucleotides (216). However, since this inhibition is competitive with respect to the coenzymes, it seems likely that it results from binding of purine nucleotides to the active site (197).

a. *Purine Nucleotides.* In the case of the bovine liver enzyme, activators of glutamate oxidation, which are bound to the purine nucleotide site, are inhibitors of alanine oxidation and vice versa (217,218). Purine nucleotides and NAD are bound to this site. In general, adenosine derivatives (ADP, ATP, 5′-AMP, 3′-AMP, adenosine, adenosine diphosphoribose, and dATP) activate the glutamate dehydrogenase reaction (2′-AMP and 3′,5′-cyclic AMP have no effect), while guanosine and inosine derivatives (GDP, GTP, dGTP, IDP, and ITP) are inhibitors. Pyrimidines (UDP, CDP) do not modify the reaction (214). The order of specificity of binding to this site (in the case of the bovine liver enzyme in the absence of coenzyme) is ADP, GTP, ATP, ITP, IDP, and GDP (214). Binding of nucleotides is not competitive with respect to coenzyme (at the active site). In the presence of coenzyme the dissociation constant of activators is increased and that of inhibitors is decreased. Therefore the order of specificity in the presence of coenzyme (NADP or NADPH) is GTP, GDP, ITP, ADP, IDP, ATP, NAD (214). The values of the dissociation constants of these purine nucleotides are low (0.4–50 μM), and metals are not required for binding (214). Purine deoxyribonucleotides, as dATP and dGTP, have dissociation constants about 10-fold higher than their respective ribose analogs (214). NADP, unlike NAD, is not bound to this site (214) except at a high pH (219).

Most mammalian enzymes (bovine liver and brain, rabbit muscle, and pig kidney) are qualitatively similar in their response to purine nucleotide. The above mammalian enzymes, however, differ from rat, avian, and frog liver enzymes with respect to both the effect and the binding constants of the nucleotide (197,220). For example (in the presence of α-ketoglutarate and NH_4^+), ADP activates NADPH oxidation about 4-fold with the bovine liver, 10-fold with the rat liver (197), and has no effect on NADPH oxidation with the frog and tadpole liver enzymes (197,220,221). ATP activates NADH oxidation (when the concentration of NADH is low) with the bovine liver enzyme (214) but under the same conditions has very little effect on the chicken, frog, and tadpole enzymes (197,220,221). ATP has only a slight activating effect on NADPH oxidation with the bovine liver enzyme (214), activates the rat liver enzyme 6-fold (197), and has no effect on the frog and tadpole liver enzymes (197,220,221). ADP and ATP are bound tightly to all of these enzymes in both the presence and absence of coenzyme.

b. *Reduced Pyridine Nucleotides.* Since reduced pyridine nucleotides are not displaced from the enzyme by purine nucleotides and, in some cases, binding of the reduced pyridine nucleotide is enhanced by purine nucleotides (GTP or GDP), it is believed that reduced pyridine nucleotides are bound to a site different from the purine nucleotide site (214). Binding of a reduced pyridine nucleotide to this site inhibits both glutamate and alanine dehydrogenase activity (222). As with the case of the purine nucleotide site, reduced diphosphopyridine nucleotides are bound more tightly than are reduced triphosphopyridine nucleotides (203,214).

c. *Other Modifiers.* The effects of many other compounds discussed below are completely reversed by ADP, and do not seem to be competitive with coenzyme at the active site. This suggests that these modifiers are also bound to the purine nucleotide site. In many of these cases, however, rigorous proof of the nature of the binding site is not available.

i. *Hormones.* Some sterols and diethylstilbesterol have been found to inhibit glutamate and activate alanine dehydrogenase activity (223–226). Under rather specific conditions, corticosterone, hydrocortisone, and progesterone can activate NAD reduction with glutamate (227–229). Thyroxine is known to inhibit glutamate dehydrogenase activity (230–232).

ii. *Metal chelators.* A number of metal chelating reagents have been observed to inhibit the bovine liver enzyme (199). In some cases the ability of these reagents to inhibit seems to be more related to the aromatic or planar nature of the compound rather than to chelation (233).

EDTA activates the glutamate dehydrogenase reaction even after the enzyme has been passed through Chelex-100 (*197*).

iii. *Sulfhydryl reagents and heavy metal ions.* Zn^{2+} inhibits glutamate and activates alanine dehydrogenase activity (*197,222,234*). Methylmercuric hydroxide (or bromide) has an opposite effect (*224,235*). Many sulfhydryl reagents, although bound to the enzyme, have little effect on activity, but silver ions inhibit (*1,199,234,236,237*).

iv. *Other compounds.* Many other compounds have been studied. N'-alkylnicotinamide chlorides inhibit glutamate and activate alanine dehydrogenase activity if the alkyl chain is long (8–12 carbon units), and have an opposite effect if the alkyl chain is short (*238*).

3. COENZYME SPECIFICITY

The bovine liver enzyme is reactive with either di- or triphosphopyridine coenzymes. The frog liver enzyme is much less reactive with triphosphopyridine nucleotides since (a) the Michaelis constants are high, (b) the maximal velocities are low, (c) concentrations of NADPH 3-fold higher than the Michaelis constant are inhibitory, and (d) purine nucleotides do not activate the reaction. These results are summarized in Table X.

TABLE X

REACTIVITY OF COENZYMES WITH FROG LIVER GLUTAMATE DEHYDROGENASE (*220*)

		Kinetic constant			
Coenzyme	Modifier	Michaelis at active site (mM)	Dissociation of coenzyme at allosteric site (mM)	V_{max} (active site saturated) (sec^{-1})	V_{max} (both sites saturated with coenzyme) (sec^{-1})
NADP	None	0.5	Very high	4	—
	10 μM ADP or ATP	0.5	Very high	4	—
NAD	None	0.02	0.3	1.4	8.6
	10 μM ATP	0.07	Very high	7.1	—
	5 μM ADP	0.02	Very high	10.7	—
NADPH	None	0.2	1	100	0
	10 μM ADP or ATP	0.2	1	100	0
NADH	None	0.03	0.06	250	0
	100 μM ADP	0.2	0.2	1,670	0

The tadpole liver enzyme is reactive almost exclusively with diphosphopyridine nucleotides (*221*). The rat liver enzyme appears to be much more reactive with triphosphopyridine nucleotides (*239,240*), especially in the presence of ADP (*197*). Nonanimal enzymes are specific for only one of the two types of coenzymes (*197*).

4. SUBSTRATE SPECIFICITY

Glutamate dehydrogenase can react with a large number of substrates (*199*). Inhibition studies and the ability of the enzyme to utilize monocarboxylic acids suggest that the α-carboxyl group is not essential for activity but does play a role in the binding of substrates to the enzyme (*213,217,232,241–244*).

The metabolic significance of slow reactions with amino acids other than glutamate is uncertain. However, the reactivity of the frog enzyme with glutamine is about one-third that of glutamate (*244*). It is also possible that *in vivo* the enzyme exists as an inhibited complex with GTP. Under these conditions some amino acids, such as norvaline, are more reactive than glutamate (*218*).

On the basis of some experimental data it would appear that glutamate reacts with a different form of the enzyme than does alanine or other monocarboxylic acids (*218*). This would explain the lack of competition between alanine and glutamate and the fact that purine nucleotide activators of glutamate oxidation are inhibitors of alanine oxidation. However, with the frog liver enzyme, alanine is apparently a competitive inhibitor of glutamate, and both alanine and glutamate oxidation are activated by ADP and ATP (*244*). The kinetic effects of amino acids are very complex. Leucine and valine can behave as activators or inhibitors of α-ketoglutarate reduction depending upon the concentration of this substrate (*245*). Glutamate and other amino acids alter the apparent Michaelis constant of the coenzyme and consequently the apparent dissociation constant of a nucleotide modifier (*203,207,214,245*). Double reciprocal plots of velocity versus glutamate concentration [in the presence of GTP and the bovine enzyme (*214*) and alanine and the frog enzyme (*244*)] are not always linear. Abortive complexes such as NADH-enzyme-glutamate can form (*209*). Therefore kinetic results with various amino acids are difficult to interpret mechanistically, and more detailed studies are required. It would be useful in some experiments to measure the rate of formation of the specific keto acid from either glutamate or the monocarboxylic amino acids and not just the total rate of formation of reduced pyridine nucleotide.

5. PHYSICAL PROPERTIES

Since the bovine liver enzyme dissociates upon dilution (*200*), precise measurements of the molecular weight have been difficult. A wide variety of procedures have been used including gel filtration (*246,247*), electron microscopy (*248,249*), and more conventional techniques, i.e., light scattering, sedimentation (*199*). According to most recent estimates the bovine liver enzyme exists as a polymer with a molecular weight of 2.0×10^6 (*250*) (under favorable conditions for complete association) which can reversibly dissociate upon dilution into enzymically active monomers with molecular weights of 4×10^5 (*197,251*). The nature of the amino acid residues involved in the association-dissociation have been studied (*252*) using the sensitive technique of difference spectroscopy.

The chicken liver enzyme has very little tendency to associate and has a molecular weight of 4.3×10^5 (*253,254*). The frog liver enzyme can exist as a dimer which reversibly dissociates upon dilution into enzymic active monomers which have a molecular weight of 2.5×10^5 (*203*). The NAD-dependent enzyme of *N. crassa* does not associate and has a molecular weight of 2.5×10^5 (*255*).

In general, coenzyme (bound to the active or to the purine nucleotide site) and activators of the glutamate reaction (bound to the purine nucleotide site) increase the molecular weight of the enzyme, while inhibitors of the glutamate reaction (bound to the purine nucleotide site) decrease the molecular weight (*214,256*). Reduced pyridine nucleotides (bound to an allosteric site) decrease the molecular weight (*214,220,256*). Many other properties of the enzyme such as optical rotary dispersion (*257–259*), stability (*199,225,260–262*), and antigenicity are also altered by modifiers (*263,264*).

6. CORRELATION OF PROTEIN STRUCTURE WITH ACTIVITY

It can be seen from the previous discussion that the enzyme is quite flexible and can be modified with respect to both its activity and its physical properties by a large number of low molecular weight compounds. Therefore it seems likely that activators (such as ADP) induce a conformational change in the enzyme so that it becomes both more reactive with glutamate and more capable of association. GTP has an opposite effect. This does not mean that only associated enzyme is active with glutamate and dissociated enzyme with alanine. At low enzyme concentrations, only monomers exist and both activities are found. This does mean that glutamate activity and association of enzyme monomers may both be affected by the same type of conformational change. This change in the conformation of the enzyme will result in an increase in molecular weight only if the enzyme concentration is sufficiently high (*214*). Therefore there is not *always* a rigorous correlation between the molecular weight of the enzyme and activity (*214,256,265, 266*). The complexity of the relationship between enzyme structure and binding of nucleotides and coenzyme can be illustrated by the following experimental results. At high enzyme concentrations in the presence of NADH, two types of GTP binding sites are detected. The results suggest a cooperative effect of GTP binding (*267*) [binding of each molecule of GTP enhances binding of additional molecules (*268*)]. Under these conditions two distinct protein peaks are detected in the ultracentrifuge (*258,267*), and the average number of moles of GTP bound per mole of total protein correlates with the increase in area of the slower peak. Thus, under these conditions, binding of GTP apparently produces an enzyme conformation which is more suited for the binding of additional GTP. In the absence of coenzyme or at low enzyme concentrations this particular transition is not apparent. Under the latter conditions only one type of GTP binding site is detected.

The enzyme has apparently increased in complexity with evolution. For example, nonanimal enzymes are not modified by purine nucleotides, do not seem to associate, and react with either NADP or NAD (*197,255*). The frog liver enzyme reacts mainly with NAD, associates only to the level of a dimer, and purine nucleotides mainly effect the reaction with NAD (*203,220*). The tadpole liver enzyme is even more exclusively reactive with NAD (*221*). The bovine liver enzyme is reasonably reactive with both NADP and NAD, associates at least to the tetramer level, and the reaction with all four coenzymes is modified by purine nucleotides.

Antibody prepared against animal glutamate dehydrogenase (frog and bovine liver) cross-reacts with the enzyme from other animal (*215,221,269–271*), but not from nonanimal, sources (*215*). Since nonanimal enzymes do not possess purine nucleotide sites, this suggests that these sites are the antigenic part of animal enzymes. However, kinetic studies of the antigen-antibody reaction demonstrate that the antibody is not competitive with nucleotides at either active or allosteric sites (*270,271*). The frog and bovine liver enzymes are immunochemically quite similar (*271*). Therefore antigenic sites are apparently distinct from active and allosteric sites and, in contrast with these sites, have not been markedly modified with animal evolution. Binding of antibody alters both the enzyme-catalyzed reaction and the binding of nucleo-tides. Therefore antibody must influence both active and allosteric sites. It seems possible that the antibody is directed against a part of the enzyme molecule which mediates the interaction between active and allosteric sites.

7. STRUCTURE OF THE ENZYME

In addition to the previously described reversible dissociation to enzymic active monomers, the bovine liver enzyme can be irreversibly dissociated into inactive chains with detergents, guanidine, extremes of pH, and other reagents (230,231,251,272–276). All of these chains have the same, or nearly the same, molecular weight of 50,000 (273). Therefore the enzymically active monomer apparently consists of 8 polypeptide chains which according to end-group analysis all have N-terminal alanine and C-terminal threonine residues (274,277). As yet, no differences have been detected between the chains.

Estimates of the number of binding sites per mole of active bovine liver monomer are (using 4×10^5 for the molecular weight): NAD, 6 (205); NADPH 4 (278); NADH, 8 (279); GTP, 8 (267); and zinc, 0.6–0.8 (280).

Efforts to dissociate the molecule into chains with methylmercurials have been unsuccessful (224). The enzyme has been inactivated by incubating at pH 2.0 and reactivated by incubating at pH 7.6 with mercaptoethanol (281).

B. Alanine Dehydrogenase

The overall equilibrium constant of this reaction, as with the glutamate dehydrogenase reaction, is about 10^{-14} M (282). The enzyme (L-ala-nine: NAD oxidoreductase, EC 1.4.1.1) has been purified and crystallized from *Bacillus subtilis* (283). This enzyme reacts also with L-α-amino-butyrate, L-valine, L-isoleucine, L-serine, and L-norvaline, and in the opposite direction with α-ketoglutarate, glyoxylate, hydroxpyruvate, and α-ketoglutarate. This alanine dehydrogenase is reversibly inactivated by dilution; light scattering, but not sedimentation data, are consistent with dissociation of the enzyme (282).

The enzyme has also been purified (to about 65%) from vegetative forms of *B. cereus* (284). Both the *B. cereus* and *B. subtilis* enzymes have similar molecular weights [2.5×10^5 (284) and 2.3×10^5 (282), respec-tively].

C. General Amino Acid Oxidases

Numerous enzyme systems have been described in the literature that support the concept that amino acids are oxidatively deaminated in accordance with the general overall equation (Eq. 13). Since molecular

$$R—CH(NH_3^+)—COO^- + O_2 + H_2O \rightarrow R—CO—COO^- + NH_4^+ + H_2O_2 \quad (13)$$

oxygen is the hydrogen acceptor, the enzymes catalyzing the above reaction are termed amino acid oxidases. These enzymes, as far as they have been investigated, are all flavoproteins. The classical scheme for the individual steps of the overall reaction may be represented as as in Eqs. (14–16). Equation (13) is the sum of these reactions. In the absence

$$R—CH(NH_3^+)—COO^- + E\text{-flavin} \leftrightarrow$$
$$[R—C(=NH)—COO^-] + E\text{-flavin}—H_2 + H^+ \quad (14)$$

$$[R—C(=NH)—COO^-] + H_2O + H^+ \leftrightarrow R—CO—COO^- + NH_4^+ \quad (15)$$

$$E\text{-flavin}—H_2 + O_2 \rightarrow E\text{-flavin} + H_2O_2 \quad (16)$$

of catalase, the hydrogen peroxide formed may react with the α-keto acid produced as shown in Eq. (17). If catalase is present, Eq. (17) does

$$R—CO—COO^- + H_2O_2 \rightarrow R—COO^- + CO_2 + H_2O \quad (17)$$

not take place, and the overall reaction is shown in Eq. (18). Only one

$$R—CH(NH_3^+)—COO^- + \tfrac{1}{2}O_2 \rightarrow R—CO—COO^- + NH_4^+ \quad (18)$$

atom of oxygen is taken up per mole of amino acid oxidized, and the α-keto acid is not destroyed.

Indirect support for the formation of the hypothetical imino acid depicted in reaction (14) has been provided by a number of studies which exclude α,β-unsaturation in the course of the reaction. For example, it has been shown that: (a) the four isomers of isoleucine are enzymically oxidized by the appropriate amino acid oxidase to the corresponding optically active α-keto-β-valeric acids (114,285); (b) the L-isomers of β-phenylserine are converted by L-amino acid oxidase to the respective isomers of mandelic acid (286); (c) the L- and D-isomers of α-aminophenylacetic acid, which have no β-hydrogen atom, are attacked by the amino acid oxidases (8,287,288); and (d) on oxidation of L-leucine in the presence of D_2O by L-amino acid oxidase, no deuterium is found in the isolated α-ketoisocaproic acid (289).

The occurrence of free radicals in flavoprotein-catalyzed reactions has been established. The intermediate steps involved in the reduction and oxidation of flavin prosthetic group(s), as well as the specific molecular complexes formed, during amino acid oxidase-catalyzed reactions have been studied intensively. Fluorescence measurements, spectral studies, and electron spin resonance spectroscopy have provided data with regard to the properties of flavins in various oxidation states and the nature of various enzyme-substrate or inhibitor complexes. The reader is referred to several excellent reviews which have appeared (*290,291*), or will appear (*292*), for details of work in this area.

Until recently the amino acid oxidase reaction had been studied only in the direction of ammonia and α-keto acid formation. In the presence of air the reaction proceeds to completion and is essentially irreversible because of the reoxidation of the reduced flavoprotein by molecular oxygen. Meister and Radhakrishnan (*293,294*) have clearly shown that both D- and L-amino acid oxidases catalyze the anaerobic synthesis of D- and L-amino acids from the corresponding α-keto acid and ammonia. When an amino acid, ammonia, and the α-keto acid analog of a second amino acid are incubated with either amino acid oxidase under anaerobic conditions, the formation of the second amino acid is observed in accordance with the reactions in Eqs. (19) and (20). Since the overall

$$R_1—CH(NH_4{}^+)—COO^- + \text{E-flavin} + H_2O \leftrightarrow$$
$$R_1—CO—COO^- + NH_4{}^+ + \text{E-flavin}—H_2 \quad (19)$$

$$R_2—CO—COO^- + \text{E-flavin}—H_2 + NH_4{}^+ \leftrightarrow$$
$$R_2—CH(NH_3{}^+)—COO^- + \text{E-flavin} + H_2O \quad (20)$$

reaction, (Eq. 21), is equivalent to a transamination reaction, it has been

$$R_1—CH(NH_3{}^+)—COO^- + R_2—CO—COO^- \leftrightarrow$$
$$R_1—CO—COO^- + R_2—CH(NH_3{}^+)—COO^- \quad (21)$$

designated a "pseudotransamination." The reaction is markedly accelerated by the addition of ammonia; $^{15}NH_3$ leads to the formation of the ^{15}N-labeled amino acid. Spectrophotometric studies show that the reduced amino acid oxidase can be reoxidized anaerobically by the addition of ammonia and α-keto acid. These and other experiments clearly show that the reaction observed does not involve transamination and also establish that the reduction of the flavoprotein by an amino acid and the hydrolysis of the imino acid are reversible reactions.

The reactivities of different amino acids with the different amino acid oxidases are listed in Table XI.

The amino acid oxidases have been reviewed recently by Meister (*295*).

1. L-AMINO ACID OXIDASES

Four L-amino acid oxidases [L-amino acid : oxygen oxidoreductase (deaminating), EC 1.4.3.2] have been partially purified and studied from such widely different sources as snake venom, rat kidney, turkey liver, and molds.

   a. L-*Amino Acid Oxidase of Snake Venom.* Zeller and Maritz (*296*) described an enzyme, originally discovered in snake venom and later shown to be present in various tissues of both venomous and non-venomous snakes, which they termed *ophio*-L-amino acid oxidase. The enzyme is widely distributed in a large number of snake venoms (*297*). The L-amino acid oxidase from the venom of the cottonmouth mocassin (*Agkistrodon piscivorus piscivorus*) has been highly purified by Singer and Kearney (*298*) and that from the venom of the eastern diamondback rattlesnake (*Crotalus adamanteus*) has been crystallized by Wellner and Meister (*299*). Although the L-amino acid oxidases of other species of snake have been studied, they have not as yet been extensively purified.

   i. *Physical properties.* On the basis of spectrophotometric and enzymic methods, it has been established that flavin adenine dinucleotide (FAD) is the prosthetic group of the L-amino acid oxidase of both *A.p. piscivorus* (*300*) and *C. adamanteus* (*299*). The flavin content of the enzymes from mocassin and rattlesnake venoms is approximately the same; the values reported were 1 mole of FAD per 62,800 gm of protein (*298*) and 64,000 gm of protein (*299*), respectively. The sedimentation coefficients of the two enzymes are essentially the same (*298,299*). Molecular weight determinations of both enzymes by the Archibald " approach to equilibrium " method and by sedimentation and diffusion gave values between 130,000 and 150,000 (*299*). Since the minimum molecular weight on the basis of FAD content is in the neighborhood of 64,000, it was concluded that both enzymes contain 2 moles of FAD per mole (*299*). The turnover numbers reported for the mocassin and rattlesnake enzymes are 3100 (*298*) and 7000 (moles of substrate oxidized per minute per mole of enzyme) (*301*), respectively. The pH optima of both enzymes are in the range 7.0–7.5 (*298,299*).

   The enzymes from both sources were shown to be homogeneous in the ultracentrifuge (*298,299*). However, crystalline preparations of the enzyme from rattlesnake venom exhibited three components upon electrophoresis in the Tiselius apparatus (*299*). Since each component had essentially the same specific activity, they probably represent three distinct forms of the enzyme. The different forms are not spontaneously interconvertible, since the isolation of one component by column electrophoresis and subjecting it to reelectrophoresis in the Tiselius apparatus gave only a single peak. The enzyme isolation was routinely carried out with pooled venom from a large number of snakes. Therefore the enzyme was isolated and crystallized from the venom of a single snake. This preparation yielded two components upon electrophoresis;

TABLE XI

Comparison of Reactivity of Amino Acids with Different Amino Acid Oxidases (1)

Type of amino acid	D-Amino acid oxidase of sheep kidney	L-Amino acid oxidase of cobra venom	L-Amine acid oxidase of N. crassa	D-Amino acid oxidase of N. crassa	L-Amino acid oxidase of Mytilus edulis
Straight-chain aliphatic monoaminomonocarboxylic acids	3-C > 6-C > 4-C > 5-C > 8-C; 11-C, 12-C, 18-C not attacked	6-C > 8-C > 5-C; 3-C, 4-C, 11-C, 12-C, 18-C not attacked	4-C > 5-C > 6-C > 3-C > 8-C; 11-C, 12-C, 18-C not attacked	4-C > 5-C > 6-C > 3-C > 18-C; 8-C, 11-C, 12-C not attacked	6-C > 5-C > 4-C > 8-C > 3-C
Branched-chain monoaminomonocarboxylic acids	Valine > isoleucine > leucine	Only leucine oxidized	Leucine > isoleucine > valine	Leucine > isoleucine, valine	Only leucine oxidized
Aliphatic monoaminodicarboxylic acids	13-C rapidly oxidized; 4-C slowly oxidized; 5-C, 6-C, and 7-C not attacked	Only 13-C oxidized	7-C > 6-C > 13-C > 5-C > 4-C; All oxidized	5-C > 7-C > 4-C > 6-C; 13-C not attacked	C-11 > C-7 > C-6; 4-C and 5-C not attacked
Diaminomonocarboxylic acids	Ornithine > lysine	Not attacked	Ornithine > lysine	Not attacked	Lysine, ornithine > citrulline > arginine, canavanine > histidine > hydroxylysine (α-γ-diaminobutyric not attacked)
Amino acids with cyclic substituent	Tyrosine > aminophenylalanine > dimethylaminophenylalanine > phenylalanine	Tyrosine > aminophenylalanine and phenylalanine > dimethylaminophenylalanine	Dimethylaminophenylalanine > phenylalanine > aminophenylalanine > tyrosine	p-Aminophenylalanine > β-pyridyl(4)alanine > phenylalanine > tyrosine	Tyrosine > phenylalanine > tryptophan > 5-hydroxytrytophan

Dimethylaminophenylalanine consumed > 1 atom O, ε-N-sulfanilyllysine not attacked	Tyrosine, aminophenylalanine, dimethylaminophenylalanine, tryptophan, and β-furfuryl(2)alanine consumed > 1 atom O	Tyrosine, aminophenyl alanine, dimethylamino-phenylalanine, ε-N-sulfanilyllysine consumed > 1 atom O	p-(β-Aminoethyl)-phenylalanine, p-aminophenylalanine, p-dimethylaminophenylalanine, pyridyl(4)alanine, tyrosine consumed > 1 atom O	—
β-Quinolyl(2)alanine > β-quinolyl(4)alanine	Histidine and ε-N-sulfanilyllysine not attacked β-Quinolyl(4)alanine > β-quinolyl(2)-alanine	—	—	
—	—	β-Quinolyl(4)alanine and β-quinolyl(2)-alanine attacked at similar rates		
Unclassified amino acids Proline > methionine > serine > threonine > cystine	Only methionine oxidized	Serine > cystine > methionine > threonine	Only methionine attacked	Proline not attacked Methionine > ethionine > cystathionine > cystine > meso-cysteine, L-cystine, LL-djenkolic acid > homocysteine
—	—	Proline and piperidine-α-carboxylic acid not attacked	—	Cysteic acid, homocystic acid, cysteinesulfinic acid not attacked α,α′-Diaminopimelic acid (meso and LL) oxidized
—	—	Cysteine consumed > 1 atom O	—	
Total number amino acids tested 38	38	38	38	42
Total number amino acids oxidized 31	19	33	23	31

the venom from another snake gave three components. Hence it appears that there is a considerable variability in the properties of L-amino acid oxidase, not only among different species of snakes, but also even among individuals of the same species. The structural differences responsible for the differences in mobility of the three forms of the enzyme have not been established. The different forms cannot be associated with large differences in molecular weight in view of the homogeneity obtained on ultracentrifugation. The similar specific activities of the three forms of the enzyme suggest that changes at the active site of the enzyme are not involved either. Although the enzyme isolated from *A.p. piscivorus* was homogeneous on electrophoresis (*298*) at different pH values, the electrophoretic mobility of some preparations differed markedly, suggesting that more than one form of L-amino acid oxidase may occur in mocassin venom as well.

ii. *Specificity.* Early specificity studies have been reviewed by Zeller (*297*) and more recently by Meister (*295*). Additional work on the structural requirements for substrate activity has been carried out recently by Zeller *et al.* (*302*). The reactivities of different amino acids with the different amino acid oxidases are listed in Table XI. Many of the studies recorded in the literature were carried out under variable conditions of amino acid concentration, pH, and with either oxygen or air as the gas phase. Since all of the factors may affect the rates of oxidation of different amino acids to a different degree, the numerical values are of limited use. In general, L-amino acid oxidases are highly specific for the L-isomer. Zeller (*297*) has summarized the specificity requirements as follows: " The substrate must possess a free carboxyl group, an unsubstituted α-amino group, and an organic radical. A second amino or carboxyl group inhibits a substance otherwise suitable as a substrate for the enzyme." If the second amino group is acylated, or if the second carboxyl group is converted to an amide or ester, oxidation occurs. If the charged group on the terminal carbon atom of the acidic and basic amino acids is neutralized, as indicated above, the compound is oxidized; L-asparagine, L-glutamine, L-citrulline, ε-*N*-benzoyl-L-lysine, and δ-*N*-carbobenzoxy ornithine are substrates, as are the ω-ethyl esters of aspartate and glutamate (*295*). The recent observation that 3,4-dehydro-L-proline is oxidized by rattlesnake L-amino acid oxidase (*303*) indicates that the requirement for a primary α-amino group is not absolute.

A number of amino acid analogs [β-arylalanines (*304*)] and derivatives of α-L-aminodicarboxylic acids [β-aspartylalanine, β-aspartylglycine, γ-glutamylalanine, etc. (*67*)] are oxidized by snake venom L-amino acid oxidases. The rates of oxidation of the β-arylalanines are comparable to those of the corresponding naturally occurring amino acids, supporting the generalization of Zeller (*297*) that the β-group exerts relatively little influence on the substrate activity of different amino acids. However, in studies with α,β-diasymmetric amino acids as substrates for ten different L-amino acid oxidases, it was found that substrates with the α-L,β-L-configuration were more readily oxidized than were the corresponding diastereoisomers (*305,306*).

iii. *Inhibitors.* A variety of carboxylic and sulfonic acids have been studied as possible inhibitors of venom L-amino acid oxidase (*296, 307–309*). Benzoic, salicylic, mandelic, iodoacetic, some aliphatic and aromatic sulfonic acids, and sulfonamides inhibit competitively. Carbonyl reagents such as hydroxylamine and semicarbazide are reported also to be competitive inhibitors. The latter effect has suggested the presence of a carbonyl group on the enzyme surface essential for substrate activation. Riboflavin and some of its analogs were found to be potent inhibitors of L-amino acid oxidase (*300*). This inhibition is apparently not competitive with respect to either substrate or FAD.

In their original studies on venom L-amino acid oxidase, Zeller and Maritz (*296*) found that the optimal concentration of L-leucine for an enzyme from *Vipera aspis* was 6.7×10^{-3} M and higher concentrations inhibited markedly. The purified enzyme from mocassin venom was also inhibited by substrate, but it was found that in oxygen the optimal concentration of L-leucine was higher than in air (*298*). Kinetic studies, in which the reaction with L-leucine, L-methionine, and L-valine were examined in the presence of various concentrations of oxygen, have been carried out with the crystalline enzyme from rattlesnake venom (*301*). In air or in oxygen, L-leucine and L-methionine inhibited the reaction markedly in concentrations above 0.01 M. L-Valine, which is a poor substrate for the enzyme, gave little or no inhibition in air or oxygen.

iv. *Reversible inactivation.* The reversible inactivation of mocassin L-amino acid oxidase was reported by Singer and Kearney, who have made an extensive study of this phenomenon (*310*). According to these workers, the enzyme exists as an equilibrium mixture of active and inactive forms. The equilibrium between the active and inactive forms was dependent upon the pH and upon the concentration of Cl$^-$ and other monovalent anions. The dependence of the equilibrium constant of the reaction on pH resembled the dissociation curve of an acid having a pK of about 6.5. The rates of inactivation and reactivation were found to be strongly temperature-dependent. However, the rate of inactivation was independent of pH. Since no change in size, shape, electrophoretic mobility, or solubility of the enzyme was detected upon extensive inactivation, the process was interpreted as involving only a limited alteration of the configuration of the native enzyme. On the basis of the nature of the pH activity curve and the other considerations noted above, it has been suggested that the inactive form of the enzyme differs from the active form by the presence of a new acidic group, probably imidazole in nature.

Crystalline L-amino acid oxidase from rattlesnake venom has also been shown to undergo reversible inactivation (311). Studies on the sedimentation coefficients, electrophoretic mobilities, and solubilities of the two forms of the enzyme are compatible with the idea that no dissociation, aggregation, or denaturation of the enzyme occurs during the inactivation process. No differences in the two forms of the enzyme could be detected by immunochemical techniques. Changes in the spectrum and optical rotatory dispersion of the two forms were observed; these changes were completely reversed upon reactivation. There is no loss of FAD, or other cofactors, since none must be added for reactivation. The spectrum of the inactive form of the enzyme is more like that of free FAD than is the active form. It was concluded that inactivation and reactivation are associated with conformational changes at the active center of the enzyme and with changes in the mode of binding of FAD to the protein.

b. Rat Kidney L-Amino Oxidase. Early studies by Krebs (8) with homogenates of liver and kidney clearly indicated that different systems exist in animal tissues for the oxidation of D- and L-amino acids. Subsequent investigation in numerous laboratories has established that there is a very active D-amino acid oxidase in mammalian systems, but no L-amino acid oxidase with comparable activity has been demonstrated. It is probable that most of the activity observed by Krebs with L-amino acids in crude tissues may be accounted for by the coupled action of transaminases and L-glutamate dehydrogenase (see Section II). However, an enzyme which oxidizes a series of L-amino acids has been described by Green and associates (312–315) and was first brought to a high stage of purity by this group. This enzyme has been obtained recently in crystalline form from rat kidney mitochondria (316). The activity of this enzyme is too low to account for the rate of L-amino acid oxidation observed by Krebs. Moreover, the enzyme has a greater activity with a series of L-α-hydroxy acids than with L-amino acids. On this basis, it may be inappropriate to continue to consider this enzyme an L-amino acid oxidase.

i. *Physical properties.* The crystalline enzyme obtained from extracts of rat kidney mitochondria by Nakano and Danowski (316) gave one band on starch gel electrophoresis and a single symmetrical peak in the ultracentrifuge. In agreement with earlier studies (313), flavin mononucleotide (FMN) was shown by spectral and enzymic methods to be the prosthetic group. On the basis of the FMN content (0.92%) of the crystalline enzyme, the minimum molecular weight was calculated to be 49,300. Since the molecular weight of the enzyme was $88,900 \pm 1100$, as determined by sedimentation equilibrium experiments, it was concluded that there are 2 moles of FMN per mole of enzyme. The same conclusion had been reached by earlier workers (313). However, this group reported not only a monomer with a molecular weight of about 130,000, but also a tetramer with a calculated molecular weight of 552,000 containing 8 FMN groups (313,317). Both components had equal catalytic activity.

ii. *Other properties.* The enzyme catalyzes the oxidation of a variety of α-aminomonocarboxylic and α-hydroxy acids of the L-configuration. The hydroxyamino acids, serine and threonine, the dicarboxylic amino acids, aspartate and glutamate, and the diamino acids, lysine and ornithine, are not attacked (312,316). D-Amino acids and optically inactive acids, such as glycine and α-hydroxyisobutyric acid, are not oxidized.

In general, the pH optimum for the oxidation of amino acids is higher (8.7–10) than that for α-hydroxy acids (about 8) (312,316,318). The turnover numbers reported for the α-hydroxy acids are 3–4 times that reported for the amino acids (313,316). The turnover number for the oxidation of L-leucine is far below those reported for other L-amino acid oxidases and for D-amino acid oxidases.

The enzyme is low or absent in cat, dog, guinea pig, rabbit, pig, ox, or sheep tissues, the best source being rat liver or kidney (312). The limited distribution of this L-amino acid oxidase and its extremely low turnover number make it highly doubtful that this enzyme plays any major part in mammalian amino acid metabolism.

c. Avian Liver L-*Amino Acid Oxidases.* An L-amino acid oxidase which is relatively specific for the basic amino acids has been obtained from turkey liver (319–325). The enzyme has been purified approximately 1000-fold (322). It catalyzes the typical oxidative type of reaction giving rise to 1 mole of ammonia with the utilization of 1 mole of oxygen in the absence of catalase. The specificity of the enzyme is greatest for the basic amino acids (ornithine, lysine > arginine > histidine > aromatic amino acids > leucine > methionine). Crude preparations exhibited activity with alanine. The following amino acids were found to be inactive as substrates: glutamic acid, aspartic acid, α-aminoadipic acid, serine, valine, and isoleucine. In studies with ^{15}N it was clearly established that the α-amino group is removed in the oxidation of the basic amino acids (323,325).

A general L-amino acid oxidase has been purified as a particulate preparation from the microsomal fraction of chicken liver (326). A number of different L-amino acids are oxidized, but neither D-amino acids nor α-hydroxy acids are substrates for the enzyme. Although this enzyme also catalyzes the oxidation of the basic amino acids, lysine, ornithine and arginine, the greatest activity was observed with L-leucine. Hence, on the basis of relative substrate specificity, this enzyme is different from the enzyme from turkey liver.

d. *Mold* L-*Amino Acid Oxidase.* The presence of an L-amino acid oxidase in both the mycelium and culture fluid of *Neurospora crassa* was reported by Bender *et al.* (*327*) and by Bender and Krebs (*328*). The production of the enzyme is depressed in nutritionally rich media and stimulated under conditions of nutritional deprivation; L-amino acid oxidase activity is greatly increased in media limited in biotin or nitrogen (*327,329*). Purification of the enzyme has been achieved (*330*). The prosthetic group of the purified enzyme was shown to be FAD in a ratio of 1 mole of FAD per 11,000 gm of nondialyzable nitrogen. The turnover number of the preparation was 2100 moles of phenylalanine per mole of FAD per minute. Burton (*330*) observed a broad plateau between pH 6 and 9.5 in the pH activity curve using phenylalanine as substrate, whereas Thayer and Horowitz (*329*), using a narrower pH range of 5.6–7.6, found considerable variation in the pH optimum with different amino acids as substrates. The latter investigators interpreted their findings to mean that the state of ionization of the amino acid substrate was of importance for activity. Further support of this interpretation was pointed out (*329*) in the case of esterification of the carboxyl group of histidine with the resultant loss of activity.

Although retaining a high substrate specificity for the L-amino acids, *Neurospora* L-amino acid oxidase appears to have a broader substrate specificity than the other L-amino acid oxidases studied. Thus, as can be seen from Table XI, thirty-three of thirty-eight amino acids tested were active as substrates. In addition to these amino acids, the following also were found to be active as substrates: canavanine, arginine, citrulline, α,ε-diaminopimelic acid, cystathionine, α-aminohydroxy-*n*-caproic acid, glutamine, α,γ-diaminobutyric acid, and glycine (*329,330*). Substrate concentration studies revealed that this enzyme, like the L-amino acid oxidase from snake venom, is inhibited by high substrate concentrations.

Other mold L-amino acid oxidases have been demonstrated in acetone powders of the mycelia of various species of *Penicillium* and *Aspergillus* (*331*).

e. *Other* L-*Amino Acid Oxidases.* In addition to the enzymes mentioned above, L-amino acid oxidases have also been found in certain bacteria (*332*) and in mollusks (*333,334*). The substrate specificity of an

L-amino acid oxidase obtained from the digestive gland of *Mytilus edulis* is shown in Table XI. Although amino acid oxidases are widely distributed in mollusks, they are usually active with the D-isomers.

2. D-AMINO ACID OXIDASES

Although D-amino acids can no longer be considered "unnatural," they are, from a quantitative standpoint, nevertheless of limited distribution in nature. This fact makes the explanation of the widespread occurrence of highly active D-amino acid oxidases [D-amino acid : oxygen oxidoreductase (deaminating), EC 1.4.3.3] somewhat difficult. The capriciousness of nature—or of the enzymologist—is clearly revealed in the fact that a great deal is known about the properties and mechanism of action of D-amino acid oxidases but very little about the role of D-amino acids in metabolism.

a. Mammalian D-Amino Acid Oxidases. Since the initial demonstration by Krebs (*8*) of the existence of a D-amino acid oxidase in mammalian liver and kidney, much progress has been made in the purification, mechanism of action, substrate specificity, and other properties of this enzyme. The enzyme has been shown to be present in the liver and kidney of all vertebrates studied (see *295*) with the exception of mouse liver (*335*).

Highly active, crystalline preparations of the enzyme have been obtained in several laboratories. A common step in the purification procedures has been the stabilization of the enzyme toward heat by the addition of a competitive inhibitor (benzoate). The holoenzyme was crystallized by Kubo *et al.* (*336,337*) who stated that it contained about 10 atoms of iron per flavin group. However, Massey *et al.* (*338*), using a slightly modified procedure, obtained the crystalline enzyme in an iron-free form. Yagi *et al.* (*339*) have reported that the above preparations of D-amino acid oxidase contain benzoate. This group (*340,341*) has isolated the enzyme as the apoprotein by removal of the FAD by acid ammonium sulfate treatment, the holoenzyme by reconstitution of the apoenzyme with FAD, and the artificial benzoate complex by treatment of the holoenzyme with benzoate. A method for the crystallization of the apoenzyme has been described recently by Miyake *et al.* (*342*).

i. *Physical properties.* The prosthetic group of D-amino acid oxidase was shown to be FAD by Warburg and co-workers (*343,344*). The crystalline enzyme isolated by Massey *et al.* (*338*) was found to contain 1 molecule of FAD per unit weight of 45,500. Molecular-weight values for D-amino acid oxidase (*338,345,346*), like those for most enzymes that undergo reversible association and dissociation, have been difficult to determine. Charlwood *et al.* (*346*) showed that in solution the enzyme existed as an equilibrium mixture of aggregates in which molecular-weight species two and four times the unit molecular weight of 45,500 were the predominant forms. Evidence for even higher-molecular-weight species was obtained. A recent study by Antonini *et al.* (*347*) demonstrates the importance of experimental conditions on the state of enzyme association. These workers found that the extent of polymer formation is decreased as the temperature and pH are lowered and as the concentration of chloride ion is increased. Polymer formation is also affected by the form of the enzyme, increasing in the order apo-enzyme, benzoate complex, and holoenzyme. Under all conditions studied, the value of the molecular weight, as determined by light scattering, extrapolated to zero protein concentration is in the neighborhood of 90,000. Hence, since the minimum molecular weight based on FAD content is 45,500, the enzyme is concluded to have 2 moles of FAD in its monomeric form.

The effects of temperature on the properties of D-amino acid oxidase have been studied in detail by Massey *et al.* (*348*). They report that the enzyme undergoes a temperature-dependent, reversible transition in structure around 12–14°. Changes in ultraviolet absorption and fluorescence indicate that this transition results in a greater solvent interaction of protein tryptophan residues below the critical temperature than above it. The effect of temperature on the visible absorption spectrum are indicative of altered binding constants of FAD to the apoenzyme at different temperatures. The sedimentation constant of the enzyme was found to change as a function of temperature as well as the catalytic properties. Although it is possible to account for the observed changes in physical properties with temperature in terms of a temperature-dependent association, Massey *et al.* (*348*) suggest that the association phenomena are probably a secondary reflection of a primary configurational change in the protein which is induced by the temperature change. Evidence has also been presented for a change in protein conformation in the formation of holoenzyme from FAD and apoenzyme (*349*). The reaction proceeds in two stages: a rapid binding of FAD followed by a slow secondary change coincident with the appearance of catalytic activity.

ii. *Specificity.* D-Amino acids show very large differences in reactivity and affinity for the enzyme. The relative rates of oxidation of various amino acids is not only dependent upon the substrate concentration but also upon whether air or oxygen is used as the gas phase. In addition, the pH activity curves vary from one amino acid to another. Variations in the purity of the enzyme preparation, amino acid concentration, gas phase, and pH make comparisons between the results of various investigators difficult, if not impossible. Earlier studies of an extensive nature have been carried out by Bender and Krebs (*328*) and by Green-stein *et al.* (*350*). Recent studies include those reported by Dixon and Kleppe (*351*). The following generalizations may be made. The enzyme is completely specific for the D-isomers. D-Proline has a maximum velocity three times that of any other amino acid. The next most reactive substrates are D-methionine, D-alanine, D-isoleucine, and D-phenylalanine. The acidic and basic amino acids are not oxidized with the exception of D-ornithine. D-Histidine is oxidized. Like the mammalian L-amino acid oxidase, but in contrast to the L-amino acid oxidase of snake venom, D-amino acid oxidase attacks not only amino acids with a primary amino group but also those with a secondary amino group, e.g., *N*-monomethylamino acids (*8,352,353*). β-Amino acids, *N*-acyl-α-amino acids, and peptides containing α-D-amino acids are not attacked. α,β-Diasymmetric amino acids with an α-D,β-D configuration are more readily oxidized by the enzyme than are their corresponding diastereoisomers (*305,306*).

iii. *Inhibitors.* Numerous investigators have studied various compounds as inhibitors of mammalian D-amino acid oxidase. According to Hellerman and Frisell (*354*), there are three types of inhibition: (a) that which is due to the interference of the combination of FAD with the apoenzyme; (b) that which is due to a competition with the amino acid substrate or its primary oxidation product; and (c) that which is due to the sensitivity of D-amino acid oxidase to certain sulfhydryl-characterizing reagents.

Quinine, atabrine, and related substances at low concentrations ($1 \times 10^{-3}\ M$) act as effective competitive inhibitors of FAD (*355*). AMP, ADP, and FMN are considerably more potent competitive inhibitors of FAD than quinine and a number of other compounds tested (*356–358*). Adenosine and adenine are less effective inhibitors and apparently act by a different mechanism. Kearney (*359*) has demonstrated that analogs of FAD (riboflavin, isoriboflavin, dichloroflavin, and galactoflavin) are very effective inhibitors of the enzyme. Their action is in part competitive with the prosthetic group, although an irreversible combination with the protein is also manifested. Other competitive inhibitors include adenosine monosulfate (*360*), riboflavin monosulfate (*360*), chlorotetracycline (*361*), and various anions (*358*). On the basis of these inhibition studies, it has been suggested (*358,360*) that the phosphoryl group, as well as adenine and riboflavin moieties, are all involved in the binding of the prosthetic group to the apoenzyme. Additional support for this conclusion has been provided by studies with analogs of FAD, either modified in the adenylate or riboflavin moieties [for references see (291,362)]. Alkylation of the 3-NH of the isoalloxazine ring of FAD results in total inactivity of the compound as a coenzyme (362), although it has been concluded that this group is not bound to the protein since the holoenzyme exhibits fluorescence (358).

The second type of inhibition, i.e., competitive inhibitors of the amino acid substrates, is that originally observed with aromatic acids (363,364). However, on the basis of kinetic studies, Hellerman and associates (365,366) have shown that this type of inhibition is not limited to compounds with aromatic properties, but is associated in general with anions possessing conjugated double-bond systems in their structures of the type, $-\overset{|}{C}=\overset{|}{C}-X^-$, where X^- is a carboxylate or phenolic hydroxyl group. Thus acyclic and heterocyclic compounds with this constituent grouping, such as acrylates, indole and pyridine derivatives, and phenolates, are equal to or greater than benzoate in inhibitory potency. If either the double bond or the anionic portion of the structure is eliminated, the inhibitory action of the compound is decreased or eliminated. The inhibitory action of these compounds is probably due to their obvious structural similarity to that of the intermediate α-imino acids. Other inhibitors of this type are L-leucine, L-phenylalanine (367), fatty acids (368), aliphatic α-keto or α-hydroxy acids (351), alkylthio fatty acids (369), and kojic acid (370).

The sulfhydryl character of D-amino acid oxidase has been indicated by numerous investigators (354,355,371,372). Frisell and Hellerman (354) have shown that sulfhydryl reagents which have a carboxylate function, such as p-chloromercuribenzoate and o-iodosobenzoate, have a dual inhibitory action, i.e., as sulfhydryl-reactive reagents and as substrate competitive inhibitors. Phenylmercuric acetate, on the other hand, has no carboxylate function and it essentially inhibits irreversibly. Inhibition by this compound leaves a residual active protein which retains unimpaired its affinity for substrate or for authentic substrate competitive inhibitors. Their findings do not support the idea that protein sulfhydryl groups are directly concerned with the bindings of substrate, as had been suggested earlier (372). Kinetic evidence consistent with the idea that p-mercuribenzoate competes with the adenylate moiety of FAD for the protein and that this moiety combines with a protein —SH group has been presented (373,374). Recent studies neither support nor contradict this role of the sulfhydryl group (342).

    b. Mold D-Amino Acid Oxidases. Horowitz (375) observed that extracts of Neurospora contained an active D-amino acid oxidase similar in its action to that of the kidney enzyme. The enzyme was found to oxidize the D-forms of most of the amino acids tested (Table XI). In general, activity for the different amino acids was similar to that reported later by Bender and Krebs (328), with methionine being the most active. The *Neurospora* D-amino acid oxidase has a pH optimum at 8.0–8.5 and was not significantly inhibited by cyanide (0.001 *M*) or iodoacetate (0.001 *M*). Benzoate (0.01 *M*), in contrast to its effect on kidney D-amino acid oxidase, did not inhibit *Neurospora* D-amino acid oxidase. DL-α-Amino-α-methylbutyric acid was observed to act as a competitive inhibitor, although it was not itself oxidized. According to Thayer and Horowitz (329), the D-amino acid oxidase is present in young cultures of all strains of *Neurospora* examined.

The presence of D-amino acid oxidase in the following molds has been reported by Emerson *et al.* (376): *Penicillium chrysogenum, notatum,* and *roqueforti* (the latter being particularly active), and *Aspergillus niger. Penicillium sanguineum* was found to be practically free of the enzyme. These investigators were able to prepare soluble D-amino acid oxidase preparations from *P. chrysogenum* which showed a pH optimum at 8.5. The reaction involved was that shown by Eq. (18), as deduced from balance studies. The enzyme was specific for D-amino acids and was highly active for leucine, methionine, alanine, α-aminobutyric acid, norleucine, phenylalanine, and valine. Less active as substrates were isoleucine, α-aminocaprylic acid, tryptophan, lysine, threonine, and glutamic acid. Aspartic acid and glycine were not oxidized. All attempts to purify the enzyme were unsuccessful. The authors make no mention of the nature of the prosthetic group, but it would appear that this enzyme is a flavoprotein.

c. Other D-*Amino Acid Oxidases.* D-Amino acid oxidase is also present in the livers of invertebrates, e.g., *Sepia officinalis* and *Octopus vulgaris* as well as *Helix* (377,378). The substrate specificity of the enzyme from this source resembles that of sheep kidney, although certain differences were noted.

Both D- and L-amino acids are oxidized by many bacteria, e.g., *Proteus, Pseudomonas, Bacterium cadaveris,* and *Escherichia coli,* but the enzymes have not been studied in detail (see 2).

D. Specific Amino Acid Oxidases

1. D-GLUTAMIC ACID OXIDASE

Blaschko and associates first reported that the livers of certain cephalopods contain not only a general D-amino acid oxidase (378) but also a specific D-glutamic acid oxidase (*379*). The latter also attacks D-aspartic acid, but at a slower rate. A 1000-fold purification of D-glutamic acid oxidase from the hepatopancreas of *Octopus vulgaris* has been achieved (*380*). The purified enzyme attacks D-glutamic acid ($K_m = 8 \times 10^{-3}$ M) and also D-aspartic acid ($K_m = 4.5 \times 10^{-3}$ M) but at a slower rate. Although crude extracts exhibited activity with other D-amino acids, the purified enzyme did not. In view of the constant ratio of activities with the two substrates during the purification procedure it appears that one enzyme is involved. The stoichiometry of the reaction was established in the presence of catalase, and the α-keto acids were identified as the 2,4-dinitrophenylhydrazones. The optimum pH for both substrates is at 8.1–8.3. Benzoate (10^{-3} M) has been demonstrated to inhibit D-amino acid oxidase (*363*), but had no effect on this system. Veronal, a known inhibitor of glycine oxidase (*381*), and urethane inhibited the enzyme, as did L-glutamic acid. The latter inhibition was competitive in nature. The enzyme is inhibited by sulfhydryl reagents such as monoiodoacetate, *o*-iodosobenzoate, and *p*-chloromercuribenzoate. Higher activities were observed in oxygen than in air. A partial resolution of the enzyme was achieved and reactivation was obtained with FAD but not with FMN or riboflavin.

A D-glutamic acid oxidase has been purified 100-fold from *Aspergillus ustus* (*382,383*). The enzyme deaminated D-glutamic acid and D-aspartic acid but had no activity with other DL-amino acids. The enzyme resembled that prepared from the octopus in its other properties (*101*). Thus the optimum pH is at 8 for both substrates, and it appears, on the basis of inhibition and purification studies, that one enzyme is involved here as well. The enzyme is inhibited by 10^{-3} M concentration of KCN and *p*-chloromercuribenzoate but not by metal-binding agents.

2. GLYCINE OXIDASE

Ratner and co-workers (*381*) have partially purified an apoenzyme from pig kidney which upon addition of FAD oxidizes glycine or sarcosine according to Eq. (18), since their preparation contained catalase. The preparation contained D-amino acid oxidase, but evidence was presented suggesting that the two enzyme systems were different. The preparation had a Michaelis constant of the order of 0.04 M with a pH optimum at 8.3; at pH 7, the enzyme activity dropped to one-eighth that at the optimum pH. The enzyme was found to be present in the kidney and liver of all the following animals studied: cat, dog, lamb, ox, rat (liver only), pig, human beings, and rabbit. The enzyme does not attack *N*-dimethylglycine, phenylglycine, *p*-aminophenylglycine, creatine, or peptides of glycine. A detailed comparison of D-amino acid oxidase and glycine oxidase of kidney has cast some doubt on the separate existence of the two enzymes (*384*).

3. D-ASPARTIC ACID OXIDASE

Still *et al.* (*385*) reported that rabbit kidney and liver contain a soluble enzyme which catalyzes the aerobic oxidation of D-aspartate to oxalacetate plus NH_3 with the formation of hydrogen peroxide. In a later study by Still and Sperling (*386*) the D-aspartic acid oxidase was resolved and reactivated by the addition of FAD. The purified enzyme showed about one-sixth the activity with D-glutamate; this, according to these workers, is best explained by the presence of a D-glutamic acid oxidase. The activity of D-aspartic acid oxidase is higher than that of D-amino acid oxidase in rabbit kidney and liver, and they are of the same order of activity in pig kidney. In contrast to pig kidney D-amino acid oxidase, which is inhibited by benzoic acid, the D-aspartic acid oxidase was unaffected. The enzyme is not present in rat liver (*387*).

V.   NONOXIDATIVE DEAMINATION

A. Hydroxyamino Acid Dehydrases (Deaminases)

This group of enzymes catalyzes a nonoxidative deamination reaction resulting from a primary dehydration of the substrate. The reaction for serine ($R = H$) and for threonine ($R = CH_3$—) has been postulated to occur as shown in Eq. (22), and that for homoserine as depicted in

$$\underset{\underset{OH}{|}\;\underset{NH_2}{|}}{R-C-CH-COOH} \underset{-H_2O}{\rightleftharpoons} \underset{\underset{NH_2}{|}}{R-CH=C-COOH} \rightleftharpoons$$

$$\underset{\underset{NH}{||}}{R-CH_2-C-COOH} \xrightarrow{+H_2O} \underset{\underset{}{\overset{O}{||}}}{R-CH_2-C-COOH} + NH_3 \qquad (22)$$

Eq. (23). The coenzyme for these enzymes has been shown to be pyridoxal-P. The dehydration and deamination may be effected through the

$$\underset{\underset{OH}{|}\quad\underset{NH_2}{|}}{CH_2-CH_2-CH-COOH} \underset{-H_2O}{\rightleftharpoons} \underset{\underset{NH_2}{|}}{CH_2=CH-CH-COOH} \rightleftharpoons \underset{\underset{NH_2}{|}}{CH_3-CH=C-COOH}$$

$$\rightleftharpoons \underset{\underset{NH}{||}}{CH_3-CH_2-C-COOH} \xrightarrow{+H_2O} \underset{\overset{O}{||}}{CH_3-CH_2-C-COOH} + NH_3 \qquad (23)$$

formation of a Schiff base between the substrate and pyridoxal-P (see Section III, B, 1). Studies reported on the dehydrases include those in mammalian liver (*388–400*), bacteria (*401–424*), *Neuospora crassa* (*425–428*), and yeast (*429*).

1. L-SERINE, THREONINE, AND HOMOSERINE DEHYDRASES OF MAMMALIAN LIVER

L-Serine dehydrase [L-serine hydro-lyase (deaminating), EC 4.2.1.13] and L-threonine dehydrase [L-threonine hydro-lyase (deaminating), EC 4.2.1.16] have been separated and purified from sheep liver by Sayre and Greenberg (*388*). L-Serine dehydrase was found to be specific for L-serine. No activity was observed with the D-isomer or with D- or L-threonine. L-Threonine dehydrase was active with L-threonine, but not the D-isomer, and to a slight extent with L-serine. Additional studies on L-threonine dehydrase from sheep liver have been reported by Nishimura and Greenberg (*389*). The purified enzyme deaminates L-allothreonine and L-serine as well as L-threonine. The activity with L-threonine is strongly inhibited by L-serine.

L-Serine dehydrase [L-serine hydro-lyase (deaminating), EC 4.2.1.13] was first obtained in a partially purified form from rat liver by Selim and Greenberg (*390,391*), who demonstrated that their protein preparation not only catalyzed the deamination of both L-serine and L-threonine but also contained cystathionine synthetase [L-serine hydro-lyase (adding L-homocysteine), EC 4.2.1.21] activity as well. Evidence was presented subsequently by Goldstein et al. (*392*) that the L-serine and L-threonine dehydrase activities were a function of the same enzyme. Pitot and associates (*393*) have investigated the variation in activity of this enzyme under a variety of physiological conditions. They, as well as Goldstein et al. (*392*) and others (*394,395*), have established that the activity is greatly increased in livers of rats fed a high-protein diet. In view of this fact, rats maintained on a high-protein diet were used by Nagabhushanam and Greenberg (*396*) in their isolation from liver of a homogeneous preparation of the enzyme, cystathionine synthetase-L-serine, and L-threonine dehydrase.

The molecular weight of the enzyme has been estimated to be about 20,000 from its sedimentation and diffusion coefficients and from ultracentrifugal equilibrium measurements by the Archibald method (*396*). The enzyme contains one mole of pyridoxal-P per mole (*396*). The purified enzyme is free of cystathionase and homocysteine and cysteine desulfhydrases (*396*). In addition to catalyzing the synthesis of cystathionine and the deamination of L-serine and L-threonine, the enzyme deaminates DL-allothreonine and certain β-chloro-α-amino acids. The latter activity demonstrates that a β-hydroxyl group is not essential for substrate activity. Since allothreonine and both *threo-* and *allo-β-*chloro-DL-α-aminobutyrates are deaminated, the stereospecificity of the enzyme is limited to isomers involving the α-carbon and the configuration at the β-carbon is not crucial for enzymic activity. The D-isomers of the hydroxyamino acids are not substrates for the enzyme. Both L-homocysteine and L-cysteine are inhibitors (*396*). It has been reported that the activity of the enzyme is strongly inhibited by elemental sulfur (*397*). Activation with monovalent cations $(K^+ > NH_4^+ > Rb^+ > Li^+ > Na^+)$ was observed (*396*). Carbonyl and sulfhydryl reagents are inhibitory (*396*).

The highest activity of the enzyme was found in mouse, rat, and dog livers; chicken liver had lower activity (*396*). Beef, horse, hog, and sheep livers contained little or no activity. Sheep liver catalyzed only the deamination of threonine with wide variations in activity observed (*396*). This result is difficult to reconcile with the studies on serine dehydrase in sheep liver reported earlier (*388*).

An enzyme which catalyzes the deamination of L-homoserine [L-homoserine hydro-lyase (deaminating), EC 4.2.1.16] and the cleavage of L-cystathionine to L-cysteine and α-ketobutyrate plus ammonia has been obtained in crystalline form from rat liver by Matsuo and Greenberg (*398,399*) and by Kato *et al.* (*400*). The enzyme also contains cystine desulfurase and cysteine desulfhydrase activities (*400*). The molecular weight determinations by the two groups (*398–400*) give values of approximately 190,000. From an estimation of the pyridoxal-P-binding capacity, it has been calculated that each mole of enzyme contains 4 moles of pyridoxal-P (*398*).

The reaction mechanism has been investigated using ^{14}C-labeled cystathionine as the substrate (*399*). Although radioactive homoserine could not be demonstrated as a product in the cleavage reaction, it was shown to give rise to radioactive cystathionine in the reverse reaction (*399*). A significant deamination of L-serine, djenkolic acid, and lanthionine was observed (*399*). Inhibitors for the enzyme include heavy metals, sulfhydryl, carbonyl, and chelating reagents (*399*). L-Cysteine and L-cystine are competitive inhibitors of the enzyme (*397*).

2. BACTERIAL DEHYDRASES

The nonoxidative deamination of serine and threonine by *E. coli* has been studied by several groups of investigators (*401–424*). Distinct enzymes specific for the D- and L-isomers of the amino acids have been demonstrated. However, like the mammalian enzymes, the bacterial dehydrases do not exhibit strict substrate specificity for a given amino acid.

D-Serine dehydrase [D-serine hydro-lyase (deaminating), EC 4.2.1.14] of wild-type *E. coli* is an inducible enzyme. On the basis of their experiments on the induction of D-serine dehydrase, Pardee and Prestidge (*408*) concluded that D-serine and D-threonine are dehydrated by the same enzyme. The isolation by McFall (*409*) of a group of mutants which are constitutive for D-serine dehydrase provided an abundant source of the enzyme. D-Serine dehydrase has been obtained in a crystalline form from such mutants by Dupourque *et al.* (*410*) and by Labow and Robinson (*411*). The enzyme has been shown to be pure by ultracentrifugal and electrophoretic techniques. The molecular weight of the enzyme is 37,300 (*411*). One mole of pyridoxal-P is bound per mole of enzyme (*410,411*). This enzyme and the mammalian L-serine and L-threonine dehydrase-cystathionine synthetase are unusual among pyridoxal-P enzymes in that they possess only a single binding site for the coenzymes per molecule. Borohydride reduction of the holoenzyme followed by acid hydrolysis and chromatography led to the identification of ε-pyridoxyllysine (*411*). Hence, in this enzyme, as in other pyridoxal-P enzymes, the coenzyme is bound to the ε-amino group of a lysine residue. D-Serine is the preferred substrate; much lower activities were observed with D-threonine (*410,411*). No significant activity was observed with L-serine, L-threonine, D-cysteine, and DL-homoserine (*410,411*). O-Methylserine is a potent competitive inhibitor (*410*).

Two distinctly different L-threonine dehydrases [L-threonine hydro-lyase (deaminating), EC 4.2.1.16] have been demonstrated in *E. coli* (*412,413*). One of these, the "synthetic" threonine dehydrase, is a constitutive enzyme, participates in the biosynthesis of isoleucine, and is susceptible to end-product inhibition by this compound. The second one, the "catabolic" threonine dehydrase, is inducible in the absence of glucose or oxygen, is insensitive to isoleucine, appears to function in the utilization of serine or threonine as an energy source, and clearly requires 5'-AMP for activity (*402*). L-Serine is a substrate for both enzymes.

Additional studies on the "synthetic" threonine dehydrase have been carried out by Changeux (*414,415*) who has confirmed and extended Umbarger's finding that L-isoleucine is a specific inhibitor. Isoleucine has been shown to stabilize the enzyme and protect it against inactivation by urea, but does not affect the sedimentation behavior of the enzyme (*414,415*). Detailed kinetic studies on the enzyme from *Salmonella typhimurium* have been reported (*416*).

The role of 5'-AMP in the activation of "catabolic" threonine dehydrase from *E. coli* has been studied in detail (*417–419*). Evidence for the fact that 5'-AMP is a modifier controlling the activity of this enzyme and, therefore, energy metabolism under anaerobic conditions, was obtained. Physical studies indicate that this compound induces conformational changes in the protein and that the active enzyme is a dimer (*417,419*). As is true of other pyridoxal-P enzymes, the coenzyme is bound to the enzyme as a Schiff base with the ε-amino group of a lysyl residue (*418*).

The mechanism of the reaction catalyzed by "catabolic" threonine dehydrase of *E. coli* has been studied by Phillips and Wood (*418*) by isotopic methods. Tritium is lost from α-tritiothreonine to the medium, and incorporation of deuterium from D_2O into α-ketobutyrate is observed. Deuterium from D_2O and ^{18}O from $H_2^{18}O$ were incorporated into threonine during dehydration at nearly equal rates. These results show that the initial dehydration to form the unsaturated amino acid (see Eq. 22) is a reversible process. These authors obtained evidence for the finite existence of free iminobutyrate, e.g., the production of α-aminobutyrate in reaction mixtures treated with borohydride, and tritium in the α-carbon on reduction with borotritide. All the data obtained are consistent with an α,β-elimination mechanism.

L-Threonine dehydrase of *Clostridium tetanomorphum* requires ADP, rather than AMP, for activation (*420–422*). This enzyme is not inhibited *in vitro* by isoleucine. The available data indicate that ADP is a catabolic regulator of anaerobic energy production in this organism.

It is of interest to note that L-serine dehydrase activity has been observed with purified preparations of tryptophanase from *E. coli* (*423*) and with the B protein of tryptophan synthetase from the same organism (*424*).

3. OTHER DEHYDRASES

Extracts of *Neurospora crassa* catalyze the deamination of both serine and threonine (*425–428*). A specific D-serine and D-threonine dehydrase has been purified 35- to 40-fold from this mold (*427*). An absolute requirement for pyridoxal-P was demonstrated. No requirement for AMP or glutathione could be demonstrated. The preparation was not active with the L-isomers of serine and threonine or DL-homoserine and DL-homocysteine. The rate of deamination of D-threonine was very slow compared to that of D-serine. Activity was observed with D-glutamate and D-aspartate. Since other D-amino acids were not deaminated by the preparation, these results could not be due to a contamination with D-amino acid oxidase. Furthermore, when either of these amino acids was incubated in the presence of D-serine, the keto acid production was a summation of that for each substrate alone. Pyridoxal-P had no effect on keto acid formation from the dicarboxylic amino acids. It is of interest to note that D-amino acid oxidase of *N. crassa* does not attack D-serine or D-threonine (*375*).

L-Serine and L-threonine dehydrase has been partially purified from *N. crassa* extracts by Yanofsky and Reissig (*428*). This enzyme, like the D-specific one, has an absolute requirement for pyridoxal-P and is specific for the L-isomers. AMP and glutathione, activators of L-serine dehydrase of *E. coli*, did not stimulate the enzyme from *N. crassa*, nor did biotin.

An enzyme from yeast that catalyzes the deamination of both L-serine and L-threonine has been purified 20-fold (*429*). The ratio of activities with the two amino acids remains constant during this process. Activation by L-valine and L-isoleucine is observed at low concentrations $(1.6 \times 10^{-4}\ M)$; higher concentrations of L-isoleucine are inhibitory. The D-isomers of the hydroxyamino acids inhibit.

4. Other Nonoxidative Deaminases

Microorganisms and some plant tissues are capable of catalyzing a multiplicity of nonoxidative amino acid deamination reactions. It is beyond the scope of this chapter to review these systems, since for the most part they have not been studied with cell-free preparations. Two exceptions to the latter statement are the aspartase and tryptophanase systems. The latter is discussed in Chapter 15.

The aspartase reaction may be formulated as in Eq. (24). The system

$$HOOCCH_2\underset{\underset{NH_2}{|}}{C}HCOOH \rightleftharpoons HOOCCH\!\!=\!\!CHCOOH + NH_3 \qquad (24)$$

has been reviewed by Erkama and Virtanen (*375a*). A similar reaction (discussed in Chapter 15) is carried out by histidine deaminase to yield urocanic acid.

Other types of nonoxidative amino acid deamination are as shown in Eqs. (25–27). Meister (*3*) has reviewed this general topic.

$$RCHNH_2COOH + H_2O \rightarrow NH_3 + RCHOHCOOH \qquad (25)$$

$$RCHNH_2COOH + 2H \rightarrow NH_3 + RCH_2COOH \qquad (26)$$

$$RCHNH_2COOH + R_1CHNH_2COOH + H_2O \rightarrow$$
$$2NH_3 + RCOCOOH + R_1CH_2COOH \qquad (27)$$

The desulfhydrases are covered in Chapter 17.

VI. DEAMIDATION

A. Glutaminase

Glutaminase catalyzes the reaction shown in Eq. (28).

$$\begin{array}{ccc}
\text{COO}^- & & \text{COO}^- \\
| & & | \\
\text{CHNH}_3{}^+ & & \text{CHNH}_3{}^+ \\
| & & | \\
\text{CH}_2 & + \text{HOH} \rightarrow & \text{CH}_2 \qquad + \text{NH}_4{}^+ \\
| & & | \\
\text{CH}_2 & & \text{CH}_2 \\
| & & | \\
\text{CONH}_2 & & \text{COO}^-
\end{array} \qquad (28)$$

Glutaminases (L-glutamine amidohydrolase, EC 3.5.1.2) are widespread in nature, having been found in animal and plant tissues, yeast, and bacteria (430,431). The enzyme is found in the highest level in kidney (432) and is apparently a mitochondrial enzyme (184,433–435). Glutaminase has been purified about 300-fold from fresh hog and dog kidney homogenates (434–436). The enzyme is quite labile (432).

Since animal tissue are in general impermeable to glutamate, but not to glutamine (437), glutaminase may be important in converting blood-borne glutamine to intracellular glutamate. The high level of this enzyme in the kidney is undoubtedly an important factor in the regulation of ammonia excretion (431) and conservation of cations.

The enzyme from mammalian or bacterial sources shows a high degree of substrate specificity and does not react with asparagine (432,438).

The mammalian enzyme is activated by divalent ions, phosphate being the most effective (434,436,439). Borate is a competitive inhibitor of phosphate with the hog kidney enzyme (434), and of glutamine with the dog kidney enzyme (436). Both enzymes are stabilized by borate or phosphate (434,436). Since the activation constant for phosphate is about 0.05 M, it is not known if phosphate functions as an *in vivo* activator (433).

Kinetic data are consistent with the concept that glutamate, phosphate, and not glutamine can add to the free enzyme (432,434,436). Glutamate is a competitive inhibitor of phosphate and a noncompetitive inhibitor of glutamine. This inhibition is believed to be important in regulation of renal ammonia production (440).

The enzyme is inhibited by sulfhydryl reagents, phthalein dyes, and flavianic acid (1,432). The glutaminase from *C. welchii* is activated by monovalent ions, bromide being the most effective (441).

The reaction is essentially irreversible ($\Delta F = -3.42$ kcal per mole) (*441*). Studies of the reversibility of the reaction have been made with isotopic (*434*) and colorimetric methods (*442*).

B. Asparaginase

Asparaginase (L-asparagine amidohydrolase, EC 3.5.1.1) catalyzes the reaction shown in Eq. (29).

$$
\begin{array}{ccc}
\text{COO}^- & & \text{COO}^- \\
| & & | \\
\text{CHNH}_3{}^+ & & \text{CHNH}_3{}^+ \\
| & + \text{H}_2\text{O} \rightarrow & | \qquad + \text{NH}_4{}^+ \\
\text{CH}_2 & & \text{CH}_2 \\
| & & | \\
\text{CONH}_2 & & \text{COO}^-
\end{array}
\qquad (29)
$$

The enzyme is widely distributed in microorganisms, plants, and animal tissues (see *1,443*). A homogeneous enzyme with a molecular weight of 1.4×10^5 has been prepared from guinea pig serum (*444*). This enzyme has a high degree of substrate specificity (*438*).

Recent interest in the enzyme has been focused on its antilymphoma activity (*445*) (see *446* for additional references). Both purified guinea pig serum and liver enzymes (*444,446*) have such activity; however, the serum enzyme is more active in terms of antitumor activity per unit enzyme activity (*446*).

An enzyme from a *Pseudomonas* strain GG 13 has been purified several hundred fold which has both glutaminase and asparaginase activity (*447,448*). The molecular weight of the enzyme is about 2.5×10^4. Phosphate and borate activated the glutaminase and inhibited the asparaginase activity. Cyanide and nitrate ions had opposite effects. Product inhibition studies of the glutaminase reaction with glutamate and $\text{NH}_4{}^+$ were similar to those of mammalian glutaminase. L-Aspartate and $\text{NH}_4{}^+$ did not inhibit asparaginase activity. Since the two activities followed each other during purification, it is suggested that both activities are associated with the same protein. The different effects of various inhibitors on the two activities could be similar to that observed with glutamate and alanine activities with glutamate dehydrogenase.

VII. AMINO ACID RACEMASES AND EPIMERASES

Enzymes capable of interconverting the D- and L-isomers of amino acids are widely distributed in microorganisms. These enzymes, in conjunction with the D-amino acid transaminases, provide a route for the biosynthesis of several D-amino acids which occur in bacterial capsules, cell walls, and peptide antibiotics (3).

Racemases have been demonstrated for (a) alanine in *Streptococcus faecalis* and a variety of bacteria (75,449,450), and in *B. terminalis* spores (451); (b) glutamic acid in *Lactobacillus arabinosis* (452–454) and *L. fermenti* (455); (c) methionine in *S. faecalis* (456,457) and in *Pseudomonas* (458); (d) lysine in *Proteus vulgaris* and several other microorganisms (459–461); (e) proline in *Clostridium sticklandii* (462); and (f) serine in the fat body of the silkworm (463). Pyridoxal-P has been reported to be the cofactor in the alanine (449,451), glutamic acid (452), and methionine (458) racemases. It has been suggested that the role of the cofactor in enzymic racemization is similar to that for transamination, i.e., reaction of the amino acid with the pyridoxal to form a Schiff base (98) (see Fig. 2). A simple rearrangement of this intermediate would destroy the asymmetry of the α-carbon, and racemization would occur. The presence of FAD, in addition to pyridoxal-P, in the alanine racemase of *B. subtilis* has been established, and a mechanism in which the flavin coenzyme functions as a hydrogen acceptor and donor in the further reactions of the Schiff base intermediate has been suggested (464). Evidence that the glutamate racemase of *L. fermenti* is a flavoprotein has been reported as well (455). No evidence was found for the participation of pyridoxal-P in the proline racemase or reported for lysine racemase.

Epimerases for hydroxyproline and diaminopimelic acid have been demonstrated. In the pathway of hydroxyproline metabolism by an inducible strain of *Pseudomonas* (465), the initial reaction involves a reversible epimerization of hydroxy-L-proline to allohydroxy-D-proline by racemization at carbon atom 2. Hydroxyproline-2-epimerase has been obtained in a highly purified form (466). The presence of pyridoxal-P as a cofactor in this enzyme has been ruled out. An epimerase active with *meso-* or L-diaminopimelic acid, but not the D-isomer, has been demonstrated in *E. coli* and a large variety of other microorganisms (467) that synthesize lysine via the diaminopimelic acid pathway. This enzyme catalyzes the racemization of only one of the two asymmetric centers of the molecule. It is not activated by pyridoxal-P.

VIII. UREA BIOSYNTHESIS AND RELATED SYSTEMS

In preceding sections of this chapter the important metabolic reactions which yield ammonia have been discussed. Certain of these systems are capable of fixing ammonia (glutamic dehydrogenase, alanine dehydrogenase, L-amino acid oxidase, etc.). The fixation of ammonia in the glutamine synthetase system will be discussed in Chapter 16. The present section will deal with enzymes which fix ammonia to form carbamyl phosphate and enzymes which utilize carbamyl phosphate for the synthesis of arginine (and urea) and pyrimidines.

The reader is referred to reviews of earlier (468–471) and comparative biochemical aspects of urea synthesis (2).

A. Synthesis of Carbamyl Phosphate

Two different types of enzyme systems are known to be involved in the formation of carbamyl phosphate, namely, carbamyl phosphate synthetase and carbamate kinase. In the case of carbamyl phosphate synthetase, 2 moles of ATP are required per mole of carbamyl phosphate produced (472–475) from bicarbonate ions and ammonia (animal systems) (476) or glutamine (bacterial systems) (477,478). Carbamate kinase (bacterial) is characterized by utilizing a single mole of ATP per mole of carbamyl phosphate formed and uses carbamate as the substrate (479). Current evidence indicates that carbamate kinase is not a major pathway for carbamyl phosphate biosynthesis in microorganisms (477) but rather is involved in ATP generation by cleavage of citrulline. Alternative pathways for carbamyl phosphate biosynthesis have been suggested and have been reviewed recently (3,468).

1. CARBAMYL PHOSPHATE SYNTHETASE

a. Distribution and Properties of the Enzyme. The distribution of this enzyme has been reviewed extensively (2,3,468,480). The enzyme is found in the liver of all ureotelic animals, in the intestinal mucosa of the dog, rabbit, rat (481), and in earthworm gut (482). It is also found in the liver of the predominantly (but not exclusively) ammontelic African lungfish (483). The enzyme is located in the mitochondrial portion of liver (472) and the soluble portion of earthworm gut (482).

The level of the enzyme, in liver, increases during embryogenesis (484–487). In the case of the tadpole the increase in the level of the enzyme is known to be the result of *de novo* synthesis (488–490) and correlates with urea formation (491). Antibody prepared against frog liver carbamyl phosphate synthetase cross-reacts with liver extracts of all ureotelic species studied, but not with comparable extracts from non-ureotelic animals (492), suggesting that this enzyme does not exist in these tissues. Thus, while the role of this enzyme for urea formation is well established, i.e., it is only found in tissue capable of forming urea, it seems possible that an enzyme of a different nature is involved in pyrimidine synthesis in extrahepatic tissues and in various tissues in nonureotelic animals.

Recently enzymes have been fractionated from nonanimal sources which utilize glutamine (in place of ammonia) as a substrate for carbamyl group synthesis. This was first found in mushrooms (*493*) and subsequently in *E. coli* (*474,475,477,478*) and other microorganisms. Free ammonium ions do not seem to be involved in the reaction catalyzed by the mushroom enzyme, while the *E. coli* enzyme reacts better with glutamine (*475,478*). The stoichiometry of the *E. coli* enzyme is the same as that of animal enzymes (*474,475*).

The most highly purified animal enzyme is from frog liver (*473*). The enzyme from this source is reasonably stable and abundant (2.5–4.0-fold purification over the initial extract yields about a 90% pure enzyme (*473,494*). The molecular weight of the enzyme is 3.2×10^5 (*494*).

b. Nature of the Reactions Catalyzed. Animal enzymes have an obligatory cofactor requirement for an acyl derivative of a five-carbon dicarboxylic acid (*495,496*). Since *N*-acetylglutamate is found in liver (*497*) and has the highest affinity for the enzyme (*496,498,499*), it is generally believed that this is the natural cofactor. The *E. coli* and mushroom enzymes do not require acetylglutamate (*493,475,478*).

In the absence of ammonium ions the frog liver enzyme can function as a bicarbonate-dependent ATPase (*500*). This reaction is irreversible and is about 3-fold less reactive (in terms of rate of ADP formation) than the reaction in the presence of saturating concentrations of ammonium ions. In the course of the reaction, ^{18}O-bicarbonate donates an ^{18}O-atom to the P_i formed (*501*). It is believed that this reaction generates a labile, enzyme-bound, "active CO_2" (*500,502*). If the concentration of ATP is low (0.04 mM) and ammonium ions are added, the ATPase reaction is inhibited, suggesting that a more stable enzyme-bound intermediate is formed. If the concentration of ATP is high (0.4 mM), ammonium ions activate the reaction, suggesting that the stable intermediate is phosphorylated with the subsequent formation of the products, ADP, P_i, and carbamyl phosphate (*502*). Kinetic and enzyme stability data are consistent with there being two types of ATP binding sites, tight and loose. Therefore it seems likely that the two ATP cleaving reactions occur at two different interacting sites and that ATP is reactive at the loose site only in the presence of ammonium ions. In the absence of ammonium ions, ATP bound to the loose site is an allosteric activator (*499,502*).

These reactions can be summarized as follows:

Reactions at tight ATP binding site:

$$\text{Enz} + \text{HCO}_3^- + \text{ATP} \underset{}{\overset{\text{M}^{2+}}{\rightleftharpoons}} \text{Enz-X} \rightarrow \text{Enz-Y} + \text{ADP} \tag{30}$$

$$\text{Enz-Y} + \text{HOH} \rightarrow \text{HCO}_3^- + \text{P}_i + \text{Enz} \tag{31}$$

$$\text{Enz-Y} + \text{HOH} + \text{RNH}_2 \rightarrow \text{Enz-Z} + \text{ROH} + \text{P} \tag{32}$$

Reaction at loose ATP binding site:

$$\text{Enz-Z} + \text{ATP} \underset{}{\overset{\text{M}^{2+}}{\rightleftharpoons}} \text{Enz} + \text{NH}_2\text{CO}_2\text{PO}_3^{2-} + \text{ADP} \tag{33}$$

In the presence of saturating concentrations of ATP and RNH_2, the reaction as shown in Eq. (31) is inhibited, so the overall reaction is

$$\text{HOH} + 2\,\text{ATP}^{4-} + \text{HCO}_3^- + \text{RNH}_2 \rightarrow$$
$$2\,\text{ADP}^{3-} + \text{HOPO}_3^{2-} + \text{ROH} + \text{NH}_2\text{CO}_2\text{PO}_3^{2-} + \text{H}^+ \tag{34}$$

Similar reaction sequences have been proposed (*3,500,503*), and this sequence is consistent with elegant pulse labeling experiments performed with *E. coli* enzyme. In the case of the liver enzyme, Enz would be enzyme-acetylglutamate complex and R would be a hydrogen atom while, in the case of the *E. coli* enzyme, Enz would be free (uncomplexed) enzyme and R would be γ-glutamyl. As yet no enzyme-bound intermediate has been isolated in either of the above systems.

The liver (*500*) and *E. coli* (*475*) enzymes also catalyze a reverse reaction which can be represented as in Eqs. (35a) and (35b).

$$\text{H}^+ + \text{ADP}^{3-} + \text{NH}_2\text{CO}_2\text{PO}_3^{2-} + \text{Enz} \underset{}{\overset{\text{M}^{2+}}{\rightleftharpoons}} \text{ATP}^{4-} + \text{Enz-Z} \tag{35a}$$

$$\text{Enz-Z} + \text{HOH} \rightarrow \text{Enz} + \text{HCO}_3^- + \text{NH}_4^+ \tag{35b}$$

The frog liver enzyme requires acetylglutamate for this reverse reaction, and phosphate has no effect (*500*). The *E. coli* enzyme requires either glutamate or P_i and will not react in the absence of both (*475*). The products of the reverse reaction, with the exception of ATP, have not been definitely identified.

The liver enzymes are also known to catalyze a reversible reaction with acetyl or formyl phosphate (*504–506*). Studies of the stoichiometry of these and reactions (35a,b) indicate that 1 mole of ATP is produced per mole of acid phosphate consumed (*500,504,505*). Acetylglutamate is not an obligatory activator of the reaction with acetyl or formyl phosphate and ADP (*505*).

Equilibrium dialysis, heat inactivation, and kinetic experiments performed with the frog liver enzyme indicate that both acetylglutamate and ATP can be bound to free enzyme (*494,499,502*). Acetylglutamate increases the dissociation constant of ATP at the tight site and decreases this value at the loose site. Kinetic data are consistent with a rapid equilibrium derivation of the reaction sequences (30–33) with NH_4^+ and ATP (to the loose site) adding randomly to the Enz-Y complex. As can be seen from Table XII, the rapid equilibrium assumption seems to

TABLE XII

DISSOCIATION CONSTANTS OF FROG LIVER CARBAMYL PHOSPHATE
SYNTHETASE (499,502)[a]

Activator or substrate	Conditions	Value (mM) at:	
		23°C	50°C
Acetylglutamate	Absence of ATP and NH_4^+	1[b]	0.07[c]
	Saturating ATP and NH_4^+	1[b]	0.07[c]
ATP (tight site)	Absence of acetylglutamate and NH_4^+	0.05[b]	
	Absence of NH_4^+; and acetylglutamate saturating	0.2[b,d]	0.1[b,c]
	Saturating acetylglutamate and NH_4^+	0.05[b]	0.03[c]
ATP (loose site)	Saturating acetylglutamate and absence of NH_4^+	0.2[b,d]	0.1[b,c]
	Saturating acetylglutamate and NH_4^+	0.5[b]	0.5[b]
NH_4^+	Saturating acetylglutamate and ATP at tight site	0.7[b]	—
	Saturating acetylglutamate and ATP (both sites)	2[b]	—

[a] All experiments performed at pH 7.5 in K^+ and Mg^{2+}. Methods used were: (b) kinetics, (c) heat inactivation, and (d) spectroscopy.

be valid (i.e., Michaelis and dissociation constants of substrates from enzyme complexes are equal).

Kinetic constants are markedly dependent upon ionic strength and the nature of the specific cation, i.e., K^+ or Na^+ (494). Perhaps of most physiological interest is the fact that K^+ markedly decreases the Michaelis and activation constants of substrates and acetylglutamate (494). Extremely high salt concentrations raise the Michaelis constant of acetylglutamate, but do not alter the maximal velocity (507), as though high salt concentrations prevent binding of acetylglutamate.

c. *Function of Acetylglutamate.* Many properties of the frog liver enzyme; stability (499,504,508,509), optical rotatory dispersion (504), and binding of ATP (494,499,502) are altered by acetylglutamate. The enzyme is less stable in the presence of acetylglutamate at both high

(50°) (*499,508*) and low temperatures (4°) (*504,509*). Accompanying inactivation in the cold is partial dissociation of the native enzyme ($S_{20,w} = 11$) into a slower-sedimenting component ($S_{20,w} = 6$). These changes are not instantaneous but require several hours. Binding of acetylglutamate to the catalytically active site at higher temperatures does not apparently result in dissociation of the enzyme (*499,510*). Acetylglutamate protects the enzyme in the presence of mercuribenzoate (*494,511*), but enhances the inactivating effect of *N*-ethylmaleimide (*511*). ATP and Mg^{2+} reverse the effect of acetylglutamate in the presence of *N*-ethylmaleimide but not in the presence of mercuribenzoate (*511*). It is therefore believed that binding of acetylglutamate masks some and exposes other sulfhydryl groups. The groups thus exposed are masked by ATP and Mg^{2+} (*511*).

Numerous experiments have failed to demonstrate a substrate carrier role for acetylglutamate (*3,468,501,503,512*). It is therefore believed that the marked conformation change induced in the enzyme by acetylglutamate (suggested by the previously mentioned experiments) is involved in the activation process. This is consistent with experiments which show that preincubating the enzyme with acetylglutamate alone is sufficient for activation (*494*). In additional experiments (measuring the rate of ADP formation), it has been found that preincubating the enzyme with acetylglutamate results in both an increase in enzyme activity and an abolition of an initial lag. Thus, when enzyme at the concentration (E_0) and acetylglutamate at the concentration (A) (i.e., not saturating) are preincubated for 10 minutes and then diluted 2-fold and assayed, the same rate is obtained as when enzyme at the concentration (E_0)/2 is assayed directly with acetylglutamate at the concentration (A), without preincubation. Thus the important factor is the concentration of acetylglutamate during the incubation. These results suggest that the activation process is slow, and only slowly reversible (*510*).

On the basis of experiments performed with analogs of acetylglutamate, it seems that a five-carbon dicarboxylic acid is required for binding to the enzyme. Only L-isomers are bound. To be an activator, the substitution on the α-carbon atom must be of the general structure R—CO—X—, where X is either a nitrogen or an oxygen atom (*499, 512,513*). In the case of the most effective activator known, namely, acetylglutamate, R is a methyl group and X is a nitrogen atom (*496,498,499,512,513*).

It is intriguing to speculate that the *E. coli* and frog liver enzymes are quite similar, with glutamine and glutamate functioning in the same manner with the *E. coli* enzyme as acetylglutamate, ammonia, and water function with animal enzyme [as suggested by Eqs. (30–33)]. Glutamine apparently has a very low affinity for the animal enzymes (*476, 498*).

d. Relationship between Carbamyl Phosphate Synthetase and Glutamate Dehydrogenase. A problem associated with the concept of ammonia production via the glutamate dehydrogenase reaction is that the chemical equilibrium of this reaction strongly favors glutamate production (*199–203*). Also the turnover members of purified frog liver glutamate dehydrogenase (in terms of rate of glutamate oxidation) and carbamyl phosphate synthetase are both low, about 7 sec^{-1}. However, the level of carbamyl phosphate synthetase found in frog liver is about 6-fold greater than that of glutamate dehydrogenase. Also, the reaction catalyzed by carbamyl phosphate synthetase is essentially irreversible. Therefore, since both enzymes are found in liver mitochondria, it seems that any ammonia produced from glutamate oxidation can be irreversibly converted to carbamyl phosphate. Also, the other two products of the glutamate dehydrogenase reaction (NADH and α-Kg), can be rapidly metabolized. Therefore the glutamate dehydrogenase reaction may never reach equilibrium. If this is the case, then, the following modifiers of the initial velocity of glutamate oxidation via glutamate dehydrogenase are important: (a) NAD, a substrate for glutamate oxidation is also an allosteric activator of glutamate oxidation, while NADH, a product, is an allosteric inhibitor of the reverse reaction; (b) ATP, a substrate of the carbamyl phosphate synthetase reaction, is an allosteric activator of glutamate oxidation and has very little effect upon the reverse reaction; (c) ADP, a product of the carbamyl phosphate synthetase reaction, activates the glutamate dehydrogenase reaction in both directions. However, P_i, a product of the carbamyl phosphate synthetase and transcarbamylase reactions, enhances the activating effects of ADP on glutamate oxidation and inhibits these effects on the reverse reaction (*220,510*).

e. Synthesis de novo of Carbamyl Phosphate Synthetase. The level of carbamyl phosphate synthetase in tadpole liver can be increased by administering thyroxine (*491*). The increase in carbamyl phosphate synthetase activity parallels the amount of injected leucine-^{14}C incorporated into the enzyme. These experiments indicate that, during tadpole metamorphosis, carbamyl phosphate synthetase is synthesized *de novo* and is not merely assembled from preexisting proenzyme or precursors (*488*). More recently it has been found that liver slices (but as yet not a cell-free system) from thyroxine-treated tadpoles can synthesize carbamyl phosphate synthetase (*489*). Pulse-labeling experiments, performed with these liver slices and leucine-^{14}C, revealed that there was a 6-fold increase in the specific radioactivity of the enzyme, in a mitochondrial fraction, during a second period of incubation in the absence of leucine-[14]C. These results suggest that enzyme precursors are formed outside of the mitochondria and are then assembled into active mitochondrial enzyme. Additional experiments were consistent with the concept that puromycin can inhibit the synthesis of enzyme precursors, but not the final assembly of precursors, into active enzyme (*490*).

2. CARBAMATE KINASE

Carbamate kinase (EC 2.7.2.2) catalyzes the reaction

$$NH_2COO^- + ATP^{4-} \xrightleftharpoons{Mg^{2+}} NH_2CO_2PO_3{}^{2-} + ADP^{3-} \qquad (36)$$

The reaction is reversible, with an overall equilibrium constant (determined directly) of 26×10^{-3} (*514*). The turnover number of the reverse reaction is about 8-fold greater than the forward reaction (*514*), and the reverse reaction is probably of greater physiological significance in organisms capable of generating ATP by the breakdown of arginine via citrulline (*477*).

The enzyme is present in a number of microorganisms (*468*) and has been crystallized from *Streptococcus faecalis* (*514*) and *S. lactis* (*515, 516*). Kinetic studies of the *S. faecalis* enzyme are consistent with a sequential order of addition of substrates to the enzyme, the purine nucleotide adding first (*514*). Carbamyl phosphate can add to free enzyme (in the absence of purines); however, the dissociation constant of carbamyl phosphate at this site (0.15 mM) is about 15-fold higher than what would be expected if the order of addition were random. Since high concentrations of carbamyl phosphate inhibit the reverse reaction (phosphorylation of ADP) when the concentration of ADP is low, these results suggest that the mechanism is truly sequential but an abortive Enz-carbamyl phosphate complex can form. One mole of ADP is bound per 33,000 gm of enzyme (the minimal molecular weight by amino acid analysis). This enzyme, like carbamyl phosphate synthetase, can function as an ATPase in the presence of HCO_3^- and in the absence of carbamate at a rate about 3×10^{-3}-fold less than the parent reaction. The enzyme is frequently used for preparing ATP, and in this connection it should be noted that the reverse reaction is practically independent of pH over the range 5.0–8.0 (*514,517*). Values of some kinetic constants are given in Table XIII.

On the basis of sedimentation-equilibrium, the molecular weights of the *S. faecalis* and *S. lactis* enzymes have been reported to be 4.6×10^4 (*518*) and 6.6×10^4 (*516*). Both enzymes have the same sedimentation coefficient ($s_{20,w} = 4$) (*514,519*), and the minimal molecular weight of the *S. faecalis* enzyme (on the basis of amino acid composition) is 3×10^4 (*514*).

TABLE XIII

KINETIC CONSTANTS OF *S. faecalis* CARBAMATE KINASE (*514*)[a]

Compound	Constant		
	Michaelis (μM)	Dissociation (μM)	Maximal velocity (moles) (3×10^4 gm enzyme-sec)$^{-1}$
ATP	8	120[b]	Forward reaction
		70[c]	92
Carbamate	80	—	—
ADP	50	7.6[b]	Reverse reaction
		5[c]	730
		7[d]	
Carbamyl phosphate	100	—	—

[a] Methods used were: (*b*) product inhibition, (*c*) initial velocity kinetics, and (*d*) equilibrium dialysis.

B. Fate of Carbamyl Phosphate

There are two major reactions in which carbamyl phosphate participates, namely, pyrimidine biosynthesis and urea formation. Thus this is a branch point in nitrogen metabolism and is subject to regulation. Therefore it is not surprising that there seems to be a reciprocal relationship between the activities of the two enzymes, aspartate and ornithine transcarbamylase. For example, aspartate transcarbamylase increases while ornithine transcarbamylase decreases during liver regeneration (*520,521*). This may be looked upon as an example of biochemical dedifferentiation. The urea cycle represents an example of a biochemically differentiated system. In the face of a limited supply of carbamyl phosphate, regenerating liver must synthesize, among many other things, pyrimidines, and thus must utilize its supply of carbamyl phosphate for the more rapid synthesis of pyrimidine precursors. This is done by increasing the aspartate transcarbamylase activity. At the same time, ornithine transcarbamylase is decreased, thus minimizing the utilization of carbamyl phosphate. When regeneration is complete and growth has ceased, biochemical differentiation occurs with a decrease of aspartate transcarbamylase and an increase in ornithine transcarbamylase, thus permitting the operation of the biochemically differentiated or specialized pathway. The factors which regulate the levels of the transcarbamylases and their relation to growth of the liver are not completely understood as yet.

1. ASPARTIC TRANSCARBAMYLASE

Carbamyl phosphate is a substrate for a system which leads to the synthesis of carbamyl aspartate (ureidosuccinate), a pyrimidine precursor (*521–532*). This enzyme (carbamoyl phosphate : L-aspartate carbamoyltransferase, EC 2.1.3.2) catalyzes the reaction as shown in Eq. (37).

$$
\begin{array}{c}
\text{COO}^- \\
| \\
\text{HCNH}_3{}^+ \\
| \\
\text{CH}_2 \\
| \\
\text{COO}^-
\end{array}
+ \text{NH}_2\text{CO}_2\text{PO}_3{}^{2-} \rightleftharpoons
\begin{array}{c}
\text{COO}^- \\
| \\
\text{HCNHCONH}_2 \\
| \\
\text{CH}_2 \\
| \\
\text{COO}^-
\end{array}
+ \text{HOPO}_3{}^{2-} + \text{H}^+
\qquad (37)
$$

The enzyme has a high degree of substrate specificity (*533,534*), although it is capable of reacting with hydroxyaspartate (*535*). The enzyme is widely distributed (*2*) and has been shown to occur in microorganisms (*530,533,536–539*), plants (*540*), and animal tissues (*534*). The enzyme activity is particularly high in cells and tissues which are actively growing and undergoing cell division (*521,534,540,541*). Thus the enzyme activity has been shown to be relatively high in tissues such as testis (*534*), intestinal mucosa (*534*), noeplastic growths (*541*), regenerating liver (*521*), liver cells in tissue culture (*520*), and the meristem and roots of plant seedlings (*540*). It seems probable from the wide distribution of this enzyme and its activity that each cell must be capable of synthesizing carbamyl aspartate for pyrimidine synthesis. The failure to date to demonstrate carbamyl phosphate synthetase activity in animal tissues other than the liver and intestine (*481*) raises the question whether carbamyl phosphate is transported from the liver or is synthesized in the extrahepatic tissues by some other mechanism (*468,542*).

The *E. coli* enzyme has been crystallized (*543*) and has been the subject of many recent investigations. The inhibition of the synthesis of this enzyme by uracil is a classical example of repression (*539*).

The enzyme is specifically inhibited by cytidine triphosphate (*544*). Thus, if nucleoside phosphates do not accumulate but are converted to nucleic acids, they do not inhibit the enzyme. Since aspartic transcarbamylase is the first enzyme in the pathway of pyrimidine biosynthesis, this is a classical example of feedback inhibition.

The *E. coli* enzyme can be reversibly dissociated into two types of subunits (*545*). The larger, with a molecular weight of 9.6×10^4, possesses catalytic activity, does not bind CTP, and is termed the catalytic subunit. The second ($MW = 3 \times 10^4$) is enzymically inactive, possesses CTP binding sites, and is termed the regulatory subunit. The native protein ($MW = 3.1 \times 10^5$) consists of 2 catalytic subunits, 4 regulatory subunits, and possesses 8 CTP binding sites (presumably 2 per regulatory subunit).

Kinetic experiments suggest that at low aspartate concentrations the catalytic subunit has a greater affinity for aspartate and is more active than the native enzyme (*545,546*). At high aspartate concentrations the native enzyme probably has a greater affinity for aspartate than the catalytic subunit. Plots of velocity versus aspartate concentration are sigmoidal in experiments performed with native enzyme, but appear to be of the Michaelis-Menten type in experiments performed with the regulatory subunit (*547*).

These results are the most rigorous support of the concept of separate and specific sites for substrate and modifiers which interact through changes in enzyme conformation (*544,548,549,550*). Presumably the regulatory subunit inhibits the reaction by altering the conformation of the catalytic subunit. Since Michaelis-Menten kinetics are obtained only with the catalytic subunit, it would seem that the regulatory subunit also alters some substrate binding sites or permits interaction among these sites.

Inhibitory effects of pyrimidines on aspartic transcarbamylase from lettuce seedlings, *Pseudomonas fluorescens*, but not *B. subtilis*, have been found (*551*). The rat liver enzyme is known to be more inhibited by deoxyribonucleotides than by ribonucleotides (*552*).

2. ENZYMES OF THE UREA CYCLE

a. Ornithine Transcarbamylase. This enzyme (carbamoylphosphate : L-ornithine carbamoyltransferase, EC 21.3.3.), catalyzes the reaction shown in Eq. (38). It has been extensively purified from pea seedlings

$$
\begin{array}{c}
NH_3^+ \\
| \\
CH_2 \\
| \\
CH_2 \\
| \quad\quad + NH_2CO_2PO_3^{2-} \rightleftharpoons \\
CH_2 \\
| \\
CHNH_3^+ \\
| \\
COO^-
\end{array}
\quad
\begin{array}{c}
NHCONH_2 \\
| \\
CH_2 \\
| \\
CH_2 \\
| \quad\quad + H^+ + HOPO_3^{2-} \\
CH_2 \\
| \\
CHNH_3^+ \\
| \\
COO^-
\end{array}
\quad (38)
$$

(553), bovine (554,555) and rat liver (556), S. faecalis (554,557), and E. coli (558). In most cases, advantage has been taken of the heat stability of the enzyme in the presence of ornithine.

The enzyme has been found in the liver of all ureotelic animals (2,469,480), but is absent in the liver of birds and most fish. It is apparently localized in the mitochondria (559).

The equilibrium of the reaction favors citrulline synthesis and the value of the equilibrium constant at pH 7.5 and 37° is 1×10^5 (neglecting hydrogen-ion concentration (469,556). In the presence of arsenate, or high concentrations of P_i, the enzyme is capable of catalyzing the following overall reaction (560–571).

$$\text{Citrulline} + P_i \rightleftharpoons \text{ornithine} + \text{carbamyl phosphate}$$

$$\text{Carbamyl phosphate} + \text{ADP} \rightleftharpoons \text{ATP} + NH_4^+ + HCO_3^-$$

$$\text{Sum: Citrulline} + P_i + \text{ADP} \rightleftharpoons \text{ATP} + \text{ornithine} + HCO_3^- + NH_4^+$$

Ornithine transcarbamylase shows a high degree of substrate specificity (553–556) for ornithine The molecular weight of the bovine liver enzyme is about 1×10^5 (554). The sedimentation coefficient of the pea and of both rat and bovine liver enzymes is 5.5 (553,554,556), while the E. coli enzyme has an $S_{20,w}$ of 7.4 (550).

    b. *Arginosuccinate Synthetase*. This enzyme [L-citrulline : L-aspartate ligase (AMP), EC 6.3.4.5] catalyzes the following reaction:

$$
\begin{array}{c}
\text{NHCONH}_2 \\
|\\
(\text{CH}_2)_3 \\
|\\
\text{CHNH}_3^+ \\
|\\
\text{COO}^-
\end{array}
+
\begin{array}{c}
\text{COO}^- \\
|\\
\text{H}_3\text{N}^+ - \text{C} - \text{H} \\
|\\
\text{CH}_2 \\
|\\
\text{COO}^-
\end{array}
+ \text{ATP}^{4-} \xrightarrow{\text{Mg}^{2+}}
$$

$$
\begin{array}{c}
\text{NH}_2 \quad\quad \text{COO}^- \\
|\qquad\qquad\ | \\
\text{C} \overset{+}{=} \text{NH} - \text{CH} \\
|\qquad\qquad\ | \\
\text{NH} \quad\quad \text{CH}_2 + \text{AMP}^{2-} + \text{HOP}_2\text{O}_6{}^{3-} + \text{H}^+ \quad\quad (39) \\
|\qquad\qquad\ | \\
(\text{CH}_2)_3 \quad\ \text{COO}^- \\
|\\
\text{H}_3\overset{+}{\text{N}} - \text{CHCOO}^-
\end{array}
$$

It is found in the soluble fraction of tissue extract. It is present in the liver (487,572,574,575) and kidney (573) of most mammals and in yeast (576).

Argininosuccinate synthetase has been highly purified from mammalian liver and kidney (572,577). The enzyme is specific for citrulline, aspartate, and ATP (572,577,578). It is inhibited by α-methyl aspartate (9) and γ-acetylornithine (579). The equilibrium constant for the reaction is strongly in favor of synthesis of arginosuccinate (572).

c. Argininosuccinase. Argininosuccinate is cleaved by argininosuccinase (L-argininosuccinate arginine-lyase, EC 4.3.2.1) to form arginine and fumarate as shown in Eq. (40). The enzyme is widely distributed,

$$
\begin{array}{c}
\underset{\substack{|\\NH_2}}{C}\!\!=\!\!\overset{+}{N}H\!\!-\!\!\underset{\substack{|\\CH_2\\|\\COO^-}}{CH}\\[2pt]
\end{array}
\quad\rightleftharpoons\quad
\begin{array}{c}
NH_2\\
C\!\!=\!\!\overset{+}{N}H_2\\
NH\\
(CH_2)_3\\
\overset{+}{H_3N}\!\!-\!\!CH\!\!-\!\!COO^-
\end{array}
\;+\;
\begin{array}{c}
CHCOO^-\\
\parallel\\
{}^-OOC\!\!-\!\!CH
\end{array}
\qquad (40)
$$

having been demonstrated in mammalian kidney and liver (*471,572, 575,580*), anurian liver (*487*), yeast (*575*), jackbean and pea seeds (*581,582*), *Chlorella* (*581,583*), *E. coli* (*471*), and *Neurospora* (*584*). The reaction is readily reversible, with an equilibrium constant of $11 \times 10^{-3} M$ (*572*).

Argininosuccinase has been obtained recently in a homogeneous, crystalline state from beef liver (*585*). It has a high degree of substrate specificity (*572*) but can catalyze a similar reaction with canavaninosuccinate. The beef liver enzyme is unstable at 0° in Tris buffer or water (*572,585*). Loss of activity begins after 4 hours but can be restored by incubating the enzyme for a few minutes in phosphate buffer at 38°. After 24 hours at 0° in Tris buffer or water, the loss of activity is irreversible. Phosphate, arginine, and argininosuccinate prevent inactivation in the cold. Concomitant with cold inactivation is partial dissociation of the enzyme from the native protein ($s_{20,w} = 9.3$) into a second form ($s_{20,w} = 7.9$). In the presence of phosphate in the cold, the enzyme partially dissociates into a minor component (15% of total) with a $s_{20,w}$ of 5.4. *p*-Hydroxymercuribenzoate completely prevents cold inactivation, but has no effect on direct assays of enzyme activity (*585*).

The molecular weight of the enzyme is 2×10^5 (*585*). Argininosuccinate is cleaved by a trans-elimination mechanism (*586*).

d. Arginase. Arginase (L-arginineurohydrolase, EC 3.5.3.1) catalyzes the hydrolysis of arginine to ornithine and urea as shown in Eq. (41).

$$
\begin{array}{c}
NH_2\\
C\!\!=\!\!\overset{+}{N}H_2\\
NH\\
(CH_2)_3\\
CHNH_3{}^+\\
COO^-
\end{array}
\;+\;H_2O\;\xrightarrow{\;Mn^{2+}\;}\;
\begin{array}{c}
NH_3{}^+\\
(CH_2)_3\\
CHNH_3{}^+\\
COO^-
\end{array}
\;+\;NH_2CONH_2
\qquad (41)
$$

The enzyme is present in high levels in the livers of all ureotelic animals. It is found in lower amounts in kidney of all animals (587–589), mammary glands, (590), testis, and skin (591). It was formerly believed that the enzyme is not present in the liver of uricotelic animals (592); however, it has recently been found in the liver of some birds (593). The enzyme appears to be localized in the cytoplasm (594).

Arginase has been prepared in high purity from mammalian liver (587,595–597). The free guanidino and carboxyl groups of the substrate are essential for activity; however, the amino group can be substituted without loss of activity (598,599). The enzyme is activated by Co^{2+} (600) or Mn^{2+} (601). The activation process is slow and, at room temperature, may require several hours (587). These metals protect the enzyme against heat inactivation (602). The sedimentation coefficients of the horse liver enzyme, in the presence of Mn^{2+}, and of the calf liver enzyme, in the absence of Mn^{2+}, are, respectively, 5.6 (602) and 3.8 (596). These and other sedimentation results suggest that arginase is a polymer in the presence of Mn^{2+} (596,603) and that the original determination of the molecular weight of the horse liver enzyme (1.4×10^5) represents that of the polymer (587,602).

Purified arginase is remarkably stable at high concentrations, but very labile after dilution. Dilute solutions of the enzyme are stabilized by ornithine or glycine (587). Ornithine and lysine are potent competitive inhibitors of arginine, while monoamino acids are noncompetitive inhibitors (604).

The human liver and erythrocyte enzymes dissociate into two components which can be separated by centrifugation. Antibodies prepared against these two enzymes cross-react; however, the antierythrocyte antibody only precipitates the slower-sedimenting component, while the antiliver antibody precipitates both components (597).

IX. SUMMARY

It is clear from the previous discussion that there exists a variety of enzymes capable of catalyzing deamination, transamination, and deamidation reactions as a result of which amino acids, amines, and amides are freed of their nitrogen moiety. In many cases both the levels and the specific activities of these enzymes can be modified. In general, these changes are consistent with the known metabolic roles of the carbon chains and the nitrogen moieties of amino acids. Under gluconeogenic conditions, enzymes involved in the conversion of the carbon chain to glucose, i.e., transaminases, dehydrases, tend to increase.

Under conditions which lead to high rates of amino acid degradation (high-protein diets, fasting, etc.), enzymes responsible for the maintenance of low ammonia levels, i.e., urea cycle enzymes, adapt accordingly. For example, cortisone increases the levels of the soluble liver aspartate-α-ketoglutarate transaminase (*179*). Lysine and arginine increase the levels of the ornithine-α-ketoglutarate transaminase, and threonine and phenylalanine increase the level of threonine deaminase in rat liver (*393*).

Other examples can be found in the ornithine urea cycle and related systems. The tadpole must make a transition from ammonotelism to ureotelism. Therefore, consistent with the needs of the animal during this transition, it is found that the levels of urea cycle enzymes increase during the metamorphosis of tadpoles of *Rana catesbeiana* (*484,487*) and these increases correlate with urea excretion (*491*). The enzymes carbamyl phosphate synthetase, argininosuccinate synthetase, and argininosuccinase are barely detectable in the premetamorphic tadpole, and the levels of these enzymes increase rapidly when morphological metamorphosis is only beginning to be evident. If premetamorphic tadpoles are treated with thyroxine (which induces gross metamorphosis), the increase in levels of these enzymes occurs ahead of gross metamorphic changes (*491*). The ratio of enzyme activity in metamophic to premetamorphic tadpoles is essentially the same whether metamorphosis is naturally induced or thyroxine-induced. This ratio is higher in the case of enzymes involved in urea biosynthesis than it is in the case of any of the other enzymes studied (*484*). Enzymes that catalyze reactions which produce or utilize substrates for the urea cycle are also altered during tadpole metamorphosis. For example, the level of ornithine-α-ketoglutarate transaminase decreases (*605*), while that of glutamate-oxaloacetate transaminase increases in tadpole liver during metamorphosis (*606*). The latter transaminase is the only transaminase studied which increases during metamorphosis. It is interesting to note that the level of this transaminase also parallels urea formation in *Xenopus laevis* (*607*). Glutamate dehydrogenase is the only dehydrogenase studied which increases markedly during metamorphosis; this is in keeping with its key role in ureotelism (*221,608*).

High-protein diets are known to increase the levels of ornithine urea cycle enzymes in rat liver (*609*). In the case of arginase it is known that this change in level of enzyme activity is a balance between enzyme synthesis and degradation (*610*). In HeLa cells, arginine decreases the level of argininosuccinate synthetase and argininosuccinase, and increases the level of activity of arginase (*611*). The latter cells cannot convert ornithine to citrulline but can convert citrulline to arginine, while in liver the major function of the arginine-synthesizing enzymes is urea production.

Many other regulatory factors have been discussed in the sections dealing with specific enzymes. A complete review of the many excellent genetic, metabolic, and hormonal studies of this system would be a chapter in itself.

X. ADDENDUM

A. Transamination

1. GLUTAMATE-OXALOACETATE TRANSAMINASE

Detailed structural studies have been performed on the pig heart mitochondrial enzyme (612). The enzyme is a dimer (MW $= 1 \times 10^5$) composed of two similar monomers. Apparently some pyridoxal-P is bound to sites other than the two active centers. The N-terminal amino acid of the mitochondrial monomer is serine, and that of the soluble monomer is alanine. There at least two electrophoretically distinct forms of the mitochondrial enzyme.

Photooxidation of the soluble enzyme results in loss of enzyme activity. This is related to the destruction of one histidine residue (613). New data have been obtained from temperature jump studies (614). Only the α-subform of the soluble enzyme was used in these experiments. For this reason and apparently as the result of improved instrumentation, these results are consistent with previous spectrophotometric and kinetics studies.

2. GLUTAMATE-PYRUVATE TRANSAMINASE

The soluble pig heart enzyme has been prepared to a high state of purity (615). Formate activates this enzyme, and pyruvate can be bound to the phosphopyridoxal form of the enzyme (616). The rat liver enzyme has been prepared to a homogeneous state and characterized physically (617). Previous treatment of rats with glucocorticoids increases the level of this enzyme, but there are no apparent differences between the enzyme isolated from treated and nontreated rats.

3. OTHER TRANSAMINASES

The phosphoserine aminotransferase has been purified 500-fold from sheep brain (618). Glutamate is the most reactive amino acid with this enzyme. The tyrosine aminotransferase has been purified and crystallized from rat liver (619).

B. Glutamate Dehydrogenase

Recently experiments have been performed with the acetylated bovine liver enzyme (*620,621*). The acetylated enzyme does not associate and has a molecular weight of 4×10^5, thus confirming this value as the molecular weight of the monomer. While purine nucleotides can be bound to the acetylated enzyme, there is essentially no interaction between the active coenzyme sites and the purine nucleotide sites of the acetylated enzyme. On the basis of kinetic experiments performed at high and low enzyme concentrations and direct binding experiments it seems that GTP in the presence of coenzyme is preferentially bound to the monomeric form of the enzyme (*622*). Therefore at enzyme concentrations greater than 0.1 mg per ml (where a mixture of monomers and polymers is present) the preferential binding gives rise to an apparent cooperative interaction between GTP binding sites. At lower enzyme concentrations there no such effects, presumably because no polymer is present. ADP in the presence of NADH binds preferentially to the polymer. In the absence of nucleotides the specific activity of the enzyme is independent of enzyme concentration.

The mechanism of NADH binding to the bovine liver enzyme has been studied with equilibrium dialysis (*623*). The results suggest that there is no interaction between the eight NADH binding sites. In these experiments ADP decreased, and GTP increased, NADH binding. This indicates that NADH has an equal affinity for both the monomeric and the polymeric forms of the enzyme and that ADP and GTP modify the affinity of NADH for these forms. It leaves unexplained the substrate inhibition produced by NADH in kinetic experiments. It should be mentioned that a buffer different from that in the purine nucleotide binding experiments was used in the equilibrium dialysis experiments.

Pyridoxal-P inhibits both glutamate and alanine dehydrogenase activity (*624*). After treatment with pyridoxal-P (like acetylation and reaction with antibody) there is diminished interaction between active coenzyme and purine nucleotide sites. Also, the enzyme after treatment with pyridoxal-P does not associate. Inhibition by pyridoxal-P apparently is the result of the formation of a Schiff base with a lysyl residue. The reaction is reversible, unless the Schiff base is reduced.

The electron transport inhibitor rotenone inhibits the rate of NADH oxidation but not the rate of NAD reduction (at pH 8.0) (*625*).

The enzyme has been crystallized from dogfish liver (*626*). This enzyme has allosteric properties similar to other animal glutamate dehydrogenase; however, it does not tend to associate and apparently its molecular weight (3.3×10^5) is not altered by purine nucleotides (*627*).

The *Neurospora* (NADP) enzyme is inactive at a pH below 8.0. At a lower pH (7.2) the enzyme is activated by substrates and many dicarboxylic acids (*628*).

C. Deamidation

Two asparaginases have been purified from *E. coli* (*629,630*). Only one is a potent antilymphoma agent. Neither requires α-keto acids nor is affected by phosphate ions.

D. Urea Biosynthesis and Related Systems

1. CARBAMYL PHOSPHATE SYNTHETASE

Apparently the enigma about the mechanism of formation of carbamyl phosphate in nonhepatic animal tissues has been resolved. A mouse spleen enzyme has been purified which uses glutamine as the ammonia donor more effectively than ammonia (*631*). The enzyme is labile but stabilized by glycerol. It is not activated by acetylglutamate. In this tissue the activity of aspartate transcarbamylase is comparatively high and not regulated by nucleotides. The carbamyl phosphate synthetase is present at rather low levels and is inhibited by UTP. Consequently it is believed that regulation of pyrimidine biosynthesis is at the carbamyl phosphate synthetase level. Activity of this enzyme has also been detected in mouse thymus, testis, stomach, and small intestine (*632*).

A similar enzyme has been found in Ehrlich ascite tissue (*633*) and a soluble fraction of fetal rat and pigeon liver (*634*). The Michaelis constant for glutamine (10 μM) with these enzymes is about 100-fold lower than that of ammonia. They are not activated by acetylglutamate. In the case of rat liver the levels of aspartate transcarbamylase and the glutamine-dependent carbamyl phosphate synthetase (which are both found in the soluble fraction) decrease while the levels of the acetyl-glutamate-activated carbamyl phosphate synthetase and ornithine transcarbamylase (which are found in the insoluble mitochondrial fractions) increase during gestation.

Pyrimidine nucleotides (uridine) inhibit, and purine nucleotides activate, the glutamine, requiring *E. coli* enzyme in both cases by altering the apparent affinity for ATP (*635*). Therefore these results are consistent with the concept that the activity of this enzyme is regulated by feedback from nucleotide levels.

The partial reversal of the carbamyl phosphate synthetase reaction (synthesis of 1 mole of ATP from 1 mole of ADP and carbamyl phosphate) with the *E. coli* enzyme does not require glutamate or phosphate as previously believed (*636*). Apparently the discrepancy resulted from using suboptimal concentrations of potassium ion in previous experiments (*475*). It is of interest that an analog of glutamine (L-2-amino-4-oxo-5-chloropentanoic acid) irreversibly inhibits this enzyme when glutamine, but not ammonium, ions are the nitrogen source (*637*). Inhibition probably results from blocking the glutamine site. Similarly, hydrazine and hydroxylamine are competitive inhibitors of the reaction with ammonium ions and the frog liver enzyme (*638*).

There are apparently three stages involved in the induction of the frog liver enzyme (*639*). First there is an early response to thyroxine which results in the new formation of ribosomal RNA. This is followed by an exponential increase in the level of carbamyl phosphate synthetase under the continuous influence of thyroxine. Also there is an enhancement of enzyme synthesis by some factors such as temperature which are independent of thyroxine.

2. ASPARTATE AND ORNITHINE TRANSCARBAMYLASES

A new method of preparing the *E. coli* aspartate transcarbamylase in higher yields has been developed (*640*). Recent structural work confirms the concept that the native enzyme consists of four catalytic subunits (each of which is a dimer composed of two identical chains) and four regulator subunits (each of which is a single chain). The amino terminal residue of the regulatory subunit is threonine and the catalytic subunit is alanine (*641*).

Ornithine transcarbamylase has been obtained in pure and crystalline form from *Streptococcus* D10 and characterized (*642*).

REFERENCES

1. P. P. Cohen and H. J. Sallach, *in* "Metabolic Pathways" (D. M. Greenberg, ed.), 2nd ed., Vol. II, pp. 1–78. Academic Press, New York, 1961.
2. P. P. Cohen and G. W. Brown, Jr., *in* "Comparative Biochemistry" (M. Florkin and H. Mason, eds.), Vol. II, pp. 161–244. Academic Press, New York, 1960.
3. A. Meister, "Biochemistry of the Amino Acids," 2nd ed. Academic Press, New York, 1965.
4. R. Schoenheimer, "The Dynamic State of Body Constituents," Harvard Univ. Press, Cambridge, Mass., 1949.
5. H. B. Vickery, G. W. Pucher, R. Schoenheimer, and D. Rittenberg, *J. Biol. Chem.* **135**, 531 (1940).
6. W. C. Rose, *Federation Proc.* **8**, 546 (1949).

7. H. A. Krebs, *Z. Physiol. Chem.* **217**, 191 (1933).
8. H. A. Krebs, *Biochem. J.* **29**, 1620 (1935).
9. A. E. Braunstein, *Advan. Enzymol.* **19**, 335 (1957).
10. K. H. Bässler and C. H. Hammar, *Biochem. Z.* **330**, 555 (1958).
11. A. L. Sinitayna, *Biokhimiya* **19**, 80 (1954).
12. E. V. Rowsell, *Biochem. J.* **64**, 235 (1956).
13. F. J. R. Hird and M. A. Marginson, *Arch. Biochem. Biophys.* **115**, 247 (1966).
14. E. V. Rowsell, *Biochem. J.* **64**, 246 (1956).
15. H. A. Krebs and K. Henseleit, *Z. Physiol. Chem.* **210**, 33 (1932).
16. A. E. Braunstein, *in* "The Enzymes" (P. D. Boyer, H. A. Lardy, and K. Myrbäck, eds.), 2nd ed., Vol. II, pp. 113–184. Academic Press, New York, 1960.
17. A. Meister, *in* "The Enzymes" (P. D. Boyer, H. A. Lardy, and K. Myrbäck, eds.), 2nd ed., Vol. VI, pp. 193–217. Academic Press, New York, 1962.
18. E. E. Snell, *in* "Comprehensive Biochemistry" (M. Florkin and E. H. Stotz, eds.), Vol. 15, pp. 138–199. Elsevier, Amsterdam, 1964.
19. Report of the Commission on Enzymes of the International Union of Biochemistry, Rev. Ed., *in* "Comprehensive Biochemistry" (M. Florkin and E. H. Stotz, eds.) pp. 82–85. Elsevier, Amsterdam, 1964.
20. A. E. Braunstein and M. G. Kritzmann, *Enzymologia* **2**, 129 (1937).
21. E. C. C. Lin, M. Civen, and W. E. Knox, *J. Biol. Chem.* **233**, 1183 (1958).
22. E. C. C. Lin and W. E. Knox, *J. Biol. Chem.* **233**, 1186 (1958).
23. I. W. Sizer and W. T. Jenkins, *in* "Methods in Enzymology" (S. P. Colowick and N. O. Kaplan, eds.), Vol. V, p. 677. Academic Press, New York, 1962.
24. D. Rudman and A. Meister, *J. Biol. Chem.* **200**, 591 (1953).
25. P. S. Cammarata and P. P. Cohen, *J. Biol. Chem.* **187**, 439 (1950).
26. E. V. Rowsell and K. Corbett, *Biochem. J.* **70**, 7P (1958).
27. P. P. Cohen and G. L. Hekhuis, *J. Biol. Chem.* **140**, 711 (1941).
28. H. G. Albaum and P. P. Cohen, *J. Biol. Chem.* **149**, 19 (1943).
29. H. C. Lichstein and P. P. Cohen, *J. Biol. Chem.* **157**, 85 (1945).
30. N. Rautanen, *J. Biol. Chem.* **163**, 687 (1946).
31. M. J. K. Leonard and R. H. Burris, *J. Biol. Chem.* **170**, 701 (1947).
32. L. I. Feldman and I. C. Gunsalus, *J. Biol. Chem.* **187**, 821 (1950).
33. J. M. Wiame and R. Storck, *Biochim. Biophys. Acta* **10**, 268 (1953).
34. Y. S. Halpern and N. Grossowicz, *Bull. Research Council Israel* **6E**, 21 (1956).
35. J. R. S. Fincham and A. B. Boulter, *Biochem. J.* **62**, 72 (1956).
36. D. G. Wilson, K. W. King, and R. H. Burris, *J. Biol. Chem.* **208**, 863 (1954).
37. D. H. Hug and C. H. Werkman, *Arch. Biochem. Biophys.* **72**, 369 (1957).
38. A. Meister, *J. Biol. Chem.* **195**, 813 (1952).
39. J. Youatt, *Biochem. J.* **68**, 193 (1958).
40. A. Meister and S. V. Tice, *J. Biol. Chem.* **187**, 173 (1950).
41. J. A. Jacquez, R. K. Barclay, and C. C. Stock, *J. Exptl. Med.* **96**, 499 (1952).
42. C. B. Thorne and D. M. Molnar, *J. Bacteriol.* **70**, 420 (1955).
43. H. J. Sallach, *J. Biol. Chem.* **223**, 1101 (1956).
44. Y. Nishizuka, M. Takeshita, S. Kuno, and O. Hayaishi, *Biochim. Biophys. Acta* **33**, 591 (1959).
45. J. E. Willis and H. J. Sallach, *Phytochemistry* **2**, 23 (1963).
46. H. I. Nakada, *J. Biol. Chem.* **239**, 468 (1964).
47. S. P. Bessman, J. Rossen, and E. C. Layne, *J. Biol. Chem.* **201**, 385 (1953).
48. E. Roberts and H. M. Bregoff, *J. Biol. Chem.* **201**, 393 (1953).
49. E. Roberts, *Arch. Biochem. Biophys.* **48**, 395 (1954).

50. A. Meister, *J. Biol. Chem.* **206**, 587 (1954).
51. F. P. Kupiecki and M. J. Coon, *J. Biol. Chem.* **229**, 743 (1957).
52. O. Hayaishi, Y. Nishizuka, M. Tatibana, M. Takeshita, and S. Kuno, *J. Biol. Chem,* **236**, 781 (1961).
53. C. F. Baxter and E. Roberts, *J. Biol. Chem.* **233**, 1135 (1958).
54. J. K. Miettilinen and A. I. Virtanen, *Acta Chem. Scand.* **7**, 1243 (1953).
55. E. M. Scott and W. B. Jacoby, *J. Biol. Chem.* **234**, 932 (1959).
56. R. A. Salvador and R. W. Albers, *J. Biol. Chem.* **234**, 922 (1959).
57. A. Waksman and E. Roberts, *Biochemistry* **4**, 2132 (1965).
58. H. J. Vogel, *Proc. Natl. Acad. Sci. U.S.* **39**, 578 (1953).
59. A. M. Albrecht and H. J. Vogel, *J. Biol. Chem.* **239**, 1872 (1964).
60. J. R. S. Fincham, *Biochem. J.* **53**, 313 (1953).
61. C. Peraino and H. C. Pitot, *Biochim. Biophys. Acta* **73**, 222 (1963).
62. H. J. Strecker, *J. Biol. Chem.* **240**, 1225 (1965).
63. N. Katunuma, Y. Matsuda, and I. Tomino, *J. Biochem. (Tokyo)* **56**, 499 (1964).
64. W. I. Sher and H. J. Vogel, *Proc. Natl. Acad. Sci. U.S.* **43**, 796 (1957).
65. R. H. Vogel and M. J. Kopac, *Biochim. Biophys Acta* **37**, 539 (1960).
66. B. Peterkofsy and C. Gilvary, *J. Biol. Chem.* **236**, 1432 (1961).
67. T. T. Otani and A. Meister, *J. Biol. Chem.* **224**, 137 (1957).
68. H. Wada and E. E. Snell, *J. Biol. Chem.* **237**, 133 (1962).
69. W. R. Dempsey and E. E. Snell, *Biochemistry* **2**, 1414 (1963).
70. M. Fujioka and E. E. Snell, *J. Biol. Chem.* **240**, 3044, 3050 (1965).
71. K. Kim and T. T. Tchen, *Biochem. Biophys. Res. Commun.* **9**, 99 (1962).
72. K. Haase and G. Schmid, *Biochem. Z.* **337**, 69 (1963).
73. K. Kim, *J. Biol. Chem.* **239**, 783 (1964).
74. B. N. Ames and B. L. Horecker, *J. Biol. Chem.* **220**, 113 (1956).
75. C. B. Thorne, C. G. Gomez, and R. D. Housewright, *J. Bacteriol.* **69**, 357 (1955).
76. P. Meadow and E. Work, *Biochim. Biophys. Acta* **28**, 596 (1958).
77. H. K. Kuramitsu and J. E. Snoke, *Biochim. Biophys. Acta* **62**, 114 (1962).
78. M. Martinez-Carrion and W. T. Jenkins, *J. Biol. Chem.* **240**, 3538, 3547 (1965)
79. D. E. Green, L. F. Leloir, and V. Nocito, *J. Biol. Chem.* **161**, 559 (1945).
80. P. S. Cammarata and P. P. Cohen, *J. Biol. Chem.* **193**, 53 (1951).
81. D. H. Cruickshank and F. A. Isherwood, *Biochem. J.* **69**, 189 (1958).
82. B. Bulos and P. Handler, *J. Biol. Chem.* **240**, 3283 (1965).
83. Z. N. Canellakis and P. P. Cohen, *J. Biol. Chem.* **222**, 53 (1956).
84. Z. N. Canellakis and P. P. Cohen, *J. Biol. Chem.* **222**, 63 (1956).
85. G. A. Jacoby and B. N. La Du, *J. Biol. Chem.* **239**, 419 (1964).
86. T. P. Singer and E. B. Kearney, *Biochim. Biophys. Acta* **11**, 276 (1953).
87. W. T. Jenkins and I. W. Sizer, *Federation Proc.* **15**, 283 (1956).
88. R. T. Taylor and W. T. Jenkins, *J. Biol. Chem.,* **241**, 4396 (1966).
89. A. Ichihara and E. Koyama, *J. Biochem. (Tokyo)* **59**, 160 (1966).
90. V. Nurmikko and R. Raunio, *Acta Chem. Scand.* **15**, 1263 (1961).
91. W. B. Jacoby and D. M. Bonner, *J. Biol. Chem.* **221**, 689 (1956).
92. " Chemical and Biological Aspects of Pyridoxal Catalysis " (E. E. Snell, P. M. Fasella, A. E. Braunstein, and A. Rossi-Fanelli, eds.), Pergamon, Oxford, 1963.
93. E. E. Snell, *J. Am. Chem. Soc.* **67**, 194 (1945).
94. D. E. Metzler, M. Ikawa, and E. E. Snell, *J. Am. Chem. Soc.* **76**, 648 (1954).
95. D. E. Metzler and E. E. Snell, *J. Am. Chem. Soc.* **74**, 979 (1952).
96. J. B. Longenecker and E. E. Snell, *Proc. Natl. Acad. Sci. U.S.* **42**, 221 (1956).
97. D. E. Metzler and E. E. Snell, *J. Biol. Chem.* **198**, 353 (1952).

98. J. Olivard, D. E. Metzler, and E. E. Snell, *J. Biol. Chem.* **199**, 669 (1952).
99. D. E. Metzler, J. B. Longenecker, and E. E. Snell, *J. Am. Chem. Soc.* **75**, 2786 (1953); **76**, 639 (1954).
100. A. E. Braunstein and G. Y. Vilenkina, *Dokl. Akad. Nauk SSSR* **66**, 243 (1949).
101. M. A. Karasek and D. M. Greenberg, *J. Biol. Chem.* **227**, 191 (1957).
102. A. E. Braunstein and M. M. Shemyakin, *Biokhimiya* **18**, 393 (1953).
103. E. E. Snell, *Vitamins Hormones* **16**, 77 (1958).
104. D. E. Metzler, *J. Am. Chem. Soc.* **79**, 485 (1957).
105. Y. Matsuo, *J. Am. Chem. Soc.* **79**, 2016 (1957).
106. G. D. Kalyankar and E. E. Snell, *Nature* **180**, 1069 (1957).
107. T. C. Bruice and R. M. Topping, *J. Am. Chem. Soc.* **84**, 2448 (1962).
108. T. C. Bruice and R. M. Topping, *J. Am. Chem. Soc.* **85**, 1480, 1488, 1493 (1963).
109. F. Schlenk and E. E. Snell, *J. Biol. Chem.* **157**, 425 (1945).
110. H. C. Lichstein, I. C. Gunsalus, and W. W. Umbreit, *J. Biol. Chem.* **161**, 311 (1945).
111. A. Meister, H. A. Sober, and E. A. Peterson, *J. Biol. Chem.* **206**, 89 (1954).
112. S. W. Tanenbaum, *J. Biol. Chem.* **218**, 733 (1956).
113. M. A. Hilton, F. W. Barnes, Jr., and T. Enns, *J. Biol. Chem.* **219**, 833 (1956).
114. A. Meister, *J. Biol. Chem.* **190**, 269 (1951).
115. D. B. Sprinson and D. Rittenberg, *J. Biol. Chem.* **184**, 405 (1950).
116. S. Grisolia and R. H. Burris, *J. Biol. Chem.* **210**, 109 (1954).
117. N. Tamiya and T. Oshima, *J. Biochem.* (*Tokyo*) **51**, 78 (1962).
118. A. Nisonoff, F. W. Barnes, Jr., T. Enns, and S. von Suchling, *Bull. Johns Hopkins Hosp.* **94**, 117 (1954).
119. W. T. Jenkins and I. W. Sizer, *J. Biol. Chem.* **234**, 1179 (1959).
120. W. T. Jenkins and R. T. Taylor, *J. Biol. Chem.* **240**, 2907 (1965).
121. A. E. Evangelopoulos and I. W. Sizer, *J. Biol. Chem.* **240**, 2983 (1965).
122. W. T. Jenkins and I. W. Sizer, *J. Biol. Chem.* **235**, 620 (1960).
123. H. Lis, P. M. Fasella, C. Turano, and P. Vecchini, *Biochim. Biophys. Acta* **45**, 529 (1960).
124. W. T. Jenkins and L. D'Ari, *Biochem. Biophys. Res. Commun.* **22**, 376 (1966).
125. W. T. Jenkins and L. D'Ari, *J. Biol. Chem.* **241**, 2845 (1965).
126. H. Wada and E. E. Snell, *J. Biol. Chem.* **237**, 127 (1962).
127. S. F. Velick and J. Vavra, *J. Biol. Chem.* **237**, 2109 (1962).
128. S. F. Velick and J. Vavra, in "The Enzymes" (P. D. Boyer, H. Lardy, and K. Myrbäck, eds.), Vol. 6, 219. Academic Press, New York, 1962.
129. R. C. Hughes, W. T. Jenkins, and E. H. Fischer, *Proc. Natl. Acad. Sci. U.S.* **48**, 1615 (1962).
130. J. Olivard and E. E. Snell, *J. Biol. Chem.* **213**, 203, 215 (1955).
131. P. Fasella, G. G. Hammes, and B. L. Vallee, *Biochim. Biophys. Acta* **65**, 142 (1962).
132. E. E. Snell, in "Chemical and Biological Aspects of Pyridoxal Catalysis" (E. E. Snell, P. M. Fasella, A. E. Braunstein, and A. Rossi Fanelli, eds.), p. 1. Pergamon Press, Oxford, 1963.
133. W. T. Jenkins, *Federation. Proc.* **20**, 978 (1961).
134. E. H. Cordes and W. P. Jencks, *Biochemistry* **1**, 773 (1962).
135. W. P. Jencks and E. Cordes, in "Chemical and Biological Aspects of Pyridoxal Catalysis" (E. E. Snell, P. M. Fasella, A. E. Braunstein, and A. Rossi Fanelli, eds.), p. 57. Pergamon, Oxford, 1963.
136. B. E. C. Banks, A. A. Deamantes, and C. A. Vernon, *J. Chem. Soc.* 4235 (1961).
137. W. T. Jenkins, S. Orlowski, and I. W. Sizer, *J. Biol. Chem.* **234**, 2657 (1959).
138. C. P. Henson and W. W. Cleland, *Biochemistry* **3**, 338 (1964).

139. J. S. Nisselbaum and O. Bodansky, *J. Biol. Chem.* **241**, 2661 (1966).
140. Y. Morino, H. Itoh, and H. Wada, *Biochem. Biophys. Res. Commun.* **13**, 348 (1963).
141. J. S. Nisselbaum and O. Bodansky, *J. Biol. Chem.* **239**, 4232 (1964).
142. S. Hopper and H. L. Segal, *J. Biol. Chem.* **237**, 3189 (1962).
143. V. Bloomfield, L. Peller, and R. A. Alberty, *J. Am. Chem. Soc.* **84**, 4367 (1962).
144. T. I. Diamonstone and G. Litwack, *J. Biol. Chem.* **238**, 3859 (1963).
145. G. Hammes and P. Fasella, *J. Am. Chem. Soc.* **84**, 4644 (1962).
146. A. Nisonoff, F. W. Barnes, Jr., and T. Enns, *J. Biol. Chem.* **204**, 957 (1953).
147. H. A. Krebs, *Biochem. J.* **54**, 82 (1953).
148. G. A. Fleischer, C. S. Potter, K. G. Wakim, M. Pankow, and D. Osborne, *Proc. Soc. Exptl. Biol. Med.* **103**, 229 (1960).
149. J. W. Boyd, *Biochem. J.* **81**, 434 (1961).
150. P. Borst and E. M. Peeters, *Biochim Biophys. Acta* **54**, 188 (1961).
151. R. H. Hooker and C. S. Vestling, *Biochim. Biophys. Acta* **65**, 358 (1962).
152. N. Katunuma, T. Matsuzawa, and A. Fujino, *J. Vitaminol.* (*Kyoto*) **8**, 74 (1962).
153. J. W. Boyd, *Biochim. Biophys. Acta* **113**, 302 (1966).
154. H. Wada and Y. Morino, *Vitamins Hormones* **22**, 411 (1964).
155. W. T. Jenkins, D. A. Yphantis, and I. W. Sizer, *J. Biol. Chem.* **234**, 51 (1959).
156. Y. Morino, H. Kagamiyama, and H. Wada, *J. Biol. Chem.* **239**, PC 943 (1964).
157. F. Schlenk and A. Fisher, *Arch. Biochem.* **12**, 69 (1947).
158. D. E. O'Kane and I. C. Gunsalus, *J. Biol. Chem.* **170**, 433 (1947).
159. H. Lis, *Biochim. Biophys. Acta* **28**, 191 (1958).
160. H. Lis and P. Fasella, *Biochim. Biophys. Acta* **33**, 567 (1959).
161. C. Turano, A. Giartosio, F. Riva, and P. Vecchini *in* "Chemical and Biological Aspects of Pyridoxal Catalysis" (E. E. Snell, P. M. Fassella, A. E. Braunstein, and A. Rossi Fanelli, eds.), p. 149. Pergamon Press, Oxford, 1963.
162. G. Marino, R. Zito, and V. Scardi, *Boll. Soc. Ital. Biol. Sper.* **40**, 720 (1964).
163. G. Marino, A. M. Greco, V. Scardi, and R. Zito, *Biochem. J.* **99**, 589 (1966).
164. G. Marino, V. Scardi, and R. Zito, *Biochem. J.* **99**, 595 (1966).
165. G. Marino, V. Scardi, and R. Zito, *Boll. Soc. Ital. Biol. Sper.* **42**, 168 (1966).
166. O. L. Polyanovsky and V. I. Ivanov, *Biokhimiya* **29**, 728 (1964).
167. O. L. Polyanovsky, *Biochim. Biophys. Res. Commun.* **19**, 364 (1965).
168. P. Fasella and G. Hammes, *Biochemistry* **4**, 801 (1965).
169. Y. M. Torchinsky and L. G. Koreneva, *Biochim. Biophys. Acta* **79**, 426 (1964).
170. M. Martinez-Carrion, F. Riva, C. Turano, and P. Fasella, *Biochem. Biophys. Res. Commun.* **20**, 206 (1965).
171. G. Schreiber, M. Eckstein, A. Oeser, and H. Holzer, *Biochem. Z.* **340**, 13 (1964).
172. G. Schreiber, M. Eckstein, G. Maass, and H. Holzer, *Biochem. Z.* **340**, 21 (1964).
173. F. J. R. Hird and E. V. Rowsell, *Nature* **166**, 517 (1950).
174. E. V. Rowsell, *Biochem. J.* **64**, 635 (1956).
175. E. Kafer and J. K. Pollak, *Exptl. Cell Res.* **22**, 120 (1961).
176. N. Katunuma, K. Mikumo, M. Makato, and M. Okada, *J. Vitaminol. Kyoto* **8**, 68 (1962).
177. E. V. Rowsell, K. V. Turner, and J. A. Carnie, *Biochem. J.* **89**, 65P (1963).
178. N. Katunuma, S. Matsuda, and M. Isumi, *in Symp. Enzyme Chem.* **16**, 70 (1962); pub. (1963).
179. H. L. Segal and Y. S. Kim, *Proc. Natl. Acad. Sci. U.S.* **50**, 912 (1963).
180. S. Hopper and H. L. Segal, *Arch. Biochem. Biophys.* **105**, 501 (1964).
181. Y. Takeda, A. Ichihara, H. Tanioka, and H. Inoue, *J. Biol. Chem.* **239**, 3590 (1964).
182. R. W. Swick, P. L. Barnstein, and J. L. Stange, *J. Biol. Chem.* **240**, 3334, 3341 (1965).

183. W. T. Jenkins *in* "Chemical and Biological Aspects of Pyridoxal Catalysis" (E. E. Snell, P. M. Fasella, A. E. Braunstein, and A. Rossi Fanelli, eds.), p. 139. Pergamon Press, Oxford, 1963.
184. M. Errera and J. P. Greenstein, *J. Biol. Chem.* **178**, 495 (1949).
185. R. Richterich-van Baerle and L. Foldstein, *Experentia* **13**, 30 (1957).
186. A. Meister, *J. Biol. Chem.* **200**, 571 (1953).
187. A. Meister, *J. Biol. Chem.* **210**, 17 (1954).
188. A. Meister and S. V. Tice, *J. Biol. Chem.* **206**, 561 (1954).
189. A. Meister and P. E. Fraser, *J. Biol. Chem.* **210**, 37 (1954).
190. A. Meister, *Science* **120**, 43 (1954).
191. A. Meister, H. A. Sober, S. V. Tice, and P. E. Fraser, *J. Biol. Chem.* **197**, 319 (1952).
192. H. L. Luschensky, *Arch. Biochem. Biophys.* **31**, 132 (1952).
193. A. B. Kilby and E. Neville, *Biochim. Biophys. Acta* **19**, 389 (1956).
194. L. S. Olenicheva, *Biokhimiya* **20**, 165 (1955).
195. C. Monder and A. Meister, *Biochim Biophys. Acta* **28**, 202 (1958).
196. E. Schmidt, F. Schmidt, H. D. Horn, and U. Gerlach, *in* "Methods of Enzymatic Analysis" (H. U. Bergmeyer, ed.), p. 658. Academic Press, New York, 1963.
197. C. Frieden, *J. Biol. Chem.* **240**, 2028 (1965).
198. J. H. Copenher, Jr., W. H. McShan, and R. K. Meyer, *J. Biol. Chem.* **183**, 73 (1950).
199. C. Frieden, *in* "The Enzymes" (P. D. Boyer, H. A. Lardy, and K. Myrbäck, eds.). 2nd Ed., Vol. 7, p. 18. Academic Press, New York, 1963.
200. J. Olson and C. B. Anfinsen, *J. Biol. Chem.* **197**, 67 (1952).
201. H. J. Strecker, *Arch. Biochem. Biophys.* **32**, 448 (1951).
202. J. E. Snoke, *J. Biol. Chem.* **233**, 271 (1956).
203. L. A. Fahien, B. O. Wiggert, and P. P. Cohen, *J. Biol. Chem.* **240**, 1083 (1965).
204. K. Wallenfels, H. Sund, and H. Diekmann, *Biochem. Z.* **239**, 48 (1957).
205. H. Kubo, T. Yamano, M. Iwatsubo, H. Watari, T. Soyama, J. Siraiski, S. Sawada and N. Nawashima, *in* "Proc. Intern. Symp. Enzyme Chem. Tokyo and Kyoto" (K. Ichikara ed.), p. 345. Academic Press, New York, 1957.
206. T. Sato, *Osaka Daigaku Igaku Zasshi* **9**, 949 (1957) [*C.A.* **52**, 4725 (1958)].
207. C. Frieden, *J. Biol. Chem.* **234**, 2891 (1959).
208. H. F. Fisher, *J. Biol. Chem.* **235**, 1830 (1960).
209. H. F. Fisher and D. G. Cross, *Biochem. Biophys. Res. Commun.* **20**, 120 (1965).
210. R. F. Chen, *Biochem. Biophys. Res. Commun.* **17**, 141 (1964).
211. S. Grisolia, C. L. Quijada, and M. Fernandez, *Biochim. Biophys. Acta* **81**, 61 (1964).
212. H. R. Levy and B. Vennesland, *J. Biol. Chem.* **228**, 85 (1957).
213. H. F. Fisher and L. L. McGregor, *J. Biol. Chem.* **236**, 791 (1961).
214. C. Frieden, *J. Biol. Chem.* **238**, 3286 (1963).
215. N. Talal and G. M. Tompkins, *Science* **146**, 1309 (1964).
216. C. S. Stachow and B. D. Sanwal, *Biochem. Biophys. Res. Commun.* **17**, 368 (1964).
217. G. M. Tompkins, K. L. Yielding, and J. Curran, *Proc. Natl. Acad. Sci. U.S.* **47**, 270 (1961).
218. G. M. Tompkins, K. L. Yielding, J. F. Curran, M. R. Summers, and M. W. Bitensky, *J. Biol. Chem.* **240**, 3793 (1965).
219. G. di Prisco, S. M. Arfin, and H. J. Strecker, *J. Biol. Chem.* **240**, 1611 (1965).
220. L. A. Fahien, B. O. Wiggert, and P. P. Cohen, *J. Biol. Chem.* **240**, 1091 (1965).
221. B. O. Wiggert and P. P. Cohen, *J. Biol. Chem.* **241**, 210 (1960).
222. K. L. Yielding, G. M. Tompkins, and D. S. Trundle, *Biochim. Biophys. Acta* **85**, 342 (1964).
223. K. L. Yielding and G. M. Tompkins, *Proc. Natl. Acad. Sci. U.S.* **46**, 1483 (1960).

224. M. W. Bitensky, K. L. Yielding, and G. M. Tompkins, *J. Biol. Chem.* **240**, 663, 668 (1965).
225. S. Grisolia, M. Fernandez, R. Amelunxen and C. L. Quijada, *Biochem. J.* **85**, 568 (1962).
226. K. L. Yielding, G. M. Tompkins, J. S. Manday, and J. F. Curran, *Biochem. Biophys. Res. Commun.* **2**, 303 (1960).
227. L. L. Engel, J. F. Scott and R. F. Colman, *Federation Proc.* **19**, 159 (1960).
228. L. L. Engel, J. F. Scott and R. F. Colman, *Recent Progr. Hormone Res.* **16**, 79 (1960).
229. C. A. Villee, J. M. Loring, and J. M. Spencer, *Federation Proc.* **20**, 231 (1961).
230. J. Wolff, *J. Biol. Chem.* **237**, 230 (1962).
231. J. Wolff, *J. Biol. Chem.* **237**, 236 (1962).
232. W. S. Caughey, J. D. Smiley, and L. Hellermann, *J. Biol. Chem.* **224**, 591 (1957).
233. K. L. Yielding and G. M. Tompkins, *Biochim. Biophys. Acta* **62**, 327 (1962).
234. K. L. Yielding, G. M. Tompkins, and D. S. Trundle, *Biochim. Biophys. Acta* **77**, 703 (1963).
235. L. Hellerman, K. A. Schellenberg, and O. K. Reiss, *J. Biol. Chem.* **233**, 1468 (1958),
236. J. A. Olson and C. B. Anfinsen, *J. Biol. Chem.* **202**, 841 (1953).
237. G. Pfleiderer, D. Teckel, and T. Wieland, *Biochem. Z.* **328**, 187 (1956).
238. B. M. Anderson and M. L. Reynolds, *J. Biol. Chem.* **241**, 1688 (1966).
239. J. M. Tager and S. Papa, *Biochim. Biophys. Acta* **99**, 570 (1965).
240. M. Klingberg and W. Slenczka, *Biochem. Z.* **331**, 486 (1959).
241. K. H. Bässler and C. H. Hammar, *Biochem. Z.* **330**, 446 (1958).
242. J. Struck and I. W. Sizer, *Arch. Biochem. Biophys.* **86**, 260 (1960).
243. J. Wolff and C. E. Wolff, *Biochim. Biophys. Acta* **26**, 387 (1957).
244. B. O. Wiggert and P. P. Cohen, *J. Biol. Chem.* **240**, 4790 (1965).
245. E. Kun and B. Achmatowicz, *J. Biol. Chem.* **240**, 2619 (1965).
246. P. Andrews, *Biochem. J.* **96**, 595 (1965).
247. K. S. Rogers, L. Hellerman, and T. E. Thompson, *J. Biol. Chem.* **240**, 198 (1965).
248. G. P. Greeville and R. W. Horne, *J. Mol. Biol.* **6**, 506 (1963).
249. C. E. Hull, *J. Biophys. Biochem. Cytol.* **7**, 613 (1960).
250. H. Sund, *Acta Chem. Scand. Suppl.* **1**, 102 (1963).
251. M. E. Magar, *Biochim. Biophys. Acta* **99**, 275 (1965).
252. D. G. Cross and H. F. Fisher, *Biochemistry* **5**, 880 (1966).
253. C. Frieden, *Biochim. Biophys. Acta* **62**, 421 (1961).
254. K S. Rogers, P. J. Geiger, P. J. Thompson, and L. Hellerman, *J. Biol. Chem.* **238**, PC 481 (1963).
255. R. W. Barratt and W. N. Strickland, *Arch. Biochem. Biophys.* **102**, 66 (1963).
256. G. M. Tompkins, K. L. Yielding, N. Talal, and J. F. Curran, in *Cold Spring Harbor Symp. Quant. Biol.*, **28**, 461 (1963).
257. M. E. Magar, *Biochim. Biophys. Acta* **96**, 345 (1965).
258. P. M. Bayley and G. K. Radda, *Biochem. J.* **94**, 31P (1965).
259. P. M. Bayley and G. K. Radda, *Biochem. J.* **98**, 105 (1966).
260. C. Frieden, *J. Biol. Chem.* **238**, 146 (1963).
261. J. C. Warren, D. O. Carr, and S. Grisolia, *Biochem. J.* **93**, 409 (1964).
262. M. W. Bitensky, K. L. Yielding, and G. M. Tompkins, *J. Biol. Chem.* **240**, 1077 (1965).
263. N. Talal, G. M. Tompkins, J. F. Muskinski, and K. L. Yielding, *J. Mol. Biol.* **8**, 46 (1964).
264. N. Talal and G. M. Tompkins, *Biochim. Biophys. Acta* **89**, 226 (1964).

265. H. F. Fisher, D. G. Cross, and L. L. McGregor, *Biochim. Biophys. Acta* **99**, 165 (1965).
266. H. F. Fisher, D. G. Cross, and L. L. McGregor, *Nature* **96**, 895 (1962).
267. R. F. Colman and C. Frieden, *Biochem. Biophys. Res. Commun.* **22**, 100 (1966).
268. A. J. Hill, *Biochem. J.* **7**, 471 (1913).
269. A. J. Bollet, F. S. Davis, and F. O. Hurt, *J. Exptl. Med.* **116**, 109 (1962).
270. L. Corman and N. O. Kaplan, *Biochemistry* **4**, 2175 (1965).
271. L. A. Fahien, H. G. Steinman, and R. McCann, *J. Biol. Chem.* **241**, 4700 (1966).
272. C. Frieden, *J. Biol. Chem.* **237**, 2396 (1962).
273. E. Masler and C. Tanford, *J. Biol. Chem.* **239**, 4217 (1964).
274. B. Jirgenson, *J. Am. Chem. Soc.* **83**, 3161 (1961).
275. H. F. Fisher, L. L. McGregor, and U. Power, *Biochem. Biophys. Res. Commun.* **8**, 402 (1962).
276. H. F. Fisher, L. L. McGregor, and U. Power, *Biochim. Biophys. Acta* **65**, 175 (1962).
277. E. Appella and G. M. Tompkins, *J. Mol. Biol.* **18**, 77 (1966).
278. C. Frieden, *Biochim. Biophys. Acta* **47**, 428 (1961).
279. G. P. Pfleiderer and F. Auricchio, *Biochem. Biophys. Res. Commun.* **6**, 53 (1964).
280. S. J. Aldenstein and B. L. Vallee, *J. Biol. Chem.* **233**, 589 (1958).
281. H. Sund, *Angew Chem. Intern. Ed. Engl.* **3**, 802 (1964).
282. A. Yoshida and E. Freese, *Biochim. Biophys. Acta* **96**, 248 (1965).
283. A. Yoshida and E. Freese, *Biochim. Biophys. Acta* **92**, 33 (1964).
284. N. G. McCormick and H. O. Halvorson, *J. Bacteriol.* **87**, 68 (1964).
285. A. Meister, *Nature* **168**, 1119 (1951).
286. W. S. Fones, *Arch. Biochem. Biophys.* **36**, 486 (1952).
287. A. Meister, L. Levintow, R. M. Kingsley, and J. P. Greenstein, *J. Biol. Chem.* **192**, 535 (1951).
288. G. K. Radda, *Nature* **203**, 936 (1966).
289. C. Frieden and S. F. Velick, *Biochim. Biophys. Acta* **23**, 439 (1957).
290. H. Beinert and G. Palmer, *Advan. Enzymol.* **27**, 105 (1965).
291. H. Beinert, *in* "The Enzymes" (P. D. Boyer, H. A. Lardy, and K. Myrbäck, eds.,) Vol. 2 (part a), p. 339. Academic Press, New York, 1960.
292. "Symposium on Flavin and Flavoproteins" (E. C. Slater, ed.), Elsevier, Amsterdam, 1966.
293. A. N. Radhakrishnan and A. Meister, *J. Am. Chem. Soc.* **79**, 5828 (1957).
294. A. N. Radhakrishnan and A. Meister, *J. Biol. Chem.* **233**, 444 (1958).
295. A. Meister, *in* "The Enzymes" (P. D. Boyer, H. A. Lardy, and K. Myrbäck, eds.,) Vol. VII, p. 609. Academic Press, New York, 1963.
296. E. A. Zeller and A. Maritz, *Helv. Chim. Acta* **27**, 1888 (1944).
297. E. A. Zeller, *Advan. Enzymol.* **8**, 459 (1948).
298. T. P. Singer and E. B. Kearney, *Arch. Biochem.* **29**, 190 (1950).
299. D. Wellner and A. Meister, *J. Biol. Chem.* **235**, 2013 (1960).
300. T. P. Singer and E. B. Kearney, *Arch. Biochem.* **27**, 348 (1950).
301. D. Wellner and A. Meister, *J. Biol. Chem.* **236**, 2357 (1961).
302. E. A. Zeller, G. Ramachander, G. A. Fleisher, T. Ishimaru, and V. Zeller, *Biochem, J.* **95**, 262 (1965).
303. H. Weissbach, A. V. Robertson, B. Witkop, and S. Udenfriend, *Anal. Biochem.* **1**, 286 (1960).
304. E. Frieden, L. T. Hsu, and K. Dittmer, *J. Biol. Chem.* **192**, 425 (1951).
305. M. Winitz, L. Bloch-Frankenthal, N. Izumiya, S. M. Birnbaum, C. G. Baker, and J. P. Greenstein, *J. Am. Chem. Soc.* **78**, 2423 (1956).

306. M. Winitz, S. M. Birnbaum, and J. P. Greenstein, *J. Am. Chem. Soc.* **77**, 3106 (1955).
307. E. A. Zeller and A. Maritz, *Helv. Chim. Acta* **28**, 365 (1945).
308. E. A. Zeller and A Maritz, *Helv. Chim. Acta* **28**, 1615 (1945).
309. E. A. Zeller, B. Islin, and A. Maritz, *Helv. Physiol. Pharmacol. Acta* **4**, 233 (1946).
310. T. P. Singer and E. B. Kearney, *Arch. Biochem.* **33**, 377, 397, 414 (1951).
311. D. Wellner, *Biochemistry* **5**, 1585 (1966).
312. M. Blanchard, D. E. Green, V. Nocito, and S. Ratner, *J. Biol. Chem.* **155**, 421 (1944).
313. M. Blanchard, D. E. Green, V. Nocito, and S. Ratner, *J. Biol. Chem.* **161**, 583 (1945).
314. D. E. Green, V. Nocito, and S. Ratner, *J. Biol. Chem.* **148**, 461 (1943).
315. D. E. Green, D. H. Moore, V. Nocito, and S. Ratner, *J. Biol. Chem.* **156**, 383 (1944).
316. M. Nakano and T. S. Danowski, *J. Biol. Chem.* **241**, 2075 (1966).
317. D. H. Moore, *J. Biol. Chem.* **161**, 597 (1945).
318. W. K. Paik and S. Kim, *Biochim. Biophys. Acta* **96**, 66 (1965).
319. P. Boulanger and R. Osteux, *Compt. Rend.* **234**, 1409 (1952).
320. P. Boulanger and R. Osteux, *Compt. Rend.* **235**, 524 (1952).
321. P. Boulanger and R. Osteux, *Compt. Rend.* **238**, 406 (1953).
322. P. Boulanger and R. Osteux, *Compt. Rend.* **241**, 126, 613 (1955).
323. P. Boulanger, J. Coursaget, J. Bertrand, and R. Osteux, *Compt. Rend.* **244**, 2255 (1957).
324. P. Boulanger and R. Osteux, *Biochim. Biophys. Acta* **21**, 552 (1956).
325. P. Boulanger, J. Bertrand, and R. Osteux, *Biochim. Biophys. Acta* **26**, 143 (1957).
326. J. Struck and I. W. Sizer, *Arch. Biochem. Biophys.* **90**, 22 (1960).
327. A. E. Bender, H. A. Krebs, and N. H. Horowitz, *Biochem. J.* **45**, xxi (1949).
328. A. E. Bender and H. A. Krebs, *Biochem. J.* **46**, 210 (1950).
329. P. S. Thayer and N. H. Horowitz, *J. Biol. Chem.* **192**, 755 (1951).
330. K. Burton, *Biochem. J.* **50**, 258 (1952).
331. S. G. Knight, *J. Bacteriol.* **55**, 401 (1948).
332. P. K. Stumpf and D. E. Green, *J. Biol. Chem.* **153**, 387 (1944).
333. J. Roche, P. E. Glahn, P. Manchon, and N. Van Thoai, *Biochim. Biophys. Acta* **35** 111 (1959).
334. H. Blaschko and D. B. Hope, *Biochem. J.* **62**, 355 (1956).
335. J. Shack, *J. Natl. Cancer Inst.* **3**, 398 (1943).
336. H. Kubo, T. Yamano, M. Iwatsubo, H. Watari, T. Soyama, J. Shiraishi, S. Sawada, N. Kawashima, S. Metani, and K. Ito, *Bull. Soc. Chim. Biol.* **40**, 431 (1958).
337. H. Kubo, T. Yamano, M. Iwatsubo, H. Watari, T. Shita, and A. Isomoto, *Bull. Soc. Chim. Biol.* **42**, 569 (1960).
338. V. Massey, G. Palmer, and R. Bennett, *Biochim. Biophys. Acta* **48**, 1 (1961).
339. K. Yagi, T. Ozawa, and H. Harada, *Nature* **184**, 1938 (1959); **188**, 745 (1960).
340. K. Yagi and T. Ozawa, *Nature* **193**, 483 (1961).
341. K. Yagi and T. Ozawa, *Biochim. Biophys. Acta* **62**, 397 (1962).
342. Y. Miyaki, K. Aki, S. Hashimito, and T. Yamano, *Biochim. Biophys. Acta* **105**, 86 (1965).
343. O. Warburg, W. Christian, and A. Greise, *Biochem. Z.* **297**, 417 (1938).
344. O. Warburg and W. Christian, *Biochem. Z.* **298**, 150 (1938).
345. K. Yagi, T. Ozawa, and T. Ooi, *Biochim. Biophys. Acta* **54**, 191 (1961).
346. P. A. Charlwood, G. Palmer, and R. Bennett, *Biochim. Biophys. Acta* **50**, 17 (1961).
347. E. Antonini, M. Brunori, M. R. Bruzzesi, E. Chiancone, and V. Massey, *J. Biol. Chem.* **241**, 2358 (1966).
348. V. Massey, B. Curti, and H. Ganther, *J. Biol. Chem.* **241**, 2347 (1966).
349. V. Massey and B. Curti, *J. Biol. Chem.* **241**, 3417 (1966).

350. J. P. Greenstein, S. M. Birnbaum, and M. C. Otey, *J. Biol. Chem.* **204**, 307 (1953).
351. M. Dixon and K. Kleppe, *Biochim. Biophys. Acta* **96**, 368 (1965).
352. P. Karrer and R. Appenzeller, *Helv. Chim. Acta* **26**, 808 (1943).
353. P. Handler, F. Bernheim, and J. R. Klein, *J. Biol. Chem.* **138**, 203 (1941).
354. W. R. Frisell and L. Hellerman, *J. Biol. Chem.* **225**, 53 (1957).
355. L. Hellerman, A. Lindsay, and M. R. Bovanick, *J. Biol. Chem.* **163**, 553 (1946).
356. K. Burton, *Biochem. J.* **48**, 458 (1951).
357. O. Walaas and E. Walaas, *Acta Chem. Scand.* **8**, 1104 (1954).
358. E. Walaas and O. Walaas, *Acta Chem. Scand.* **10**, 122 (1956).
359. E. B. Kearney, *J. Biol. Chem.* **194**, 747 (1952).
360. F. Egami and K. Yagi, *J. Biochem. (Tokyo)* **43**, 153 (1956).
361. K. Yagi, J. Okuda, T. Ozawa, and K. Okada, *Biochim. Biophys. Acta* **34**, 372 (1959).
362. B. M. Chassy and D. B. McCormick, *Biochim. Biophys. Acta* **110**, 91 (1965).
363. J. R. Klein and H. Kamin, *J. Biol. Chem.* **138**, 507 (1941).
364. G. R. Bartlett, *J. Am. Chem. Soc.* **70**, 1010 (1948).
365. L. Hellerman, F. P. Chinard, and V. R. Dietz, *J. Biol. Chem.* **147**, 443 (1943).
366. W. R. Frisell, H. J. Lowe, and L. Hellerman, *J. Biol. Chem.* **223**, 75 (1956).
367. S. Edlbacher and O. Wiss, *Helv. Chim. Acta* **27**, 1831 (1944).
368. W. T. Brown and P. G. Scholefield, *Proc. Exptl. Biol. Med.* **82**, 34 (1953).
369. W. T. Brown and P. G. Scholefield, *Biochem. J.* **58**, 368 (1954).
370. J. R. Klein and N. S. Olsen, *J. Biol. Chem.* **170**, 151 (1947).
371. T. P. Singer and E. S. G. Barron, *J. Biol. Chem.* **157**, 241 (1945).
372. T. P. Singer, *J. Biol. Chem.* **174**, 11 (1948).
373. K. Yagi and T. Ozawa, *Nature* **184**, 1227 (1959).
374. K. Yagi and T. Ozawa, *Biochim. Biophys. Acta* **42**, 381 (1960).
375. N. H. Horowitz, *J. Biol. Chem.* **154**, 141 (1944).
375a. J. Erkama and A. I. Virtanen, in "The Enzymes" (J. B. Sumner and K. Myrbäck, eds.), 1st Ed., Vol. I, Part 2, pp. 1244–1249. Academic Press, New York, 1951.
376. R. L. Emerson, M. Puziss, and S. G. Knight, *Arch. Biochem.* **25** 299, (1950).
377. H. Blaschko and J. Hawkins, *Biochem. J.* **49**, xliv (1951).
378. H. Blaschko and J. Hawkins, *Biochem. J.* **52**, 306 (1952).
379. H. Blaschko and J. M. Himms, *J. Physiol. (London)* **128**, 7P (1955).
380. E. Rocca and F. Ghiretti, *Arch. Biochem. Biophys.* **77**, 336 (1958).
381. S. Ratner, V. Nocito, and D. E. Green, *J. Biol. Chem.* **152**, 119 (1941).
382. K. Izaki, S. Mizushima, H. Takahashi, and J. Kakaguchi, *Symp. Enzyme Chem. (Tokyo)* **11**, 143 (1956).
383. S. Mizushima and J. Sakaguchi, *Bull Agr. Chem. Soc. Japan* **20**, 131 (1956).
384. A. H. Neims and L. Hellerman, *J. Biol. Chem.* **237**, P976 (1962).
385. J. L. Still, M. V. Buell, W. E. Knox, and D. E. Green, *J. Biol. Chem.* **179**, 831 (1949).
386. J. L. Still and E. Sperling, *J. Biol. Chem.* **182**, 585 (1950).
387. H. I. Nakada and S. Weinhouse, *J. Biol. Chem.* **187**, 663 (1950).
388. F. W. Sayre and D. M. Greenberg, *J. Biol. Chem.* **220**, 787 (1956).
389. J. S. Nishimura and D. M. Greenberg, *J. Biol. Chem.* **236**, 2684 (1961).
390. A. S. M. Selim and D. M. Greenberg, *J. Biol. Chem.* **234**, 1474 (1959).
391. A. S. M. Selim and D. M. Greenberg, *Biochim. Biophys. Acta* **42**, 211 (1960).
392. L. Goldstein, W. E. Knox, and E. J. Behrman, *J. Biol. Chem.* **237**, 2855 (1962).
393. C. Peraino, R. L. Blake, and H. C. Pitot, *J. Biol. Chem.* **240**, 3039 (1965).
394. R. A. Freedland and E. H. Avery, *J. Biol. Chem.* **239**, 3357 (1964).
395. E. Ishikawa, T. Ninagawa, and M. Suda, *J. Biochem. (Tokyo)* **57**, 506 (1965).
396. A. Nagabhushanam and D. M. Greenberg, *J. Biol. Chem.* **240**, 3002 (1965).

397. A. Kato, M. Ozura, and M. Suda, *J. Biochem.* (*Tokyo*) **59**, 40 (1966).
398. Y. Matsuo and D. M. Greenberg, *J. Biol. Chem.* **230**, 545, 561 (1958).
399. Y. Matsuo and D. M. Greenberg, *J. Biol. Chem.* **234**, 507, 516 (1959).
400. A. Kato, M. Ogura, H. Kimura, and M. Suda, *J. Biochem.* (*Tokyo*) **59**, 34 (1966).
401. E. Chargaff and D. B. Sprinson, *J. Biol. Chem.* **151**, 273 (1943).
402. W. A. Wood and I. C. Gunsalus, *J. Biol. Chem.* **181**, 171 (1949).
403. E. F. Gale and M. Stephenson, *Biochem. J.* **32**, 392 (1938).
404. D. E. Metzler and E. E. Snell, *J. Biol. Chem.* **198**, 363 (1952).
405. H. C. Lichstein and W. W. Umbreit, *J. Biol. Chem.* **170**, 423 (1947).
406. H. C. Lichstein and J. F. Christman, *J. Biol. Chem.* **175**, 649 (1948).
407. F. Binkley, *J. Biol. Chem.* **150**, 261 (1943).
408. A. B. Pardee and L. S. Prestidge, *J. Bacteriol.* **70**, 667 (1955).
409. E. McFall, *J. Mol. Biol.* **9**, 746 (1964).
410. D. Dupourque, S. A. Newton, and E. E. Snell, *J. Biol. Chem.* **241**, 1233 (1966).
411. R. Labow and W. G. Robinson, *J. Biol. Chem.* **241**, 1239 (1966).
412. H. E. Umbarger and B. Brown, *J. Bacteriol.* **71**, 443 (1955); **73**, 105 (1956).
413. H. E. Umbarger and B. Brown, *J. Biol. Chem.* **233**, 415 (1958).
414. J. P. Changeux, *Bull. Soc. Chim. Biol.* **46**, 927, 947 (1964).
415. J. P. Changeux, *Bull. Soc. Chim. Biol.* **47**, 115, 267 (1965).
416. P. Maeba and B. D. Sanwal, *Biochemistry* **5**, 525 (1966).
417. M. Hirata, M. Tokushige, A. Inagaki, and O. Hayaishi, *J. Biol. Chem.* **240**, 1711 (1965).
418. A. T. Phillips and W. A. Wood, *J. Biol. Chem.* **240**, 4703 (1965).
419. A. T. Phillips and W. A. Wood, *Biochem. Biophys. Res. Commun.* **15**, 530 (1964).
420. O. Hayaishi, M. Gefter, and H. Weissbach, *J. Biol. Chem.* **238**, 2040 (1963).
421. M. Tokushige, H. R. Whiteley, and O. Hayaishi, *Biochem. Biophys. Res. Commun*, **13**, 380 (1963).
422. H. R. Whiteley and O. Hayaishi, *Biochem. Biophys. Res. Commun.* **14**, 143 (1964).
423. W. A. Newton, Y. Morino, and E. E. Snell, *J. Biol. Chem.* **240**, 1211 (1965).
424. I. P. Crawford and J. Ito, *Proc. Natl. Acad. Sci. U.S.* **51**, 390 (1964).
425. C. Yanofsky, *J. Biol. Chem.* **194**, 279 (1952).
426. C. Yanofsky, *J. Biol. Chem.* **198**, 343 (1952).
427. J. L. Reissig, *Arch. Biochem. Biophys.* **36**, 234 (1952).
428. J. L. Reissig and C. Yanofsky, *J. Biol. Chem.* **202**, 567 (1953).
429. M. Boll and H. Holzer, *Biochem. Z.* **343**, 504 (1965).
430. G. A. Zittle, in "The Enzymes" (J. B. Sumner and K. Myrbäck, eds.), 1st ed., Vol. I, Part 2, pp. 922–945. Academic Press, New York, 1951.
431. A. Meister, *Physiol. Rev.* **36**, 103 (1956).
432. E. Roberts, *in* "The Enzymes" (P. D. Boyer, H. Lardy, and K. Myrbäck, eds.), 2nd ed., Vol. 4, Part A, pp. 285–300. Academic Press, New York, 1960.
433. M. C. Otey, S. M. Birnbaum, and J. P. Greenstein, *Arch. Biochem. Biophys.* **49**, 245 (1954).
434. J. D. Klingman and P. Handler, *J. Biol. Chem.* **232**, 369 (1958).
435. J. A. Shepherd and G. Kalnitsky, *J. Biol. Chem.* **192**, 1 (1951).
436. F. W. Sayre and E. Roberts, *J. Biol. Chem.* **233**, 1128 (1958).
437. P. Schwerin, S. P. Bessman, and H. Waelsch, *J. Biol. Chem.* **184**, 37 (1950).
438. A. Meister, L. Levintow, R. E. Greenfield, and A. Abendschein, *J. Biol. Chem.* **215**, 411 (1955).
439. J. P. Greenstein and C. E. Carter, *J. Natl. Cancer Inst.* **7**, 57, 433 (1947).
440. L. Goldstein, *Am. J. Physiol.* **210**, 661 (1966).

441. D. E. Hughes and D. H. Williamson, *Biochem. J.* **51**, 45 (1952).
442. T. Benzinger and R. Hems, *Proc. Natl. Acad. Sci. U.S.* **42**, 896 (1959).
443. J. E. Varner, *in* "The Enzymes" (P. D. Boyer, H. Lardy, and K. Myrbäck, eds.), 2nd ed., Vol. 4, Part A, p. 244. Academic Press, New York, 1960.
444. T. O. Yellin and J. C. Wriston, Jr., *Biochemistry* **5**, 1605 (1966).
445. J. D. Broome, *Nature* **191**, 1114 (1961).
446. H. M. Suld and P. A. Herbut, *J. Biol. Chem.* **240**, 2234 (1965).
447. M. E. Ramadan, F. El Asmer, and D. M. Greenberg, *Arch. Biochem. Biophys.* **108**, 143 (1964).
448. M. E. Ramadan, F. El Asmer, and D. M. Greenberg, *Arch. Biochem. Biophys.* **108**, 150 (1964).
449. W. A. Wood and I. C. Gunsalus, *J. Biol. Chem.* **190**, 403 (1951).
450. U. Roze and J. L. Strominger, *Mol. Pharmacol.* **2**, 92 (1966).
451. B. T. Stewart and H. O. Halvorson, *Arch. Biochem. Biophys.* **49**, 168 (1954).
452. S. A. Narrod and W. A. Wood, *Arch. Biochem. Biophys.* **35**, 462 (1952).
453. P. Ayengar and E. Roberts, *J. Biol. Chem.* **197**, 453 (1952).
454. L. Glaser, *J. Biol. Chem.* **235**, 2095 (1960).
455. M. Tanaka, Y. Kato, and S. Kinoshita, *Biochem. Biophys. Res. Commun.* **4**, 114 (1961).
456. G. D. Shockman and G. Toennies, *Arch. Biochem. Biophys.* **50**, 1 (1954).
457. G. D. Shockman and G. Toennies, *Arch. Biochem. Biophys.* **50**, 9 (1954).
458. R. E. Kallio and A. D. Larson, *Symp. Amino Acid Metab., Baltimore, 1954, Johns Hopkins Univ., McCollum-Pratt Inst., Contrib.* **105**, 616 (1955).
459. H. T. Huang, D. A. Kita, and J. W. Davisson, *J. Am. Chem. Soc.* **80**, 1006 (1958).
460. H. T. Huang and J. W. Davisson, *J. Bacteriol.* **76**, 495 (1958).
461. P. S. Thayer, *J. Bacteriol.* **78**, 150 (1959).
462. T. C. Stadtman and P. Elliott, *J. Biol. Chem.* **228**, 983 (1957).
463. N. G. Srinivasan, J. J. Corrigan, and A. Meister, *J. Biol. Chem.* **240**, 796 (1965).
464. W. F. Diven, R. B. Johnston, and J. J. Scholz, *Biochim. Biophys. Acta* **67**, 161 (1963).
465. E. Adams, *J. Biol. Chem.* **234**, 2083 (1959).
466. E. Adams and I. L. Norton, *J. Biol. Chem.* **239**, 1525 (1964).
467. M. Antia, D. S. Hoare, and E. Work, *Biochem. J.* **65**, 448 (1957).
468. P. P. Cohen, *in* "The Enzymes" (P. D. Boyer, H. Lardy, and K. Myrbäck, eds.), 2nd ed., Vol. 6, pp. 477–494. Academic Press, New York, 1962.
469. P. P. Cohen and M. Marshall, *in* "The Enzymes" (P. D. Boyer, H. Lardy, and K. Myrbäck, eds.), 2nd ed., Vol. 6, pp. 327–338. Academic Press, New York, 1962.
470. H. A. Krebs, *in* "The Enzymes" (J. B. Sumner and K. Myrbäck, eds.), 1st ed., Vol. II, Part 2, pp. 866–885. Academic Press, New York, 1951.
471. S. Ratner, *Advan. Enzymol.* **15**, 319 (1954).
472. R. L. Metzenberg, L. M. Hall, M. Marshall, and P. P. Cohen, *J. Biol. Chem.* **229**, 1019 (1957).
473. M. Marshall, R. L. Metzenberg, and P. P. Cohen, *J. Biol. Chem.* **233**, 102 (1958).
474. P. M. Anderson and A. Meister, *Biochemistry* **4**, 2803 (1965).
475. S. M. Kalman, P. H. Duffield, and T. Brzozowski, *J. Biol. Chem.* **241**, 1871 (1966).
476. S. Grisolia and P. P. Cohen, *J. Biol. Chem.* **191**, 189 (1951).
477. A. Piérard and J. M. Wiame, *Biochem. Biophys. Res. Commun.* **15**, 76 (1964).
478. S. M. Kalman, P. H. Duffield, and T. Brzozowski, *Biochem. Biophys. Res. Commun.* **18**, 530 (1965).
479. M. E. Jones and F. Lipmann, *Proc. Natl. Acad. Sci U.S.* **46**, 1194 (1960).
480. E. Baldwin, *Comp. Biochem. Physiol.* **1**, 24 (1960).

481. L. M. Hall, R. C. Johnson, and P. P. Cohen, *Biochim. Biophys. Acta* **37**, 144 (1960).
482. S. H. Bishop and J. W. Campbell, *Science* **142**, 1583 (1963).
483. P. A. Janssens and P. P. Cohen, *Science* **152**, 358 (1966).
484. P. P. Cohen, *Harvey Lecture Ser.* **60**, 119 (1966).
485. A. L. Kennan and P. P. Cohen, *Develop. Biol.* **1**, 511 (1959).
486. G. W. Brown, Jr., and P. P. Cohen, *in* "A Symposium on the Chemical Basis of Development" (W. D. McElroy and H. B. Glass, eds.), p. 495. Johns Hopkins Press, Baltimore, Maryland, 1958.
487. G. W. Brown, Jr., W. R. Brown, and P. P. Cohen, *J. Biol. Chem.* **234**, 1775 (1959).
488. R. L. Metzenberg, M. Marshall, W. K. Paik, and P. P. Cohen, *J. Biol. Chem.* **236**, 162 (1961).
489. M. Tatibana and P. P. Cohen, *J. Biol. Chem.* **239**, 2905 (1964).
490. M. Tatibana and P. P. Cohen, *Proc. Natl. Acad. Sci. U.S.* **53**, 104 (1965).
491. W. K. Paik and P. P. Cohen, *J. Gen. Physiol.* **43**, 683 (1960).
492. M. Marshall and P. P. Cohen, *J. Biol. Chem.* **236**, 718 (1960).
493. B. Levenberg, *J. Biol. Chem.* **237**, 2590 (1962).
494. M. Marshall, R. L. Metzenberg, and P. P. Cohen, *J. Biol. Chem.* **236**, 2229 (1961).
495. P. P. Cohen and S. Grisolia, *J. Biol. Chem.* **182**, 747 (1950).
496. S. Grisolia and P. P. Cohen, *J. Biol. Chem.* **204**, 753 (1953).
497. L. M. Hall, R. L. Metzenberg, and P. P. Cohen, *J. Biol. Chem.* **230**, 1013 (1958).
498. M. Grassl and S. J. Bach, *Biochim. Biophys. Acta* **42**, 154 (1960).
499. L. A. Fahien, J. M. Schooler, G. A. Gehred, and P. P. Cohen, *J. Biol. Chem.* **239**, 1935 (1964).
500. R. L. Metzenberg, M. Marshall, and P. P. Cohen, *J. Biol. Chem.* **233**, 1560 (1958).
501. M. E. Jones and L. Spector, *J. Biol. Chem.* **235**, 2897 (1960).
502. L. A. Fahien and P. P. Cohen, *J. Biol. Chem.* **239**, 1925 (1964).
503. M. E. Jones, *Ann. Rev. Biochem* **34**, 381 (1965).
504. S. Grisolia and L. Raijman, *Advan. Chem. Ser.* **44**, 128 (1964).
505. L. Raijman and S. Grisolia, *J. Biol. Chem.* **239**, 1272 (1964).
506. W. B. Novoa and S. Grisolia, *J. Biol. Chem.* **237**, PC2711 (1962).
507. J. Kennedy and S. Grisolia, *Biochim. Biophys. Acta* **96**, 102 (1965).
508. J. Caravaca and S. Grisolia, *Biochem. Biophys. Res. Commun.* **1**, 94 (1959).
509. L. Raijman and S. Grisolia, *Biochem. Biophys. Res. Commun.* **4**, 262 (1961).
510. L. A. Fahien and P. P. Cohen, unpublished experiments.
511. W. B. Novoa, H. A. Tigier, and S. Grisolia, *Biochim. Biophys. Acta* **113**, 84 (1966).
512. C. M. Allen and M. E. Jones, *Arch. Biochem. Biophys.* **114**, 115 (1966).
513. J. M. Schooler, L. A. Fahien, and P. P. Cohen, *J. Biol. Chem.* **238**, PC1909 (1963).
514. M. Marshall and P. P. Cohen, *J. Biol. Chem.* **241**, 4197 (1966).
515. S. H. Bishop and S. Grisolia, *Biochim. Biophys. Acta* **118**, 211 (1966).
516. S. H. Bishop, *Federation Proc.* **25**, 523 (1966).
517. M. E. Jones, *in* "Methods in Enzymology" (S. P. Colowick and N. O. Kaplan, eds.), Vol. 6, p. 645. Academic Press, New York, 1963.
518. S. M. Kalman and P. H. Duffield, *Biochim. Biophys. Acta* **92**, 498 (1964).
519. S. Grisolia, P. Harmon, and L. Raijman, *Biochim. Biophys. Acta* **62**, 293 (1962).
520. S. Kim and P. P. Cohen, *Arch. Biochem. Biophs.* **109**, 421 (1965).
521. E. Calva and P. P. Cohen, *Cancer Res.* **19**, 679 (1959).
522. J. R. Bronk and R. B. Fisher, *Biochem. J.* **64**, 106 (1956).
523. J. R. Bronk and R. B. Fisher, *Biochem. J.* **64**, 111 (1956).
524. J. R. Bronk and R. B. Fisher, *Biochem. J.* **64**, 118 (1956).
525. A. G. Gornall and A. Hunter, *J. Biol. Chem.* **147**, 593 (1943).

526. S. J. Bach and M. Smith, *Biochem. J.* **64**, 417 (1956).
527. F. Cedrangolo, R. Piazza, and G. Rametta, *Ital. J. Biochem.* **6**, 56 (1957).
528. F. Cedrangolo, *Enzymologia* **19**, 335 (1958).
529. L. D. Wright, K. A. Walentik, D. S. Spicer, J. W. Huff, and H. R. Skeggs, *Proc. Soc. Exptl. Biol. Med.* **75**, 293 (1950).
530. I. Lieberman and A. Kornberg, *J. Biol. Chem.* **207**, 911 (1954).
531. L. W. Weed and D. W. Wilson, *J. Biol. Chem.* **207**, 439 (1954).
532. P. Reichard, H. L. Smith, Jr., and G. Hanshoff, *Acta Chem. Scand.* **9**, 1010 (1955).
533. P. Reichard, *Acta Chem. Scand.* **10**, 548 (1956).
534. J. M. Lowenstein and P. P. Cohen, *J. Biol. Chem.* **220**, 57 (1956).
535. H. J. Sallach, *J. Biol. Chem.* **234**, 900 (1959).
536. P. Reichard and G. Hanshoff, *Acta Chem. Scand.* **9**, 519 (1955).
537. R. A. Yates and A. B. Pardee, *J. Biol. Chem.* **221**, 743 (1956).
538. R. A. Yates and A. B. Pardee, *J. Biol. Chem.* **221**, 757 (1956).
539. R. A. Yates and A. B. Pardee, *J. Biol. Chem.* **227**, 677 (1957).
540. L. I. Stein and P. P. Cohen, *Arch. Biochem. Biophys.* **109**, 429 (1965).
541. E. Calva, J. M. Lowenstein, and P. P. Cohen, *Cancer Res.* **19**, 101 (1959).
542. L. H. Smith, Jr., and P. Reichard, *Acta Chem. Scand.* **10**, 1024 (1956).
543. M. Shepherdson and A. B. Pardee, *J. Biol. Chem.* **227**, 677 (1957).
544. J. C. Gerhart and A. B. Pardee, *J. Biol. Chem.* **237**, 891 (1962).
545. J. C. Gerhart and H. K. Schachman, *Biochemistry* **4**, 1054 (1965).
546. J. C. Gerhart and A. B. Pardee, *Federation Proc.* **23**, 727 (1964).
547. J. C. Gerhart, *in* D. E. Atkinson, *Ann. Rev. Biochem.* **35**, Pt. I, 85 (1966).
548. J. P. Changeux, *Brookhaven Sym. Biol.* **17** [BNL 869 (C–40)], 232 (1964).
549. J. Monod, J. P. Changeux, and F. Jacob, *J. Mol. Biol.* **6**, 306 (1963).
550. J. Monod, J. Wyman, and J. P. Changeux, *J. Mol. Biol.* **12**, 88 (1965).
551. J. Neumann and M. E. Jones, *Arch. Biochem. Biophys.* **104**, 438 (1964).
552. E. Bresnick, *Biochim. Biophys. Acta* **61**, 598 (1962).
553. K. Kleczkowski and P. P. Cohen, *Arch. Biochem. Biophys.* **107**, 271 (1964).
554. M. Marshall and P. P. Cohen, personal communication, 1966.
555. G. H. Burnett and P. P. Cohen, *J. Biol. Chem.* **229**, 337 (1957).
556. P. Reichard, *Acta Chem. Scand.* **11**, 523 (1957).
557. J. M. Ravel, M. L. Grona, J. S. Humphreys, and W. Shive, *J. Biol. Chem.* **234**, 1452 (1959).
558. P. Rogers and G. D. Novelli, *Biochim. Biophys. Acta* **33**, 423 (1959).
559. J. Caravaca and S. Grisolia, *J. Biol. Chem.* **235**, 684 (1960).
560. H. A. Krebs, L. V. Eggleston, and V. A. Knivett, *Biochem. J.* **59**, 185 (1955).
561. H. A. Krebs, P. K. Jensen, and L. V. Eggleston, *Biochem. J.* **70**, 397 (1959).
562. V. A. Knivett, *Biochem. J.* **50**, xxx (1952).
563. H. D. Slade, *Arch. Biochem. Biophys.* **42**, 204 (1953).
564. E. L. Oginsky and R. F. Gehrig, *J. Biol. Chem.* **204**, 721 (1953).
565. V. A. Knivett, *Biochem. J.* **56**, 602 (1954).
566. M. Korzenovsky and C. H. Werkman, *Arch. Biochem. Biophys.* **46**, 174 (1953).
567. H. D. Slade, C. C. Doughty, and W. C. Slump, *Arch. Biochem. Biophys.* **48**, 338 (1954).
568. V. A. Knivett, *Biochem. J.* **58**, 480 (1954).
569. E. L. Oginsky, *Symp. Amino Acid Metab., Baltimore, 1954, Johns Hopkins Univ., McCollum-Pratt Inst. Contrib.* **105**, 300 1955.
570. M. Korzenovsky, *Symp. Amino Acid Metab. Baltimore, 1954, Johns Hopkins Univ., McCollum-Pratt Inst., Contrib.* **105**, 309, (1955).

571. H. D. Slade, *Symp. Amino Acid Metab., Baltimore, 1954, Johns Hopkins Univ., McCollum-Pratt Inst., Contrib.* **105**, 321 (1955).
572. S. Ratner, *in* "The Enzymes" (P. D. Boyer, H. Lardy, and K. Myrbäck, eds.), 2nd ed., Vol. 6, p. 495. Academic Press, New York, 1962.
573. S. Ratner and B. Petrack, *J. Biol. Chem.* **191**, 693 (1951).
574. S. Ratner and B. Petrack, *J. Biol. Chem.* **200**, 161 (1953).
575. S. Ratner, *Symp. Amino Acid Metab., Baltimore, 1954, Johns Hopkins Univ., McCollum-Pratt Inst., Contrib.* **105**, 231 (1955).
576. S. Ratner and B. Petrack, *Arch. Biochem. Biophys.* **65**, 582 (1956).
577. A. Schuegraf, S. Ratner, and R. C. Warner, *J. Biol. Chem.* **235**, 3597 (1960).
578. B. Petrack and S. Ratner, *J. Biol. Chem.* **233**, 1494 (1958).
579. H. Tigier, J. Kennedy, and S. Grisolia, *Biochim. Biophys. Acta* **110**, 423 (1965).
580. S. Ratner, W. P. Anslow, Jr., and B. Petrack, *J. Biol. Chem.* **204**, 115 (1953).
581. J. B. Walker and J. Myers, *J. Biol. Chem.* **203**, 143 (1953).
582. D. C. Davison and W. H. Elliott, *Nature* **169**, 313 (1952).
583. J. B. Walker, *Proc. Natl. Acad. Sci. U.S.* **38**, 561 (1952).
584. J. R. S. Fincham and J. B. Boylen, *Biochem. J.* **61**, xiii (1955).
585. E. A. Havir, H. Tamir, S. Ratner, and R. C. Warner, *J. Biol. Chem.* **240**, 3079 (1965),
586. H. D. Hoberman, E. A. Havir, O. Rochovansky, and S. Ratner, *J. Biol. Chem.* **239**, 3818 (1964).
587. D. M. Greenberg, *in* "The Enzymes" (P. D. Boyer, H. Lardy, and K. Myrbäck, eds.), 2nd ed., Vol. 4, Part A, p. 257. Academic Press, New York, 1960.
588. A. Hunter and J. A. Dauphinee, *Proc. Roy. Soc. (London)* **97B**, 227 (1924).
589. A. Hunter, *J. Biol. Chem.* **81**, 505 (1929).
590. S. J. Folley and A. L. Greenbaum, *Biochem. J.* **40**, 46 (1945).
591. E. Roberts and S. Frankel, *Cancer Res.* **9**, 231 (1949).
592. A. Clementi, *Atti Accad. Lincei (5)* **23**, 612 (1914).
593. G. W. Brown, Jr., *Arch Biochem. Biophys.* **114**, 184 (1966).
594. D. Rosenthal, B. Gottlieb, J. D. Gorry, and H. M. Vars, *J. Biol. Chem.* **223**, 469 (1956).
595. D. M. Greenberg, *in* "Methods in Enzymology" (S. P. Colowick and N. O. Kaplan, eds.), Vol. 1, p. 268. Academic Press, New York, 1955.
596. W. Graasmann, H. Hörmann, and O. Janowsky, *Z. Physiol. Chem.* **312**, 273 (1958).
597. J. Cabello, V. Prajoux, *Biochim. Biophys. Acta* **105**, 583 (1965).
598. A. Hunter, *Biochem. J.* **32**, 826 (1938).
599. A. Hunter and H. E. Woodward, *Biochem. J.* **35**, 1298 (1941).
600. A. Hunter and C. E. Downs, *J. Biol. Chem.* **155**, 173 (1944).
601. M. S. Mohamed and D. M. Greenberg, *Arch. Biochem. Biophys.* **8**, 349 (1945).
602. D. M. Greenberg, A. E. Bagot, and D. A. Roholt, Jr., *Arch. Biochem. Biophys.* **62**, 446 (1956).
603. S. J. Bach and J. D. Killip, *Biochim. Biophys. Acta* **29**, 273 (1958).
604. A. Hunter and C. E. Downs, *J. Biol. Chem.* **157**, 427 (1945).
605. H. Nakagawa and P. P. Cohen, personal communication.
606. S. K. Chan and P. P. Cohen, *Arch. Biochem. Biophys.* **104**, 325 (1964).
607. P. A. Janssens, *Comp. Biochem. Physiol.* **13**, 217 (1964).
608. N. DeGroot and P. P. Cohen, *Biochim. Biophys. Acta* **59**, 588 (1962).
609. R. T. Schimke, *in Cold Spring Harbor Sym. Quant. Biol.* **26**, 363 (1961).
610. R. T. Schimke, *J. Biol. Chem.* **239**, 3808 (1964).
611. R. T. Schimke, *J. Biol. Chem.* **239**, 136 (1964).
612. M. Martinez-Carrion and D. Tiemeir, *Biochemistry* **6**, 1715 (1967).

613. M. Martinez-Carrion, C. Turano, F. Riva, and P. Fasella, *J. Biol. Chem.* **242**, 1426 (1967).

614. P. Fasella and G. G. Hammes, *Biochemistry* **6**, 1798 (1967).

615. M. H. Saier, Jr., and W. T. Jenkins, *J. Biol. Chem.* **242**, 91 (1967).

616. M. H. Saier, Jr., and W. T. Jenkins, *J. Biol. Chem.* **242**, 101 (1967).

617. P. W. Gatehouse, S. Hopper, L. Schatz, and H. L. Segal, *J. Biol. Chem.* **242**, 2319 (1967).

618. H. Hirsch and D. M. Greenberg, *J. Biol. Chem.* **242**, 2283 (1967).

619. S. Hayashi, P. K. Granner, and G. M. Tompkins, *J. Biol. Chem.* **242**, 3998 (1967).

620. R. F. Colman and C. Frieden, *J. Biol. Chem.* **241**, 3652 (1966).

621. R. F. Colman and C. Frieden, *J. Biol. Chem.* **241**, 3661 (1966).

622. C. Frieden and R. F. Colman, *J. Biol. Chem.* **242**, 1705 (1967).

623. K. L. Yielding and B. B. Holt, *J. Biol. Chem.* **242**, 1079 (1967).

624. B. M. Anderson, C. D. Anderson and J. E. Churckich, *Biochemistry* **5**, 2893 (1966).

625. R. A. Butlow, *Biochemistry* **6**, 1088 (1967).

626. L. Corman, L. M. Prescott, and N. O. Kaplan, *J. Biol. Chem.* **242**, 1383 (1967).

627. L. Corman and N. O. Kaplan, *J. Biol. Chem.* **242**, 2840 (1967).

628. D. J. West, R. W. Tuveson, R. W. Barratt, and J. R. S. Fincham, *J. Biol. Chem.* **242**, 2134 (1967).

629. H. A. Campbell, L. T. Mashburn, E. A. Boyse, and L. J. Old, *Biochemistry* **6**, 721 (1967).

630. H. Cedar and J. H. Schwartz, *J. Biol. Chem.* **242**, 3753 (1967).

631. M. Tatibana and K. Ito, *Biochem. Biophys. Res. Commun.* **26**, 221 (1967).

632. M. Tatibana, K. Ito, M. Terada, and A. Inagaki, *Abstr. G-17, 7th Intern. Congr. Biochem., Tokyo, 1967.*

633. S. E. Hager and M. E. Jones, *J. Biol. Chem.* **242**, 5667 (1967).

634. S. E. Hager and M. E. Jones, *J. Biol. Chem.* **242**, 5674 (1967).

635. P. M. Anderson and A. Meister, *Biochemistry* **5**, 3164 (1966).

636. P. M. Anderson and A. Meister, *Biochemistry* **5**, 3157 (1966).

637. E. Khedouri, P. M. Anderson, and A. Meister, *Biochemistry* **5**, 3552 (1966).

638. S. McKinley, C. D. Anderson, and M. E. Jones, *J. Biol. Chem.* **242**, 3381 (1967).

639. H. Nakagawa, K. Kim, and P. P. Cohen, *J. Biol. Chem.* **242**, 635 (1967).

640. J. C. Gerhart and H. Holoubach, *J. Biol. Chem.* **242**, 2886 (1967).

641. G. L. Hervé and G. R. Stark, *Biochemistry* **6**, 3743 (1967).

642. S. H. Bishop and S. Grisolia, *Biochim. Biophys. Acta* **139**, 344 (1967).

CHAPTER 15 (Part I)

Carbon Catabolism of Amino Acids

David M. Greenberg

I. SCOPE OF THE CHAPTER

The metabolism of the α-amino group of the amino acids and of the remaining carbon moiety, in general, follows independent pathways once the amino groups are removed. The metabolism of the amino nitrogen is discussed in Chapter 14. It is the purpose of this chapter to follow the fate of the carbon moiety and of those nitrogen groups that continue to remain attached to the carbon residue to the point where they are lost.

In the majority of cases the carbon residue of an amino acid in the course of its catabolism is transformed into a short-chain di- or mono-carboxylic acid; in the latter case this may be a straight or a branched chain. The problem then becomes one of the catabolic pathways of these different mono- or dicarboxylic acids. Their pathways will be traced in this chapter to the point where they converge into well-known products whose metabolism is discussed elsewhere in this work.

Many of the clues to the metabolic pathways of the carbon residues of the amino acids were obtained by the investigation of the ketogenic, glycogenic, and antiketogenic properties of the different amino acids. The early literature on this subject has been reviewed by Dakin (1). Other important evidence was secured from the excretion of incompletely metabolized products in the urine, particularly when a radical difficult to degrade, such as the phenyl group, was combined in the molecule (1), or in subjects with hereditary defects causing "inborn errors of metabolism" (2), which led to the accumulation and excretion of incompletely metabolized products.

In the course of time it has been possible to induce conditions in experimental animals simulating the genetic anomalies of metabolism, thus providing conditions favorable for the study of the catabolism of the amino acids. This type of experiment has been raised to a much more exact level by the experimental production of numerous mutants of microorganisms with specific and serial metabolic defects, starting with the bread mold, *Neurospora crassa* (3), but now extended to a variety of bacteria and yeasts (4). However, the greatest advances have been made by tracer experiments with isotopically labeled molecules and the employment of enzyme systems ever more purified and better defined. The new advances in the methodology of metabolic experiments, in conjunction with the highly refined chromatographic procedures for the isolation and identification of even very minute amounts of the intermediate products of metabolism, have made possible very rapid advances in identifying at least the major chemical intermediates of the metabolism of the amino acids, as also of the other food components.

As a result the gross metabolic pathways are fairly well known for most of the amino acids.

Great advances have been made in the purification of many of the enzymes concerned in the catalysis of individual reactions of the degradative pathways of many of the amino acids. The purified enzymes also have permitted detailed studies of the reaction mechanisms. Among the most significant of the areas of development in recent years has been elucidation of the regulatory processes by which organisms adjust the individual reaction rates of their numerous metabolic processes to conform to the organisms' physiological requirements.

II. AMINO ACIDS LINKED WITH THE CITRIC ACID (TCA) CYCLE

The major catabolic pathways of the carbon chains of the amino acids alanine, aspartic, and glutamic acids appear to be readily apparent once these amino acids lose their amino groups. When this occurs, alanine is converted to pyruvate, aspartic acid to either oxalo-acetic or fumaric acid, and glutamic acid to α-ketoglutaric acid. All of the above acids are integral members of the citric acid cycle, and the subsequent degradation of each one has been adequately explained in terms of the operation of this cycle (see Volume I, Chapter 4).

However, other degradative pathways have been discovered for aspartic and glutamic acids, as will be described below.

A. Alanine

The only known pathways for the dissimilation of alanine are via transamination or oxidation to yield pyruvic acid. This is subsequently oxidized to CO_2 by means of the TCA cycle. Nonspecific L-amino acid oxidoreductases that deaminate L-alanine occur in snake venom, certain bacteria, and in the kidney (see Chapter 14). Under certain conditions, L-alanine can be oxidized by glutamate dehydrogenase. A dehydro-genase that preferentially oxidizes L-alanine has been discovered and purified from certain bacteria (5–8). This enzyme [L-alanine : NAD oxidoreductase (deaminating), EC 1.4.1.1] was crystallized from *Bacillus subtilis*. It has a molecular weight of about 228,000. The enzyme also catalyzes the oxidation of L-α-aminobutyrate, L-valine, L-isovaline, L-serine, and L-norvaline at decreasing rates from 15% to 5% of the rate of oxidation of L-alanine. It can catalyze the reductive amination of a variety of α-keto acids. The optimum activity for deamination is at pH 8.8–9.0, for reamination at pH 10.0–10.5. K_m values of the various substrates were as follows: L-alanine 1.7 mM, NAD$^+$ 1.8 × 10^{-4} M at pH 10.0; pyruvate 5.4 × 10^{-4} M, NADH 2.3 × 10^{-5} M, NH$_4^+$ 10^{-2} M at pH 8.0. The equilibrium constant for the reaction represented by the equation:

$$\frac{[\text{Pyruvate}][\text{NADH}][\text{NH}_4^+][\text{H}^+]}{[\text{Alanine}][\text{NAD}^+][\text{H}_2\text{O}]} = 3 \times 10^{-4}$$

The L-alanine dehydrogenase is inhibited by most divalent cations and by PCMB. Inhibition by the latter can be reversed by cysteine.

D-Alanine is decomposed by D-amino acid dehydrogenase, but it appears to be metabolized mainly by conversion to L-alanine through the action of alanine racemase.

B. Aspartic Acid

In the degradation of aspartic acid the major pathway in the vertebrate organism appears to be transamination to oxaloacetic acid; the latter is then oxidized by means of the TCA cycle (see Chapter 14).

A number of alternative pathways exist for the metabolism of aspartic acid. It can couple with citrulline, forming argininosuccinic acid, which is subsequently cleaved to fumarate and arginine (see Chapter 14).

The carbamyl group of citrulline can also be transferred to L-aspartate, it is believed through carbamyl phosphate as an intermediate, to form carbamyl-L-aspartate (see Chapter 14). This in turn is the precursor of dihydroorotic acid and ultimately of the pyrimidines.

A specific D-aspartate dehydrogenase was discovered in rabbit kidney and liver by Still et al. (9).

Other pathways of aspartic acid dissimilation are via the aspartase reaction and β-decarboxylation. They are discussed below.

1. Aspartase (L-Aspartate ammonia-lyase, EC 4.3.1.1)

Aspartase is found in microorganisms and plants, but not in vertebrates. This enzyme reversibly catalyzes the reaction shown in Eq. (1).

$$\text{HOOCCH}_2\text{CHNH}_2\text{COOH} \rightleftharpoons \text{HOOCCH}=\text{CHCOOH} + \text{NH}_3 \qquad (1)$$
$$\text{L-Aspartic acid} \qquad\qquad\qquad \text{Fumaric acid}$$

In this reaction, ammonia is either removed directly from aspartic acid or added to the fumaric acid in a manner similar to the removal or addition of water by fumarase.

The name aspartase for this enzyme was proposed by Woolf (10). Preparation of cell-free extracts and fractionations of the enzyme were first carried out by Gale on *Escherichia coli* (*11*). He observed activation of some ammonium sulfate fractions by adenylate and formate. Subsequently various degrees of purification of aspartase from a variety of bacterial species have been achieved by a number of investigators (*12–16*). Extraction of the enzyme is improved by *n*-butanol (*13*). The aspartase preparations have generally contained other enzymes, in particular fumarase, which results in the formation of malate in addition to fumarate. Aspartase preparations free of this enzyme were obtained by Williams and McIntyre (*15*) and by Dupue and Moat (*16*). The protection of the enzyme against inactivation and also some degree of enhancement of its activity by formate has been confirmed by Williams and McIntyre (*15*). These investigators attributed the activating effect to a prevention of SH group oxidation. The stimulating effect of AMP proved to be quite small in more highly purified preparations.

Ellfolk (*13*) observed the inhibition of aspartase by EDTA, citrate, oxalate, and PP_i. Reactivation can be achieved most effectively by Ca^{2+} and Mn^{2+}. Other metal ions require higher concentrations. Ichihara *et al.* (*17*) have reported the activation of aspartase by Co^{2+} and folic acid. Activation by Co^{2+} has been confirmed (*16*) but not activation by folic acid. Physical-chemical constants of aspartase are as follows: optimum pH at 7.0–7.5, K_m for aspartase is $2–3 \times 10^{-2} M$, and the equilibrium constant is 2×10^{-2} at 37° and pH 6.8 in 0.05 M phosphate buffer.

Highly purified aspartase from *Enterobacter aerogenes* has a molecular weight of $\sim 180,000$. It is cleaved into four subunits by PCMB. The tetramer can be regenerated by treatment with mercaptoethanol (*17a*).

Emery (*18*) has shown that aspartase also catalyzes the condensation of fumarate with NH_2OH to yield *N*-hydroxyaminoaspartate. It was not possible to crystallize the latter, but its presence was confirmed by various tests. One of the most definitive tests was catalytic hydrogenation of the enzymic product to L-aspartate.

Other evidence that the NH_2OH reaction is catalyzed by aspartase is that the ratios of activity of the enzyme on aspartate and on hydroxyaminoaspartate formation remain constant during purification. Further evidence is that NH_2OH is competitive with aspartate and vice versa. The optimum pH was found to be at 7.0 and the K_m 0.03 M for NH_2OH in this reaction.

a. Stereospecificity. The stereospecificity of the enzyme for the addition of the amino group of aspartate is shown by the fact that only L-aspartate is a substrate and D-aspartate is inactive in the aspartase reaction. It has also been established that aspartase is stereospecific for the hydrogen at the β-carbon (*19,20*). Upon incubation of fumarate and

NH_3 in D_2O, only one molecule of D is found per mole of aspartate. Fumarate formed from the D-labeled aspartate lacks the isotope. If the reaction were not stereospecific, one would expect to find 3 molecules of stably bound D atoms per molecule of aspartate and 2 atoms per molecule of fumarate after the reaction components had become completely equilibrated.

The addition of the D atom in aspartate has been shown to be *cis* to the amino group by magnetic resonance spectrophotometry in the same manner as this has been shown for the addition of water in the formation of malate, catalyzed by fumarase.

According to Halpern and Umbarger (*21*), the aspartase reaction is relatively unimportant for the biosynthesis of amino acids in *E. coli*.

2. DECARBOXYLATION OF ASPARTATE

Dissimilation of aspartate in certain microorganisms can proceed via β-decarboxylation (Eq. 2). Aspartate is the only amino acid in which

$$HOOCCHNH_2CH_2COOH \rightarrow HOOCCHNH_2CH_3 + CO_2 \qquad (2)$$

enzymic cleavage of the distal carboxyl group has been observed; decarboxylation of other amino acids occurs via the α-carboxyl group. Since the discovery of this reaction (*22,23*) the properties of the responsible enzyme (L-aspartate: 4-carboxyl-lyase, EC 4.1.1.12) has been extensively studied by a number of investigators in a variety of bacterial species (*24–31*). A preparation of high purity has been obtained from *Alkaligenes faecalis* (*28,29*), and the enzyme has been crystallized from *Achromobacter* sp. (*30,31*).

A very interesting property of the aspartate β-decarboxylase is that activation of the enzyme activity occurs not only with pyridoxal-P, but also with a variety of α-keto acids and with certain aldehydes. Combination of pyridoxal-P and carbonyl compounds produces the greatest degree of activation and allows enzyme activity to be continued for much longer periods of time.

Study of the activating effect of carbonyl compounds first led to the formulation of the hypothesis to explain the mechanism of the reaction that the carbonyl compound displaces the binding of the formyl group of pyridoxal-P to an ϵ-amino group of a lysine residue in the protein (*27*). This reaction, it was presumed, would enable L-aspartate and certain amino acid analogs of aspartate (DL-*erythro*-β-hydroxyaspartate) to combine more readily with the aldehyde group of pyridoxal-P and thus promote the subsequent decarboxylation reaction.

However, the discovery by Meister and co-workers (*28,29*) that aspartate β-decarboxylase is also a nonspecific transaminase of very low activity indicates a very different kind of mechanism. Support for this has been obtained by isolation of trace amounts of alanine-^{14}C and glutamate-^{14}C in experiments with labeled pyruvate and α-ketoglutarate.

These authors propose that the enzyme becomes inactivated by conversion of the pyridoxal-P to pyridoxamine-P by some of the aspartate undergoing transamination, in addition to the more rapid decarboxylation. The α-keto compounds react with the inactive pyridoxamine-P to restore the active pyridoxal-P.

It has also been observed that addition of pyridoxal-P to the inactivated enzyme leads to a restoration of enzyme activity in the absence of an α-keto acid. This is explained by Meister *et al.* as a consequence of the ready dissociation of the pyridoxamine-P from the holoenzyme and its replacement by pyridoxal-P.

Aspartate β-decarboxylase is a very large protein, of approximately 800,000 molecular weight. It combines with 13 pyridoxal-P molecules per molecule of enzyme. It is maximally active over the wide range of pH 3.5–6.0. The K_m for L-aspartate is 80 μM. An interesting property of the enzyme is that it has an absorption maximum at 360 mμ, instead of the usual peak of pyridoxal-P-dependent enzymes at 410–430 mμ. Addition of L-aspartate causes an immediate shift of this peak to 325 mμ. This is reversed by pyruvate (*26,31*). In addition to the inhibition by certain amino acid analogs of aspartate (β-hydroxyaspartate, β-methylaspartate), the enzyme is inhibited by the dicarboxylic acids, maleate and succinate. An interesting finding is that combination of the enzyme with β-hydroxyaspartate prevents the reduction of the pyridoxal-P by borohydride.

C. Glutamic Acid

1. Via Glutamic Acid Dehydrogenase

L-Glutamate is oxidized by a specific enzyme which has been found to be widely distributed in bacteria, yeast, plants, and mammalian tissues. This enzyme is unique among the enzymes acting on amino acids in that disphosphopyridine nucleotide (NAD) serves as its coenzyme.

The equation for the reaction is:

$$\text{L-Glutamate} + H_2O + NAD^+ \rightarrow \alpha\text{-ketoglutarate} + NADH + NH_4^+ \qquad (3)$$

L-Glutamate dehydrogenase was discovered independently by von Euler *et al.* (*32*) and Dewan (*33*). Crystallization of this enzyme was achieved by Strecker (*34*) and by Olsen and Anfinsen (*35*) from beef liver, by Snoke (*36*) from chicken liver, and more recently from other sources (*37*).

The glutamate dehydrogenase of plants and most animal tissues exhibits a high degree of preference for NAD. NADP-specific glutamate dehydrogenases occur in yeast and *Escherichia coli* (*37*).

The properties of glutamate dehydrogenase are discussed in detail in Chapter 14.

2. The Decarboxylation Pathway

An important alternative pathway of the metabolism of glutamate is via its decarboxylation to γ-aminobutyrate, transamination of the γ-aminobutyrate to succinic semialdehyde, and oxidation of the latter to succinate (Fig. 1). Succinate is utilized in the TCA cycle.

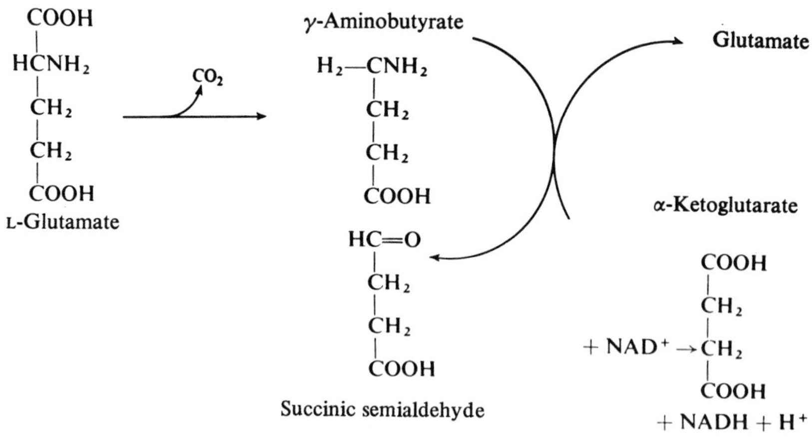

FIG. 1. Decarboxylation pathway of L-glutamate dissimilation.

γ-Aminobutyrate and the enzyme that catalyzes its formation, glutamate decarboxylase, were discovered in the nervous system in vertebrates (*38,39*). The enzymes that catalyze the dissimilation of γ-aminobutyrate, the transaminase and succinic semialdehyde dehydrogenase, occur in certain other tissues of the body in addition to the nervous system. The γ-aminobutyrate system occurs mainly in the gray matter of nerves (*38,42*), although trace amounts of γ-aminobutyrate have been detected in the white matter of the cerebral cortex (*43*). Recently it has been reported that the largest proportion of γ-aminobutyrate and glutamate decarboxylase is found mainly in nerve ending fragments (*44*).

It has been established that the γ-aminobutyrate enzyme system increases progressively with the development of the central nervous system (45,46).

The exact neurological function of γ-aminobutyrate has not been clearly established (47), but its chemical distribution indicates an inhibitory function since it occurs in high concentration in inhibitory axons and in trace amounts in excitatory axons (48,49).

a. *Glutamate Decarboxylase* (L-glutamate-1-carboxylase, EC 4.1.1.15). The activity of glutamate decarboxylase is high in brain. Even crude extracts of acetone-dried brain powder exhibit high decarboxylase activity. Attempts to purify the enzyme from this source have not been successful because of its association with the particulate matter of nervous tissue and its extreme lability without stabilizing agents.

It has been reported that the glutamate decarboxylase is a bound component of nerve cell mitochondria (50,51), although contrary findings have been reported (46). Roberts and Simonson (52) found that mouse brain glutamate decarboxylase could be stabilized by addition of reduced glutathione (0.01 M) or other thiols and pyridoxal-P ($5 \times 10^{-4} M$).

The brain enzyme has optimum activity at pH 6.8. There is abundant evidence that it is a pyridoxal-P-dependent enzyme and that it requires free protein SH groups for activity. Its lability to a considerable degree is due to the oxidizability of the essential SH groups, e.g., by O_2.

An extensive study of a variety of inhibitors of mouse brain glutamate decarboxylase and the mechanism of their action has been published by Roberts and Simonson (52).

Glutamate decarboxylase is also found in numerous bacteria (53-55), and in plants (56) where obviously γ-aminobutyrate has physiological functions other than those in brain. The conditions necessary for the induction of decarboxylases in bacteria have been established by Gale (53). One of them is growth on an acid medium.

Highly purified glutamate decarboxylase preparations have been prepared from several bacteria (54,55), where this enzyme is a soluble protein. The most highly purified of these preparations from bacteria was obtained by Shukuya and Schwert (54) from a strain of E. *coli* with a specific activity of 18,000 μl/CO_2/10 min/mg protein. The enzyme preparation exhibited one large, sharp boundary with a small amount of a second peak on sedimentation in an ultracentrifuge. It had an $s_{20,w}$ of 12.8 and an estimated molecular weight of 300,000. Chloride was found to enhance the bacterial enzyme activity by increasing the maximum velocity without altering the K_m. The optimum activity of this preparation is at pH 3.8, and the K_m at the above pH and 36° is 8.2×10^{-4} M.

In previous work on a less purified enzyme preparation of the enzyme from *E. coli* (*55*) the pH optimum was observed at 4.8–5.0 and the K_m 4.5 × 10^{-3} M at pH 5.0. The *E. coli* decarboxylase was estimated to bind 2 pyridoxal-P molecules per molecule of enzyme. It exhibited an absorption maximum at 415 mμ between pH 4.5 and 5.7.

An interesting property of the glutamate decarboxylase from *E. coli* is that it is less stable at 0° than at 25°. Shukuya and Schwert (*54*) interpret this effect as resulting from increased dissociation of the pyridoxal-P and a change in conformation of the apoenzyme that hinders its reassociation with the coenzyme at higher temperatures. Protection against the decreased stability at 25° is afforded by high salt concentration or agents with a high dielectric constant.

Glutamate decarboxylase is also present in higher plants (see *56* for review). The maximal activity for the enzyme from carrots was found to be at pH 5.5–5.8 and the K_m 3.6 mM.

The decarboxylase of sunflower cotyledons has been dissociated into apo- and coenzyme by treatment with acid-$(NH_4)_2SO_4$ and reconstituted by addition of pyridoxal-P (*57*). Ohno and Okunuki (*58*) found that a number of monocarboxylic and dicarboxylic acids inhibited the enzyme.

b. γ-Aminobutyrate Aminotransferase (4-aminobutyrate : 2 oxoglutarate aminotransferase, EC 2.6.1.19). γ-Aminobutyrate aminotransferase was first discovered in nervous tissue (*59,60*) where its distribution parallels that of γ-aminobutyrate and glutamate decarboxylase (*61,62*). In the mammal this enzyme also occurs in liver (*60*). Extracts of acetone powders of beef brain (*63*) showed optimum activity for transamination at pH 8.2. K_m values of 3 mM for γ-aminobutyrate and 4 mM for α-ketoglutarate were obtained. The enzyme activity was enhanced 20–130% by addition of pyridoxal-P. The reaction was specific for α-ketoglutarate, but transamination occurred with β-alanine and a number of other homologs of γ-aminobutyrate; the transaminase is inhibited by various carbonyl and SH reagents. The inhibition by NH_2OH is reversed by pyridoxal-P. Microorganisms (*63-66*) and plants (*67,68*) that contain glutamate decarboxylase also contain γ-aminobutyrate aminotransferase. The plant γ-aminobutyrate aminotransferase exhibited a higher activity with pyruvate than with α-ketoglutarate (*67,68*).

The most highly purified preparations of γ-aminobutyrate aminotransferases have been prepared from bacteria (*65,66*). A purification of 57-fold was obtained from aqueous extracts of a strain of *Pseudomonas fluorescens* to give a specific activity of 3.4 μmoles of substrate transaminated/min/mg protein. The purified enzyme was found to be specific for γ-aminobutyrate and α-ketoglutarate. The optimum activity ranged between pH 8.5 and 9.0. The equilibrium constant was estimated to be 0.1 over a temperature range of 20–38° and a pH range of 7.4–8.8. The bacterial enzyme was not influenced by pyridoxal-P, but it was inhibited by carbonyl reagents. From the influence of pH on the reaction it was deduced that the zwitterion of γ-aminobutyrate is the enzymically reactive species.

It was also established that the reaction does not involve a transcarboxylation by an experiment with glutamate-2-^{14}C and succinic semialdehyde. The γ-aminobutyrate produced in the reaction was not radioactive.

The best manner for assaying the transaminase is to couple the reaction with succinic semialdehyde oxidoreductase. By determining the NADH formed fluorometrically the assay is made extremely sensitive (62).

c. *Succinate Semialdehyde Dehydrogenase* (succinate-semialdehyde: NAD(P) oxidoreductase, EC 1.2.1.16). This, the third enzyme in the pathway of L-glutamate dissimilation via γ-aminobutyrate, has been purified from monkey brain (69), *Pseudomonas fluorescens* (65,66), and guinea pig kidney (62).

The dehydrogenase was purified 150-fold from monkey brain to a specific activity of 3 μmoles/hour/mg protein at 20° and pH 8.1. The reaction was NAD-dependent, NADP being inactive. The succinate semialdehyde dehydrogenase was associated with the particle fraction of brain, but could be solubilized by repeated freezing and thawing. The enzyme preparation from guinea pig kidney was enriched 1000-fold (62).

Activation of the dehydrogenase occurred with SH compounds and EDTA, while γ-aminobutyrate inhibited it. After fractionation with DEAE-cellulose and Sephadex the enzyme specifically catalyzed the oxidation of succinic semialdehyde only, with NAD as the hydrogen acceptor.

The pH activity curve of the dehydrogenase was unusual in that it was S-shaped with maximum activity at pH 9.0–9.4. K_m values were 7.8×10^{-5} M for succinic semialdehyde and 5×10^{-5} M for NAD at 38°.

The succinate semialdehyde dehydrogenase from *Pseudomonas fluorescens* (65,66) differs from the mammalian dehydrogenase in having 8 times greater activity with NADP than with NAD. It has been purified about 50-fold to a specific activity of 3.6 μmoles/minute/mg protein at room temperature and pH 7.9. The enzyme was inducible when the organism was grown on γ-aminobutyrate or pyrrolidine. Maximal activity was at pH 8.5, but the enzyme was most stable at pH 6.5–7.1. Protection of the enzyme against decay is afforded by

2-mercaptoethanol. Loss of enzyme activity could be reversed by incubation with SH compounds.

3. FERMENTATION OF GLUTAMIC ACID

   a. *The γ-Aminobutyrate Pathway.* In certain bacteria, γ-aminobutyrate is fermented to acetate, butyrate, and ammonia. In this pathway succinic semialdehyde is also an intermediate compound, but it is reduced to γ-hydroxybutyrate instead of being oxidized to succinate.

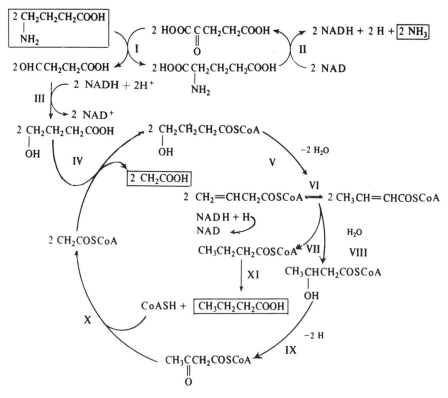

FIG. 2. Fermentation of D-glutamate via γ-aminobutyrate (*71*).

The fermentation of γ-aminobutyrate has been extensively studied with cell-free extracts of *Clostridium aminobutyricum* by Hardman and Stadtman (*70–72*). The individual reaction steps in this metabolic pathway are given in Fig. 2. The overall reaction can be represented by Eq. (4).

$$2 \text{ γ-Aminobutyrate} + 2H_2O \rightarrow 2 \text{ NH}_3 + 2 \text{ acetate} + \text{butyrate} \qquad (4)$$

The enzymes that catalyze the initial stages of the process from γ-aminobutyrate to γ-hydroxybutyrate are γ-aminobutyrate aminotransferase, glutamate dehydrogenase, and γ-hydroxybutyrate dehydrogenase.

Study of the fermentation sequence with crude, dialyzed extracts of *Cl. aminobutyricum* established that the following cofactors were required to obtain maximal activity; α-ketoglutarate, CoA, NAD NADP, P_i, and 2-mercaptoethanol. Proof of the formation of succinic semialdehyde could be obtained by preventing the oxidation of the NADH formed in the glutamate dehydrogenase reaction (Fig. 2, II). This was accomplished best by the addition of iodoacetamide, which prevented the further reaction. In such experiments there was a stoichiometric formation of succinic semialdehyde from γ-aminobutyrate.

Under the usual anaerobic conditions of incubation the reaction products consisted of γ-hydroxybutyrate, acetate, butyrate, and glutamate, with a molar oxidation-reduction index of 1.05.

Reduction of succinic semialdehyde to γ-hydroxybutyrate was demonstrated with dialyzed extracts of *Cl. aminobutyricum* and shown to be NADH-dependent. Partial purification of the enzyme γ-hydroxybutyrate dehydrogenase* (4-hydroxybutyrate:NAD oxidoreductase, EC 1.1.1.61) has been achieved from several bacterial species (*72,73*). The dehydrogenase preparations were found to be specific for γ-hydroxybutyrate and NAD. The enzyme from *Cl. aminobutyricum* attained a specific activity of 5.28 μmoles/min/mg protein. Its pH optimum was at 8.5, and values of K_m of $1.1 \times 10^{-2}\ M$ and $2.65 \times 10^{-4}\ M$ were estimated for γ-hydroxybutyrate and NAD, respectively.

The γ-hydroxybutyrate dehydrogenase of *Cl. aminobutyricum* is a divalent cation and SH-dependent enzyme. It is activated most effectively by Zn^{2+} ($10^{-4}\ M$), but other divalent cations are also effective. Its thiol nature is shown by the inhibiting effect of various SH reagents. The dehydrogenase from a *Pseudomonas* was found to have optimum activity at pH 7.0 and to be resistant to inhibition by SH reagents (*73*).

The individual steps of the reductive portion of the second phase of fermentation (Fig. 2, IV or XI) were deduced from the following experimental evidence: CoA or acetyl-CoA is not required in the formation of γ-hydroxybutyrate, but is essential for the subsequent conversion to acetate and butyrate. The transfer of CoA from acetyl-CoA to γ-hydroxybutyrate was measured by the characteristic that decomposition of acetyl-CoA is readily catalyzed by phosphotransacetylase in the

---

* Brain extracts have been found to have weak succinate semialdehyde reducing activity and the occurrence of γ-hydroxybutyrate in normal brain tissue has been reported (*73,74*). The subsequent fate of the latter is unknown.

presence of arsenate, whereas γ-hydroxybutyryl-CoA is not affected. Residual acetyl-CoA can then be estimated by the decrease in absorption at 232 mμ. Next it was determined that γ-hydroxybutyryl-CoA is a substrate for the oxidation of NADH by cell-free extracts of *Cl. aminobutyricum*, but γ-hydroxybutyrate does not react.

The inclusion of vinylacetyl-CoA and crotonyl-CoA in the reaction sequence was based on the results of a number of experiments. On incubation of γ-hydroxybutyryl-pantetheine with enzyme in the absence of an electron donor formation of crotonyl pantetheine could be measured by its strong absorption at 203 mμ.* A large increase in absorption was observed in these experiments. Confirming evidence was obtained by isolating crotonyl carbamate and identifying it by paper chromatography.

Direct evidence of the formation of vinylacetyl-CoA was not obtained, but it was established that the isomerization could be made to take place with the bacterial extracts.

In the presence of NADH, catalysis of the reduction to butyrate by the enzyme preparation proceeded both with crotonyl-CoA and crotonyl-pantetheine, the latter reacting much more slowly.

The deacylation of the thiol esters of butyrate and acetate with the accompanying formation of ATP was shown to occur by coupling fermentation of γ-aminobutyrate with the enzymes CoA-transferase, phosphotransacetylase, and acetokinase (75). This yielded 1 mole of ATP per 2 moles of γ-aminobutyrate. A second mole of ATP was assumed to be derived from the reoxidation of NADH by the usual electron transfer chain. Formation of ATP explains the requirement for P_i observed in the fermentation.

  b. β-Methylaspartate-Mesconate Pathway. In *Clostridium tetanomorphum*, glutamate is metabolized fermentatively by the pathway shown in Fig. 3 with the eventual production of acetate and butyrate. This pathway is of special interest because it involves the vitamin B_{12}-coenzyme dependent reversible conversion of L-glutamate to L-*threo*-β-methylaspartate. This dependency was discovered by Barker and co-workers (see *76–78*) because treatment with charcoal removes the coenzyme from the enzyme system. The reaction is catalyzed by the enzyme methylaspartate mutase.

Study of the reaction by tracer methods with isotopic compounds has established the intramolecular transfer of a glycine moiety between C-3 and C-4 of glutamate, with a simultaneous movement of a hydrogen

---

* Crotonyl-CoA cannot be estimated similarly because of the interference in the UV absorption by the adenine moiety.

atom in the opposite direction. Free glycine, acrylic acid, and α-keto-glutarate have been shown not to be intermediates, nor is a free proton involved in the reaction. Two enzymes and a third protein (protein S) are required to catalyze the conversion of L-glutamate to mesaconate.

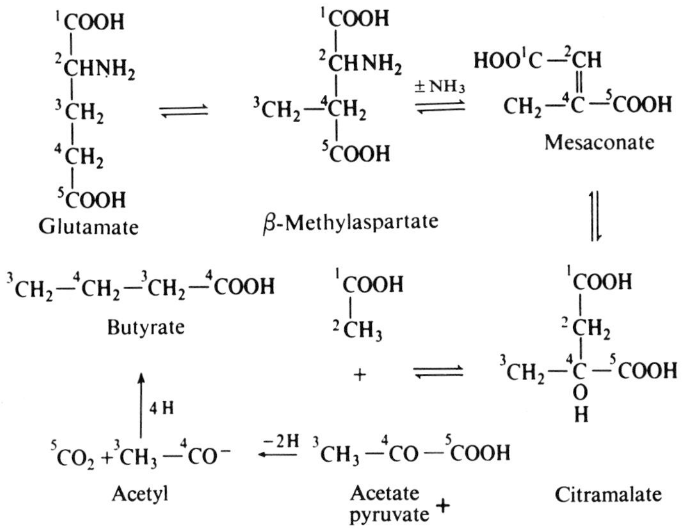

FIG. 3. L-Glutamate fermentation via β-methylaspartate (77).

Glutamate mutase (L-*threo*-3-methylaspartate carboxyaminomethyl-mutase, EC 5.4.99.1). This enzyme, which catalyzes the first reaction, has been purified to the extent of being largely freed of β-methylaspartate ammonia-lyase and protein S. The procedure consists of treatment of *Cl. tetanomorphum* extracts with charcoal, protamine sulfate, and $(NH_4)_2SO_4$, and finally column chromatography on hydroxylapatite (79). The S protein is obtained by successive acid precipitation at pH values from 6.1 to 3.6 and subsequent salting out with $(NH_4)_2SO_4$ (78).

The most accurate assay method is to determine anaerobically the conversion of β-methylaspartate to glutamate. Some Ca^{2+} is added to inhibit any residual β-methylaspartate ammonia-lyase. With this system it can be demonstrated that the required reaction components are a cobamide coenzyme and an SH compound (2-mercaptoethanol is most active). When the reaction is performed aerobically in the opposite direction, the S protein (reduced) enhances the rate about 3-fold. Optimum activity of this reaction is at pH 8.5. The purified enzyme is specific for L-glutamate and L-*threo*-β-methylaspartate. Approximate K_m values are 1–2 mM for glutamate and 0.5 mM for β-methylaspartate at

$38°$. Measurement of the equilibrium constant (L-glutamate)/(β-methylaspartate) yielded the value 10.6 at pH 8.2 and $38°$.

More recently the two components of glutamate mutase have both been highly purified (79a). One, component E, is a protein of molecular weight about 128,000. The second one, component S, is a small protein of molecular weight 17,000. Component E is comparatively stable to storage, but is rapidly inactivated by basic buffers (Tris). The presence of L-glutamate or methylaspartate protects against inactivation. Component S contains five SH groups, two of which are essential for enzyme activity. The activating and protective effect of mercaptoethanol is dependent upon its reducing function. The combination of components E and S is required for glutamate mutase activity.

The specificity for the cobamide coenzyme is comparatively low. One absolute requirement is the presence of a 5-deoxyadenosine group in the molecule. The most potent form of the coenzyme is the one containing the benzimidazole nucleotide. The K_m of this is 2.3×10^{-7} M.

β-Methylaspartate ammonia-lyase (EC 4.3.1.2). This enzyme was first crystallized by Bright and Ingraham (80). The formation of mesaconate, catalyzed by this enzyme, is estimated by the increase in absorbancy at 240 mμ. It is a stable enzyme with optimum activity at pH 9.7. In addition to L-threo-β-methylaspartate, the enzyme is capable of deaminating L-erythro-β-methylaspartate and L-aspartate. It has been shown to be distinct from aspartate ammonia-lyase. Estimates of K_m values gave 6.5×10^{-4} M for β-methylaspartate and 2.3 mM for aspartate at pH 9.76 and $25°$ (81). The equilibrium constants of the reaction were estimated to be 0.306 and 0.238 at pH 9.7 and 7.9, respectively (81).

Both a divalent and a monovalent cation are required for activity of the enzyme. The most potent are Mg^{2+} and K^+, although other related cations can function, but less effectively (81–84). Ca^{2+} is a potent inhibitor of this enzyme, acting competitively against Mg^{2+}.

Studies on the stereospecificity of methylaspartate ammonia-lyase from the exchange reaction with deuterium of the solvent has resulted in a number of interesting conclusions regarding the mechanism involved.

It has been determined that the substrate preferred by the enzyme is threo-2-L-3-methylaspartate, which is 100 times more reactive than the corresponding erythro compound, and that the preferred geometry of the elimination and addition of ammonia from and to the substrates is trans (82–85).

The deuterium exchange reaction at C-3 with the solvent is increased by the presence of NH_3, requires Mg^{2+}, and is inhibited by Ca^{2+} and Ag^+ in the same fashion as the overall reaction catalyzed by methylaspartate ammonia-lyase (86,87). This indicates that the overall reaction proceeds by the H^+ at C-3 becoming detached and combining with an S group in the active center of the enzyme, converting the β-methyl-aspartate to a carbanion.

The exchange reaction has a bell-shaped pH rate curve that at its maximum is 1.5 pH units lower than the pH maximum of the overall reaction. This is taken to indicate that the overall reaction is determined by two ionizable groups, an enzyme-bound SH and the amino group of the enzyme-bound methylaspartate carbanion (86–87).

Evidence for participation of an SH group in the β-methylaspartate-lyase reaction has been obtained by Williams and Libano (87). This evidence consists of the observation that the rate of loss of enzyme activity by photooxidation in the presence of methylene blue parallels the rate of destruction of SH groups on photooxidation. Also, enzyme activity is inhibited by SH alkylating agents, e.g., N-ethylmaleimide. Protection against the enzyme inactivation by both photooxidation and SH reagents could be obtained by the presence of substrates and of analogs structurally similar at the C-3 of methylaspartate.

Bright and Silverman (86) concluded from a study of the initial rates of the reaction with the free and the complexed substrate, employing asparate as a model compound, that the Mg^{2+} is complexed with the enzyme and that free methylaspartate is the true substrate. Bright (88) also concluded from a kinetic analysis that the binding of the metal-ion activator and β-methylaspartate to the β-methylaspartate ammonia-lyase was by a random-order, rapid, equilibrium mechanism.

The final steps in the fermentation of glutamate by Cl. tetanomorphum are hydration of mesaconate to citramalate and cleavage of the latter by a reversible aldolase-type reaction to acetate and pyruvate (Fig. 2, IV). This reaction is catalyzed by the enzyme mesaconitate hydratase (88–91). The enzyme has been purified about 14-fold from sonic extracts of Pseudomonas fluorescens (90). It converted mesaconate to (+)-citra-malate stoichiometrically until equilibrium concentrations were reached. The equilibrium ratio of (citramalate)/(mesaconate) has the value 6.0 at pH 8.2, the optimum pH for the reaction. The enzyme requires Fe^{2+} for activity. The K_m of citramalate is 5×10^{-4} M at the optimum pH.

That the reaction product was the (+)-citramalate was shown by Barker et al. (89) by enzymic synthesis.

III. GLYCINE

Glycine is utilized in numerous synthetic processes such as the forma-tion of purines, porphyrins, creatine, ethanolamine, choline, and gluta-thione. The reactions for these syntheses are discussed in other chapters of this work. This section is devoted to a narration of the pathways of the catabolism of glycine.

Two major routes of dissimilation of glycine have become apparent. One involves its deaminative conversion to glyoxylate and the subsequent metabolism of glyoxylate by a number of pathways. The second route is by the direct decomposition of glycine to yield a monocarbon adduct of tetrahydrofolate, CO_2 and NH_3. In addition, there are a number of other special pathways that appear to be of lesser importance for the catabolism of glycine.

A. Historical Development

Reviews of the earlier work on the metabolism of glycine are contained in the previous editions of this work and in a symposium on Amino Acid Metabolism in the McCollum-Pratt Institute in 1955 (92) by Weinhouse, Sakami, and Mackenzie. Weinhouse and co-workers demonstrated the formation of glyoxylate from glycine-2-^{14}C in liver by the addition of a pool of unlabeled glyoxylate and by showing that the label was retained in C-2. The conversion of the α-carbon of glycine to formate was first observed by Sakami (93) and by Siekevitz and Greenberg (94). It was subsequently shown that the compound formed from the C-2 of glycine was at the oxidation level of formaldehyde rather than of formate, and that the decomposition reaction was dependent on a reduced pteridine coenzyme (see Sakami in 92).

A requirement for pyridoxal-P in the metabolism of glycine and serine was first discovered by Lascelles and Woods (95) with washed cells of *Streptococcus faecalis R*, and subsequently in liver preparations (see Sakami in 92).

These early findings have been confirmed by numerous investigators in vertebrates, bacteria, molds, and higher plants.

B. Metabolism via Glyoxylate

Conversion of glycine to glyoxylate has been shown to occur mainly by transamination, although a flavoprotein that oxidizes glycine has been found in liver and kidney (96). This enzyme, it now appears, is D-amino acid oxidase. A number of nonspecific aminotransferases can catalyze the transamination of glycine. More recently, relatively specific glycine aminotransferases have been prepared from liver (97), bacteria (97,98), and higher plants (99).

The equilibrium of the reaction strongly favors the formation of glycine from glyoxylate and L-glutamate. The latter is the most active amino donor. Conversion of glyoxylate to glycine by transamination also proceeds readily nonenzymically (*100*).

Weinhouse and co-workers also demonstrated that, in liver, glyoxylate can be oxidized to formate and to oxalate (*92*). The overall reactions are represented by the following equations:

$$NH_2CH_2COOH \rightarrow O=CH-COOH \quad \text{(transamination)}' \qquad (5)$$

$$O=CH-COOH + \tfrac{1}{2} O_2 \rightarrow HCOOH + CO_2 \qquad (6)$$

$$O=CH-COOH + \tfrac{1}{2} O_2 \rightarrow HOOCCOOH \qquad (7)$$

At low concentrations of glyoxylate, formation of formate is favored, with no production of oxalate; at relatively high concentrations (≥ 0.005 M), formation of oxalate is favored. The oxidation of glyoxylate to oxalate is catalyzed by xanthine dehydrogenase (*101*). The reaction also takes place in pigeon liver homogenate, which is reported to contain no typical xanthine dehydrogenase (*92*). Oxalic acid cannot be an important intermediate in the catabolism of vertebrates because of its inertness to further oxidation. Apparently any oxalate formed in the animal body is eliminated by excretion in the urine.

The dissimilation of glyoxylate is not simple and can follow two general pathways, particularly in microorganisms. One of these pathways is initiated by condensation of glyoxylate with an acyl-CoA, the best known being acetyl-CoA. This reaction is a fundamental one in the glyoxylic acid cycle discovered by H. L. Kornberg. The alternative pathway involves an aldol condensation of glyoxylate with a carbonyl compound. The subject of glyoxylate metabolism has been reviewed by Kornberg and Elsden (*102*) and by Kornberg in the section on the Glyoxylate Cycle in Volume I, Chapter 4, of this series.

Glycine, in all probability, is only a minor source of glyoxylate. The latter can be formed by cleavage of isocitrate to glyoxylate and succinate, catalyzed by the enzyme isocitrate lyase. Glyoxylate is also formed from carbohydrate precursors via the pentose phosphate cycle.

1. DISSIMILATION OF GLYOXYLATE

a. Condensation with Acyl-CoA. As mentioned above, glyoxylate may be metabolized by condensation with acetyl-CoA, catalyzed by the enzyme, malate synthetase. The reaction proceeds according to Eq. (8).

$$CH_3COSCoA + O=CHCOOH + H_2O \rightarrow \begin{array}{c} HO-CH-COOH \\ | \\ CH_2COOH \end{array} + CoASH \qquad (8)$$

b. Aldol Condensation Pathways. The alternative pathway of dissimilation of glyoxylate is through condensation with a variety of carbonyl compounds. It includes condensation of 2 molecules of glyoxylate to form tartronic semialdehyde, condensation with pyruvate to yield α-keto-β-hydroxybutyrate, and condensation with α-ketoglutarate to form α-keto-β-hydroxyadipate.

The significance of the tartronic semialdehyde pathway for the oxidation of glyoxylate was established by Kornberg and Goto (103). The reactions involved are represented below:

$$2\,O{=}CHCOOH \rightarrow O{=}CHCHOH{-}COOH + CO_2 \qquad (9)$$
$$\text{Glyoxylic acid} \qquad\qquad \text{Tartronic semialdehyde}$$

$$O{=}CHCHOHCOOH + NADH + H^+ \rightarrow CH_2OH{-}CHOH{-}COOH + NAD^+ \quad (10)$$
$$\text{Glycerate}$$

The condensation of glyoxylate to tartronic semialdehyde was discovered by Krakow and Barkulis (104), and the enzyme glyoxylate carboligase catalyzing the reaction (Eq. 9) was partially purified and characterized by Krakow et al. (105). It was demonstrated that glyoxylate carboligase has a requirement for both thiamine pyrophosphate (TPP) and Mg^{2+}. Optimum pH is at 6.5. The identification of tartronic semialdehyde as the product of the enzymic reaction has been proved by comparison of the enzymically and chemically prepared compounds (106).

Tartronic semialdehyde reductase (EC 1.1.1.60), the enzyme that catalyzes the reduction of tartronic semialdehyde to D-glycerate (Eq. 10) has been crystallized from a Pseudomonad by Goto and Kornberg (107) after a 200-fold purification. It has a broad plateau of optimum activity between about pH 6.0 and 8.0. The molecular weight is about 91,000. Estimates of K_m are 2×10^{-4} M for tartronic semialdehyde and 2×10^{-5} M for NADH.

Glyoxylate also condenses with α-ketoglutarate to form 2-oxo-3-hydroxyadipate (108). This reaction is catalyzed by a carboligase, which may be the same enzyme that catalyzes the formation of tartronic semialdehyde. This enzymic reaction, like the glyoxylate condensation reaction, has a requirement for TPP and Mg^{2+} (or Mn^{2+}) (109).

2-Oxo-3-hydroxyadipate is decarboxylated to yield 2-oxo-3-hydroxyglutarate. The latter in turn is hydrogenated to form α-ketoglutarate. This series of reactions provides a cyclic mechanism for the oxidation of glyoxylate (109). The enzyme system for this process was discovered in *Rhodopseudomonas spheroides*, where it is highly active. More recently it has been shown to occur in liver mitochondria (110). The α-ketoglutarate-glyoxylate carboxyligase has been purified about 80-fold from rat liver mitchondria acetone powder (110).

C. Tetrahydrofolate-Dependent Pathway of Glycine Metabolism

A considerable body of evidence has accumulated that the α-carbon of glycine is converted to methylene-folate $\cdot$ H_4 and the carboxyl group to CO_2 without the intermediate formation of glyoxylate or free formaldehyde or formate as an intermediary in the reaction in liver (*111–113*), various bacteria (*114,115*), and plants (*116*). Tetrahydrofolate (or its polyglutamate congeners) and pyridoxal-P have been shown to be cofactors for this reaction pathway.

The evidence ruling out glycolate, glyoxylate, and formaldehyde as intermediates in this pathway of glycine dissimilation was that addition of these compounds to systems oxidizing glycine-^{14}C did not lead to incorporation of the label into the above compounds, nor to dilution of the label in the C-3 of serine, nor in the methyl groups of thymine and other methylated compounds derived from C-2 of glycine.

More explicit information on the details of the folate $\cdot$ H_4-dependent pathway of glycine metabolism was obtained from studies with bacteria, particularly *Pepticoccus glycinophilus*. In this organism, glycine is fermented to acetate, CO_2, and NH_3. The acetate is derived from the C-2 and the CO_2 from the C-1 of glycine. Sagers and Gunsalus (*117*) and Klein and Sagers (*118*) proposed the accompanying reaction scheme to explain the route from glycine to acetate.

$$NH_2-CH_2-COOH + \text{folate} \cdot H_4 \xrightarrow{\text{PLP, NAD}} CH_2-\text{folate} \cdot H_4 + CO_2 + NH_3 + 2H \quad (11)$$

$$CH_2-\text{folate} \cdot H_4 + \text{glycine} \rightarrow \text{serine} + \text{folate} \cdot H_4 \quad (12)$$

$$\text{Serine} \rightarrow \text{pyruvate} + NH_3 \quad (13)$$

$$\text{Pyruvate} \rightarrow \rightarrow \rightarrow \text{acetate} \quad (14)$$

The enzyme system catalyzing the decomposition of glycine by this pathway has been fractionated by Klein and Sagers (*119,120*) using the exchange of $H^{14}CO_3^-$ with the carboxyl group of glycine as a test method.

This work has resulted in the separation of the enzyme system into two protein fractions. One of the proteins of comparatively high molecular weight was found to be a pyridoxal-P protein, which was destroyed by heating. The second protein of low molecular weight, purified 23-fold, was heat-stable. Maximal CO_2 exchange was observed at pH 7.0. The reaction required the addition of SH compounds and was enhanced by the presence of EDTA. Mercaptoethanol was the most effective SH compound. The reaction was inhibited by formaldehyde, glyoxylate, PCMB, and Hg^{2+} and Cu^{2+} ions. Both proteins were required for the decarboxylation of glycine to proceed.

Michaelis constants for the CO_2 exchange reaction were determined to be 4.6×10^{-6} M for pyridoxal-P, 0.32 M for glycine, and 0.031 M for HCO_3^-. The pyridoxal-P containing enzyme was characterized by its absorption properties, which coincided with those of known pyridoxal-P enzymes, and by the fact that treatment with cysteine resulted in loss of the CO_2 exchange activity, accompanied by a sharp decrease in the absorption at 430 mμ. The activity and absorption peak could be restored by incubation with low concentrations of pyridoxal-P.

Baginsky and Huennekens (121) and Klein and Sagers (122) fractionated this enzyme system from *Pepticoccus glycinophilus* and isolated four proteins designated P_1, P_2, P_3, and P_4. Of these, P_1 and P_2 catalyze the $^{14}CO_2$ exchange reaction. P_1 is the pyridoxal-P containing protein. It is yellow-green in color and has an absorption maximum at 430 mμ. P_2 is a heat-stable, colorless protein of low molecular weight. P_3 is a flavoprotein. It has been shown that P_1 and P_2 are required for the labilization of the carboxyl group of glycine. To achieve the reduction of the flavoprotein, P_3 requires the participation of P_1, P_2, and P_4. Baginsky and Huennekens (121) obtained evidence that P_3 is a lipoyl dehydrogenase. According to Klein and Sagers (122), the precise role of P_4 and its possible association with folate $\cdot$ H_4, and of P_2 and its possible function in the release of glycine carboxyl groups has not been adequately established. A tentative scheme for the cleavage of glycine is offered by the sequence of Eqs. (15–17).

$$NH_2CH_2\overset{*}{C}OOH + HCO_3^- \underset{PLP}{\overset{P_1P_2}{\rightleftharpoons}} NH_2CH_2COOH + H\overset{*}{C}O_3^- \qquad (15)$$

$$NH_2\overset{*}{C}H_2COOH + FH_4 + P_3\text{---}FAD \overset{P_1P_2P_4}{\rightleftharpoons}$$

$$FH_4\text{---}\overset{*}{C}H_2 + CO_2 + NH_3 + P_3\text{---}FADH_2 \qquad (16)$$

$$P_3\text{---}FADH_2 + DPN^+ \rightleftharpoons P_3\text{---}FAD + DPNH + H^+ \qquad (17)$$

D. Glycine Catabolism via Aminoketones

1. Δ-Aminolevulinate

A number of CoA acylates have been found to be capable of condensing with glycine to form aminoketones. These may then be further enzymically decomposed to complete the cyclic oxidation of glycine. The most important of these reactions is the condensation of succinyl-CoA with glycine to form δ-aminolevulinic acid. The latter is the precursor of the porphyrins. The details of porphyrin synthesis are discussed in Chapter 18.

The evidence for the oxidation of glycine via δ-aminolevulinate has been provided by Shemin *et al.* (*123,124*).

This evidence consisted of the demonstration that the C-5 of δ-amino-levulinate, derived from glycine, was converted to the ureido carbon of uric acid and that it also forms formate.

Equal amounts of radioactivity were isolated in added carrier formate from the urine of rats fed either δ-aminolevulinate-5-^{14}C or glycine-2-[14]C. The enzyme catalyzing the formation of δ-aminolevulinate is not highly specific; it can also catalyze the condensation of acetyl-CoA and of propionyl-CoA with glycine (*125*).

2. AMINOACETONE

A second aminoketone that appears to be of considerable significance in the catabolism of glycine is aminoacetone formed from the condensation of glycine and acetyl-CoA with extracts of erythrocytes (*125*), certain bacteria (*126,127*), and mitochondria from guinea pig liver (*128*). Urata and Granick also observed formation of aminoacetone from malonyl-CoA. This occurs, presumably, by the prior decarboxylation of the malonyl-CoA to acetyl-CoA. Enzymic decompositon of amino-acetone to 2-oxopropanal (methylglyoxal, pyruvyl aldehyde) was demonstrated by Elliott (*129*). An enzyme catalyzing this reaction was observed in beef plasma. From the observation that monoamine oxidase inhibitors inhibited the oxidation of α-aminoacetone, Urata and Granick (*128*) suggested that this was the enzyme probably catalyzing the oxidation of aminoacetone to 2-oxopropanal. Since 2-oxopropanal is converted to D-lactate by the enzyme glyoxylase, a cycle for the catabolism of glycine analogous to the δ-aminolevulinate cycle can be envisioned, as was first proposed by Elliott (*127*). The reactions of this cycle are shown in Eqs. (18–20).

$$CH_3COSCoA + HOOC-CH_2-NH_2 \rightarrow CH_3COCH_2NH_2 + CO_2 + CoASH \quad (18)$$
$$\text{Aminoacetone}$$

$$CH_3COCH_2NH_2 + O \xrightarrow{\text{monamine oxidase}} CH_3COCHO + NH_3 \quad (19)$$
$$\text{2-Oxopropanal}$$

$$CH_3CO-CHO + H_2O \xrightarrow{\text{glyoxylase}} CH_3CHOH-COOH \quad (20)$$
$$\text{D-Lactic acid}$$

D-Lactate* can be oxidized to pyruvate, which can be converted to acetyl-CoA.

In their study of aminoacetone formation by guinea pig liver mitochondria, Urata and Granick observed a 140% increase in yield in the presence of iodoacetamide. This was shown to be due to the acylation of CoASH by the iodoacetamide. CoASH was observed to inhibit synthesis of aminoacetone.

In vertebrates, the aminoacetone cycle proceeds directly from 2-oxopropanal to pyruvate (see Section IV, C, 3).

IV. THE HYDROXYAMINO ACIDS

From the observation of Dakin (1) that phenylserine is oxidized in the body to benzoic acid, Knoop (130) had the insight to propose that the catabolism of β-hydroxy-α-amino acids leads to the formation of glycine. This has been proven correct by more recent investigations and a number of enzymes have been discovered that catalyze an aldol cleavage of these amino acids which is reversible, as shown in Eq. (21).

$$RCH_2OH\!-\!CHNH_2COOH \rightleftharpoons RCHO + CH_2NH_2COOH \qquad (21)$$

A. Phenylserine

The benzaldehyde portion of DL-*threo*-β-phenylserine labeled at C-3 with ^{14}C has been shown to be incorporated into the benzoyl moiety of urinary hippuric acid (131). Phenylserine is the compound that instituted the classic investigations on the β-hydroxyamino acids. An enzyme has been found that occurs in liver and kidney of mammals and induces the cleavage of phenylserine to benzaldehyde and glycine (132). This enzyme, phenylserine aldolase (133), has been shown to be distinct from threonine aldolase, although similar in properties, by its partial separation from the latter.

---

* A lactate racemase (EC 5.1.2.1) has been observed in a number of microorganisms [D. Dennis and N. O. Kaplan, Biochem. Z. **38**, 485 (1963); T. Hiyama, S. Fukui, and K. Kitihara, J. Biochem. (Tokyo) **64**, 99 (1968)]. The lactate racemase of Lactobacillus saki (Hiyama et al.) has been highly purified and found to have a molecular weight of 25,000. Its optimum activity is at pH 5.8–6.2 and K_m values of 1.7×10^{-2} M and 8×10^{-2} M were estimated for D- and L-lactate, respectively.

Some bacteria contain D-lactate dehydrogenase (EC 1.1.1.28) [E. M. Tarmy and N. O. Kaplan, J. Biol. Chem. **243**, 2579 (1968)]. The D-lactate dehydrogenase of E. coli has a molecular weight of 115,000 and contains 12 reduced cysteinyl residue per molecule of enzyme. D-Lactate dehydrogenase has also been found in invertebrates [G. L. Long and N. O. Kaplan, Science **162**, 685 (1968)].

The enzyme is specific for the L-*threo* form of the amino acid. As found with other enzymes of this type, pyridoxal-P is required as a coenzyme and free sulfhydryl groups are required for its activity.

The synthesis of phenylserine from benzaldehyde and glycine has been observed by a number of authors (*132,133*). Burns and Fiedler (*132*) observed that both the L-*threo* and L-*erythro* amino acids are formed in the synthesis.

It is interesting that the same enzyme preparation also catalyzes the splitting of L-*threo*-β-thienylserine to thiophene-2-aldehyde and glycine, and the synthesis of L-*erythro*-thienylserine from the breakdown products.

B. Serine

1. ALDOL CLEAVAGE

Serine can be decomposed by an aldol cleavage to yield glycine, but not free formaldehyde. The enzymic reaction is more complicated in this case, requiring not only pyridoxal-P, but also folate · H_4 as cofactors. Elwyn *et al.* (*134*) have concluded that conversion to glycine is probably the major pathway for L-serine metabolism.

In bacteria the metabolic importance of the decomposition of serine to glycine is shown by the observation that an *E. coli* mutant lacking serine hydroxymethyltransferase required glycine, but not serine, for growth (*135*).

The reversible conversion of serine to glycine is catalyzed by the enzyme serine hydroxymethyltransferase (L-serine: tetrahydrofolate-510-hydroxymethyltransferase, EC 2.1.2.1). Developments in the knowledge of the properties of this enzyme up to 1960 have been reviewed by a number of authors (*136,137*) and will not be further considered here.

An essentially pure preparation of this enzyme from frozen rabbit liver has been isolated by Schirch and Mason (*138*). Its molecular weight determined by the sedimentation equilibrium method is 331,000. Four moles of pyridoxal-P were present per mole of the isolated enzyme. However, after reduction with sodium borohydride, which completely abolishes the enzyme activity, addition of pyridoxal-P gave a restoration of one-third of the original specific activity. This observation, in conjunction with the fact that the activity of the purified enzyme can be increased by one-third by addition of pyridoxal-P to the incubation medium, suggests that the fully active enzyme would contain 6 moles of pyridoxal-P per mole of enzyme.

Upon acid hydrolysis of the borohydride-reduced enzyme, there was isolated ε-pyridoxyllysine. This establishes that the formyl group of pyridoxal-P is combined as a Schiff base with the ε-amino groups of lysine residues.

At 10- to 15-fold the enzyme concentration required to react with L-serine, rabbit liver hydroxymethyltransferase slowly catalyzes the decomposition of α-methylserine to D-alanine and methylenefolate · H_4.

The absorption maximum characteristic of pyridoxal-P in the enzyme at 430 mμ is shifted to a lower wavelength of 424–427 mμ on addition of serine or glycine. Addition of folate · H_4 to the glycine-containing enzyme preparation leads to the formation of a large absorption peak at 492 mμ and a smaller one at 460 mμ.

In *Clostridium cylindrosporum* a hydroxymethyltransferase has been studied which requires tetrahydropteroyl polyglutamate for activity. Other essential components are NAD, Mn^{2+}, and P_i for this enzyme system (*139*).

2. Hydrolytic Deamination

From a study of the distribution of the label of serine-^{14}C in glycogen and in certain amino acids in the intact rat (*140,141*), Koeppe and co-workers determined that conversion to pyruvate is an important pathway of L-serine degradation, while a product of unknown nature is formed as an intermediate from D-serine. Further evidence for the enzymic dehydration and deamination of serine to pyruvic acid is provided by the observation of Lien and Greenberg (*142*) that isotopic alanine was the amino acid formed in greatest amount from serine-3-^{14}C by liver mitochondrial preparations. The alanine could be formed from pyruvate by transamination.

Preparations of serine dehydratase have been isolated from liver and from a variety of microorganisms. In microorganisms, dehydratases specific for D- and L-serine have been discovered. The reaction of the hydroxyamino acid dehydratases, of which there are a large number, can be represented by the following general equation:

$$R \cdot CHOH{-}CHNH_2COOH + H_2O \rightleftharpoons R{-}CH_2CO{-}COOH + NH_3 \qquad (22)$$

Pyridoxal-P is the coenzyme in all instances. The enzymes are usually not specific for a single hydroxyamino acid, and they may catalyze other α, β, or γ elimination or condensation reactions as well.

Serine-deaminating enzymes occur in many bacterial species (*E. coli, Pseudomonas aeruginosa, Proteus* Ox_{19}, *Clostridium welchii* (*143–146*), and *Neurospora* (*147,148*).

The studies on serine deamination indicated the presence of two distinct enzymes, one active on the L-form and the other on the D-form in *E. coli* (*143,144*) and *Neurospora* (*147,148*).

In mammalian liver several enzymes with activity on L-serine appear to exist. Sayre and Greenberg (*149*) separated the deaminating activities for L-serine and L-threonine in sheep liver.

An enzyme was isolated from rat liver that appeared to be a homogeneous protein by a variety of criteria (*150,151*), and which deaminated both L-serine and L-threonine, and, in addition, catalyzed the synthesis of cystathionine from homocysteine and L-serine.* Brown *et al.* (*152*), however, have reported that they have been able to separate the cystathionine-synthesizing activity from the serine-deaminating activity of rat livers. Nakagawa *et al.* (*152a*) reported that they have crystallized the serine dehydratase of rat liver and that it exhibits no cystathionine-synthesizing activity. The ratio of enzyme activity of serine : threonine was found to be about 1.45. The molecular weight was estimated to be 63,500, instead of 21,000 as reported by Nagabhushanam and Greenberg. The highly purified enzyme had about equal specific activity for L-serine and L-threonine. This enzyme appears to account for the total deaminating activity of these two hydroxyamino acids in the mouse and the rat, but not in sheep, dog, or chicken (*151*). An enzyme catalyzing deamination of L-threonine, but not of L-serine, also appears to be present in human liver (*153*).

β-Chloralanine has also been shown to serve as a substrate for purified serine dehydratase (*151*).

a. Mechanism of the Reaction. Experimental evidence that the mechanism of the deamination involves a dehydration is provided by the inability of serine dehydratase to deaminate phosphoserine and O-ethers of serine (*154*). The theoretical basis for the mechanism of action of pyridoxal-P-dependent enzymes is discussed in detail in Chapter 14.

The reaction for the transamination of serine to hydroxypyruvate is discussed in Chapter 16.

b. Control Mechanisms of Serine Metabolism. Experimental studies of recent years have revealed the presence of important nutritional and hormonal factors that regulate the levels of serine dehydratase in rat liver, and, consequently, the metabolism of serine. In the rat, which has been the species studied mainly, the serine dehydratase of liver also

---

* More recently, a cystathionine synthetase preparation devoid of hydroxyamino acid deaminating activity has been isolated in the author's laboratory. The error in the earlier work resulted from a faulty method of assay (S. Kashiwamata and D. M. Greenberg, publication pending).

appears to possess the major, if not exclusive, threonine deaminating activity. The activity of this enzyme has been shown to decrease greatly on a high-carbohydrate, low-protein diet (155,156). Its activity is increased moderately by fasting (157) and to a large degree by high-protein administration or injection of a mixture of the ten nutritionally essential amino acids (158). Ishikawa et al. (159) observed that this enzyme is increased in alloxan diabetes and this is counteracted by insulin. In adrenalectomized rats there was no increase in enzyme activity, but an increase could be produced in adrenalectomized diabetic rats by administration of hydrocortisone. The activity of this enzyme has also been shown to be enhanced by thyroxine administration (160). Generally, the response to high protein and corticosteroids is lost in liver hepatomas (161), although some minimal deviation tumors are exceptions. Pitot and co-workers (162) have established that the increase in enzyme activity results from an increase in enzyme protein content. Using actinomycin D to inhibit protein synthesis, they obtained estimates of the template lifetimes of serine dehydratase. This was 6–8 hours in normal rat liver, less than 2 hours in several rat hepatomas, and longer than 2 weeks in the Morris hepatoma 5123.

C. Threonine

Three pathways are known in higher animals for the enzymic degradation of threonine. One involves the conversion of L-threonine to α-ketobutyric acid. The enzyme catalyzing the reaction, threonine dehydratase [L-threonine hydrolyase (deaminating), EC 4.2.1.16] has been extensively purified and its properties studied. A second pathway involves the cleavage of L-threonine and preferably L-allothreonine to glycine and acetaldehyde. This enzyme has been named threonine aldolase (L-threonine acetaldehyde-lyase, EC 4.1.2.5), or allothreonine aldolase (L-allothreonine acetaldehyde-lyase, EC 4.1.2.6). A third pathway involves the dehydrogenation and decarboxylation of L-threonine to yield aminoacetone. The equations for the different pathways of threonine dissimilation are shown in Fig. 4.

These three pathways have also been demonstrated to occur in various microorganisms.

The amino group of threonine, like that of lysine, is not available for reversible transfer reactions. Thus the ^{15}N from other amino acids is not found in threonine (163), and the $^{14}C : ^{15}N$ ratio of threonine isolated from body tissues shows only a small difference in the dilution of the two labels (163).

1. HYDROLYTIC DEAMINATION

Conversion of threonine to 2-oxobutyric acid is catalyzed by the pyridoxal-P-dependent enzyme, threonine dehydratase. Two species of the enzyme appear to be present in mammalian liver, and a number of different species in bacteria.

One of the mammalian threonine dehydratase species studied in sheep liver (*164–165a*) is characterized by being inactivated by serine at pH 7.0 after a short period of activity and by being altered to a form with lower activity at pH 9.* In addition to L-threonine and L-serine, L-allothreonine is also a substrate for this enzyme, although it is cleaved at about 0.2 of the rate of L-threonine.

The threonine dehydratase of sheep liver has been purified to a specific activity of 13 μmoles L-threonine decomposed/min/mg protein (*164*). Its optimum activity is pH 8.2–8.6. The Michaelis constant for L-threonine was estimated to be 8.0 mM. The variation of this constant with respect to pH has been studied by Davis and Metzler (*165*). The enzyme activity is enhanced by monovalent cations, the most effective being K^+ and NH_4^+.

An interesting observation is that the activity of this enzyme in liver is highly variable from animal to animal.

The second type of threonine dehydratase has been studied in rat liver (*151,167*). It was reported to be primarily a serine dehydratase by Selim and Greenberg (*150*) and also exhibited cystathionine synthetase activity. Goldstein et al. (*166*) observed that the activity ratios of the two amino acids remained constant (serine:threonine = 1.4:1) over an 8-fold range of purification. Subsequently Nagabhushanam and Greenberg (*151*), taking advantage of the observations of Pitot and co-workers (*156,158*) that the enzyme activity is greatly enhanced on a high-protein intake, purified the enzyme to an apparently homogeneous state. The molecular weight of this preparation is about 20,000. Michaelis constants for this enzyme are 52 and 59 mM for L-serine and L-threonine, respectively.

Bacterial threonine dehydratases differ greatly in properties and functions from those of the vertebrates. Two kinds of biodegradative

---

* To explain the inactivation of threonine dehydratase, McLemore and Metzler (*165a*) suggest that an oxazolidine ring derivative is formed between serine and the enzyme-bound pyridoxal-P. A. T. Phillips (*165b*), on the contrary, proposes that this occurs by alkylation by aminoacrylate (a presumed intermediate of serine deamination) of an essential thiol group in the catalytically active center of the enzyme. In support of this theory it was observed that N-ethylmaleimide-treated enzyme was not inactivated by serine. N-Ethylmaleimide presumably alkylates the same essential SH group. It yields an enzyme that has a residual activity of 10% of the fully active normal enzyme.

threonine dehydratases and a biosynthetic enzyme that catalyzes the first step in isoleucine biosynthesis have been discovered.

The biosynthetic enzyme is allosterically inhibited by isoleucine and is protected against inhibition by valine (*167–170a*). This type of enzyme has been highly purified from *Rhodopseudomonas spheroides* (*170*). The properties of the biosynthetic threonine dehydratase will be discussed in Chapter 16 in relation to the biosynthesis of isoleucine.

One of the biodegradative threonine dehydratases occurs in *E. coli*. It is stimulated by AMP and is induced by either threonine or serine in *E. coli* grown anaerobically in the absence of glucose (*171–173b*). This enzyme is considered to play an important role in anaerobic energy production by supplying α-keto acids for oxidoreduction. It controls regeneration of NAD by reduction of the α-keto acids formed from L-threonine or L-serine. The enzyme is most strongly activated by 5′-AMP; other mononucleotides, such as GMP, CMP, and dAMP, also stimulate but much less effectively. Activation by AMP occurs at all concentrations of L-threonine. AMP at a concentration of 10 mM decreased the K_m value of L-threonine and of L-serine 5-fold and increased the V_{max} 10-fold. The presence of AMP protects the enzyme against inactivation by heat, dilution, PCMB, and hydroxylamine. D-Threonine enhances the protective effect of AMP. A typical Michaelis-Menten type concentration-velocity curve was obtained even in the absence of AMP. Even so, it is concluded that AMP is an allosteric effector in controlling the rate of threonine deamination. The mononucleotides probably exert their action by favoring dimer formation of the enzyme. Phillips and Wood (*172*) determined by sucrose gradient centrifugation the molecular weights of the monomer and dimer of the enzyme to be 82,000 and 155,000, respectively.

From a kinetic study of the effect of AMP analogs on the activation of *E. coli* threonine dehydratase, Nakazawa *et al.* (*173a,173b*) found that the base portion of the ribonucleotide was relatively unspecific, but that adenine contributed to the binding to the enzyme. Major determinants of the activation potential were the 2′-hydroxy group of the ribose and the 5′-phosphate.

A different biodegradative L-threonine dehydratase is found in *Clostridium tetanomorphum* (*174–176a*). This enzyme is stimulated by ADP, not AMP. The enzyme activity is enhanced 10-fold by 10^{-4} M ADP, owing to an increase in the affinity for L-threonine in the presence of the dinucleotide. The V_{max} is not increased. The kinetics of this enzyme is different from that of the one in *E. coli* in that a sigmoid activity-substrate concentration curve occurs in the absence of ADP and this changes to a hyperbolic curve with ADP. No change in molecular weight is found as a result of the presence of ADP. The latter protects the enzyme against inactivation just as AMP protects the *E. coli* enzyme. The degradative pathway of L-threonine deamination is linked to ATP synthesis as shown by the series of reactions in Eqs. (23–25). As the ATP is utilized

$$\text{L-Threonine} + H_2O \xrightarrow{\text{deaminase}} \alpha\text{-ketobutyrate} + NH_3 \qquad (23)$$

$$\alpha\text{-Ketobutyrate} + O + P_i \rightarrow \text{propionyl-P} + CO_2 + H_2O \qquad (24)$$

$$\text{Propionyl-P} + ADP \xrightarrow{\text{acyl kinase}} \text{propionate} + ATP \qquad (25)$$

for energy-requiring reactions, liberating ADP, the latter transmits a signal that accelerates the activity of the threonine dehydratase. This in turn leads to the formation of more keto acid.

To explain the kinetics of the enzyme, two substrate sites are postulated, a catalytic site and an activating site.

This enzyme has now been crystallized (*176a*). The pure enzyme displays an absorption maximum at 415 mμ in neutral solution, indicative of bound pyridoxal-P. The enzyme dissociates into subunits in the absence of ADP or L-threonine. Reaggregation takes place in the presence of the above compounds. Only ADP increases binding of L-threonine by lowering its K_m value.

2. Cleavage to Glycine and Acetaldehyde

The enzymic formation of glycine from threonine was first reported by Braunstein and Vilenkina (*177*). This finding has been confirmed by Metzler and Sprinson (*178*) and the laboratories of the writer (*179–183*). Lien and Greenberg (*179*) isolated glycine by ion-exchange chromatography upon incubating DL-threonine-1,2-^{14}C with rat liver mitochondria, or after injecting the amino acid into the intact animal. Formation of glycine in the intact animal was also shown by isolating hippuric acid from the urine after administration of isotopic threonine or allothreonine along with benzoic acid (*180*).

Considerable information is now available about the properties of the enzyme(s)* that catalyzes this cleavage reaction. Enzyme preparations have been purified to a considerable degree from sheep and rat liver (*181–183*). The enzyme preparations catalyze the cleavage of only the L-forms of threonine and allothreonine. L-Allothreonine is decomposed much more rapidly than is L-threonine.

* Schirch and Gross [*J. Biol. Chem.* **243**, 5651 (1968)] identified the threonine aldolases as being serine transhydroxymethylase in rabbit liver.

A number of findings suggested that separate enzymes for L-threonine and L-allothreonine might be present in sheep liver (*181*). Thus, (a) the ratio of enzyme activity on threonine to allothreonine changed in the process of purification; (b) compounds with functional groups sterically related to threonine and allothreonine, such as serinol and serine, inhibited the decomposition of the two substrates differently; (c) different responses in the decomposition of the two substrates were obtained with different enzyme preparations. A threonine aldolase specific for L-threonine has been found in *Cl. pasteurianum* (*182a*).

The contrary conclusion was reached in the study of rat liver threonine aldolase (*183*). In these preparations the ratios of allothreonine activities remained constant during purification and partial inactivation by various procedures. Furthermore, the rates were not additive when both substrates were assayed together.

With the rat liver enzyme preparation, which was purified 175-fold, K_m values were obtained of 20 mM for L-threonine and 0.217 mM for L-allothreonine. At saturating levels of the substrates, the rate of decomposition was about 2-fold greater for L-allothreonine.

Pyridoxal-P was demonstrated to be the coenzyme for both the sheep and rat enzymes. The pyridoxal-P is bound very much more strongly to the rat liver enzyme than to the sheep liver enzyme. Removal of the coenzyme from the former is quite difficult.

The reversibility of the reaction has been demonstrated by a number of investigators (*133,181,183*). A careful study by Malkin and Greenberg (*183*) indicated that L-allothreonine is the sole synthetic product formed in the enzymic reaction.

3. THE AMINOACETONE PATHWAY

An aminoacetone cycle, proposed by Green and Elliott (*184*), is shown in Fig. 4.

Production of aminoacetone from L-threonine by washed cell suspensions of *Staphylococcus aureus* was first reported by Elliott (*185*). Shortly afterward Neuberger and Tait (*186*) demonstrated the oxidation of L-threonine with the concomitant reduction of NAD by cell-free preparations of *Rhodopseudomonas spheroides*. Aminoacetone formation from L-threonine by mitochondria of guinea pig liver was reported by Urata and Granick (*128*). Decomposition of aminoacetone by blood plasma preparations, presumably by monoamine oxidase, to 2-oxopropanal (methylglyoxal) was discovered by Elliott (*187*).

L-Threonine dehydrogenase (L-threonine : NAD oxidoreductase) has been partially purified from *Staph. aureus* (*184*) and from solubilized bullfrog liver mitochondria (*188*). Amphibian livers were found to have higher enzyme activity than mammalian livers.

The bacterial enzyme has optimum activity at pH 8.0–8.5. Potassium ions both activated the enzyme activity (maximal at 0.5 M) and protected it against heat inactivation (*184*). Other cations of the akali elements acted similarly. The enzyme is specific for both L-threonine and NAD.

The bullfrog liver threonine dehydrogenase is quite similar in properties to that of the bacterial enzyme with respect to substrate specificity and region of optimum pH activity. No activation by K^+ was observed. The enzyme was inhibited by sulfhydryl reagents and could be partially reactivated by 0.001 M glutathione. The K_m values were found to be approximately the same for both the bacterial (8.7 mM) and the liver (7.5 mM) enzyme preparations. The K_m obtained for NAD with the bacterial enzyme was 47 μM.

Only a single enzyme activity has been observed in the decomposition of L-threonine to aminoacetone. Consequently, decarboxylation of the presumed intermediate, 2-amino-3-oxobutyrate, must occur spontaneously. This is supported by the known instability of β-keto acids.

It was mentioned previously that Elliott (*185*) observed the deamination of aminoacetone to 2-oxopropanal by blood plasma preparations containing monoamine oxidase. Both Urata and Granick (*128*) and Green and Elliott (*184*) observed the decomposition of aminoacetone by liver breis. This suggests the deamination of aminoacetone by transamination. The reaction for conversion of 2-oxopropanal to D-lactate catalyzed by the enzyme glyoxylase (Fig. 4) is well known. An alternative reaction reported by Monder (*188a*) and confirmed by Higgins *et al.* (*188b*) is the oxidation of 2-oxopropanal to pyruvate. This reaction utilizes either pyridine nucleotide as the hydrogen acceptor. Monder partially purified the oxidase from sheep liver with a 90-fold enhancement of activity. The reaction product was identified as pyruvate by reaction with muscle lactic dehydrogenase and identity of the 2,4-dinitrophenylhydrazone with the authentic compound. The K_m values for the oxopropanal with this enzyme was 2.7 mM when NAD was the coenzyme and 0.21 mM when NADP was the coenzyme. The liver enzyme was found to be inhibited by carbonyl and thiol reagents.

Various α-ketoaldehydes are substrates for the enzyme. Higgins *et al.* (*188b*), employing cell-free extracts of *B. subtilis*, studied the transamination of aminoacetone and oxidation of the product. It was found that α-ketoglutarate, oxaloacetate, and pyruvate all could serve as amino group acceptors for transamination with the crude enzyme preparation. Oxidation of the 2-oxopropanal, presumably formed by transamination, proceeded if NAD was present in the incubation mixture.

Pyruvate, of course, can readily be reduced if lactic dehydrogenase is available, yielding L-lactic acid. This reaction couples the aminoacetone pathway more directly with the glycolytic pathway.

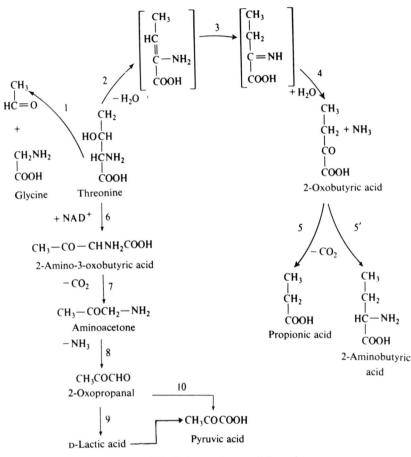

FIG. 4. Dissimilation pathways of threonine.

The identification of the different enzymes and reaction intermediates in the pathway from threonine to pyruvate, and the discovery of the formation of aminoacetone from glycine and acetyl-CoA lends great credence to the proposed aminoacetone cycle (Fig. 4). The aminoacetone pathway may very well be a substantial route for the catabolism of threonine in vertebrate organisms.

V. HISTIDINE

Histidine is unique among the amino acids in the diversity of its biochemical functions. It has been established as being a component of the catalytic center of many enzymes, owing to the special structural properties of its imidazole group. This group complexes with a variety of metal ions. Thus it is a ligand for the binding of Fe^{2+} in homogentisate oxidase and probably other enzymes whose activity is dependent on the transition bivalent metal cations. Iron porphyrin in proteins also is bonded in part with the imidazole of histidine.

Histidine is an essential nutrient for growth in the young rat and mouse, but the adult man can be maintained in nitrogen balance even if it is omitted from the diet (189). The compounds carnosine (β-alanyl-histidine) and anserine (β-alanyl-1-methylhistidine) occur in muscle. Their particular functions remain obscure, although there appears to be some relationship between these compounds and vitamin E. In vitamin E deficiency there appears to be decreased synthesis of these compounds.

Histidine is converted to histamine in many body tissues. The latter is a compound that has very potent pharmacological effects on smooth muscle. Ergothionine (3-thiolhistidine betaine) is synthesized by certain molds. It accumulates in animal erythrocytes, but is not formed by vertebrates.

Structural formulas of some of these compounds are shown below.

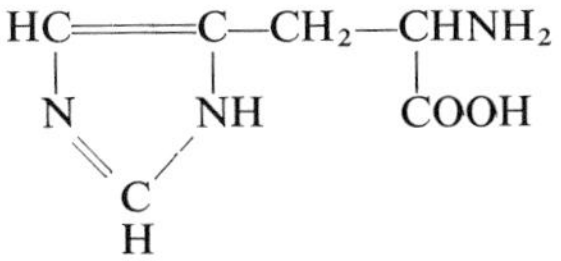

$$\begin{array}{ccc}
HC\!\!=\!\!\!=\!\!\!=\!\!C-CH_2-CHNH_2 \\
| \qquad | \qquad\quad | \\
N \quad\ NH \qquad COOH \\
\diagdown\ \diagup \\
C \\
H
\end{array}$$

Histidine

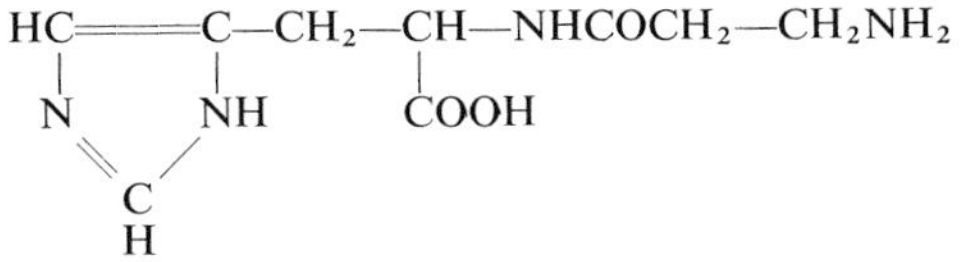

$$\begin{array}{ccc}
HC\!\!=\!\!\!=\!\!\!=\!\!C-CH_2-CH-NHCOCH_2-CH_2NH_2 \\
| \qquad | \qquad\quad | \\
N \quad\ NH \qquad\quad COOH \\
\diagdown\ \diagup \\
C \\
H
\end{array}$$

Carnosine

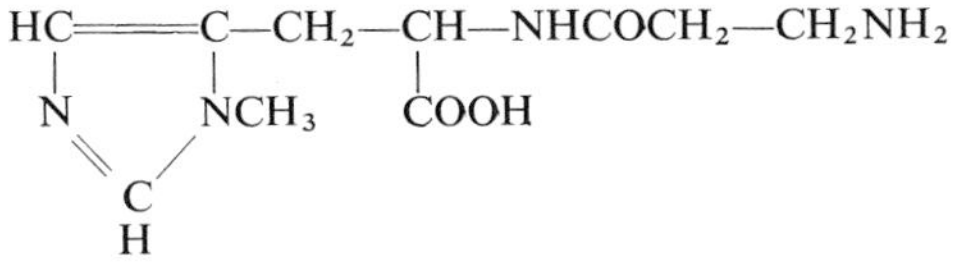

$$\begin{array}{ccc}
HC\!\!=\!\!\!=\!\!\!=\!\!C-CH_2-CH-NHCOCH_2-CH_2NH_2 \\
| \qquad | \qquad\quad | \\
N \quad\ NCH_3 \qquad COOH \\
\diagdown\ \diagup \\
C \\
H
\end{array}$$

Anserine

$$\begin{array}{ccc}
HC\!\!=\!\!\!=\!\!\!=\!\!C-CH_2-CH_2NH_2 \\
| \qquad | \\
N \quad\ NH \\
\diagdown\ \diagup \\
C \\
H
\end{array}$$

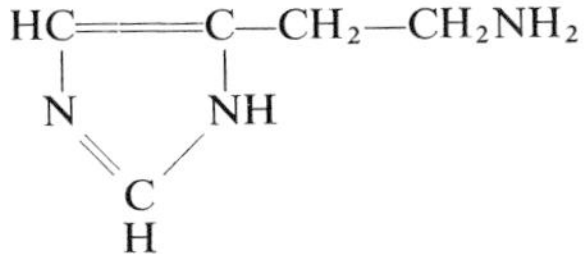

Histamine

There have been many surprising developments resulting from the study of histidine dissimilation by different organisms. The known results reveal a complex series of processes with a variety of enzymic and nonenzymic reactions and intermediate compounds. The presently available evidence has led to the scheme for the stepwise dissimilation of histidine via urocanic acid shown in Fig. 5.

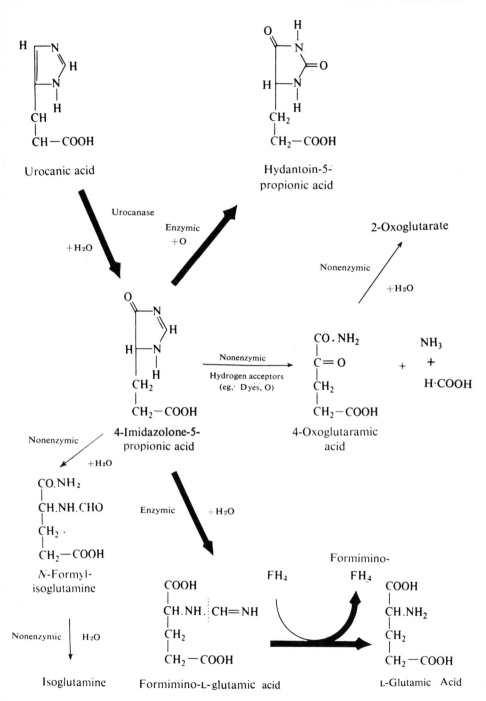

FIG. 5. Enzymic and nonenzymic pathways involved in the metabolism of urocanic acid. FH₄ represents tetrahydrofolic acid (234).

A. Historical Development

The decomposition of histidine by liver brei was observed independently by Edlbacher (*190*) and Györgi and Rothler (*191*). Edlbacher (*192*) demonstrated that histidine was decomposed by liver extracts to NH_3, formic acid, and a derivative that yields glutamic acid upon further treatment with alkali. The enzyme causing this reaction was described as capable of liberating 1 mole of NH_3 from histidine and producing a compound that yielded a second mole of NH_3 on treatment with alkali. This enzyme activity was shown to occur only in liver and to be present in the livers of all vertebrates tested. Only L-histidine was decomposed.

Urocanic acid had been isolated from the urine of various animals fed L-histidine (*193*). None was obtained from D-histidine (*194*). However, the sluggish metabolism of exogenous urocanic acid in the intact animal caused most investigators to doubt that it had a central role in histidine catabolism.

The hydrolytic splitting of urocanic acid by liver extracts was first observed by Sera and Yada (*195*) and confirmed by others (*196,197*). The enzyme that catalyzes this decomposition was named urocanase.

The Japanese investigators proposed that histidine was first deaminated to urocanic acid by histidine ammonia-lyase (EC 4.3.1.3), and the urocanic acid, in turn, was converted to a product that yielded glutamate and formate. This hypothesis has been shown to be correct.

B. Isotopic Tracer Studies

Histidine is readily oxidized to CO_2 in the animal body. Borsook *et al.* (*198*) observed that 36% of the label of an intravenous dose of L-histidine-2-^{14}C was expired in 1 hour in the rat. Similar findings have been reported by others (*199*).

1. METABOLISM IN BACTERIA

The demonstration that urocanate was an intermediate in the decomposition of histidine and proof of the reaction pathway to glutamate and formate was first established with cell-free extracts of *Pseudomonas fluorescens*.

The formation of L-glutamate was established by isolating the crystalline hydrochloride from the incubation mixture and establishing its identity (*200*). Proof that urocanate was an intermediate in the reaction sequence was obtained by employing preparations of histidine labeled with ^{15}N in the 2-position of the side chain and N-1, or with ^{14}C at C-2 of the imidazole ring (see Fig. 6) (*201*). On incubating these variously labeled histidines with the bacterial enzyme preparations and utilizing appropriate dilution techniques, isotopically labeled urocanate was isolated that retained the label in the imidazole ring and in which the ^{15}N of the α-amino group was converted to NH$_3$. The histidine labeled with ^{15}N in the N-1 position yielded glutamate in which over 90% of the excess ^{15}N was contained in the amino group. Incubation of histidine labeled with ^{14}C at C-2 of the imidazole ring yielded labeled formate.

These results with bacteria were possible because cell-free extracts of *P. fluorescens* metabolize histidine directly to glutamate, formate, and NH$_3$. *Aerobacter aerogenes* extracts yield glutamate and formamide (*202*).

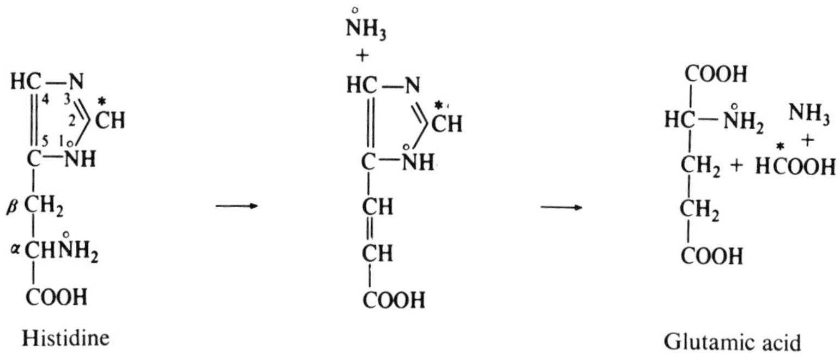

$$
\begin{array}{ccccc}
\text{HC}-\text{N} & & \overset{\circ}{\text{NH}_3}^+ & & \text{COOH} \\
\underset{5\ 1}{\overset{4\ 3}{||}}\overset{*}{\underset{2}{\diagup}}\text{CH} & & \text{HC}-\text{N} & & | \\
\text{C}-\text{NH} & & ||\quad\overset{*}{\diagdown}\text{CH} & & \text{HC}-\overset{\circ}{\text{NH}_2}\quad\text{NH}_3 \\
| & & \text{C}-\text{NH} & & |\qquad\qquad\overset{*}{+} \\
\beta\ \text{CH}_2 & \longrightarrow & | & \longrightarrow & \text{CH}_2 + \text{HCOOH} \\
| & & \text{CH} & & | \\
\alpha\ \overset{\circ}{\text{CH}}\text{NH}_2 & & || & & \text{CH}_2 \\
| & & \text{CH} & & | \\
\text{COOH} & & | & & \text{COOH} \\
& & \text{COOH} & & \\
\text{Histidine} & & & & \text{Glutamic acid} \\
& & \text{Urocanic acid} & &
\end{array}
$$

Fig. 6. Scheme of catabolism established by experiments with isotopically labeled histidine.

2. Metabolism in the Mammal

Liver preparations, as discovered by Edlbacher *et al.*, produce a material which forms the products of Eq. (26) only on alkaline hydrolysis.

$$\text{Histidine} + 4\,\text{H}_2\text{O} \rightarrow 2\,\text{NH}_3 + \text{glutamate} + \text{formate} \qquad (26)$$

This compound exhibits two acidic groups of pK' 2.4 and 4.7 and a basic group of pK' 11.11. On the basis of the titration data, Walker and Schmidt (*203*) first, and Borek and Waelsch (*204*) later, proposed that this unknown compound was formiminoglutamic acid.

Formiminoglutamic acid, it was later discovered, was excreted in the urine of rodents maintained on a folic acid-deficient diet (205). The amount excreted was increased on feeding histidine. Furthermore, it was shown that histidine was the precursor of formiminoglutamate by the administration of histidine-[15]N. The glutamate derived from the isolated formiminoglutamate contained the calculated quantity of the label. Subsequently crystalline formiminoglutamic acid was isolated (205), its composition determined, and its identity proved by synthesis (206,207).

It was also found that the formiminoglutamate was formed by diluted extracts of the bacteria mentioned previously.

Later work showed that formiminoglutamate was not directly formed from urocanate by a single enzyme reaction. The action of urocanase was to produce a very labile compound, imidazolonepropionic acid. This then was hydrolyzed through the activity of a second enzyme, imidazolonepropionate hydrolase, to formiminoglutamate (208,209).

C. Hydrolytic Dissimilation

Study of the enzymic reactions of histidine catabolism has been aided tremendously by the strong ultraviolet absorption of urocanate with a maximum at 277 mμ and a molar extinction coefficient of 18,000 (210,211). This property renders it very convenient to follow both the formation of urocanate from histidine and its disappearance spectrophometrically. The observation that in the enzymic decomposition of urocanate a product with an absorption peak at 264 mμ is formed also is of great importance in following the further sequence of histidine catabolism (208,209).

1. L-HISTIDINE AMMONIA-LYASE (EC 4.3.1.3)

In studying the properties of histidine ammonia-lyase, the alkaline pH of 9.3 is commonly used for incubation in order to minimize the interference of urocanase. The latter has a very low activity at the above pH.

Histidine ammonia-lyase from *Pseudomonas* has been purified considerably (212–213a), but the enzyme from liver has not. This is because the bacterial enzyme is present initially in higher concentration in the cell and it is considerably more stable. The enzyme is active over a wide pH range, with optimum activity at pH 9.5. It has been purified to a specific activity of 35 μmoles urocanate formed/min/mg protein. The molecular weight of this preparation was estimated to be 918,000 by sucrose gradient centrifugation (213a). The enzyme becomes inactivated on aging or dialysis, presumably through oxidation of an essential SH group. The requirement for a free SH group for activity has been established by inhibition experiments with SH reagents. Reactivation can be achieved by a variety of thiol compounds (204,210,211,214). There also may be a requirement for a metal ion. The evidence for this is that histidine ammonia-lyase is inhibited by a variety of metal-ion sequestering reagents (204,215,216). The inhibition by the sequestering agents is competitive with histidine, which also strongly binds metal ions. The exact physiological metal-ion activator is in doubt. Activation has been reported by Zn^{2+}, Cd^{2+}, and Hg^{2+} (215,216).

Studies with isotopically labeled compounds showed that the purified histidine ammonia-lyase catalyzes a hydrogen and urocanate exchange, but not an ammonia exchange (213), in addition to the irreversible deamination of histidine. The hydrogen exchange reaction was shown to involve the hydrogen on the β-carbon of histidine (213). From the above findings, Peterkofsky (213) proposed that the mechanism of histidine deamination involves the intermediate formation of an amino enzyme that can either reversibly form histidine or that decomposes to NH_3 and free enzyme.

Previous investigators have concluded that the deamination of histidine is an irreversible reaction; however, Williams and Herons (213a) reported that synthesis of histidine could be obtained by prolonged incubation (161 hours) in Tris-acetate buffer, at pH 8.0, comparatively high in Mg^{2+}. From the synthesis experiments the equilibrium constant was estimated as given in the accompanying equation. Williams and

$$\frac{[\text{Urocanate}][\text{NH}_4{}^+]}{[\text{Histidine}]} = K_{eq} = \sim 3$$

Herons concluded that the establishment of the reversibility of the histidase reaction renders unnecessary the obligate amino enzyme intermediate mechanism proposed by Peterkofsky. Either formation of a carbanion or a concerted reaction mechanism is consistent with the experimental results.

The K_m values for histidine have been estimated to vary between 9 and 24 mM with increasing pH (213).

a. Histidinemia. In this genetic disease there is an elevated blood histidine concentration, and histidine and imidazolepyruvate are excreted in the urine. It has been demonstrated that children suffering from this condition lack histidine ammonia-lyase in the stratum corneum of the skin (217). Blocking of the urocanate pathway favors transamination of histidine to imidazolepyruvate and conversion of the latter to imidazolelactate and imidazoleacetate (218–220).

2. UROCANASE

Crude liver extracts and dilute bacterial preparations decompose urocanic acid to N-formimino-L-glutamate (α-formamido-L-glutamate, L-α-formamidinoglutarate) (204,205,208).

More recently it has been demonstrated that urocanase preparations consist of two enzymes (208,209). A separation of them can be effected, so that formiminoglutamate is no longer produced in the reaction. The product formed by the first enzyme, for which the name urocanase has been retained, has been deduced to be 4(5)-imidazolone-5(4)-propionic acid. Even though this has not been proved by crystallization of the compound and verification of the structure by standard methods, there is abundant indirect evidence that the deduction is correct.

Formation of the product of urocanase activity, 4(5)-imidazolone-5(4)-propionate, is detected by the disappearance of the absorption peak of urocanate at 277 mμ and the simultaneous appearance of an absorption band with a maximum at 264 mμ at pH 7.2–7.6. The extinction coefficient of this material at 264 mμ (E_{max}) was estimated to be 3980 + 300 at pH 7.2 in contrast to 18,000 for urocanate at 277 mμ.

Acidification of the material shifts the absorption maximum to 234 mμ. The acidification stabilizes the compound, and it can be kept in the cold for long periods, particularly in the absence of oxygen, without decomposition. Brown and Kies (221) and Rao and Greenberg (222) have separated the compound by chromatography on a Dowex 50 (H$^+$) column, but have not succeeded in crystallizing it.

Imidazolonepropionic acid ("264-mμ material") is unstable in alkali and is very sensitive to light and oxidizing agents. Addition of H_2O_2 or persulfate causes the absorption at 264 mμ to disappear instantly. No well-defined products of the oxidation could be detected

The evidence that the product of the urocanase reaction is 4(5)-imidazolone-5(4)-propionic acid is based primarily on the similarity of its spectral characteristics in acid, alkaline, and neutral solutions to synthetic 4(5)-imidazolone and 4(5)-imidazolone-5(4)-acetic acid, which have been prepared by Witkop and co-workers (223,224). The above compounds are degraded by bacterial enzymes to formiminoglycine and formininoaspartic acid in a manner similar to the formation of formiminoglutamic acid in the reaction with crude urocanase preparations. Brown and Kies (221) determined that imidazolonepropionic acid has an acid group of pK′ 4.4.

a. Purification of Urocanase. Feinberg and Greenberg (208) were able to separate beef liver urocanase from the subsequent hydrolyzing enzyme by a procedure that resulted in a 30-fold increase in specific activity. Urocanase was subsequently purified several hundred-fold in our laboratory (222), and Robinson (225) has claimed an 800-fold purification in an abstract that contains no details. Urocanase is inhibited by hydroxylamine, cyanide, and a number of other carbonyl reagents. Gupta and Robinson (226) have reported that urocanase is a pyridoxal-P enzyme, but this has not been confirmed.

Urocanase appears to be an SH enzyme. It is inhibited by PCMB, and this can be reversed in part by glutathione. The K_m of urocanic acid was calculated to be 1.5×10^{-5} M.

3. IMIDAZOLONEPROPIONIC ACID HYDROLASE

This enzyme has been separated from urocanase and partially purified from rat liver (227) and *Pseudomonas fluorescens* extracts (228). Imidazolonepropionate prepared enzymically (221) and stored under nitrogen provides a stable substrate. The assay method is to determine the decrease in the absorption maximum of imidazolonepropionate at 260-264 mμ. The product of the enzymic hydrolysis was demonstrated to be formiminoglutamate by isolation and comparison with the infrared spectra of the synthetic compound and other chemical tests. Imidazolonepropionate hydrolase was shown to occur in the livers of all mammals tested and to have a 5- to 10-fold higher activity in *P. fluorescens* and *A. aerogens* grown on histidine. The findings with the bacteria indicate the inducible nature of this enzyme.

The hydrolase from rat liver was inhibited by PCMB; that from *P. fluorescens* was not. K_m values for imidazolonepropionate were estimated to be 7×10^{-5} M with the rat liver enzyme and 2.0×10^{-4} M with the bacterial enzyme. The enzymically formed formiminoglutamate was optically active, demonstrating that its precursor, the imidazolonepropionic acid, is also optically active.

4. FORMIMINOGLUTAMATE FORMIMINOTRANSFERASE

This enzyme converts formiminoglutamic acid to glutamic acid (229,230). The formimino group is transferred to form 5-formimino-FH_4 (Eq. 27). This, in turn, is converted to 5,10-methenyl $\cdot$ FH_4, which can undergo a variety of reactions with other enzymes. The enzyme was purified 100-fold from acetone powder extracts of hog liver (230).

$$\text{Formiminoglutamate} + FH_4 \rightarrow \text{5-formimino} \cdot FH_4 + \text{glutamate} \qquad (27)$$

Conversion to glutamate completes the enzymic steps specific for the dissimilation of histidine. The metabolic reactions of glutamate have already been discussed (see Section II).

D. Nonenzymic Formation of Formylisoglutamine

Imidazolonepropionic acid also decomposes nonenzymically. Feinberg and Greenberg (*208*) provisionally identified the end product to be formylisoglutamine. Revel and Magasanik (*209*) isolated the nonenzymic hydrolytic product after forming imidazolonepropionate by reacting urocanate with bacterial urocanase, and identified it as' a mixture of L- and DL-formylisoglutamine. Rao and Greenberg (*228*) isolated and identified formylisoglutamine from the products of the liver urocanase reaction and made a careful study of the nonenzymic reaction. Formylisoglutamine is formed by a first-order reaction. The half-life of formation at pH 7.2 and 27° was estimated to be 22 minutes and the first-order velocity constant 0.031 min^{-1}. On fractionating the reaction products by column chromatography on Dowex 1 and crystallizing the compound from ethanol, it was determined that it was identical with N-formyl-L-isoglutamine by the criteria of melting point, optical rotation, and infrared spectrum. On recrystallization from boiling ethanol, the optical rotation dropped and there was some decomposition to isoglutamine.

The enzymic cleavage of the imidazolone ring of imidazolonepropionate occurs by a nucleophilic attack of OH at C-4, whereas the nonenzymic cleavage occurs by a similar nucleophilic attack at C-2. Revel and Magasanik explain the presence of the D-formylisoglutamine by formation of the D-imidazolonepropionate from the L-compound via an enolization step (see Fig. 7). This, in turn, is hydrolyzed to the D-formylisoglutamine. Since the L-formylisoglutamine predominates, it suggests that the spontaneous hydrolysis is more rapid than the racemization.

Isolation of formyl-DL-isoglutamine and of isoglutamine from incubations of histidine and urocanate has been reported frequently. The discovery of the nonenzymic reaction now explains its source and helps to resolve a long-standing controversy of the dissimilation pathway of histidine.

E. Oxidative Dissimilation of Urocanate

Ichihara *et al.* (*231–233*) reported that urocanate is oxidized by an enzyme system found in liver and the bacterium *Pseudomonas aeruginosa*. The enzymes from both sources have been considerably purified by isoelectric precipitation, $(NH_4)_2SO_4$ fractionation, and adsorption and elution from calcium phosphate gel. The oxidative activity usually is markedly increased by addition of EDTA

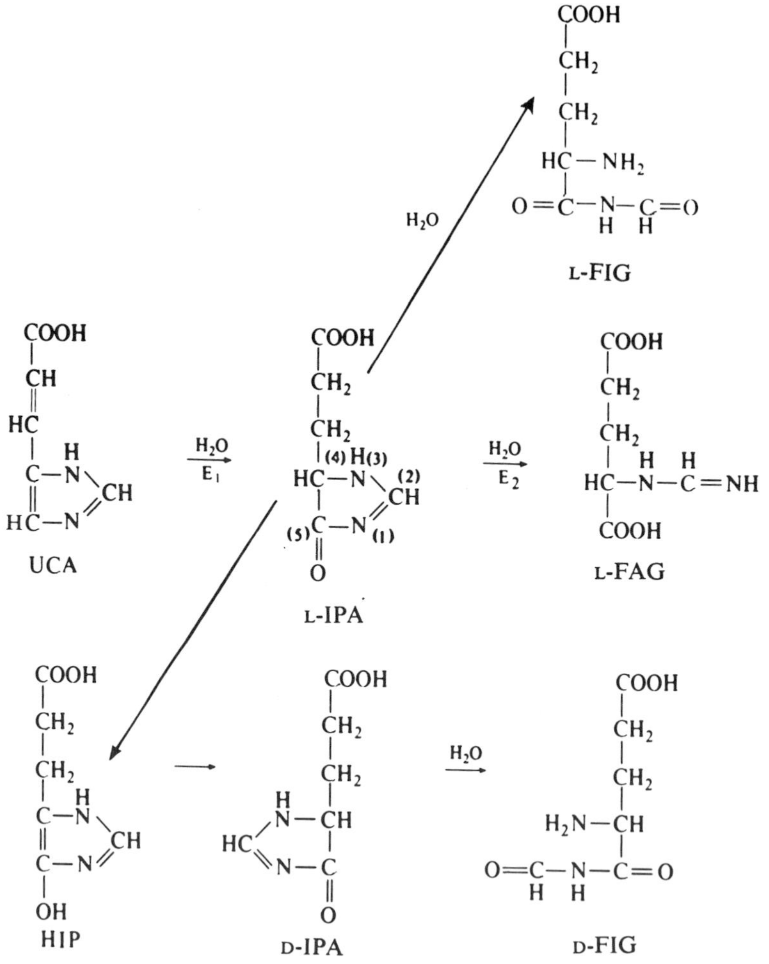

FIG. 7. Mechanism of formation of D- and L-formylisoglutamine from urocanic acid. UCA, urocanic acid; IPA, imidazolonepropionic acid; FAG, formamidinoglutaric acid; FIG, formylisoglutamine; HIP, hydroxyimidazolepropionic acid; E_1, urocanase; E_2, imidazolonepropionate-hydrolase (209).

Study of the reaction with these enzyme preparations have indicated a ratio of oxygen absorption to urocanate decomposition of a mole per mole. Products of the oxidative reaction reported to have been isolated are succinic monoureide, hydantoinacrylic acid, and α-ketoglutaric acid amide (232,233). The authors reached the conclusion that the first step in the reaction sequence was the hydrolytic formation of imidazolone-propionate by urocanase, and the latter was the immediate oxidative substrate.

More recently Brown and Kies (*221*) reported the formation from histidine and the excretion in the urine of hydantoin-5-propionic acid, and also its formation by liver extracts of guinea pig and the rat. The substrate for the oxidation was shown to be, very probably, imidazolonepropionate. No oxidation or formation of hydantoinpropionate could be demonstrated if the urocanase activity was first destroyed. The L-hydantoin-5-propionic acid was isolated by chromatography and crystallized. Its identity was unequivocally established. Brown and Kies determined that hydantoinpropionic acid was excreted unchanged when administered to experimental animals.

It is reasonable to assume that imidazolonepropionate is the substrate for the oxidation to hydantoinpropionate, because of its known susceptibility to oxidation. The nonenzymic oxidation of this compound by a variety of oxidizing agents had been observed by Miller and Waelsch (*229*) and by Feinberg and Greenberg (*208*), as will be described below. Hassall and Greenberg (*234*) proved that the chemical oxidation product is 4-oxyglutaramic acid. From the data for the enzymic formation of hydantoinpropionic acid the reaction can be formulated as shown in Eq. (28).

$$\text{Imidazolonepropionic acid} + \tfrac{1}{2}O_2 \xrightarrow{\text{oxidase}} \text{Hydantoin-5-propionic acid} \qquad (28)$$

Imidazolonepropionic acid   Hydantoin-5-propionic acid

In further study of the enzymic oxidative reaction, Hassall and Greenberg (*235*) observed that milk xanthine oxidase could catalyze the oxidation of imidazolonepropionate to hydantoinpropionate.

The oxidase in guinea pig liver that catalyzes oxidation of imidazolonepropionate to hydantoinpropionate does not, however, exhibit the properties of a typical xanthine oxidase. This enzyme has been purified about 200-fold by Hassall and Greenberg (*236*). By the assay method of Avis *et al.* (*237*) there was no oxidation of xanthine or hypoxanthine. The active enzyme fraction, like xanthine oxidase, contained FAD, but this could not be dissociated from the enzyme without the irreversible loss of activity. Enzyme activity is inhibited by reagents that inhibit FAD-containing enzymes, namely, cyanide and atabrine. The oxygen requirement for the formation of hydantoinpropionate was established to involve the net uptake of 1 μatom of oxygen per μmole of product. Determination of this value was complicated by the fact that imidazolonepropionate readily undergoes spontaneous oxidation to 4-oxoglutaramic acid in the presence of various hydrogen acceptors, including oxygen.

The pH optimum for hydantoinpropionate formation was found to be at 5.8. The K_m for imidazolonepropionate at the above pH was estimated to be 5×10^{-4} M.

More recently Payes and Greenberg (*238*) have shown that the imidazolonepropionate oxidizing enzyme has the characteristic properties of an aldehyde oxidase (*239–241*). Although oxidation of hypoxanthine and xanthine by molecular oxygen is extremely sluggish, the enzyme readily catalyzes the oxidation of these compounds by hydrogen acceptors such as 2,6-dichlorophenolindophenol, methylene blue, and cytochrome c. This enzyme, which has been extensively purified and studied from hog and rabbit liver, catalyzes the oxidation of a series of aldehydes, quinine compounds (*242*), N^1-methylnicotinamide, purine, and 4-hydroxypyrimidine. The guinea pig liver enzyme readily catalyzed oxidation of purine, hypoxanthine, xanthine, purine, 4-hydroxypyrimidine, and *p*-hydroxybenzaldehyde, with dichlorophenolindophenol as the oxidizing agent. The broad specificity of this enzyme removes much of the mystery surrounding the formation of hydantoinpropionate in the mammal. As previously mentioned, the latter is a terminal metabolite in the mammal, and was found to be excreted unchanged in the urine when administered to the rat, monkey, and man. Formation of hydantoinpropionate is probably fortuitous and only incidental to the major catalytic action of the enzyme on other substrates.

1. NONENZYMIC OXIDATION TO 4-OXOGLUTARAMATE

As mentioned above, imidazolonepropionate is a readily oxidizable compound (*208*). The nature of the nonenzymic oxidation product eluded investigators for many years. It was demonstrated to be 4-oxoglutaramic acid by Hassall and Greenberg (*234*). Imidazolonepropionate was prepared by the action of urocanase on urocanic acid and oxidized with dichlorophenolindophenol. The product was isolated by chromatography on a Dowex 1-X 8 column and crystallized. The compound was then identified to be 4-oxoglutaramic acid by elemental analysis, titration with NaOH, which indicated the presence of one carboxyl group, and by various degradations. Hydrolysis with HCl yielded NH_3 and 2-oxoglutaric acid. Catalytic hydrogenation of the 2,4-dinitrophenylhydrazones of the amide and the hydrolysis product gave isoglutamine and glutamate, respectively. The chemical characterization left little doubt that the nonenzymic oxidation product is 4-oxoglutaramic acid. This also harmonizes with the anticipated oxidative decomposition of imidazolonepropionate.

2. METABOLISM OF L-HYDANTOIN-5-PROPIONATE BY BACTERIA

While hydantoinpropionate is not further metabolized in the mammal, it can be metabolized by certain bacteria. Akamatsu (243) observed that the latter could be utilized as the sole source of nitrogen by a soil organism. With a gram-negative coccus isolated from soil by the enrichment culture technique with L-hydantoinpropionate as the sole carbon source and $(NH_4)_2SO_4$ for additional nitrogen, hydrolysis of the hydantoinpropionate to carbamylglutamate (ureidoglutarate) and then to glutamate was demonstrated by Hassall and Greenberg (244). The reaction proceeds according to Eq. (29).

L-Hydantoin-5-propionic acid   N-Carbamyl-L-glutamic acid

Hydantoinpropionate-^{14}C was incubated with cell-free extracts of the isolated organism, and the reaction products isolated by paper chromatography. The radioactive compounds were identified by cochromatography and radioautography with authentic unlabeled substances in a number of solvent systems. With the cell-free preparations and using L-hydantoin-5-propionate-2-^{14}C, the only radioactive products observed were carbamylglutamate and CO_2. With uniformly ^{14}C-labeled hydantoinpropionate there were also obtained glutamate, α-ketoglutarate, and succinate. These results indicate that the glutamate is further metabolized via the TCA cycle.

The story of histidine metabolism reported in this section relates the successful solution of a complicated problem that has been under investigation for many years and was marked by a series of controversies. Earlier investigators were misled by the fact that both enzymic and nonenzymic hydrolytic and oxidative reactions may be involved in the catabolism of urocanic acid. Furthermore, the inertness of exogenous urocanate resulted in serious doubt that it was on the main pathway of histidine metabolism.

F. Minor Pathways of Histidine Catabolism

1. DECARBOXYLATION OF HISTIDINE

Mammalian tissue contains a specific histidine decarboxylase (245–250) and a nonspecific one that decarboxylates various aromatic amino acids (251). The specific histidine decarboxylase (L-histidine carboxylase, EC 4.1.1.22) is present, apparently, in the mast cells. A high content of the enzyme has been found to occur in mice and rat embryos (245), in the rat stomach (247), and in mast cell tumors (246). Some degree of purification has been achieved with the enzyme from embryonic tissue (245) and from the mast cell tumor (246). This decarboxylase has been shown to be inducible; the stimulating factor being stress, e.g., epinephrine administration (247). The enzyme requires pyridoxal-P for activity. The pH optimum varies with the histidine concentration, being pH 6.4 at 5×10^{-3} M. The apparent K_m value is 2×10^{-5} M at pH 7.8.

According to Hakanson (245) the anion of the amino acid is the substrate for the enzyme, and calculations based on the anion form present gives a constant value of K_m of 6×10^{-7} M. The mast cell histidine decarboxylase is specifically inhibited by the α-hydrazine analog of histidine and by 4-homo-3-hydroxybenzoxylamine (252). The nonspecific aromatic amino acid decarboxylase is not inhibited by the above compounds.

Whereas histidine decarboxylase activity is comparatively low in the mammal, it is very high in certain bacteria, e.g., Lactobacillus 30a (253) and Clostridium welchii (254). The bacterial enzyme shows no requirement for pyridoxal-P. The decarboxylase has been obtained in crystalline form by Rosenthaler et al. (253) from Lactobacillus 30a. The molecular weight is 195,000, pH optimum 4.5–6.0, and the K_m 0.9 mM. The bacterial enzyme is specific for histidine. It is inhibited noncompetitively by CN^- and PCMB.

The catabolism of histamine has been reviewed quite recently (255). This occurs through its oxidation to imidazole acetaldehyde and further oxidation to imidazoleacetic acid (256). A subject of dispute is whether the enzyme catalyzing the first step in the oxidation is specific for histamine or whether it is the less specific enzyme, diamine oxidase (256,257).

A highly purified histaminase (EC 1.4.3.6) preparation, apparently homogeneous, has been prepared from hog kidney (258). Optimum activity is obtained at pH 6.8. The enzyme contains bound FAD, which is liberated by 0.1 N HCl. Histaminase also contains pyridoxal-P.

Partial inactivation of the enzyme by dialysis can be reversed by the addition of FAD and pyridoxal-P. The concentration of pyridoxal-P for activation is critical. In excess amounts it inhibits the enzyme. In addition to histamine, 1-methylhistamine is also a substrate for histaminase. This is formed through the catalytic activity of imidazole-N-methyltransferase (259) with adenosylmethionine as the methyl donor. The further oxidation of the aldehydes formed from histamine and methylhistamine is catalyzed by xanthine oxidase type of enzymes. The reactions involved are represented by Eq. (30).

$$
\begin{array}{ccc}
HC \!=\! C \cdot CH_2 \!-\! CH_2 \!-\! NH_2 & \xrightarrow{\text{histaminase}} & HC \!=\! C \!-\! CH_2 \!-\! CHO \\
\end{array}
$$

(30)

$$
\xrightarrow{\text{xanthine oxidase}} \quad HC \!=\! C \!-\! CH_2 \!-\! COOH
$$

Imidazoleacetic acid is also capable of being converted to the corresponding ribonucleotide. The presence of this compound has been detected in tissues, and imidazolacetate riboside is excreted in the urine.

2. TRANSAMINATION OF HISTIDINE TO IMIDAZOLEPYRUVATE

A very active histidine-pyruvate aminotransferase occurs in liver (260–263). The transaminase activity is about twice as high as that of urocanase and four times that of histidine ammonia-lyase in rat liver (260). Methods for assay of this transaminase have been devised by Lin et al. (263) and improved by Spolter and Baldridge (260). The enzyme is specific for pyruvate and for L-histidine. In the rat, two forms of the enzyme occur, one in the mitochondria, the other in the cell cytoplasm (261,262).

The histidine transaminase is found in the rat fetus on the 17th day of gestation and increases in activity by one-third on birth. At 24 hours after birth the enzyme level attains that found in the adult rat (261). In histidinemia, where there is an absence of histidine ammonia-lyase, transamination becomes the major route of histidine catabolism (217). The imidazolepyruvate formed is further oxidized to imidazoleacetate. The subsequent oxidative route of this compound is not known.

Studies on the kinetic properties of histidine-pyruvate aminotransferase (262) have shown that the pH optimum is at 8.5 for the mitochondrial enzyme and at pH 8.8–9.4 for the soluble enzyme. K_m values of

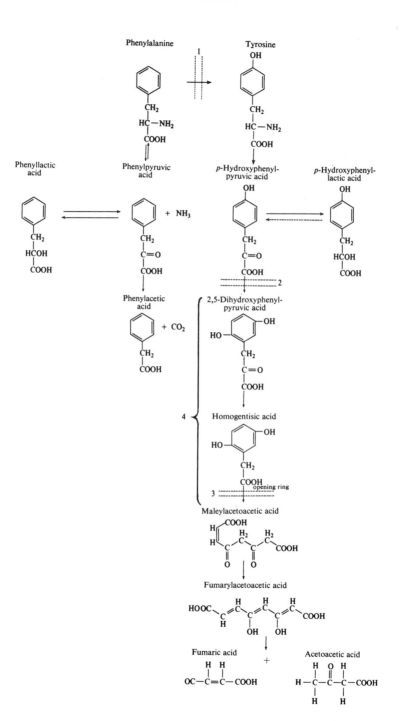

histidine are $1.5 \times 10^{-2} M$ for the mitochondrial enzyme and $3 \times 10^{-3} M$ for the cytoplasmic enzyme. Pyruvate has a much higher K_m, about $1 \times 10^{-2} M$.

VI. THE AROMATIC AMINO ACIDS

Much of the earlier knowledge of the metabolic pathways of phenylalanine and tyrosine catabolism was obtained by study of the defects in the hereditary diseases alcaptonuria, albinism, phenylketonuria, and tyrosinosis. Widespread interest in this subject dates from the publication of Garrod's "Inborn Errors of Metabolism" (2). The metabolic blocks at particular steps in the degradative pathways of phenylalanine and tyrosine for the disorders mentioned above are shown in Fig. 8. The knowledge of the metabolism of the aromatic amino acids has been rapidly advanced in recent years by isotopic tracer and enzyme studies.

Both phenylalanine and tyrosine are readily transaminated to their corresponding keto acids (see Chapter 14). The keto and hydroxy analogs of these amino acids can effectively replace the requirement for their respective amino acids in the diet of the growing rat.

A. Phenylalanine

The conversion of phenylalanine to tyrosine was clearly indicated by the feeding experiments of Womack and Rose (264), who showed that tyrosine is not essential in the diet when sufficient phenylalanine is available.

Definite proof of the conversion was obtained by isotopic experiments by Moss and Schoenheimer (265). These authors fed deutero-DL-phenylalanine to the rat and isolated the correspondingly labeled L-tyrosine. The conversion of phenylalanine to tyrosine is ireversible, as tyrosine cannot replace phenylalanine in the diet.

Fruitful enzyme studies on the hydroxylation of phenylalanine commenced with the work of Udenfriend and Cooper (266). These investigators prepared a highly unstable, soluble enzyme system from liver that catalyzed the conversion of phenylalanine to tyrosine. Enzyme activity was found in livers of all mammals tested and also in chicken liver, but in no other tissues.

FIG. 8. Metabolic pathways of phenylalanine and tyrosine. The numbered dashed lines have the following significance: 1 = block in phenylpyruvic oligophrenia; 2 = block in tyrosinosis; 3 = block in alcaptonuria.

Subsequent work has shown that the hydroxylation requires a two-enzyme system and that the hydroxyl oxygen is derived from O_2 (*267*). One of the enzymes utilizes a tetrahydropteridine (biopterin) as a coenzyme, the other enzyme requires NADPH (*268,269*) to reduce the dihydropteridine formed. The enzyme system is not so specific as was formerly believed, and it will catalyze the hydroxylation of tryptophan and a number of fluorophenylalanines (*270*). The reaction proceeds only to the formation of tyrosine with purified enzymes.

The mechanism of the hydroxylation reaction can be represented by Eqs. (31–33).

$$\text{Phenylalanine} + O_2 + \text{biopterin} \cdot H_4 \xrightarrow{\text{Enzyme I}} \text{tyrosine} + \text{biopterin} \cdot H_2 + H_2O \tag{31}$$

$$\text{Biopterin} \cdot H_2 + \text{NADPH} + H^+ \xrightarrow{\text{Enzyme II}} \text{biopterin} \cdot H_4 + \text{NADP}^+ \tag{32}$$

$$\textit{Sum:}\ \text{Phenylalanine} + O_2 + \text{NADPH} + H^+ \xrightarrow{\text{Pterin}\cdot H_4} \text{tyrosine} + H_2O + \text{NADP}^+ \tag{33}$$

Rat liver was found to be a convenient source of Enzyme I, which requires a reduced pteridine for the hydroxylation reaction. The source of Enzyme II in these experiments was sheep liver. Reduced biopterin has been shown to be the natural cofactor for hydroxylation by isolating and identifying it from rat liver extracts (*271*). A closely related compound, sepia pteridine, was shown to be very active as the coenzyme (*272*). Prior to the isolation of the natural coenzyme model, reduced pteridine compounds were utilized, particularly 2-amino-4-hydroxy-6,7-dimethyl tetrahydropteridine, even though they are much less active than the natural cofactors.

In the course of the hydroxylation it was observed that a tautomeric form of the dihydropteridine was formed that was spontaneously converted to 7,8-dihydropteridine. The latter is the normal substrate for reduction by dihydrofolate reductase. This transformation was catalyzed by P_i. The initial product which is enzymically active was deduced to be a 5,6-dihydro tautomer. It could be formed chemically from the tetrahydro compound by oxidation with 2,6-chlorophenolindophenol. Evidence for the 5,6-dihydro form of the pteridine was obtained from studies with tritiated compounds (*273*). The initially formed dihydropteridine can function in the enzymic hydroxylation reaction; the converted compound is inert.

Kaufman has obtained indirect evidence that the dihydropteridine formed during the hydroxylation has a quininoid structure.

The dihydropteridine is reduced by the sheep liver enzyme to the tetrahydro derivative, with NADPH as the hydrogen donor. This reaction is typical of the catalytic activity of dihydrofolate reductase. However, that a specific enzyme is involved is indicated by the ability of the 6,7-dimethylpteridine to serve in the phenylalanine hydroxylation reaction. The latter is not a substrate for sheep liver dihydrofolate reductase (274). An enzyme has now been prepared from rat liver that catalyzes the reduction of sepia pteridine to dihydropterin (275), with NADPH as the hydrogen donor. Dihydrofolate reductase can serve to catalyze reduction of dihydrobiopterin in the phenylalanine hydroxylation in place of the new enzyme.

The ability of the enzyme system to catalyze the hydroxylation of 4-fluorophenylalanine to tyrosine makes a simple hydroxylation mechanism for the reaction untenable. The removal of the fluorine by an alternative mechanism requires the ultilization of an additional electron per molecule of tyrosine formed, and leads to a predicted ratio of NADPH/tyrosine of 2 when NADPH is the ultimate electron donor. The experimental finding has been even higher, with ratios of 3–4. To account for the experimental findings, it is proposed that an enzyme-bound oxygen-containing derivative of the substrate is formed. This could be an oxyfluorophenylalanine, which could undergo decomposition to yield tyrosine and F^- with the concomitant oxidation of a second mole of the pteridine $\cdot$ H_4 to pteridine $\cdot$ H_2 (270).

The enzymic defect in phenylketonuria has been traced to a lack of Enzyme I of the hydroxylating enzyme system (276,277). Homogenates of fresh liver biopsy (276) or autopsy specimens (277) of phenylketonuric patients produced no detectable conversion of phenylalanine to tyrosine under conditions where liver homogenates from control individuals were active. Addition of Enzyme I of rat liver to the inactive phenylketonuric liver homogenate strikingly activated the conversion to tyrosine.

In addition to *para* hydroxylation, evidence for an *ortho* and *meta* hydroxylation of phenylalanine has accumulated through the isolation of various *m*- and *o*-hydroxyphenyl acids in the urine of phenylketonurics and of normal man and animals (278,279).

An enzyme has been demonstrated in liver that hydroxylates phenylpyruvic acid at the *ortho* position (280).

B. Tyrosine

The major pathway for the catabolism of tyrosine is through C-2 and C-5 hydroxylation of the phenyl ring, giving rise to the intermediate homogentisic acid and the cleavage of the phenyl ring of this compound eventually to form fumaric and acetoacetic acids. The route through hydroxylation at C-3 and C-4 to yield 3,4-dihydroxyphenylalanine, the adrenalin compounds, and melanin is quantitatively less important. This pathway is discussed in Chapter 16.

Five enzyme steps have been demonstrated to be required in the conversion of tyrosine to fumaric and acetoacetic acids. They consist of a transamination to p-hydroxyphenylpyruvate, a simultaneous oxidation and migration of the side chain and decarboxylation to form homogentisic acid, oxidation of the latter to maleylacetoacetic acid, isomerization of this compound to fumarylacetoacetic acid, and hydrolysis of this acid to fumarate and acetoacetate.

The isotope data support the theory proposed long ago that a migration of the side chain takes place (*281,282*) and that the original position of the side chain is replaced by a second hydroxyl group, as is found in homogentisic acid.

The genetic disease alcaptonuria has played an important part in promoting the earlier developments of the understanding of the metabolism of tyrosine. Tyrosine was shown to be the precursor of the homogentisic acid excreted by alcaptonurics, and much information on possible intermediates was gleaned by feeding experiments on patients with this condition (see *288* for review).

Knowledge of the pathway of dissimilation of tyrosine has been advanced greatly by isotopic tracer experiments and by the availability of enzyme systems that catalyze isolated reactions in the overall oxidation of tyrosine.

Incubating liver slices with isotopically labeled phenylalanine or tyrosine gave results that suggested that the fate of each of the carbon atoms of tyrosine was as is shown in the accompanying diagram.

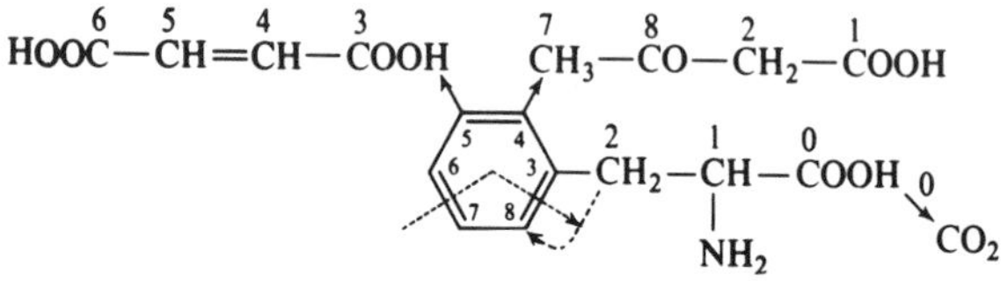

Early difficulties in the enzymic study of the catabolism caused by the instability of many of the metabolic intermediates were resolved through the discovery of a requirement for α-ketoglutarate and ascorbic acid for the oxidation of tyrosine. Starting about 1950, enzyme preparations were obtained that specifically oxidized tyrosine, p-hydroxyphenylpyruvate, and homogentisic acid to acetoacetic acid. The individual enzymes were then separated and purified, and detailed knowledge obtained of each step of the degradation process.

1. TRANSAMINATION OF TYROSINE

The initial evidence that the catabolism of tyrosine is initiated by transamination is that only 4 atoms of oxygen are required to oxidize 1 molecule of tyrosine to acetoacetate. α-Ketoglutarate stimulates the oxidation, no free ammonia is produced in the reaction, the oxidation is decreased upon removal of the transaminase coenzyme by ammonium sulfate precipitation or by dialysis, and there is an increase in the rate of tyrosine oxidation upon addition of pyridoxal-P. When p-hydroxyphenylpyruvate was used as the substrate, addition of keto acids did not increase the rate of oxidation (283,284).

The tyrosine transaminase was subsequently obtained in a highly purified form and the reaction products identified. Its properties are described in detail in Chapter 14.

Tyrosine-α-ketoglutarate transaminase is an inducible enzyme much like tryptophan pyrrolase. The ability to increase the transaminase activity in liver by tyrosine administration depends on the integrity of adrenal cortical function.

2. p-HYDROXYPHENYLPYRUVIC ACID HYDROXYLASE [p-Hydroxyphenylpyruvate ascorbate : oxygen oxidoreductase (hydroxylating), EC 1.14.2.2]

The substrate directly oxidized to homogentisic acid is p-hydroxyphenylalanine. 2,5-Dihydroxyphenylalanine, although it is converted to homogentisic acid in the alcaptonuric patient, is not attacked by the soluble enzyme system. Neither is p-hydroxyphenylacetate nor 2,5-dihydroxyphenylpyruvate.

These results indicate that the hydroxylation of p-hydroxyphenylpyruvate and migration of the side chain occur simultaneously. No separation of these two processes has been observed. Neither has it been possible to separate the decarboxylation to homogentisate from the above reactions. The p-hydroxyphenylpyruvate oxidase system has been extensively purified and its properties studied by LaDu and Zannoni (285) and by Knox et al. (286–288). In most instances, enzyme assays were performed by determining oxygen utilization manometrically. Lin et al. (289) have developed a rapid spectrophotometric assay method based on the formation of an enol borate complex of enol p-hydroxypyruvate that absorbs strongly in the ultraviolet with a maximum at 310 mμ and a molar extinction coefficient of 12,400. Use of this assay procedure requires the presence of an enzyme that catalyzes the keto-enol tautomerism of aromatic keto acids and, in the presence of borate, favors formation of the enol tautomer.

Oxidation of p-hydroxyphenylpyruvate could be limited to the formation of homogentisic acid by inhibiting oxidation of the latter with α,α-dipyridyl (0.001 M). The enzyme source used by LaDu and Zannoni was acetone powder extracts of dog liver. One mole of oxygen was consumed and 1 mole of CO_2 was liberated per mole of homogentisate formed as required by the reaction.

Excess substrate inhibited the oxidation of the p-hydroxyphenyl-pyruvate, as did also benzoquinone acetic acid. The inhibition could be prevented by addition of a variety of reducing agents characterized by p-catechol and ascorbic acid-like structures. Reduced coenzyme Q_{10} and ascorbic acid were the most active natural compounds (290–291). Folic acid maintained tyrosine metabolism in vitamin C-deficient guinea pigs *in vivo* by protecting liver p-hydroxylphenypyruvate oxidase from inhibition. The folic acid had no protective effect *in vitro* (292).

The oxidation of p-hydroxyphenylpyruvate was also strongly inhibited by diethyldithiocarbamate, and this can be partially reversed by small amounts of Cu^{2+}. The apparent requirement for Cu^{2+} was supported by the observation that the livers of copper-deficient dogs had about half the p-hydroxyphenylpyruvate oxidase activity of littermate controls. However, this lowered activity due to copper deficiency could not be increased by the addition of exogenous Cu^{2+}.

The results of LaDu and Zannoni on the dog liver enzyme were substantiated by Hager *et al.* (287) on enzyme preparations from beef and pig liver. These authors suggest that p-hydroxyphenylpyruvate hydroxylase is a copper enzyme with properties similar to tyrosinase. Additional support for this is the relative insensitivity of the enzyme to CO and CN and the very high affinity for O_2.

An interesting aspect of the oxidation of tyrosine has been the role of ascorbic acid (293,294). Incomplete tyrosine metabolism with the excretion of homogentisic acid and other incomplete oxidation products of tyrosine is found in scurvy. One site of this action of vitamin C is on L-hydroxyphenylpyruvate hydroxylase. The activating and protective action of ascorbic acid on this enzyme has been found not to be specific. As already mentioned, p-catechol and diketone type compounds can serve in place of ascorbic acid *in vitro*.

LaDu and Zannoni (285) fractionated dog liver p-hydroxyphenyl-pyruvate hydroxylase into two protein fractions both of which were required for activity. One of these required fractions, it turned out, was rich in catalase. This particular liver fraction could be replaced by pure catalase regardless of the source. However, no H_2O_2 is formed in the oxidation of p-hydroxyphenylpyruvate, so the catalase does not act by protecting the enzyme from inactivation by H_2O_2. Similarly it has not been possible to demonstrate any coupled reaction promoted by catalase, such as occurs with H_2O_2 and ethanol.

Use of dog liver in the experiments was fortunate because it has not been possible to separate the two enzyme components in rat and pig liver.

In the purified enzyme system, ascorbic acid becomes progressively less effective while 2,6-dichlorophenolindophenol gains in efficiency (285). Ascorbic acid probably serves, as do the other effective compounds, by forming an oxidation-reduction couple of suitable potential (293).

It has been established that, in the genetic disease alcaptonuria, the biochemical lesion is an absence of p-hydroxyphenylpyruvate hydroxylase. This was shown to be true in autopsy material in both the liver and kidney (295,296).

3. HOMOGENTISATE OXIDASE (Homogentisate oxidoreductase, EC 1.13.1.5)

The next enzymic step in the dissimilation of tyrosine, oxidation of homogentisic acid was clearly established by the work of Ravdin and Crandall (297). These investigators isolated two enzyme fractions from a rat liver homogenate, one of which catalyzed the oxidation of homo-genetisic acid to an open-chain diketone-dicarboxylic acid. In their hands the product isolated was 4-fumarylacetoacetic acid. Subsequent work has shown that the initial product is 4-maleylacetoacetic acid and that an isomerase is present which converts this to the fumarylacetoacetate (298). The second enzyme of Ravdin and Crandall (297), fumarylaceto-acetic acid hydrolase, hydrolytically cleaves this compound to fumarate and acetoacetate.

The above information shows that the logical position of the rupture of the benzene ring is between carbon atoms 3 and 8 as is shown in the diagram in Section VI, B.

An enzyme preparation that catalyzes the opening of the benzene ring of homogentisic acid was obtained from rabbit liver by Suda and Takeda (299) and named homogentisicase. It was observed that this enzyme is strongly inhibited by α,α'-dipyridyl, leading to the inference that Fe^{2+} is an essential component of this enzyme. This observation has been confirmed frequently (300,301). α,α'-Dipyridyl has been used to limit the oxidation of tyrosine to homogentisic acid and also to cause a chemically induced alcaptonuria in experimental animals (302).

Homogentisate oxidase has been purified about 50-fold from extracts of acetone powders of beef liver (303) and 136-fold from calf liver (304). The optimum activity of the Fe^{2+} activated enzyme is at pH 5.8.

Homogentisate oxidase has been shown to be an SH enzyme. Evidence has been obtained that the Fe^{2+} is bound to the SH as mercaptides. One ligand of the Fe^{2+} is presumed to be bound to SH, and the other to an imidazole group.

Kinetic evidence was obtained that organic mercurials inhibit the enzyme by competing with Fe^{2+}, but not with the substrate for a common enzymic binding site (304). Fe^{2+}, various organic mercurials, and reducing agents were found to protect the enzyme from irreversible aerobic oxidation. The K_m of homogentisate was estimated to be 4.0×10^{-4} M, and 1 mM for O_2 (303).

4. 4-MALEYLACETOACETATE-*cis*, *trans*-ISOMERASE (EC 5.2.1.2)

This enzyme was discovered by Knox and Edwards (298). The cleavage of homogentisate yields maleylacetoacetate, as would logically be expected. Because of the presence generally of the isomerase, fumarylacetoacetate was the product that commonly accumulated. Knox and Edwards were able to separate the isomerase from preparations containing fumarylacetoacetate hydrolase, present in the supernatant of the homogentisate oxidase, by ethanol fractionation. The reaction catalyzed is shown in Eq. (34).

$$ \text{(34)} $$

Formation of maleylactoacetate upon incubation with homogentisate oxidase is detected by the difference in its absorption spectrum from that of fumarylacetoacetate. The former at pH 3 has an absorption maximum at 230 mμ, while the latter compound has its maximum at 350 mμ. At pH 1, fumarylacetoacetate exhibits maximum absorption at 310 mμ, while maleylacetoacetate has almost no absorption. The difference in spectra is accounted for by the fact that fumarylacetoacetate exists mainly in the enol form in acid solutions, while maleylacetoacetate is largely in the keto form. Enol formation is enhanced by borate, which leads to the formation of an enol-fumarylacetoacetate-boron complex.

The isomerase requires glutathione as a coenzyme. In addition the enzyme contains essential SH groups which are inactivated by the usual thiol reagents.

The inhibition can be reversed by the addition of glutathione.

5. 4-FUMARYLACETOACETATE FUMARYL HYDROLASE (EC 3.7.1.2)

This enzyme appears to be identical with previously discovered enzymes that hydrolyse triacetic acid (*305*) and acylpyruvate (*306*). Fumarylacetoacetase is not affected by reducing agents, metal ions, or SH reagents. The products of this enzyme reaction are fumarate and acetoacetate. The oxidation of these compounds is completed through the operation of the TCA cycle.

VII. TRYPTOPHAN

A. Introduction

Tryptophan is an amino acid that has had a great fascination for biochemists because of its many important chemical and biological reactions and products. It was first isolated in 1901 (*307*) and was among the first of the amino acids that were shown to be nutritionally indispensable (*308*).

Early biochemical interest in tryptophan was generated in connection with the problem of the putrefactive production of indole and skatole by the bacteria of the intestinal tract and the relation of putrefaction and of these compounds to the health of the individual. In 1942 it was established that the vitamin pyridoxine (*309*) was important for its metabloism, and in 1945 that tryptophan could replace nicotinic acid to maintain the growth of the rat (*310*). Soon afterwards it was established that tryptophan was converted to nicotinic acid. This developed great interest in this amino acid on the part of nutritionists.

Botanists have been interested in the plant hormone heteroauxin, which is indoleacetic acid. It occurs in the urine and has been shown to be a tryptophan metabolite in plants (*311*). Another focus of general biological interest in tryptophan is that it is the precursor of the brown eye pigments of insects, ommochrome (*311*).

Tryptophan has also assumed medical importance as a result of the pharmacological action of serotonin, 5-hydroxytryptamine, and the possibility that the aromatic amines resulting from the degradation of tryptophan might be carcinogens (*311*).

The mechanism of the transformation of the indole ring of tryptophan to the pyridine ring of nictotinic acid is of singular interest to both the organic chemist and the biochemist. This has recently been solved by the brilliant discovery of Nishizuka and Hayaishi (*312*).

In the determination of the metabolic pathways of tryptophan, valuable aid has come from the ability to produce mutants in *Neurospora* with specific metabolic defects, and by means of the induction phenomenon in bacteria. This technique yields strains with metabolic characteristics highly desirable for the study of intermediate metabolism. The occurrence of certain intermediate products of tryptophan metabolism in the urine of mammals in certain instances also has provided important clues toward the unraveling of the different reaction steps in the metabolic scheme.

The outcome of the different lines of investigation is that a number of pathways of tryptophan metabolism have been established. In the vertebrate organism the two well-known pathways are the kynurenine-hydroxyanthranilic acid and the serotonin pathways. Studies with *Pseudomonas* bacteria led Stanier and Hayaishi (*313*) to propose two pathways for the dissimilation of the products of tryptophan metabolism starting at the level of kynurenine. One of them is through anthranilic acid and catechol, referred to as the aromatic pathway, and the other through kynurenic acid, named the quinoline pathway.

A diagram of the different products of tryptophan dissimilation by the several known pathways is given in Fig. 9.

Pertinent past reviews of various aspects of tryptophan metabolism are contained in (*311,314,315*).

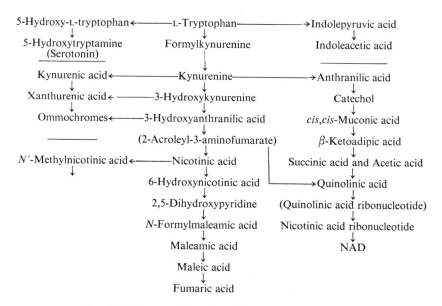

FIG. 9. Pathways and products of tryptophan catabolism.

B. Isotopic Experiments on Dissimilation of Tryptophan

Experiments with tryptophan-^{14}C have established that this amino acid is fairly readily oxidized to completion in the intact animal (*316,317*) –about as readily as is phenylalanine, according to Dalgleish and Tabechian (*317*). This is true for the carbon atoms of the aromatic ring as well as of the side chain. A large fraction of the administered ^{14}C in the tissues remained largely as tryptophan, according to Gholson *et al.* (*318*). A considerable fraction of the label is always found in the urine, since a variety of incompletely oxidized products are invariably produced in the metabolism of tryptophan (*315*). They will be discussed below.

C. The Kynurenine-Anthranilic Acid-Niacin Pathway

The kynurenine-anthranilic acid pathway is not only the route for niacin formation, it is also the main pathway for oxidation of tryptophan to CO_2 in the mammalian liver.

Gholson *et al.* (*318*) suggested that 3-hydroxyanthranilic acid is an obligate intermediate in the major pathway for the complete oxidation *in vivo* of the indole nucleus of tryptophan. The basis for this conclusion is that the rate and extent of conversion of 3-hydroxyanthranilate-1-^{14}C was that which would be expected if this compound were an intermediate in the conversion of tryptophan to acetate.

Subsequently, Henderson and Hankes (*319*) and Gholson and co-workers (*320*) demonstrated the formation of glutamic acid from both tryptophan-7a-^{14}C and 3-hydroxyanthranilic acid-1-^{14}C in the intact rat.

An illuminating experiment is given in Table I showing the products

TABLE I

RADIOACTIVE REACTION PRODUCTS OF
3-HYDROXYANTHRANILATE-1-[14]C[a]

Product	mμmoles	%
CO_2	229	6.1
Glutamic acid	171	4.5
Acetic acid	16	0.4
α-Ketoadipic acid	37	1.0
Picolinic acid	1433	38.3
Quinolinic acid	894	23.9
Substrate remaining	824	23.0

[a] Modified from (*321*).

formed on incubating cat liver with 3-hydroxyanthranilate-^{14}C (*321*). In this experiment the substrate consisted of 3.74 μmoles of 3-hydroxy-anthranilate. The incubation was performed with cat liver homogenate treated with charcoal and Dowex 1-Cl$^-$. The incubation mixture was fortified with NAD and coenzyme A. Kynureninase from a *Pseudomonas* was included to ensure cleavage of kynurenine and hydroxykynurenine to the corresponding anthranilates.

The oxidation schemes of tryptophan deduced from this and other experiments are given in Figs. 9 and 10.

The initial degradation products by this pathway are formylkynurenine, kynurenine, 3-hydroxykynurenine, anthranilic acid, or 3-hydroxy-anthranilic acid and alanine.

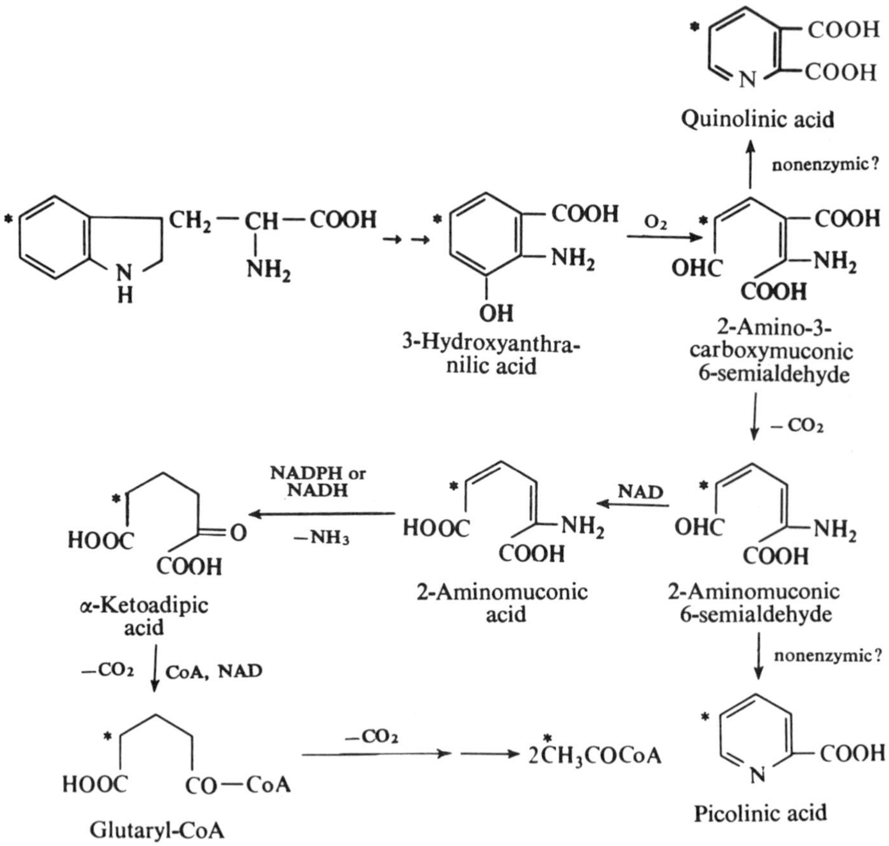

FIG. 10. Proposed pathway of metabolism of the benzene ring of tryptophan in mammalian livers (*321*).

The first of the enzymes catalyzing this sequence of reactions is named tryptophan oxygenase, better known as tryptophan pyrrolase (EC 1.13.1.12). The reaction promoted is shown in Eq. (35).

$$\text{Tryptophan} + O_2 \longrightarrow \text{N-Formylkynurenine} \tag{35}$$

Tryptophan   N-Formylkynurenine

1. TRYPTOPHAN OXYGENASE (EC 1.13.1.12)

Formation of kynurenine from tryptophan was discovered in liver by Kotake and Masayama (*322*). These authors proposed the name tryptophan pyrrolase for the enzyme. Knox and Mehler (*323*) subsequently showed that the formation of kynurenine consisted of two enzyme reactions, an initial oxidation to formylkynurenine followed by hydrolysis to kynurenine. Because the reaction was stimulated by H_2O_2 produced *in situ*, it was assumed that there was an intermediate formation and utilization of peroxide in the reaction. For this reason the enzyme was first named tryptophan oxidase-peroxidase by Knox. Experiments with ^{18}O showed that molecular O_2 was incorporated into the reaction products (*324*), but not the ^{18}O from $H_2{}^{18}O$. This observation led Tanaka and Knox (*325*) to return to the use of the original name, tryptophan pyrrolase. The recommendation of the International Union of Biochemistry is that the latter be replaced by the trivial name, tryptophan oxygenase.

It has been determined that the enzyme is an iron porphyrin protein. Tryptophan oxygenase occurs in microorganisms, the livers of mammals, amphibia and birds, and in insects. Highly purified preparations of the enzyme have been obtained from liver (*325–328*) and from a *Pseudomonas* (*327a,329*). Purification is achieved easily, requiring only acid precipitation and chromatography of the enzyme on a DEAE-cellulose column.

Four forms of the enzyme have been observed to occur in liver: the complete active holoenzyme; the apoenzyme; a form that is combined with hematin and is activated by reduction (*328*); and a fourth form that requires prolonged incubation (*330*) or addition of small amounts of cell particulates to be capable of being activated by hematin (*331*).

In a study of purification of the tryptophan oxygenase from rat liver, Tokuyama observed the complete separation of the prosthetic heme group from the rat liver enzyme, but only the partial separation from the

Pseudomonas enzyme, indicating greater affinity for the prosthetic group by the latter. The bacterial enzyme is more stable than the rat liver enzyme, this too being dependent on the ease of separation of the apoenzyme. The enzymes from both sources combine with hematin or heme to reconstitute the holoenzyme. Combination with hematin requires its reduction to yield active enzyme. Whether reconstituted with heme or hematin followed by reduction, a lag period in initiation of the oxidation reaction is observed following addition of substrate. The enzymes from both sources exhibited a requirement for an SH group(s) on the protein. The SH group on the bacterial enzyme is protected by the prosthetic group and undergoes inactivation by PCMB only, following its reduction to heme and dissociation to yield the apoenzyme. The rat liver enzyme is inhibited directly and readily by PCMB.

The molecular weight of the enzymes from both sources have been determined by sucrose gradient centrifugation to be 103,000. Optimum activity is at pH 5.5–8.0 and pH 7.0–7.4 for the bacterial and liver enzyme, respectively. During reaction of the substrate, spectral changes are observed in the Soret region of the spectrum, possibly indicating reversible oxidation and reduction of the prosthetic group.

Tryptophan pyrrolase is inducible in both microorganisms and in liver.

In *Pseudomonas*, Pelleroni and Stanier (*332*) have shown that the true inducer is L-kynurenine. L-Tryptophan also is effective, but this is a consequence of the presence of small amounts of tryptophan oxygenase and formamidase in the bacterium that cause the conversion of the tryptophan to kynurenine. The latter enzymes are induced coordinately by the increasing concentrations of the kynurenine, while the induction of the rest of the enzymes in the oxidation sequence in the anthranilate pathway proceeds sequentially.

In the mammal, increased tryptophan oxygenase activity results from a combination of factors. The chief inducing agents are tryptophan and certain adrenal cortical steroids (*328–331,333–338*).

Administration of tryptophan induces an increase in the combination of the apoenzyme with available hematin (*328*); it stabilizes the enzyme reducing the rate of degradation (*338*); and it also induces *de novo* synthesis of new enzyme protein (*334,385*). The increase in enzyme activity results from a summation of all three factors.

The corticoid steroids, such as hydrocortisone, have a greater effect in promoting protein synthesis than does tryptophan. A considerable portion of the new enzyme is in the latent form that requires incubation to be activated (*331,333*). Higher tryptophan oxygenase activities are obtained by the combined administration of tryptophan and hydrocortisone than of either one alone.

Proof of the stimulation of protein synthesis by the several inducing agents has been demonstrated by a variety of techniques. Greengard and Feigelson (328) demonstrated an increase in the precipitin reaction with tryptophan oxygenase after enzyme induction in liver. The normal increase can be inhibited by administration of puromycin or actinomycin-D(334,335).

The mechanism of activation of tryptophan oxygenase apoenzyme produced by different types of inducers has been considerably clarified by recent work of Schimke et al. (338) and Piras and Knox (336,337). Activation of the enzyme induced by hydrocortisone consists of two sequential steps, namely, conjugation of the apoenzyme with hematin to form the oxidized holoenzyme followed by reduction of the oxidized holoenzyme. The first-step reaction requires the presence of L-tryptophan or certain analogs of the latter. The best of these analogs is α-methyltryptophan. The conjugation reaction is inhibited by thiol reagents or by globin. The reduction of the oxidized holoenzyme is promoted by L-tryptophan and by ascorbate. It is reversed by oxidation in air in the absence of L-tryptophan. Only the second reaction-step is required for enzyme induced by tryptophan. Conjugation is the more rapid of the two reactions and, for the overall enzyme activation, reduction is the limiting step.

It has been deduced that different sites occur on the apoenzyme and holoenzyme that react with L-tryptophan for conjugation and for catalysis. The apoenzyme site involved in conjugation has a wider specificity with analogs and a greater affinity for L-tryptophan than the catalytic site of the holoenzyme. The K_m of L-tryptophan for conjugation is estimated to be 26 μM, that for the catalytic reaction is about 3×10^{-4} M.

Knox and Piras (336,337) observed that analogs of tryptophan that promoted the conjugation of apotryptophan oxygenase with hematin in vivo or in vitro were inducers of the enzyme. These inducers react specifically with the apoenzyme at a site having high affinity for L-tryptophan and α-methyltryptophan, but may have little or no affinity for the catalytic site. The tryptophan analogs apparently increase the tryptophan oxygenase levels both by synthesis of new protein and by slowing the rate of its disappearance. Rosenfeld and Fielgelson (339) obtained evidence that the tryptophan oxygenase of Pseudomonas contains a regulatory site different from the catalytic site that influences the functional capacity of the preexisting enzyme. The evidence for this is that the enzyme exhibits a sigmoidal saturation curve with increasing concentrations of tryptophan that is eliminated by α-methyltryptophan. The latter compound is neither a substrate nor an inhibitor of tryptophan oxygenase.

Warburg *et al.* (*340*) concluded that tryptophan oxygenase is similar to, but not the same as, the O_2 transfering enzyme of respiration.

The tryptophan oxygenase has also been partially purified from *Drosophila* and shown to be activated by vitamin C or 2-mercapto-ethanol (*341*).

Higuchi and Hayaishi (*342*) observed that extracts of rabbit ileum catalyzed the conversion of D-tryptophan to D-kynurenine, presumably via D-formylkynurenine. The enzyme preparation consumed 1 mole of O_2 per mole of tryptophan oxidized and per mole of kynurenine formed. The enzyme activity of this preparation was stimulated by methylene blue and a DPNH-generating system. The optimum reaction was at pH 8.0–8.5. L-Tryptophan was inhibitory.

Tryptophan oxygenase is subject to feedback control by nicotinyl derivatives (*343*) of which the most potent is NADPH. NADPH may be considered the ultimate distal product of the conversion of tryptophan to niacin derivatives. Analysis of the inhibition of tryptophan oxygenase by NADPH resulted in the discovery of a special type of " allosteric " inhibition (*344*). Heating of purified tryptophan oxygenase in the presence of tryptophan yields an enzyme no longer inhibited by NADPH but which still possesses considerable tryptophan oxidase activity. Fractionation of the enzyme on Sephadex G-200 yields an active form of the tryptophan oxygenase sensitive to NADPH eluted with the void volume and an inactive form eluted later. From this result it is surmised that NADPH converts a polymeric active form of the enzyme to a monomeric relatively inactive form by its action at a specific inhibitory site.

2. ARYLFORMYLAMINE AMINOHYDROLASE; FORMAMIDASE (EC 3.5.1.9)

This enzyme catalyzes the hydrolysis of formylkynurenine to kynure-nine, as shown in Eq. (36).

$$\tag{36}$$

The properties of this enzyme have been studied in rat liver (*323*), *Drosophila* (*345*), bacteria (*346*), and *Neurospora* (*347*). When the hydro-lysis is performed in the presence of $H_2{}^{18}O$, one equivalent of ^{18}O is found in the formate formed. This enzyme, as is indicated by its systema-tic name, is not specific for formylkynurenine, but will catalyze the hydrolysis of a variety of arylformylamines, but shows a preference for formylkynurenine. Formamidase has been purified 15- to 20-fold from *Neurospora* (*347*). The enzyme has a broad pH optimum at 7.3–7.8 in phosphate buffer. The Michaelis constant for formylkynurenine was estimated to be 1.1×10^{-4} *M*.

3. KYNURENINE-3-HYDROXYLASE (EC 1.14.1.2)

Interest in hydroxykynurenine in relation to niacin formation was aroused when it was observed that it yielded niacin in a *Neurospora* mutant with a genetic block after kynurenine (*348*). The compound was first isolated from the larvae of several insects (*349*). Its importance in the kynurenine pathway in the mammal was established when it was shown that it was converted to niacin and could replace the latter as a growth factor (*350,351*).

Conversion of kynurenine to 3-hydroxykynurenine was found to be catalyzed by liver and kidney mitochondria (*352*). More recently the enzyme has been shown to exist in the outer mitochondrial membrane (*353*).

The only published attempt at purifying the enzyme is that of Saito *et al.* (*354*). Several-fold purification was achieved by salting out the solubilized enzyme with 0.35–0.55 saturation of $(NH_4)_2SO_4$. These investigators determined that O_2 and NADPH were required for the hydroxylation. The experimental work established that the reaction proceeds according to Eq. (37).

$$\text{L-Kynurenine} + \text{NADPH} + \text{H}^+ + \text{O}_2 \rightarrow \text{3-hydroxykynurenine} + \text{NADP}^+ + \text{H}_2\text{O} \quad (37)$$

L-Kynurenine is the specific substrate for the enzyme, and hydroxylation occurs only at position 3.

The activity of kynurenine hydroxylase is increased 3- to 10-fold by the monovalent anions Cl^-, Br^-, CN^-, or N_3^-. The most potent of these are Cl^- and N_3^-. The optimum activity of the enzyme is at pH 8, and its Michaelis constants are 2.5×10^{-5} *M* for NADPH and 2.3×10^{-5} *M* for kynurenine.

a. Role of Riboflavin. A number of reports have appeared that suggested that a riboflavin coenzyme might be involved in the hydroxylation of kynurenine. Thus Stevens and Henderson (*355*) found a marked decrease in the activity of kynurenine hydroxylase prepared from the livers of riboflavin-deficient rats. Addition of riboflavin coenzymes or boiled liver extract failed to restore the enzyme activity. The available evidence is too indefinite to judge whether or not a riboflavin coenzyme is associated with kynurenine hydroxylase.

4. L-KYNURENINASE

a. L-Kynurenine Hydrolase (EC 3.1.3). This is a key enzyme of the kynurenine-niacin pathway. The reaction catalyzed is shown in Eq. (38). In this equation R = H or OH. The products are either anthranilic or 3-OH-anthranilic acids.

$$\underset{R}{\underset{NH_2}{\nearrow}}\!\!\!\!\overset{O}{\overset{\|}{C}}-CH_2-\underset{NH_2}{\underset{|}{CH}}-COOH \;+\; H_2O \;\longrightarrow\; \underset{R}{\underset{NH_2}{\nearrow}}\!\!\!COOH \;+\; alanine \quad (38)$$

Kynureninase has been studied in liver (*356*), *Neurospora* (*357*), and *Pseudomonas fluorescens* (*352*). It is able to catalyze the cleavage of a variety of compounds containing the $COCH_2CHNH_2COOH$ group. Liver kynureninase (*356*) splits 3-hydroxy-kynurenine about twice as rapidly as kynurenine; the enzyme from *Pseudomonas* (*358*) hydrolyzes kynurenine five times as rapidly as the hydroxy compound. Jacoby and Bonner (*357*) observed that kynureninase also hydrolyzes formyl-kynurenine.

Braunstein *et al.* (*359*) observed that alanine was produced by the scission of the side chain of kynurenine and that pyridoxal-P was required as a coenzyme for the reaction. These investigators showed that the kynureninase activity was greatly reduced in the livers of pyridoxine-deficient animals, and could be restored *in vitro* by the addition of pyridoxal-P. This finding has been amply confirmed by other workers (*360–362*).

The mechanism of the reaction follows a pattern similar to that proposed for the role of pyridoxal-P in the decomposition of other amino acids (see Chapter 14). This involves formation of a Schiff base between kynurenine and pyridoxal-P-enzyme. The shift of electrons labilizes the carbonyl group of the kynurenine which is cleaved to form anthranilic acid and the Schiff base of alanine.

In *Neurospora*, kynureninase activity was increased 600-fold by addition of L-tryptophan to the culture medium. The enzyme has been purified to a considerable degree from liver (*363*) and *Neurospora* mycelia (*357*). The enzyme from *Neurospora* was calculated to have values of K_m of 6×10^{-6} M for kynurenine and 3×10^{-6} M for hydroxy-kynurenine. Except for differences in the region of the pH of optimum activity, kynureninase was very similar in properties from all the sources studied.

5. 3-HYDROXYANTHRANILATE OXYGENASE (EC 1.13.1.6)

The reaction catalyzed by this enzyme results in the cleavage of the benzene ring and initiates the chains of reactions that lead either to the synthesis of niacin, formation of picolinic acid, or complete oxidation of hydroxyanthranilate to CO_2 in vertebrates. The generally accepted sequence of reactions leading to the various products is shown in Fig. 10. 3-Hydroxyanthranilate oxygenase* has been highly purified in recent years (364–369).

Earlier work had already shown that the enzyme required O_2, Fe^{2+}, and free SH groups for activity.

One mole of O_2 was consumed and one atom of ^{18}O was found in the picolinic or quinolic acid isolated from the reaction mixture, and no ^{18}O was found in the products when the incubation was conducted in $H_2^{18}O$ (370).

Enhancement up to 3500-fold in specific activity of 3-hydroxyanthranilate oxygenase has been obtained with extracts of beef liver (367,368). In the course of purification, enzyme activity is greatly increased by $(NH)_2SO_4$, acid precipitation, and heating. The maximal increase in activity, up to 10-fold, can be obtained by heat treatment in the absence of O_2 alone. To obtain maximum activity, the enzyme needs to be purified in the presence of Fe^{2+} and reduced glutathione. More recently, 3-hydroxyanthranilate oxygenase has also been purified from beef kidney (371, 371a). The properties of the kidney enzyme are very similar to those from liver. 3-Hydroxyanthranilate oxygenase, previously not known to exist in microorganisms, has now been demonstrated in the yeast Saccharomyces cerevisiae (372).

Oxygen destroys enzyme activity and the hydroxyanthranilate oxygenase loses activity in the course of oxidizing the substrate. This is believed to result from oxidation of the bound Fe^{2+} to Fe^{3+}.

Mitchell et al. (369) concluded that the enzyme contains tightly bound Fe which is involved in the catalytic reaction. Added Fe^{2+} is assumed to function only by serving as a reducing agent. It can be replaced by other reducing agents, ascorbic acid in particular. The latter is reported to activate the hydroxyanthranilate oxygenase independently of Fe^{2+}. Ascorbic acid also is the best protective agent in maintaining activity and gives the greatest yield of product.

---

* An enzyme associated with microsomes that hydroxylates anthranilic acid to 5-hydroxyanthranilate prepared from rabbit liver has been reported (371). The optimal pH is at 6.8 and the K_m for the substrate is 3.8×10^{-4} M. NADPH is required as a cofactor.

3-Hydroxyanthranilate oxygenase is strongly inhibited by Fe^{3+} and by Cu^{2+}. It is also inhibited noncompetitively by aminophenols, amino-benzoic acids, and hydroxybenzoic acids (368). The K_m for 3-hydroxy-anthranilate was estimated to be 6.3 μM. The optimum activity is at pH 7.0.

The immediate oxidation product of 3-hydroxyanthranilate is generally accepted as being 2-amino-3-carboxymuconic-6-semialdehyde (2-acroleyl-3-aminofumarate). It has not been isolated, but evidence for its existence is the isolation of the 2,4-dinitrophenylhydrazone by Wiss and Bettendorf (373) and the fact that its properties are consistent with the proposed structure (374).

6. QUINOLINIC ACID

2-Amino-3-carboxymuconic-6-semialdehyde reacts nonenzymically by cyclizing spontaneously to form quinolinic acid. This, as will be described below, undergoes a series of reactions to form niacin ribonu-cleotides and niacin.

The intermediate formed on oxidation of 3-hydroxyanthranilate can be converted quantitatively to quinolinic acid by heating at 100°, and nearly complete conversion is obtained on incubation of hydroxy-anthranilate with liver slices or a high concentration of anthranilate oxygenase (375,376).

7. PICOLINIC ACID

On the other hand, the intermediate (2-amino-3-carboxymuconic-6-semialdehyde) is the substrate for an enzyme that catalyzes its conversion to 2-aminomuconic-6-semialdehyde. The latter spontaneously cyclizes to form picolinic acid (375,377). This enzyme, picolinic acid carboxylase, was discovered by Mehler (378). The enzyme was demonstrated in extracts of fresh or frozen liver of various species, the most consistent activity being found in guinea pig liver. In experiments with ^{14}C-hydroxy-anthranilic acid, Mehler and May (375) demonstrated the cleavage of the carboxyl group. The enzyme is inhibited by CN^- and appears to require free SH. Picolinate carboxylase has been purified about 200-fold from cat liver with a 15% yield (379). It exhibits a broad range of maximal activity and stability between pH 6 and 9.5. The K_m value for 2-amino-3-carboxymuconic-6-semialdehyde was of the order of 10^{-6} M. The enzyme activity was inhibited by PCMB, and this was prevented by L-cysteine.

8. ENZYMIC DECOMPOSITION OF 2-AMINO-3-CARBOXYMUCONIC-6-
SEMIALDEHYDE (2-Acroleyl-3-aminofumarate)

As shown in Fig. 10, the decarboxylation product of carboxymuconic-6-semialdehyde is 2-aminomuconic-6-semialdehyde. This is further oxidized to 2-muconic acid catalyzed by an enzyme named 2-hydroxymuconic-6-semialdehyde dehydrogenase (371).

α-Muconic acid is further deaminated to α-ketoadipic acid, and the latter undergoes decarboxylation and oxidation by an α-keto acid oxidase to yield glutaryl-CoA.

9. 2-HYDROXYMUCONIC-6-SEMIALDEHYDE DEHYDROGENASE

The decarboxylation product of 2-amino-3-carboxymuconic-6-semialdehyde is extremely unstable and has not been isolated in incubation products. However, by the combined action of picolinate carboxylase and the 2-hydroxymuconic-6-semialdehyde dehydrogenase, formation of 2-aminomuconic acid can be demonstrated. Because of the instability of 2-aminomuconic-6-semialdehyde, the related compound, 2-hydroxymuconic-6-semialdehyde, was utilized as a model substrate.

2-Amino-3-carboxymuconic-6-semialdehyde is converted nonenzymically to 2-hydroxymuconic-6-semialdehyde by acidification to pH 3. 2-Hydroxymuconic-6-semialdehyde has been shown to be oxidized to 2-hydroxymuconic acid by an NAD-linked specific aldehyde dehydrogenase obtained from o-cresol-adapted Pseudomonas cells (380) and by a similar enzyme found in liver and kidney (379). This enzyme has been purified 50- to 100-fold from cat and rat livers and from the Pseudomonas mentioned above. In addition to catalyzing the oxidation of 2-hydroxymuconic-6-semialdehyde and 2-aminomuconic-6-semialdehyde, this enzyme also acts on a number of simple aliphatic aldehydes, such as formaldehyde, n-butyraldehyde, and to a very slight extent on acetaldehyde. The optimum activity is between pH 7.5 and 8.0. The K_m value for 2-hydroxymuconic-6-semialdehyde and NAD were 1.6×10^{-5} and 1.9×10^{-5} M, respectively.

The enzyme was inhibited by PCMB, and this inhibition was reversed by various SH compounds. This suggests that the dehydrogenase might be an SH enzyme. Other inhibitors are the metal ions: Co^{2+}, Ni^{2+}, Hg^{2+}, and Cu^{2+}.

The evidence that 2-aminomuconic-6-semialdehyde is formed from 2-amino-3-carboxymuconic-6-semialdehyde was obtained by incubating the latter with picolinate carboxylase and 2-hydroxymuconic-6-semialdehyde dehydrogenase in the presence of NAD. This led to the formation of a new compound with an absorption maximum at 325 mμ. This new compound was converted to 2-hydroxymuconic acid by acid. The latter was identified by various chemical tests.

The 2-aminomuconic acid has been shown to be reductively deaminated to 2-ketoadipic acid in the presence of either NADH or NADPH. This keto acid is then oxidatively decarboxylated to glutaryl-CoA (*381*).

10. Conversion of Quinolinic Acid to Niacin

Both quinolinic acid and picolinic acid of themselves are metabolically inert and are not attacked by tissue preparations. This inertness puzzled investigators of tryptophan metabolism for many years, since there was abundant evidence that tryptophan could serve as a precursor of niacin, and the most likely pyridine precursor from chemical consideration was quinolinic acid. This paradox was solved by the discovery of Nishizuka and Hayashi (*312*) that 3-hydroxyanthranilic acid incubated with 5-phosphoribosyl-1-pyrophosphate and a soluble enzyme system from rat liver is converted to niacin ribonucleotide. Quinolinic acid ribonucleotide is an intermediate in this reaction sequence. Furthermore, niacin ribonucleotide could be obtained by incubating quinolinic acid-^{14}C with PP-ribose-P and the liver enzyme with the concomitant liberation of [14]CO$_2$.

a. Quinolinate Phosphoribosyltransferase. This enzyme has been purified to a considerable degree from liver (*382,383*) and has been crystallized from a Pseudomonad isolated by enrichment culture (*384*). The liver enzyme has optimum activity at pH 6.2 and K_m values of 6×10^{-5} M for quinolinate and 5×10^{-5} M for P-ribose-PP, respectively.

The Pseudomonad grown on quinolinate has a content of this enzyme of 9% of its total protein, and it requires only a 10-fold enrichment to be sufficiently pure to crystallize. The bacterial enzyme has a molecular weight of 165,000 and can be dissociated into inactive protein units of molecular weight 54,000. The optimum activity of the bacterial enzyme is at pH 7.1. Study of the kinetics of the enzyme reaction gave results consistent with the involvement of a ternary complex of enzyme, quinolinic acid, and P-ribose-PP.

D. Quinolinic Acid Pathway of Tryptophan Dissimilation

This pathway occurs in an incomplete form in the vertebrate organism. In certain bacteria it is an important pathway for the total oxidation of tryptophan. The quinolinic acid pathway arises by transamination of kynurenine and hydroxykynurenine. The resulting products spontaneously cyclize to form kynurenic and xanthurenic acids.

Kynurenic acid was discovered in 1853 by Liebig (*385*) before tryptophan was known. Xanthurenic acid was discovered in urine by Musajo in 1935 (*386*). It was isolated from the urine of pyridoxine-deficient rats by Lepkovsky and co-workers (*387*) because of the intense green color of the chelate it forms with Fe^{3+}. Excretion of xanthurenic acid is increased in pyridoxine deficiency, particularly after feeding of tryptophan. It is also found in the urine of normal animals (*388*). Metabolism of kynurenic and xanthurenic acid does not proceed further to any significant degree in the mammal. In experiments with xanthurenic-^{14}C acid, Rothstein and Greenberg (*389*) could detect no formation of $^{14}CO_2$. Injected xanthurenic acid was found to be excreted quantitatively in the urine of the rat in the form of free xanthurenic acid and certain conjugates. The conjugates contained glucuronide residues at the 8- or 4- and 8-positions of xanthurenate, respectively, and the amino acid serine attached to the carboxyl group. The rabbit excretes xanthurenic acid as an ethereal sulfate.

Price and co-workers (*390,391*) reported that kynurenic acid may be dehydroxylated to quinaldic acid, and that the latter is excreted in the urine of men and rats. As much as 29 % of a single oral dose of kynurenic acid was recovered in the urine as quinaldic acid in human subjects. The methylation of the 8-hydroxy group of xanthurenic acid was also reported from the same laboratory (*391*)

Quinaldic acid is not degraded further (*392*). It may be excreted in the urine as the free acid or as the glycine conjugate (*392*).

1. KYNURENINE AMINOTRANSFERASE

This enzyme has been studied in liver and kidney (*393,394*), in *Pseudomonas* (*395*), and in *Neurospora* (*396*).

The enzyme is active on both L-kynurenine and L-3-hydroxykynurenine. The D-isomers are inert. It requires pyridoxal-P as a coenzyme. The most effective amino-group acceptor is α-ketoglutarate, but other α-keto acids can also function. *o*-Aminobenzoylpyruvic acid, the expected product of the action of kynurenine transaminase on kynurenine, has not been identified. This is probably because the spontaneous cyclization to form kynurenic and xanthurenic acids occurs with extreme rapidity.

The kynurenine aminotransferase has been considerably purified by Ueno *et al.* (*397*) from horse kidney. The apoenzyme was found to be activated by either pyridoxal-P or pyridoxamine-P. The optimum activity is at pH 6.5. K_m values obtained were 1.8 mM for L-kynurenine, 1×10^{-4} M for pyridoxal-P, and 2.6×10^{-6} M for pyridoxamine-P.

In vitamin B_6-deficient rats there was a greater decrease of kynureninase than of the aminotransferase (398). It was concluded that this tended to favor formation of xanthurenic acid. This agrees with the observation that excretion of xanthurenic acid is greatly increased in vitamin B_6 deficiency.

2. OXIDATIVE PATHWAY OF KYNURENIC ACID IN BACTERIA

By the use of enzyme preparations from tryptophan-adapted cells of *Pseudomonas fluorescens*, Hayaishi and co-workers obtained the conversion of kynurenic acid to glutamate, alanine, acetate, and CO_2 (399,400). Study of the intermediate steps with separated enzyme fractions led to the formulation of the scheme of kynurenic acid dissimilation shown in Fig. 11.

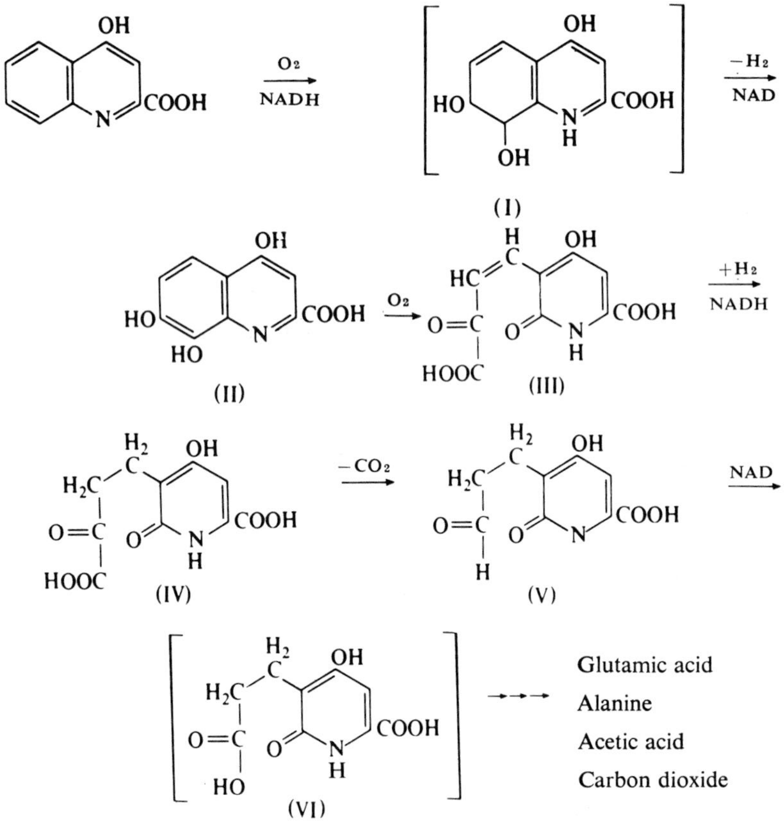

FIG. 11. Bacterial oxidation products of kynurenic acid.

The first product, compound (I), is formed in the presence of O_2 and NADH on incubation with an $(NH_4)_2SO_4$-precipitated fraction of the bacterial extract containing kynurenic acid hydroxylase (*401*). It has not been isolated in crystalline form. This compound, isolated on Dowex 1, is decomposed to 8-hydroxyxanthurenic acid on drying. From its chemical and reaction properties, compound (I) is deduced to be 7,8-dihydroxykynurenic acid-7,8-diol. Formation of the latter does not appear to be a one-step reaction, and it is postulated that a peroxide or epoxide of kynurenic acid is formed first. Compound (I) is the substrate for an NAD-linked dehydrogenase which catalyzes its conversion to 7,8-dihydroxykynurenic acid, compound (II). This enzyme has been named tentatively 7,8-dihydroxykynurenic acid-7,8-diol dehydrogenase. For this reaction O_2 is not required if NAD is regenerated.

The rate of formation of 7,8-dihydroxykynurenic acid is increased by Fe^{2+} or by boiled enzyme extract.

Compound (II) was synthesized by Behrman and Tanaka (*402*) and proposed as a probable intermediate in kynurenic acid oxidation. This was confirmed by Taniuchi and Hayaishi (*401*), although some doubt developed concerning this owing to the observation that 7- and 8-hydroxykynurenic acids are also substrates, but much poorer ones, for the subsequent oxidation step.

7,8-Dihydroxykynurenic acid is converted to an intensely yellow compound (III) with an absorption maximum at 390 mμ, catalyzed by 7,8-dihydroxykynurenic acid oxygenase. The structure of this compound (III) is deduced to be 5-(γ-carboxy-γ-oxopropenyl)-4,6-dihydroxypicolinic acid (*403*).

Kynurenic acid hydroxylase is inactivated by O_2 and requires the presence of an SH compound to retain activity. It is inhibited by PCMB and can be reactivated by glutathione or cysteine. The optimum pH for the reaction is at 7.0.

Compound (III) is reduced to compound (IV) by an NADPH-linked dehydrogenase. Compound (IV) has been obtained in crystalline form and identified, and so has its decarboxylation product, compound (V). Compound (V) is indicated, by preliminary results, to be further oxidized by an NAD-linked dehydrogenase to compound (VI).

E. The Aromatic Pathway of Tryptophan Dissimilation

Oxidation by this pathway commences with the conversion of anthranilic acid to catechol. The reaction is postulated to proceed according to the scheme shown in Eq. (39).

$$\text{[anthranilic acid]} \xrightarrow{O_2} \text{[intermediate]} + NADH + H^+ \longrightarrow \text{[catechol]} +$$

$$CO_2 + NH_4^+ + NAD^+ \qquad (39)$$

Anthranilate hydroxylase has been partially purified and studied by a number of investigators (404–406). The instability of the enzyme and the complexity of the reaction has retarded elucidation of the reaction mechanism. Taniuchi et al. (406) established definitely the formation of catechol by isotopic experiments with anthranilic acid-2-^{14}C. This had been difficult to do because of the presence of pyrocatechase in the enzyme preparations, which catalyzes oxidation of catechol to cis,cis-muconic acid. These authors also established a requirement for reduced pyridine nucleotide as a cofactor; NADH is the more active. The reaction requires O_2, and enzyme activity is lost during the incubation in O_2. Anthranilate hydroxylase has a requirement for Fe^{2+} (407). It is inhibited by PCMB and Hg^{2+}. This inhibition can be partially reversed by SH compounds. Optimal activity is at pH 7.5. In experiments with ^{18}O, Kobayashi et al. established that both atoms of O in the catechol were derived from O_2 (407). This led to a revision of the previously proposed mechanism (406) to that shown in Eq. (39).

1. OXIDATION OF CATECHOL

Two enzymes are known that catalyze oxidation of catechol with the cleavage of the benzene ring. The earlier and better studied enzyme is pyrocatechase which induces the formation of cis,cis-muconic acid, Eq. (40); the more recently discovered enzyme is metapyrocatechase which induces formation of 2-hydroxymuconic semialdehyde, Eq. (42).

2. PYROCATECHASE (Catechol: oxygen, 1,2-oxidoreductase, EC 1.13.1.1)

Pyrocatechase was discovered by Hayaishi and Hashimoto (408). It has been purified and studied from acetone-dried Micrococcus ureae cells (409) and from tryptophan-adapted Pseudomonas fluorescens cell extracts (410,410a). The enzyme has a molecular weight of about 100,000 and contains two atoms of Fe per molecule. It also contains one naked and two masked SH groups per molecule. Glutathione and SH compounds stimulate the activity of pyrocatechase, either by binding heavy metals or by reducing o-benzoquinone, which is produced from catechol spontaneously. By use of ^{18}O it was shown that the reaction utilizes one mole of O_2 per mole of catechol, and the product formed has been shown to be *cis,cis*-muconic acid *(411,412)* (Eq. 40).

$$\text{(structure: benzene ring with two OH groups)} + O_2 \longrightarrow \text{(open chain with two COOH groups)} \tag{40}$$

Pyrocatechase has a red color with a broad absorption peak at 400–600 $m\mu$ and a shoulder at 322 $m\mu$ *(412)*.

Study of the enzyme by electron spin resonance spectroscopy (ESR) *(413)* indicated that there is at least one atom of Fe^{3+} in native pyrocatechase, and the latter undergoes reduction and reoxidation during the enzyme-catalyzed oxygenation of catechol. The red color of the enzyme disappeared, as did the ESR signal for Fe^{3+}, when sodium dithionite was added to the enzyme maintained under N_2. The ESR signal also was greatly decreased when catechol was added to the enzyme. The changes of the signal and the color were reversible. On the basis of the above findings, Nakazawa et al. *(413)* proposed the mechanism for the reaction shown in Eq. (41).

$$Fe^{3+} \xrightarrow{S} Fe^{2+}S^{\circ} \xrightarrow{O_2} \overset{\overset{\displaystyle S^{\circ}}{|}}{\underset{\underset{\displaystyle O_2}{\backslash}}{Fe^{2+}}} \longrightarrow \overset{\overset{\displaystyle S^{\circ}}{|}}{\underset{\underset{\displaystyle O_2^{-}}{\backslash}}{Fe^{3+}}} \longrightarrow SO_2 + Fe^{3+} \tag{41}$$

$$\text{(I)} \qquad \text{(II)} \qquad \text{(III)} \qquad \text{(IV)}$$

In this equation S represents the substrate. The essence of the proposed scheme is that the substrate reduces the Fe^{3+} in the enzyme. The product then reacts with the O_2 to form a ternary complex. The transfer of an electron from Fe^{2+} to O_2 (III) yields a highly reactive compound (IV) in which the two free radicals react to form an oxygenated end product.

3. METAPYROCATECHASE

This oxygenase was first described by Dagley and Stopher *(414)*. The reaction differs from that catalyzed by pyrocatechase in that the reaction product formed is 2-hydroxymuconic semialdehyde, Eq. (42). When

$$\text{(structure: benzene ring with two OH groups)} + O_2 \longrightarrow \text{(open chain product)} \tag{42}$$

first isolated, this proved to be an extremely unstable enzyme that was inactivated by O_2 *(415)*. The activity could be preserved if the enzyme was kept under N_2, and the inactivated enzyme could have its activity restored when treated with reducing agents such as sodium borohydride (416).

An interesting observation that enabled the complete purification and crystallization of metapyrocatechase was that the enzyme activity could be stabilized by low concentrations of organic solvents, such as acetone and ethanol (417,418).

Utilizing *Pseudomonas arvilla*, which was found to have a very high content of the enzyme, a preparation of very high activity could be obtained with a 30-fold enrichment that readily crystallized from an $(NH_4)_2SO_4$ solution.

The molecular weight of metapyrocatechase was found to be 140,000. It contains one Fe per molecule. It is strongly inhibited by PCMB, Hg^{2+}, and Ag^+.

Homocatechol was found to be about equally as good a substrate as catechol.

Studies with ESR led Hayaishi and co-workers to propose the reaction mechanism shown in Eq. (43) (413,418). The reaction postulates that the

(43)

Fe^{2+} at the active center reacts first with O_2 to form a perferryl ion complex which establishes the equilibrium state

$$Fe^{2+}O_2 \rightleftharpoons Fe^{3+}O_2{}^-$$

This then binds with catechol, forming a ternary complex in which both O_2 and catechol are activated and react together to form an intermediate catechol peroxide.

4. LACTONIZING AND DELACTONIZING ENZYMES

The enzymes which convert *cis,cis*-muconic acid to β-ketoadipic acid were isolated by Sistrom and Stainier (411) and Evans *et al.* (419,420). One of these enzymes eliminates the olefinic bonds by lactone formation and hydrolysis. The lactonizing enzyme converts *cis,cis*-muconic acid to (+)-γ-carboxymethyl-Δ^α-butenolide. This enzyme requires Mn^{2+} for activity. It catalyzes the reversible formation of the γ-lactone, thus eliminating the γ,δ-double bond and creating a center of asymmetry at the γ-carbon atom. The delactonizing enzyme converts the γ-lactone to the enol form of β-ketoadipic acid. The enzyme is specific for the dextrorotatory γ-lactone. No cofactor requirement was discovered. The reaction sequence is shown in Eq. (44).

cis,cis-Muconic acid   (+)-γ-Carboxymethyl-

Δ^α-butenolide   (44)

(enol) β-Ketoadipic acid   (keto)

The lactonizing enzyme can also catalyze the reversible reaction between *cis,trans*-muconic acid and the levorotatory lactone.

The β-ketoadipic acid is further cleaved to succinate and acetate by a variety of microorganisms (*421*).

F. The 5-Hydroxyindole Pathway

From a study of the enterochromaffin cells (skin, intestinal mucosa), Erspamer (*422*) postulated the occurrence of a physiologically potent vascular hormone. From chemical and pharmacological studies it was concluded that this material, named "enteramine," was an indole derivative carrying a phenolic group(s) and a side chain with a terminal primary or secondary amino group.

Subsequently a compound with vasoconstrictor activity was isolated from serum by Rapport and co-workers (*423*) and identified as 5-hydroxytryptamine. This compound was named "serotonin." Enteramine and serotonin were soon shown to be one and the same substance.

The exact physiological functions of serotonin are not known. It has been associated with gastrointestinal activity, nervous system function, kidney function, anaphylaxis, etc.

The physiological and pharmacological properties of serotonin have been the subjects of several reviews (*424,425*).

1. Biogenesis of Serotonin

Evidence that tryptophan is the precursor of serotonin has been amply demonstrated by isotopic tracer experiments (*425,426*).

a. Hydroxylation. The first step in the formation of serotonin is the hydroxylation of tryptophan to form 5-hydroxytryptophan. This compound has been isolated from the urine of a number of mammalian species, and its formation has been demonstrated in liver slices (*427*) and in human metastatic carcinoid tumor slices (*428*).

Enzyme preparations that form 5-hydroxytryptophan have been isolated from liver (*429–432*), brain stem (*433,434*), and mouse mast cell tumors (*435,436*).

The soluble liver enzyme was subsequently shown to be the same enzyme that catalyzes hydroxylation of phenylalanine to tyrosine (*429,430*). Phenylalanine is a considerably better substrate for this enzyme than is tryptophan, and the K_m for phenylalanine is about 10-fold lower than that for tryptophan. Freedland (*432*) has pointed out certain parallelisms and also certain differences in the action of the liver aromatic acid hydroxylase on the two amino acids. Thus a high-phenylalanine diet suppresses both activities equally, while L-tryptophan and cortisone increase the activity on the two amino acids to the same degree. On the other hand, hydroxylation of tryptophan, but not of phenylalanine, is increased by Fe^{2+} and vitamin C. Tryptophan, but not phenylalanine, hydroxylation was found to be inhibited by EDTA.

Since until recently it was assumed that phenylalanine hydroxylation occurred only in liver, doubt arose that this was the enzyme normally concerned in formation of serotonin since this compound is formed in great amounts in tissues other than liver. This doubt appears to be dissipated by the properties of the hydroxylase prepared from mouse mast cell cancer (*438*). This enzyme, too, hydroxylates phenylalanine, and at about twice the rate of tryptophan. This enzyme preparation was found to be dependent on Fe^{2+} and a tetrahydropteridine for activity. A tetrahydropteridine requirement is also characteristic of the liver enzyme.

Certain differences of importance characterize the mast cell tumor enzyme. It has optimum activity at pH 6.7, whereas that of the liver enzyme is at 7.4, and the K_m for tryptophan is of the same magnitude as that for phenylalanine.

The tryptophan hydroxylase of brain stem has a very low degree of activity and has not been so well characterized as that of liver and the mast cell tumor.

Formation of 5-hydroxytryptophan has been demonstrated with

Chromobacterium violaceum (437,438). Incubation of L-tryptophan with resting cells of this organism leads to the formation of L-5-hydroxytryptophan, 5-hydroxyindolepyruvic acid, and 5-hydroxyindoleacetic acid. No serotonin is formed because this organism lacks the decarboxylating enzyme. The final product of the reaction in the bacteria is the formation of the purple pigment violacein. Rensen (439) has reported that 5-hydroxytryptophan hydroxylase is induced by growing the organism on L-tryptophan. The enzyme is dependent on Fe for activity.

   b. Decarboxylation of 5-Hydroxytryptophan. An enzyme that decarboxylates 5-hydroxytryptophan has been observed in kidney, liver, gastric mucosa, intestine, and lung (440–442). The enzyme has been purified as much as 100-fold from guinea pig kidney, and a requirement for pyridoxal-P has been established (441,442). This enzyme was considered to be specific for L-hydroxytryptophan for a number of years. More recently, evidence has been obtained that it can decarboxylate a large number of aromatic L-amino acids (443–446). The most active substrates in decreasing order of activity are: 3,4-dihydroxyphenylalanine, *o*-tyrosine, *m*-tyrosine, 5-hydroxytryptophan, tryptophan, and phenylalanine (443). Because of the wide range of substrate activity, Lovenberg *et al.* (446) proposed that the name of the enzyme be broadened to aromatic L-amino acid decarboxylase. In addition to guinea pig kidney, the decarboxylase has also been partially purified from dog brain stem and shown to have properties similar to the liver enzyme. The pH optimum of the enzyme is at about 9.0. 5-Hydroxytryptophan was found to have the lowest K_m value ($2 \times 10^{-5}\ M$); the magnitude for the other amino acid substrates being 20- to 1000-fold greater.

2. METABOLISM OF D-TRYPTOPHAN

   It has recently been shown that D-tryptophan is also vigorously metabolized by the mammal (447,448). A large amount of 5-hydroxytryptamine was found in the urine of the rat after D-hydroxytryptophan administration.

   It is proposed that the D-hydroxytryptophan undergoes oxidative deamination to 5-hydroxyindolepyruvate, catalyzed by D-amino acid oxidase. This compound, it is suggested, is subsequently transaminated to form L-5-hydroxytryptophan and the latter is decarboxylated to serotonin by kidney aromatic amino acid decarboxylase (448). Evidence for this was obtained by injecting rats with glutamic acid-^{15}N and D-5-hydroxytryptophan. Serotonin with a significant atom percent excess of ^{15}N was isolated from the urine of rats (448).

3. OXIDATION OF SEROTONIN

The first stage in the oxidation of serotonin is probably catalyzed by monoamine oxidase. This enzyme is present in mitochondria (449) and also as a soluble protein (450). The soluble fraction from guinea pig liver, purified 10- to 15-fold, converted serotonin to 5-hydroxyindoleacetic acid. The enzyme requires DPN and another protein fraction to form the latter product. The overall reaction is believed to proceed as shown:

Serotonin $\xrightarrow[\text{oxidase}]{\text{monoamine}}$ 5-hydroxyindole acetaldehyde

$\xrightarrow[\text{NAD}]{\text{aldehyde dehydrogenase}}$ 5-hydroxyindoleacetic acid

In the absence of the aldehyde dehydrogenase it has been possible to demonstrate formation of small amounts of the 5-hydroxyindole acetaldehyde. With an excess of aldehyde dehydrogenase and NAD, the reduction of NAD can be used to assay monoamine oxidase.

Considerable information about serotonin metabolism has been learned from the action of monoamine oxidase inhibitors. One of the first of these compounds to be discovered was 1-isonicotinyl-2-isopropyl hydrazide (Marsilid) (451). Many other compounds have since been discovered that inhibit monoamine oxidase both in vitro and in vivo (438). The outcome of the studies with monoamine oxidase inhibitors has been to show that, when the enzyme is inhibited, there is a diminished formation of 5-hydroxyindoleacetic acid and serotonin is converted to the O-glucuronide and excreted in rodents. In man there is little formation of serotonin glucuronide. The fate of serotonin in man under these conditions is not known definitely, although serotonin-O-sulfate has been observed in carcinoid patients. Serotonin may be methylated to yield N-methylserotonin (438).

G. The Tryptophanase Pathway

Various bacteria, E. coli in particular, cleave the side chain of tryptophan to yield indole (452). This putrefactive reaction is the source of indole and skatol in man. In earlier times these compounds were of great interest because they were believed to represent a serious health problem. Studies of this reaction with the enzyme tryptophanase have shown that indole, ammonia, and pyruvic acid are the reaction products (453) and that the coenzyme for the reaction is pyridoxal-P (454). Potassium or ammonium ions are required for enzyme activity (455).

In recent years highly purified preparations of the enzyme have been isolated from *Bacillus alvei* (*454,455*), and the crystalline enzyme from *E. coli* (*456,457*).

The enzyme from *E. coli* was found to be capable of catalyzing a wide range of reactions, including deamination of L-serine and L-cysteine and synthesis of tryptophan from indole and L-serine or the former and L-cysteine. The *B. alvei* enzyme has a more restricted specificity. It does not catalyze synthesis of tryptophan, but does catalyze the deamination of L-serine in addition to the tryptophanase reaction. The enzyme from *E. coli* is extremely unstable; that from *B. alvei* is quite stable and will withstand heating to 80°.

Stabilization of the *E. coli* enzyme could be achieved by storing the enzyme in the cold in 11% $(NH_4)_2SO_4$ and adding an SH compound. Crystallization yields the apoenzyme, even when the procedure is carried out in the presence of pyridoxal-P. In addition to pyridoxal-P, tryptophanase requires K^+ for activity. Only about a 9-fold enrichment is required to obtain the pure crystalline enzyme from derepressed cells of *E. coli*, whereas the enzyme from *B. alvei* required a 3400-fold purification to yield a homogeneous protein.

In the ultracentrifuge, the enzyme from both bacteria exhibited two peaks. The slower-moving peak had an s_{20} value of 9.65 for the *E. coli* enzyme (*457,458*) and 10.8 for the *B. alvei* enzyme (*455*). The molecular weight was estimated to be between 220,000 and 280,000. The optimum activity was at about pH 8.8 for the enzyme from either source. The K_m for tryptophan was approximately the same for each at about 0.3 μM. Hoch and DeMoss (*455*) found that 1 mole of pyridoxal-P was bound to 125,000–130,000 gm of the *B. alvei* enzyme, and that 2 moles of the coenzyme were required to be bound to each active site to give maximal activity. Dissociation constants of the two pyridoxal phosphates were estimated to be 1.14 and 14.4 μM, respectively. Subsequently, Gobenthan and DeMoss (*459*) found that their tryptophanase was dissociated into inactive subunits of one-half the size of the active enzyme in the presence of Tris buffer. Aggregation of the enzyme and full recovery of activity could be achieved by solution in potassium phosphate buffer.

More recently it has been shown that tryptophanase consists of four polypeptide subunits of 55,000 molecular weight each. Each subunit contains two identical peptide chains joined by disulfide linkage. Cleavage of the disulfide yields polypeptides of 28,000–30,000 molecular weight (*460*).

In a kinetic study of tryptophan-catalyzed reactions, Morino and Snell (*461*) developed the kinetic scheme shown in Fig. 12. In this scheme, E represents enzyme; T, tryptophan; C, an allosubstrate (i.e., an amino acid substrate other than tryptophan; e.g., serine, cysteine, etc.); ET, the enzyme-tryptophan complex; EC, the tryptophan-allosubstrate complex; EA, the tryptophan-aminoacrylate complex; I, indole; and RH, the substituent on the β-position of the allosubstrate ($-OH$, $-SH$, etc.) plus a hydrogen atom. The source of the tryptophanase was the crystalline enzyme from *E. coli* (*456,457*).

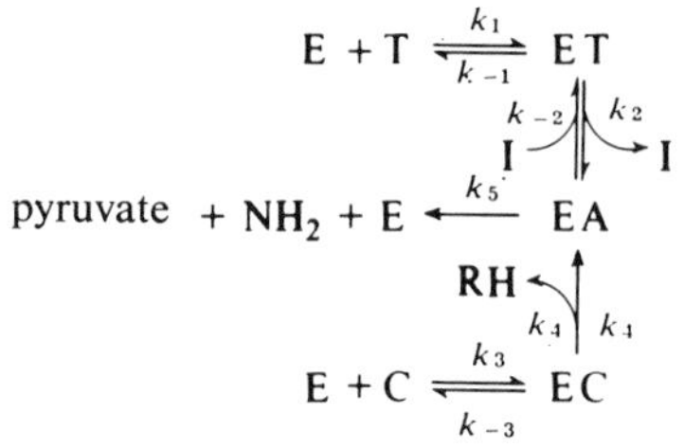

$$
\begin{array}{c}
E + T \underset{k_{-1}}{\overset{k_1}{\rightleftharpoons}} ET \\
k_{-2} \Big\updownarrow k_2 \\
I \quad \quad I \\
k_5 \\
\text{pyruvate} + NH_2 + E \longleftarrow EA \\
RH \\
k_4 \Big| k_4 \\
k_3 \\
E + C \underset{k_{-3}}{\overset{}{\rightleftharpoons}} EC
\end{array}
$$

FIG. 12. Kinetic scheme of reactions catalyzed by tryptophanase (*460*).

The reactions catalyzed by *E. coli* tryptophanase represent three kinetic cases: (a) formation of pyruvate from tryptophan, (b) formation of pyruvate from other substrates, and (c) formation of tryptophan from other substrates and indole. It was deduced from the kinetic data that the rate-determining step in each of the tryptophanase-catalyzed reactions was the removal of the β-substituent from the substrate.

In a second paper, Morino and Snell (*462*) observed that holotryptophanase in the presence of K^+ exhibited two pH-dependent absorption bands with maxima at 337 and 420 mμ. The authors concluded that the spectral change probably respresents principally the loss of a single proton from the chromophore. By plotting absorbance against pH, it was determined that the pK value of the ionization was 7.2. From a variety of evidence, it was deduced that the deprotonated form is the active form of the enzyme.

The coenzyme, pyridoxal-P, was demonstrated to be combined with the ϵ-amino group of lysine. By isolation of ϵ-pyridoxyllysine it was determined that 4 moles of pyridoxal-P were combined with 1 mole of enzyme (MW = 220,000).

By exchange experiments with deuterated or tritiated water, evidence was obtained that the initial step in the reaction of an amino substrate, following its interaction with the coenzyme, is the labilization of the α-hydrogen of the amino acid. This is followed by elimination of the β-substituent to yield an enzyme-aminoacrylate complex.

REFERENCES

1. H. D. Dakin, "Oxidations and Reductions in the Animal Body," 2nd ed. Longmans, Green, London, 1922.
2. A. E. Garrod, "Inborn Errors of Metabolism," 2nd ed. H. Frowde and Hodder & Stoughton, London, 1923.
3. G. W. Beadle, *Chem. Rev.* **37**, (1945); *Physiol. Rev.* **25**, 643 (1945).
4. J. R. S. Fincham, *Advan. Enzymol.* **22**, 1 (1960).
5. A. Pierard and J. M. Waime, *Biochem. Biophys. Acta* **37**, 490 (1960).
6. R. J. O'Connor and H. O. Halvorson, *J. Bacteriol.* **82**, 706 (1961).
7. N. G. McCormick and H. O. Halvorson, *J. Bacteriol.* **87**, 68 (1964).
8. A. Yoshida and E. Freese, *Biochem. Biophys. Acta* **92**, 33 (1964); **96**, 248 (1965).
9. J. L. Still, M. V. Buell, W. E. Knox, and D. E. Green, *J. Biol. Chem.* **179**, 831 (1949).
10. B. Woolf, *Biochem. J.* **23**, 472 (1929).
11. E. F. Gale, *Biochem. J.* **32**, 1583 (1938).
12. H. C. Lichstein and W. W. Umbreit, *J. Biol. Chem.* **170**, 423 (1947).
13. N. Ellfolk, *Acta Chem. Scand.* **7**, 884, 115 (1953); **8**, 151, 443 (1954).
14. P. A. Trudinger, *Australian J. Expt. Biol. Med. Sci.* **31**, 317 (1953).
15. V. R. Williams and R. T. McIntyre, *J. Biol. Chem.* **217**, 467 (1955).
16. R. H. Dupue and A. G. Moat, *J. Bacteriol.* **82**, 383 (1961).
17. K. Ichihara, H. Kanagawa, and M. Uchida, *J. Biochem. (Tokyo)* **42**, 439 (1955).
17a. V. R. Williams and D. J. Lartigue, *J. Biol. Chem.* **242**, 2973 (1967).
18. R. F. Emery, *Biochemistry* **2**, 1041 (1963).
19. S. England, *J. Biol. Chem.* **233**, 1003 (1958).
20. A. I. Krasna, *J. Biol. Chem.* **233**, 1010 (1958).
21. Y. S. Halpern and H. E. Umbarger, *J. Bacteriol.* **80**, 285 (1960).
22. S. R. Mardashev, L. A. Semina, R. N. Etinof, and A. I. Balyasnaya, *Biokhimya* **14**, 44 (1949).
23. A. Meister, H. A. Sober, and S. V. Tice, *J. Biol. Chem.* **189**, 577 (1951).
24. L. V. Crawford, *Biochem. J.* **68**, 221 (1958).
25. J. Cattaneo and J. C. Senez, "IUB Symposium: Chemical and Biological Aspects of Pyridoxal Catalysis, Rome, 1962," p. 217. Pergamon Press, 1963.
26. A. Meister, J. S. Nishimura, and A. Novogrodsky, "IUB Symposium: Chemical and Biological Aspects of Pyridoxal Catalysis, Rome, 1962," p. 229. Pergamon Press, 1963.
27. J. S. Nishimura, J. M. Manning and A. Meister, *Biochemistry* **1**, 442 (1962).
28. A. Novogrodsky, J. S. Nishimura, and A. Meister, *J. Biol. Chem.* **238**, PC1903 (1963).
29. A. Novogrodsky and A. Meister, *J. Biol. Chem.* **239**, 879 (1964).
30. E. M. Wilson, *Biochem. Biophys. Acta* **67**, 345 (1963).
31. E. M. Wilson and H. L. Kornberg, *Biochem. J.* **88**, 578 (1963).
32. H. von Euler, E. Adler, G. Gunther, and N. B. Das, *Z. Physiol. Chem.* **254**, 61 (1938).
33. J. G. Dewan, *Biochem. J.* **32**, 1378 (1938).
34. H. J. Strecker, *Arch. Biochem. Biophys.* **32**, 448 (1951); **46**, 128 (1953).
35. J. A. Olsen and C. B. Anfinsen, *J. Biol. Chem.* **197**, 67 (1952); **202**, 841 (1953).
36. J. E. Snoke, *J. Biol. Chem.* **223**, 271 (1956).
37. G. Frieden, *in* "The Enzymes" (P. D. Boyer, H. Lardy, and K. Myrbäck, eds.), 2nd ed., Vol. VII, Chap. 3. Academic Press, New York, 1963.
38. W. J. Wingo and J. Awapara, *J. Biol. Chem.* **187**, 267 (1950).
39. E. Roberts and S. Frankel, *J. Biol. Chem.* **187**, 55 (1950).
40. E. Roberts and S. Frankel, *J. Biol. Chem.* **188**, 789 (1951); **190**, 505 (1951).

41. R. W. Albers and R. O. Brady, *J. Biol. Chem.* **234**, 926 (1959).
42. R. W. Albers and R. A. Salvador, *Science* **128**, 359 (1958).
43. H. E. Hirsch and E. Robins, *J. Neurochem.* **9**, 63 (1962).
44. H. Weinstein, E. Roberts, and T. Kakefunada, *Biochem. Pharmacol.*, **12**, 503 (1965).
45. B. Siskin, K. Sano, and E. Roberts, *J. Biol. Chem.* **236**, 503 (1961).
46. C. J. van den Berg, G. M. J. van Kempen, J. P. Schade, and H. Veldstra, *J. Neurochem.* **12**, 863 (1965).
47. K. A. C. Elliott and H. H. Jasper, *Physiol. Rev.* **39**, 383 (1959).
48. D. R. Curtis and J. C. Watkins, *J. Neurochem.* **6**, 117 (1960).
49. E. A. Kravitz and D. D. Potter, *J. Neurochem.* **12**, 323 (1965).
50. N. F. Shatunova and I. A. Sytinsky, *J. Neurochem.* **11**, 701 (1964).
51. G. M. J. van Kempen, C. J. van den Berg, H. J. van der Helm and H. Veldstra, *J. Neurochem.* **12**, 581 (1965).
52. E. Roberts and D. G. Simonson, *Biochem. Pharmacol.* **12**, 113 (1963).
53. E. G. Gale, *Advan. Enzymol.* **6**, 1 (1946).
54. R. Shukuya and G. W. Schwert, *J. Biol. Chem.* **235**, 1649, 1653, 1658 (1960).
55. V. A. Najjar and J. Fisher, *J. Biol. Chem.* **206**, 215 (1954).
56. O. Schalles, *in* "The Enzymes" (J. B. Sumner and K. Myrbäck, eds.), Vol. II, Part 1, p. 233. Academic Press, New York, 1951.
57. J. E. Smith and E. R. Waygood, *Can. J. Biochem. Physiol.* **39**, 1055 (1961).
58. M. Ohno and K. Okunuki, *J. Biochem.* (*Tokyo*) **51**, 313 (1962).
59. S. P. Bessman, J. Rossen, and E. C. Layne, *J. Biol. Chem.* **201**, 385 (1953).
60. E. Roberts and H. M. Bergoff, *J Biol. Chem.* **201**, 393 (1953).
61. C. F. Baxter and E. Roberts, *J. Biol. Chem.* **233**, 1135 (1958).
62. F. N. Pitts, Jr., C. Quick, and E. Robins, *J. Neurochem.* **12**, 93 (1965).
63. E. Roberts, P. Ayengar, and I. Posner, *J. Biol. Chem.* **203**, 195 (1953).
64. F. F. Noe and W. J. Nickerson, *J. Bacteriol.* **76**, 674 (1958).
65. E. M. Scott and W. B. Jacoby, *Science* **128**, 361 (1958).
66. E. M. Scott and W. B. Jacoby, *J. Biol. Chem.* **234**, 932 (1959).
67. R. O. D. Dixon and L. Fowden, *Ann. Botany* (*London*) **25**, 513 (1961).
68. R. Pietruszko and L. Fowden, *Ann. Botany* (*London*) **25**, 491 (1961).
69. R. W. Albers and R. A. Salvador, *Science* **128**, 359 (1958).
70. J. K. Hardman and T. C. Stadtman, *J. Bacteriol.* **79**, 544 (1960).
71. J. K. Hardman and T. C. Stadtman, *J. Biol. Chem.* **238**, 2081, 2088 (1963).
72. J. K. Hardman, *in* "Methods in Enzymology" (E. P. Colowick and N. O. Kaplan eds.), Vol. V, p. 778. Academic Press, New York, 1962.
73. M. W. Nirenberg and W. B. Jacoby, *J. Biol. Chem.* **235**, 954 (1960).
74. S. P. Bessman and W. N. Fishbein, *Nature* **200**, 1207 (1963).
75. J. K. Hardman and T. C. Stadtman, *J. Bacteriol.* **85**, 1326 (1963).
76. H. A. Barker, "Bacterial Fermentations," CIBA lectures, John Wiley and Sons, New York, 1956.
77. H. A. Barker, *in* "The Bacteria" (J. C. Gunsalus and R. Y. Stanier, eds.) Vol. III, Chap. 3. Academic Press, New York, 1961.
78. H. A. Barker, F. Suzuki, A. Iodice, and V. Rooze, *Ann. N.Y. Acad. Sci.* **112**, 644 (1964).
79. H. A. Barker, V. Rooze, F. Suzuki, and A. A. Iodice, *J. Biol. Chem.* **239**, 3260 (1964),
79a. F. Suzuki and H. A. Barker, *J. Biol. Chem.* **241**, 878 (1966).
79b. R. L. Switzer and H. A. Barker, *J. Biol. Chem.* **242**, 2658 (1967).
80. H. J. Bright and L. L. Ingraham, *Biochim. Biophys. Acta*, **44**, 586 (1960).
81. H. A. Barker, R. D. Smyth, H. J. Bright, and L. L. Ingraham, *in* "Methods in

Enzymology" (S. P. Colowick and N. O. Kaplan, eds.), Vol. V, p. 827. Academic Press, New York.

82. H. J. Bright and L. L. Ingraham, *Biochim. Biophys. Acta* **44**, 586 (1961).
83. V. R. Williams and J. Selbin, *J. Biol. Chem.* **239**, 1635 (1964).
84. H. J. Bright, L. L. Ingraham, and R. E. Lundin, *Biochim. Biophys. Acta* **81**, 576 (1964).
85. H. J. Bright, R. E. Lundin, and L. L. Ingraham, *Biochemistry* **3**, ¦224 (1964).
86. H. J. Bright and R. Silverman, *Biochim. Biophys. Acta* **81**, 175 (1964).
87. V. R. Williams and W. Y. Libano, *Biochim. Biophys. Acta* **118**, 144 (1966).
88. H. Bright, *J. Biol. Chem.* **239**, 2307 (1964).
89. H. A. Barker, R. M. Wilson, and A. Munch-Petersen, *Federation Proc.* **16**, 151 (1957).
90. H. Katsuki, N. Ariga, F. Katsuki, J. Nagai, S. Egashira, and S. Tanake, *Biochim. Biophys. Acta* **56**, 545 (1962).
91. H. A. Barker and A. H. Blair, *Biochem. Prepn.* **9**, 21 (1962).
92. W. D. McElroy and H. B. Glass (eds.), "A Symposium on Amino Acid Metabolism, McCollum-Pratt Institute, Johns Hopkins Univ., Baltimore, 1955.
93. W. Sakami, *J. Biol. Chem.* **176**, 995 (1948).
94. P. Siekevitz and D. M. Greenberg, *J. Biol. Chem.* **180**, 845 (1949).
95. L. Lascelles and D. D. Woods, *Nature* **166**, 649 (1950).
96. S. Ratner, V. Nocito, and D. E. Green, *J. Biol. Chem.* **152**, 119 (1944).
97. H. I. Nakada, *J. Biol. Chem.* **239**, 1468 (1964).
98. L. L. Campbell, *J. Bacteriol.* **71**, 81 (1956).
99. S. K. Sinha and E. A. Cossins, *Can. J. Biochem.* **43**, 495 (1965).
100. H. I. Nakada and S. Weinhouse, *J. Biol. Chem.* **204**, 531 (1953).
101. L. W. Fleming and G. W. Crosbie, *Biochim. Biophys. Acta* **43**, 139 (1960).
102. H. L. Kornberg and S. R. Elsden, *Advan. Enzymol.* **23**, 401 (1961).
103. H. L. Kornberg and A. M. Goto, *Biochem. J.* **78**, 69 (1961).
104. G. Krakow and S. Barkulis, *Biochim. Biophys. Acta* **21**, 593 (1956).
105. G. Krakow, J. A. Hayashi and S. Barkulis, *J. Bacteriol.* **81**, 509 (1961).
106. R. C. Valentine, H. Drucker, and R. S. Wolfe, *J. Bacteriol.* **87**, 241 (1964).
107. A. M. Goto and H. L. Kornberg, *Biochem. J.* **81**, 273 (1961).
108. M. Okuyama, S. Tsuiki, and G. Kikuchi, *Biochim. Biophys. Acta* **110**, 66 (1966).
109. S. Tsuiki and G. Kikuchi, *Biochim. Biophys. Acta* **64**, 514 (1962).
110. H. Kawasaki, M. Okuyama, and G. Kikuchi, *J. Biochem. (Tokyo)* **59**, 419 (1966).
111. J. Koch and E. L. R. Stockstad, *Biochem. Biophys. Res. Commun.* **23**, 585 (1966).
112. D. R. Sanadi and M. J. Bennett, *Biochim. Biophys. Acta* **39**, 367 (1960).
113. D. A. Richert, R. Amberg, and M. Wilson, *J. Biol. Chem.* **237**, 99 (1962).
114. L. Jaenicke and J. Koch, *Biochem. Z.* **336**, 432 (1962).
115. G. Kohlhaw, B. Deus, and H. Holzer, *J. Biol. Chem.* **240**, 2135 (1965).
116. W. Franke and W. deBoer, *Z. Physiol. Chem.* **314**, 70 (1959).
117. R. D. Sagers and I. C. Gunsalus, *J. Bacteriol.* **81**, 541 (1961).
118. S. M. Klein and R. D. Sagers, *J. Bacteriol.* **83**, 121 (1962).
119. S. M. Klein and R. D. Sagers, *J. Biol. Chem.* **241**, 197 (1966).
120. S. M. Klein and R. D. Sagers, *J. Biol. Chem.* **241**, 206 (1966).
121. M. Baginsky and F. M. Huennekens, *Biochem. Biophys. Res. Commun.* **23**, 600, (1966); *Arch. Biochem. Biophys.* **120**, 703 (1967).
122. S. M. Klein and R. D. Sagers, *J. Biol. Chem.* **242**, 297, 300 (1967).
123. D. Shemin, *Symp. Amino Acid Metab.*, Baltimore, 1954, Johns Hopkins Univ. McCollum-Pratt Inst., Contrib. **105**, 727 (1955).

124. A. M. Nemeth, C. S. Russell, and D. Shemin, *J. Biol. Chem.* **299**, 415 (1957).
125. K. D. Gibson, W. G. Layer, and A. Neuberger, *Biochem. J.* **70**, 71 (1958).
126. W. H. Elliott, *Biochem. J.* **74**, 478 (1960).
127. W. H. Elliott, *Nature* **183**, 1051 (1959).
128. G. Urata and S. Granick, *J. Biol. Chem.* **238**, 811 (1963).
129. W. H. Elliott, *Biochem. J.* **74**, 90 (1960).
130. F. Knoop, *Z. Physiol. Chem.* **89**, 151 (1941).
131. G. L. Haberland, F. Bruns, and K. I. Altman, *Biochem. Z.* **326**, 107 (1954).
132. F. H. Bruns and L. Fiedler, *Nature* **181**, 1533 (1958); *Biochem. Z.* **330**, 324 (1958).
133. J. B. Gilbert, *J. Am. Chem. Soc.* **76**, 4183 (1954).
134. D. Elwyn, T. Ashmore, G. F. Cahill, Jr., S. Zothe, W. Welch, and A. B. Hastings, *J. Biol. Chem.* **226**, 735 (1957).
135. L. I. Pizer, *J. Bacteriol.* **89**, 1145 (1965).
136. F. M. Huennekens and M. J. Osborn, *Advan Enzymol.* **21**, 369 (1959).
137. J. C. Rabinowitz, *Enzymes*, **2**, 185 (1960).
138. L. Schirch and M. Mason, *J. Biol. Chem.* **237**, 2578 (1962); **239**, 1032 (1963).
139. B. E. Wright, *J. Biol. Chem.* **219**, 873 (1956).
140. R. J. Hill, D. C. Hobbs, and R. E. Koeppe, *J. Biol. Chem.* **230**, 169 (1958).
141. M. L. Minthorn, Jr., G. A. Mourkides, and R. E. Koeppe, *J. Biol. Chem.* **234**, 3205 (1959).
142. O. G. Lien, Jr., and D. M. Greenberg, *J. Biol. Chem.* **195**, 637 (1952).
143. W. A. Wood and I. C. Gunsalus, *J. Biol. Chem.* **181**, 171 (1949).
144. D. E. Metzler and E. E. Snell, *J. Biol. Chem.* **198**, 353, 363 (1952).
145. H. C. Lichstein and W. W. Umbreit, *J. Biol. Chem.* **170**, 423 (1947).
146. H. C. Lichstein and J. F. Christman, *J. Biol. Chem.* **175**, 649 (1948).
147. C. Yanofsky and J. L. Reissig, *J. Biol. Chem.* **198**, 343 (1952).
148. C. Yanofsky and J. L. Reissig, *J. Biol. Chem.* **202**, 567 (1956).
149. F. W. Sayre and D. M. Greenberg, *J. Biol. Chem.* **220**, 787 (1956).
150. A. S. M. Selim and D. M. Greenberg, *J. Boil. Chem.* **234**, 1474 (1959); *Biochim. Biophys. Acta* **42**, 211 (1960).
151. A. Nagabhushanam and D. M. Greenberg, *J. Biol. Chem.* **240**, 3002 (1965).
152. F. C. Brown, J. Mallady, and J. A. Roszell, *J. Biol. Chem.* **241**, 5220 (1966).
152a. H. Nakagawa, H. Kimura and S. Miura, *Biochem. Biophys. Res. Commun.* **28**, 359 (1967)
153. S. H. Mudd, J. D. Finkelstein, F. Irreverre, and L. Laster, *Biochem. Biophys. Res. Commun.* **19**, 665 (1965).
154. D. B. Sprinson and E. Chargaff, *J. Biol. Chem.* **164**, 411 (1964).
155. H. C. Pitot and C. Peraino, *J. Biol. Chem.* **238**, PC 1910 (1963).
156. C. Peraino and H. C. Pitot, *J. Biol. Chem.* **239**, 4308 (1964).
157. M. N. Goswami and F. Chatagner, *Experientia* **22**, 370 (1966).
158. C. Peraino, R. L. Blake, and H. C. Pitot, *J. Biol. Chem.* **240**, 3039 (1965).
159. E. Ishikawa, T. Ninagawa, and M. Suda, *J. Biochem.* (*Tokyo*) **57**, 506 (1965).
160. R. A. Freedland, *Federation Proc.* **23**, 434 (1964).
161. R. H. Bottomley, H. C. Pitot, and H. P. Morris, *Cancer Res.* **23**, 392 (1963).
162. H. C. Pitot, C. Peraino, C. Lamar, Jr., and A. L. Kennan, *Proc. Natl. Acad. Sci. U.S.* **54**, 845 (1965).
163. D. F. Elliott and A. Neuberger, *Biochem. J.* **46**, 207 (1950).
164. J. S. Nishimura and D. M. Greenberg, *J. Biol. Chem.* **236**, 2684 (1961).
165. L. Davis and D. E. Metzler, *J. Biol. Chem.* **237**, 1883 (1962).
165a. W. O. McLemore and D. E. Metzler, *J. Biol. Chem.* **243**, 441 (1968).
165b. A. T. Phillips, *Biochim. Biophys. Acta* **151**, 523 (1968).

166. L. Goldstein, W. E. Knox, and E. J. Behrman, *J. Biol. Chem.* **237**, 2855 (1962).
167. H. E. Umbarger and B. Brown, *J. Bacteriol.* **73**, 105 (1957).
168. H. E. Umbarger and B. Brown, *J. Biol. Chem.* **233**, 415 (1958).
169. H. E. Umbarger, *Cold Spring Harbor Symp. Quant. Biol.* **26**, 301 (1961).
170. P. Datta and L. Prakash, *J. Biol. Chem.* **241**, 5827 (1966).
170a. C. Leitzmann and R. W. Bernlohr, *Biochim. Biophys. Acta* **151**, 449, 461 (1968).
171. J. P. Changeux, *Cold Spring Harbor Symp. Quant. Biol.* **26**, 313 (1961).
172. A. T. Phillips and W. A. Wood, *Biochem. Biophys. Res. Commun.* **15**, 530 (1964); *J. Biol. Chem.* **240**, 4703 (1965).
173. M. Hirata, M. Tokushige, A. Inagaki, and O. Hayaishi, *J. Biol. Chem.* **240**, 1711 (1965).
173a. A. Nakazawa and O. Hayaishi, *J. Biol. Chem.* **242**, 1146 (1967).
173b. A. Nakazawa, M. Tokushige, O. Hayaishi, M. Ikehara, and Y. Mizuno, *J. Biol. Chem.* **242**, 3868 (1967).
174. O. Hayaishi, M. Gefter, and H. Weissbach, *J. Biol. Chem.* **238**, 2040 (1963).
175. M. Tokushige, H. R. Whiteley, and O. Hayaishi, *Biochem. Biophys. Res. Commun.* **13**, 380 (1963).
176. H. R. Whiteley and O. Hayaishi, *Biochem. Biophys. Res. Commun.* **14**, 143 (1964).
176a. A. Vanquickenborne and A. T. Phillips, *J. Biol. Chem.* **243**, 1312 (1968).
177. A. E. Braunstein and G. Y. Vilenkina, *Dokl. Acad. Nauk. SSSR* **66**, 243 (1949).
178. H. L. Metzler and D. B. Sprinson, *J. Biol. Chem.* **197**, 461 (1952).
179. O. G. Lien, Jr., and D. M. Greenberg, *J. Biol. Chem.* **200**, 367 (1953).
180. F. C. Chao, C. C. Delwiche, and D. M. Greenberg, *Biochim. Biophys. Acta* **10**, 103 (1953).
181. M. A. Karasek and D. M. Greenberg, *J. Biol. Chem.* **227**, 191 (1957).
182. S. C. Lin and D. M. Greenberg, *J. Gen. Physiol.* **38**, 181 (1954).
182a. R. H. Dainty, *Biochem. J.* **104**, 46P (1967).
183. L. I. Malkin and D. M. Greenberg, *Biochim. Biophys. Acta* **85**, 117 (1964).
184. M. L. Green and W. H. Elliott, *Biochem. J.* **92**, 537 (1964).
185. W. H. Elliott, *Nature* **183**, 1051 (1959).
186. A. Neuberger and G. H. Tait, *Biochim. Biophys. Acta* **41**, 164 (1960); *Biochem. J.* **84**, 317 (1962).
187. W. H. Elliott, *Nature* **185**, 467 (1960).
188. D. Hartshorne and D. M. Greenberg, *Arch. Biochem. Biophys.* **105**, 173 (1964).
188a. C. Monder, *Biochim. Biophys. Acta* **99**, 573 (1965); *J. Biol. Chem.* **242** 4603 (1967).
188b. I. J. Higgins, J. M. Turner, and A. J. Willetts, *Nature* **215**, 887 (1967).
189. W. C. Rose, W. J. Haines, D. T. Warner, and J. E. Johnson, Jr., *J. Biol. Chem.* **188**, 49 (1951).
190. S. Edlbacher, *Z. Physiol. Chem.* **157**, 106 (1926).
191. P. Györgi and H. Rothler, *Biochem. Z.* **173**, 334 (1926).
192. S. Edlbacher, *Ergeb. Enzymforsch.* **9**, 131 (1943).
193. W. J. Darby and H. B. Lewis, *J. Biol. Chem.* **146**, 225 (1942).
194. M. Konishi, *Z. Physiol. Chem.* **143**, 189 (1925).
195. Y. Sera and S. Yada, *J. Osaka City Med. Soc.* **38**, 1107 (1939).
196. V. Oyamada, *J. Biochem. (Tokyo)* **36**, 227 (1944).
197. S. Edlbacher and H. von Bidder, *Z. Physiol. Chem.* **273**, 163 (1942).
198. H. Borsook, C. L. Deasy, A. J. Haagen-Smit, G. Keighley, and P. H. Lowy, *J. Biol. Chem.* **187**, 839 (1950).
199. A. D. Iorio and L. P. Bouthillier, *Rev. Can. Biol.* **9**, 388 (1950).
200. H. Tabor and O. Hayaishi, *J. Biol. Chem.* **194**, 171 (1952).
201. H. Tabor, A. H. Mehler, O. Hayaishi, and J. White, *J. Biol. Chem.* **196**, 121 (1952).

202. B. Magasanik and H. R. Bowser, *J. Biol. Chem.* **213**, 571 (1955).
203. A. C. Walker and C. L. A. Schmidt, *Arch. Biochem.* **5**, 445 (1944).
204. B. A. Borek and H. Waelsch, *J. Biol. Chem.* **205**, 459 (1953).
205. H. Tabor, M. Silverman, A. H. Mehler, F. S. Daft, and H. Bauer, *J. Am. Chem. Soc.* **75**, 756 (1953).
206. H. Tabor and A. H. Mehler, *J. Biol. Chem.* **210**, 559 (1954).
207. J. E. Seegmiller, M. Silverman, H. Tabor, and A. H. Mehler, *J. Am. Chem. Soc.* **76**, 6205 (1954).
208. R. H. Feinberg and D. M. Greenberg, *Nature* **181**, 897 (1958); *J. Biol. Chem.* **234**, 2670 (1959).
209. H. R. B. Revel and B. Magasanik, *J. Biol. Chem.* **233**, 930 (1958).
210. A. H. Mehler and H. Tabor, *J. Biol. Chem.* **201**, 775 (1953).
211. D. A. Hall, *Biochem. J.* **51**, 499 (1952).
212. H. Tabor and A. H. Mehler, *in* "Methods in Enzymology" (S. P. Colowick and N. O. Kaplan, eds.), Vol. II, p. 228. Academic Press, New York, 1955.
213. A. Peterkofsky, *J. Biol. Chem.* **237**, 787 (1962).
213a. V. R. Williams and J. M. Heroms, *Biochim. Biophys. Acta* **139**, 214 (1967).
214. K. Matsuda, J. Itagaki, T. Wachi, and M. Uchida, *J. Biochem.* (*Tokyo*) **39**, 40 (1952).
215. M. Suda, K. Tomihata, A. Nakaya, and A. Kato, *J. Biochem.* (*Tokyo*) **40**, 257 (1953).
216. A. Peterkofsky and L. N. Mehler, *Biochim. Biophys. Acta* **73**, 159 (1963).
217. V. G. Zannoni and B. N. LaDu, *Biochem. J.* **88**, 160 (1963).
218. E. C. C. Lin, B. M. Pitt, M. Civen, and W. E. Knox, *J. Biol. Chem.* **233**, 668 (1958).
219. P. D. Spolter and R. C. Baldridge, *J. Biol. Chem.* **238**, 2071 (1963).
220. R. C. Baldridge and V. H. Auerbach, *J. Biol. Chem.* **239**, 1557 (1964).
221. D. D. Brown and M. W. Kies, *J. Am. Chem. Soc.* **80**, 6147 (1958); *J. Biol. Chem.* **234**, 3182 (1959).
222. D. R. Rao and D. M. Greenberg, *Biochim. Biophys. Acta* **43**, 404 (1960).
223. K. Freter, J. C. Rabinowitz, and B. Witkop, *Ann.* **607**, 174 (1957).
224. H. Kny and B. Witkop, *J. Am. Chem. Soc.* **81**, 6245 (1959).
225. W. G. Robinson, *Proc. 4th* Intern. Congr. Biochem., *Vienna, 1958* **4**, 91; pub. (1959–1960).
226. N. K. Gupta and W. G. Robinson, *Federation Proc.* **20**, 4 (1961).
227. S. Snyder, O. L. Silva, and M. W. Kies, *Biochem. Biophys. Res. Commun.* **5**, 165 (1961).
228. D. R. Rao and D. M. Greenberg, *J. Biol. Chem.* **236**, 1758 (1961).
229. A. Miller and H. Waelsch, *J. Biol. Chem.* **228**, 383, 397 (1957).
230. H. Tabor and L. Wyngarden, *J. Biol. Chem.* **234**, 1830 (1959).
231. K. Ichihara, S. Itagaki, Y. Suzuki, M. Uchida, and Y. Sakamoto, *J. Biochem.* (*Tokyo*) **43**, 603 (1956).
232. S. Ota, T. Wachi, M. Uchida, Y. Sakamoto, and K. Ichihara, *J. Biochem.* (*Tokyo*) **43**, 611 (1956).
233. K. Ichihara, Y. Sakamoto, N. Satani, N. Okada, S. Kakiuchi, T. Koizumi, and S. Ota, *J. Biochem.* (*Tokyo*) **43**, 797 (1956).
234. H. Hassall and D. M. Greenberg, *J. Biol. Chem.* **238**, 1423 (1963).
235. H. Hassall and D. M. Greenberg, *Biochim. Biophys. Acta* **67**, 507 (1963).
236. H. Hassall and D. M. Greenberg, *Arch. Biochem. Biophys.* **125**, 278 (1968).
237. F. G. Avis, F. Bergel, and R. C. Bray, *J. Chem. Soc.* 1100 (1955).
238. B. Payes and D. M. Greenberg, *Arch. Biochem. Biophys.* **125**, 911 (1968).
239. K. V. Rajagopalan, I. Fridovich, and P. Handler, *J. Biol. Chem.* **237**, 922 (1962).
240. K. V. Rajagopalan and P. Handler, *J. Biol. Chem.* **242**, 4097 (1967).
241. R. P. Igo and B. Mackler, *Biochim. Biophys. Acta* **44**, 310 (1960).

242. W. E. Knox, *J. Biol. Chem.* **163**, 699 (1946).
243. N. Akamatsu, Jr., *J. Biochem.* (*Tokyo*) **47**, 800 (1960).
244. H. Hassall and D. M. Greenberg, *J. Biol. Chem.* **238**, 3325 (1963).
245. R. Hakanson, *Biochem. Pharmacol.* **12**, 1289 (1963).
246. H. Weissbach, W. Lovenberg, and S. Udenfriend, *Biochim. Biophys. Acta* **50**, 177 (1961).
247. R. W. Schayer, *Am. J. Physiol.* **189**, 537 (1957); **198**, 1187 (1960)ₐ
248. A. M. Rothschild and R. W. Schayer, *Biochim. Biophys. Acta* **34**, 392 (1959).
249. J. D. Reid and D. M. Shepherd, *Life Sciences*, **1**, 5 (1963).
250. R. Kapeller-Adler, *Biochem. J.* **51**, 610 (1952); H. Swanberg, *Acta Physiol. Scand* **23**, 79 (1950).
251. W. Lovenberg, H. Weissbach, and S. Udenfriend, *J. Biol. Chem.* **237**, 89 (1962).
252. R. J. Levine, T. L. Sato, and A. Sjoerdsma, *Biochem. Pharmacol.* **14**, 139 (1965).
253. J. Rosenthaler, B. M. Guerard, E. W. Chang, and E. E. Snell, *Proc. Natl. Acad. Sci.* **54**, 152 (1965).
254. A. W. Rodwell, *J. Gen. Microbiol.* **8**, 224, 233 (1953).
255. R. Kapeller-Adler, *Federation Proc.* **24**, 757 (1965).
256. Symposium on "Catabolism of Histamine," *Federation Proc.* **24**, 757–780 (1965).
257. E. A. Zeller, *Federation Proc.* **24**, 766 (1965).
258. R. Kapeller-Adler and H. McFarlane, *Biochim. Biophys. Acta* **67**, 542 (1963).
259. D. D. Brown, R. Tomchick, and J. Axelrod, *J. Biol. Chem.* **234**, 2948 (1959).
260. P. D. Spolter and R. C. Baldridge, *J. Biol. Chem.* **238**, 2071 (1963).
261. R. Makoff and R. C. Baldridge, *Biochim. Biophys. Acta* **90**, 282 (1964).
262. H. Spolter and R. C. Baldridge, *Biochim. Biophys. Acta* **90**, 287 (1964).
263. E. C. Lin, B. M. Pitt, M. Civen, and W. E. Knox, *J. Biol. Chem.* **233**, 668 (1958).
264. M. Womack and W. C. Rose, *J. Biol. Chem.* **107**, 449 (1934).
265. A. R. Moss and R. Schoenheimer, *J. Biol. Chem.* **135**, 415 (1940).
266. S. Udenfriend and J. R. Cooper, *J. Biol. Chem.* **194**, 503 (1952).
267. C. Mitoma, *Arch. Biochem. Biophys.* **60**, 476 (1956).
268. S. Kaufman and B. Levenberg, *J. Biol. Chem.* **234**, 2683 (1959).
269. S. Kaufman, *J. Biol. Chem.* **230**, 931 (1958).
270. S. Kaufman, *Enzymes*, **1**, 373 (1963).
271. S. Kaufman, *Proc. Natl. Acad. Sci U.S.* **50**, 1085 (1963).
272. S. Kaufman, *J. Biol. Chem.* **237**, 2712 (1962); **230**, PC 2712 (1962).
273. S. Kaufman, *J. Biol. Chem.* **239**, 332 (1964).
274. D. R. Morales and D. M. Greenberg, *Biochim. Biophys. Acta* **85**, 360 (1964).
275. M. Matsubara, S. Katoh, M. A. Kino, and S. Kaufman, *Biochim. Biophys. Acta* **122**, 202 (1966).
276. H. W. Wallace, K. Moldave, and A. Meister, *Proc. Soc. Exptl. Biol. Med.* **94**, 632 (1957).
277. C. Mitoma, R. M. Auld, and S. Udenfriend, *Proc. Soc. Exptl. Biol. Med.* **94**, 634 (1957).
278. M. D. Armstrong, K. N. F. Shaw, and P. E. Wall, *J. Biol. Chem.* **218**, 293 (1956).
279. M. D. Armstrong and K. N. F. Shaw, *J. Biol. Chem.* **225**, 269 (1957).
280. K. Taniguchi and M. D. Armstrong, *J. Biol. Chem.* **238**, 4091 (1963).
281. E. Mayer, *Deut. Arch. Klin. Med.* **70**, 1443 (1901).
282. E. Friedmann, *Beitr. Chem. Physiol. Pathol.* **11**, 304 (1908).
283. B. N. LaDu and V. G. Zannoni, *Ann N.Y. Acad. Sci.* **92**, 175 (1961).
284. B. N. LaDu and D. M. Greenberg, *J. Biol. Chem.* **190**, 245 (1951).
285. B. N. LaDu and V. G. Zannoni, *J. Biol. Chem.* **217**, 777 (1955); **219**, 273 (1956).

286. W. E. Knox and L. M. Knox, *Biochem. J.* **49**, 686 (1951).
287. S. E. Hager, R. I. Gregerman, and W. E. Knox, *J. Biol. Chem.* **225**, 935 (1957).
288. W. E. Knox and B. M. Pitt, *J. Biol. Chem.* **225**, 675 (1957).
289. E. C. C. Lin, B. M. Pitt, M. Civen, and W. E. Knox, *J. Biol. Chem.* **233**, 668 (1958).
290. V. G. Zannoni and B. N. LaDu, *J. Biol. Chem.* **234** 2925 (1959).
291. V. G. Zannoni, *J. Biol. Chem.* **237**, 1172 (1962).
292. V. G. Zannoni, G. A. Jacoby, S. E. Malawista, and B. N. LaDu, *J. Biol. Chem.* **237**, 3506 (1962).
293. B. N. LaDu, Jr., and D. M. Greenberg, *Science* **117**, 111 (1953).
294. R. R. Sealock, R. L. Goodland, W. N. Summerwell, and J. M. Brierly, *J. Biol. Chem.* **196**, 761 (1952).
295. B. N. LaDu, V. G. Zannoni, L. Laster, and J. E. Seegmiller, *J. Bioi. Chem.* **230**, 251 (1958).
296. V. G. Zannoni, J. E. Seegmiller, and B. N. LaDu, *Nature* **193**, 952 (1962).
297. R. G. Ravdin and D. I. Crandall, *J. Biol. Chem.* **189**, 137 (1951).
298. W. E. Knox and S. W. Edwards, *J. Biol. Chem.* **216**, 479, 489 (1955); S. W. Edwards and W. E. Knox, *J. Biol. Chem.* **220**, 79 (1956).
299. M. Suda and Y. Takeda, *J. Biochem.* (*Tokyo*) **37**, 375 (1950).
300. M. Suda, Y. Takeda, K. Sujishi, and T. Tanaka, *J. Biochem.* (*Tokyo*) **38**, 297 (1951).
301. D. I. Crandall, *J. Biol. Chem.* **212**, 565 (1955).
302. B. Schepartz, *J. Biol. Chem.* **205**, 185 (1953).
303. K. Tokuyama, *J. Biochem.* (*Tokyo*) **46**, 1379, 1453, 1553 (1959).
304. W. G. Flamm and D. I. Crandall, *J. Biol. Chem.* **238**, 389 (1963).
305. W. M. Connors and E. Stotz, *J. Biol. Chem.* **178**, 881 (1949).
306. A. Meister and J. P. Greenstein, *J. Biol. Chem.* **175**, 573 (1948).
307. F. G. Hopkins and S. W. Cole, *J. Physiol.* (*London*) **27**, 418 (1901).
308. E. G. Wilcock and F. G. Hopkins, *J. Physiol.* (London) **35**, 88 (1906).
309. S. Lepkovsky and E. Nielson, *J. Biol. Chem.* **144**, 135 (1942); D. F. Ried, S. Lepkovsky, D. M. Bonner, and E. L. Tatum, *J. Biol. Chem.* **155**, 299 (1944).
310. W. A. Krehl, L. J. Teply, P. S. Sarma, and C. A. Elvehjem, *Science* **101**, 489 (1945).
311. C. E. Dalgliesh, *Advan. Protein Chem.* **10**, 31 (1955).
312. Y. Nishizuka and O. Hayaishi, *J. Biol. Chem.* **238**, PC 483 (1963); **238**, 3369 (1963).
313. R. Y. Stanier and O. Hayaishi, *Science*, **114**, 326 (1951).
314. C. E. Dalgliesh, *Quart. Rev.* (*London*) **5**, 227 (1951); **12**, 31 (1958).
315. J. M. Price, (ed.), *Symposium on Tryptophan Metabolism* (Div. Med. Chem.), Am. Chem. Soc., Atlantic City, N.J., Sept. 14, 1959.
316. R. K. Gholson, D. R. Rao, L. M. Henderson, R. J. Hill, and R. E. Koeppe, *J. Biol. Chem.* **230**, 179 (1958).
317. C. E. Dalgliesh and H. Tabechian, *Biochem. J.* **62**, 625 (1956).
318. R. K. Gholson, L. M. Henderson, G. A. Mourkides, R. J. Hill, and R. E. Koeppe, *J. Biol. Chem.* **234**, 96 (1959).
319. L. M. Henderson and L. V. Hankes, *J. Biol. Chem.* **222**, 1069 (1956).
320. R. K. Gholson, L. V. Hankes, and L. M. Henderson, *J. Biol. Chem.* **235**, 132 (1960).
321. Y. Nishizuka, A. Ichiyama, R. K. Gholson, and O. Hayaishi, *J. Biol. Chem.* **240**, 733 (1965).
322. Y. Kotake and T. Masayama, *Z. Physiol. Chem.* **243**, 237 (1936).
323. W. E. Knox and A. H. Mehler, *J. Biol. Chem.* **187**, 419 (1950).
324. O. Hayaishi, S. Rothberg, A. H. Mehler, and Y. Saito, *J. Biol. Chem.* **229**, 889 (1957).
325. T. Tanaka and W. E. Knox, *J. Biol. Chem.* **234**, 1162 (1959).
326. W. E. Knox and V. H. Auerbach, *J. Biol. Chem.* **214**, 307 (1955).

327. M. Given and W. E. Knox, *J. Biol. Chem.* **234**, 1787 (1959).
327a. K. Tokuyama, *Biochim. Biophys. Acta* **151**, 76 (1968).
328. O. Greengard and P. Feigelson, *J. Biol. Chem.* **236**, 1158 (1961); **237**, 1903 (1962).
329. P. Fiegelson and O. Greengard, *J. Biol. Chem.* **237**, 3714 (1962).
330. K. Tokuyama and W. E. Knox, *Biochim. Biophys. Acta* **81**, 201 (1964).
331. W. E. Knox, M. M. Piras and K. Tokuyama, *J. Biol. Chem.* **241**, 297 (1966); *Enzymol. Biol. Clin.* **7**, 1 (1966).
332. N. J. Pelleroni and R. Y. Stanier, *J. Gen. Microbiol.* **35**, 319 (1964).
333. W. E. Knox and M. Ogata, *J. Biol. Chem.* **240**, 2211 (1965).
334. O. Greengard, M. A. Smith, and G. Acs, *J. Biol. Chem.* **238**, 1548 (1963).
335. A. M. Nemeth and G. de la Haba, *J. Biol. Chem.* **237**, 1190 (1962).
336. M. M. Piras and W. E. Knox, *J. Biol. Chem.* **242**, 2952 (1967).
337. W. E. Knox and M. M. Piras, *J. Biol. Chem.* **242**, 2959 (1967).
338. R. T. Schimke, E. W. Sweeney, and C. M. Berlin, *Biochem. Biophys. Res. Commun.* **15**, 214 (1964); *J. Biol. Chem.* **240**, 4609 (1965).
339. P. Feigelson and H. Maeno, *Biochem. Biophys. Res. Commun.* **28**, 289 (1967).
340. O. Warburg, A. W. Geissler, and S. Lorenz, *Z. Physiol. Chem.* **348**, 899 (1967).
341. G. A. Marzluf, *Z. Vererbungslehre* **97**, 10 (1965).
342. K. Higuchi and O. Hayaishi, *Arch. Biochem. Biophys.* **120**, 397 (1967).
343. R. K. Gholson and J. Kori, *J. Biol. Chem.* **239**, PC 2399 (1964).
344. Y. S. Cho-Chung and H. C. Pitot, *J. Biol. Chem.* **242**, 1192 (1967).
345. E. Glassman, *Genetics* **41**, 566 (1956).
346. O. Hayaishi and R. Y. Stainer, *J. Bacteriol.* **62**, 169 (1951).
347. W. B. Jacoby, *J. Biol. Chem.* **207**, 657 (1954).
348. C. Yanofsky and D. M. Bonner, *Proc. Natl. Acad. Sci. U.S.* **36**, 167 (1950).
349. A. Butenandt, *Angew. Chem.* **61**, 262 (1949).
350. Y. Hirata, K. Nakanishi, and H. Kikkawa, *Science* **112**, 307 (1950).
351. P. W. Albert, B. T. Scheer, and H. J. Deuel, Jr., *J. Biol. Chem.* **175**, 479 (1948).
352. F. T. de Castro, J. M. Price, and R. R. Brown, *J. Am. Chem. Soc.* **78**, 2904 (1956).
353. H. Okamoto, S. Yamamoto, M. Nazaki, and O. Hayaishi, *Biochem. Biophys. Res. Commun.* **26**, 309 (1967).
354. Y. Saito, O. Hayaishi, and S. Rothberg, *J. Biol. Chem.* **229**, 921 (1957).
355. C. O. Stevens and L. M. Henderson, *J. Biol. Chem.* **234**, 1191 (1959).
356. O. Wiss and H. Fuchs, *Experientia* **6**, 1472 (1950).
357. W. B. Jacoby and D. M. Bonner, *J. Biol. Chem.* **205**, 393, 699, 709 (1953).
358. O. Hayaishi, *Symp. Amino Acid Metab., Baltimore, 1954, Johns Hopkins Univ., McCollum-Pratt Inst.*, 914 (1955).
359. A. E. Braunstein, E. V. Goryachenkova, and T. S. Pashkina, *Biokhimiya* **14**, 163 (1949).
360. O. Wiss, *Helv. Chim. Acta* **32**, 1694 (1949).
361. W. E. Knox, *Biochem. J.* **53**, 379 (1953).
362. J. B. Longenecker and E. E. Snell, *J. Biol. Chem.* **213**, 229 (1955).
363. O. Wiss and F. Weber, *Z. Physiol. Chem.* **304**, 232 (1956).
364. R. H. Decker, H. H. Kang, F. R. Leach, and L. M. Henderson, *J. Biol. Chem.* **236**, 3076 (1961).
365. C. O. Stevens and L. M. Henderson, *J. Biol. Chem.* **234**, 1188 (1959).
366. M. Iaccarino, E. Boeri, and V. Scardi, *Biochem. J.* **78**, 65 (1961).
367. G. di Prisco, A. Vescia, and E. Boeri, *Arch. Biochem. Biophys.* **95**, 400 (1961).
368. A. Vescia and G. di Prisco, *J. Biol. Chem.* **237**, 2318 (1962).
369. R. A. Mitchell, H. H. Kang, and L. M. Henderson, *J. Biol. Chem.* **238**, 1151 (1963).

370. O. Hayaishi, S. Rothberg, and A. H. Mehler, *Abstr. Am. Chem. Soc. 130th Meeting, Atlantic City, N.J.* 53C, (1956).
371. N. Ogasawara, J. E. Gander, and L. M. Henderson, *J. Biol. Chem.* **241**, 613 (1966).
371a. S. Kashiwamata, K. Nakashima, and Y. Kotake, *Biochim. Biophys. Acta* **113**, 244 (1966).
372. F. von Lingens and H. D. Heilmann, *Naturwiss.* **54**, 369 (1967).
373. O. Wiss and G. Bettendorf, *Z. Physiol. Chem.* **306**, 145 (1957).
374. H. S. Mason, *Advan. Enzymol.* **19**, 79 (1957).
375. A. H. Mehler and E. L. May, *J. Biol. Chem.* **223**, 449 (1956).
376. C. L. Long, H. N. Hill, I. M. Weinstock, and L. M. Henderson, *J. Biol. Chem.* **211**, 405 (1954).
377. A. H. Mehler, E. G. McDaniel, and J. M. Hundley, *J. Biol. Chem.* **232**, 323, 331 (1958).
378. A. H. Mehler, *J. Biol. Chem.* **218**, 241 (1956).
379. A. Ichiyama, S. Nakamura, H. Kawai, T. Honjo, Y. Nishizuka, O. Hayaishi, and S. Senoh, *J. Biol. Chem.* **240**, 740 (1965).
380. Y. Nishizuka, A. Ichiyama, S. Nakamura, and O. Hayaishi, *J. Biol. Chem.* **237**, PC 269 (1962).
381. R. K. Gholson, Y. Nishizuka, A. Ichiyama, H. Kawai, H. Nakamura, and O. Hayaishi, *J. Biol. Chem.* **237**, PC 2043 (1962).
382. R. K. Gholson, I. Ueda, N. Ogasawara, and L. M. Henderson, *J. Biol. Chem.* **239**, 1208 (1964).
383. S. Nakamura, M. Ikeda, H. Tsuji, Y. Nishizuka, and O. Hayaishi, *Biochem. Biophys. Res. Commun.* **13**, 285 (1963).
384. P. M. Packman and W. B. Jacoby, *J. Biol. Chem.* **240**, 4107 (1965); **242**, 2075 (1967).
385. J. Liebig, *Ann.* **86**, 125 (1853).
386. L. Musajo, *Atti Accad. Lincei* **21**, 368 (1935); *Gazz. Chim. ital.* **67**, 165, 171, 182 (1937).
387. S. Lepkovsky, E. Roboz, and A. J. Haagen-Smit, *J. Biol. Chem.* **149**, 195 (1943).
388. R. R. Brown and J. M. Price, *J. Biol. Chem.* **219**, 985 (1956).
389. M. Rothstein and D. M. Greenberg, *Arch. Biochem. Biophys.* **68**, 206 (1957).
390. H. Takahashi, M. Kaihara, and J. M. Price, *J. Biol. Chem.* **223**, 705 (1956).
391. J. M. Price and L. W. Dodge, *J. Biol. Chem.* **223**, 699 (1956).
392. M. Kaihara, *J. Biol. Chem.* **235**, 136 (1960).
393. O. Wiss, *Z. Physiol. Chem.* **293**, 106 (1953).
394. M. Mason, *J. Biol. Chem.* **211**, 839 (1954).
395. I. L. Miller, M. Tsuchida, and E. A. Adelberg, *J. Biol. Chem.* **203**, 205 (1953).
396. W. B. Jacoby and D. M. Bonner, *J. Biol. Chem.* **221**, 689 (1956).
397. Y. Ueno, O. Hayaishi, and R. Shukuya, *J. Biochem. (Tokyo)* **54**, 75 (1963).
398. N. Ogasawara, Y. Hagino, and Y. Kotake, *J. Biochem. (Tokyo)* **52**, 162 (1962).
399. O. Hayaishi, H. Taniuchi, M. Tashiro, Y. Yamada, and S. Kuno, *J. Am. Chem. Soc.* **81**, 3483 (1959).
400. O. Hayaishi, H. Taniuchi, M. Tashiro, and S. Kuno, *J. Biol. Chem.* **236**, 2492 (1961).
401. H. Taniuchi and O. Hayaishi, *J. Biol. Chem.* **238**, 283 (1963).
402. E. J. Behrman and T. Tanaka, *Biochem. Biophys. Res. Commun.* **1**, 257 (1959).
403. K. Horibata, H. Taniuchi, M. Tashiro, S. Kuno, O. Hayaishi, T. Sakan, K. Tokuyama and S. Senoh, *Symp. Enzyme Chem. (Tokyo)* **15**, 117 (1961).
404. T. Higashi and Y. Sakomoto, *J. Biochem. (Tokyo)* **48**, 147 (1960).
405. H. Hosokawa, H. Nakagawa, and T. Takeda, *J. Biochem. (Tokyo)* **48**, 155 (1960); **49**, 355 (1961).

406. H. Taniuchi, M. Masakazu, S. Kuno, O. Hayaishi, M. Nakajima, and N. Kurihara, *J. Biol. Chem.* **239**, 2204 (1964).
407. S. Kobayashi, S. Kuno, N. Itada, and O. Hayaishi, *Biochem. Biophys. Res. Commun.* **16**, 556 (1964).
408. O. Hayaishi and K. Hashimoto, *J. Biochem. (Tokyo)* **37**, 371 (1950).
409. K. Tokuyama, S. Katsuya, K. Asanuma, and M. Kashimura, *Med. J. Osaka Univ.* **6**, 969 (1956).
410. O. Hayaishi, M. Katagiri, and S. Rothberg, *J. Biol. Chem.* **229**, 905 (1957).
410a. A. Nakazawa, Y. Kojima and H. Taniuchi, *Biochim. Biophys. Acta* **147**, 189 (1967).
411. W. R. Sistrom and R. Y. Stanier, *J. Biol. Chem.* **210**, 821 (1954); *Nature* **174**, 513 (1954).
412. H. Taniuchi, Y. Kojima, A. Nakazawa, and O. Hayaishi, *Federation Proc.* **23**, 429 (1964).
413. T. Nakazawa, Y. Kojima, H. Fujisawa, M. Nazaka, and O. Hayaishi, *J. Biol. Chem.* **240**, PC 3224 (1965).
414. S. Dagley and D. A. Stopher, *Biochem. J.* **73**, 16P (1959).
415. Y. Kojima, N. Itada, and O. Hayaishi, *J. Biol. Chem.* **236**, 2223 (1961).
416. H. Taniuchi, Y. Kojima, F. Kanetsura, H. Ochiai, and O. Hayaishi, *Biochem. Biophys. Res. Commun.* **8**, 97 (1962).
417. M. Nozaki, H. Kagamiyama, and O. Hayaishi, *Biochem. Biophys. Res. Commun.* **11**, 65 (1963); *Biochem. Z.* **338**, 582 (1963).
418. O. Hayaishi, *Bacteriol. Rev.* **30**, 720 (1966).
419. W. C. Evans and B. S. W. Smith, *Biochem. J.* **49**, x (1951).
420. W. C. Evans, B. S. W. Smith, R. P. Linstead, and J. A. Elvidge, *Nature* **168**, 772 (1951).
421. M. Katagiri and O. Hayaishi, *Federation Proc.* **15**, 285 (1956); *J. Biol. Chem.* **226**, 439 (1957).
422. V. Erspamer, *Pharmacol. Rev.* **38**, 277 (1958); *J. Physiol. (London)* **127**, 118 (1955).
423. M. M. Rapport, A. A. Green and I. H. Page, *J. Biol. Chem.* **174**, 735 (1958); **176**, 1237, 1243 (1948).
424. I. H. Page, *Physiol. Rev.* **38**, 277 (1958).
425. S. Udenfriend, P. A. Shore, D. F. Bogdanski, H. Weissbach, and B. B. Brodie, *Recent Progr. Hormone Res.*, **13**, 1 (1957).
426. S. Udenfriend, E. Titus, H. Weissbach, and R. E. Peterson, *J. Biol. Chem.* **219**, 335 (1956).
427. J. Rensen, H. Weissbach, and S. Udenfriend, *J. Biol. Chem.* **237**, 2261 (1962).
428. D. G. Graham-Smith, *Biochim. Biophys. Acta* **86**, 176 (1964).
429. J. R. Cooper and I. Melcer, *J. Pharmacol. Exptl. Therap.* **132**, 265 (1961).
430. R. A. Freedland, I. M. Wadzinski, and H. A. Waisman, *Biochem. Biophys. Res. Commun.* **5**, 94 (1961).
431. J. Renson, H. Weissbach, and S. Udenfriend, *J. Biol. Chem.* **237**, 2261 (1962).
432. R. A. Freedland, *Biochim. Biophys. Acta* **73**, 71 (1963).
433. W. M. Gal, J. C. Armstrong, and B. Ginsburg, *J. Neurochem.* **13**, 643 (1966).
434. H. Green and J. L. Sarver, *Anal. Biochem.* **15**, 53 (1966).
435. W. Lovenberg, R. J. Levine, and A. Sjoerdsma, *Biochem. Pharmacol.* **14**, 887 (1965).
436. T. L. Sato, E. Jequier, W. Lovenberg, and A. Sjoerdsma, *European J. Pharmacol.* **1**, 18 (1967).
436a. S. Hosoda and D. Glick, *J. Biol. Chem.* **241**, 192 (1966).
437. C. Mitoma, H. Weissbach, and S. Udenfriend, *Arch. Biochem. Biophys.* **63**, 122 (1956).
438. S. Udenfriend, *Vitamins Hormones* **17**, 133 (1959).

439. J. Rensen, *Arch. Int. Physiol. Biochim.* **75**, 180 (1967).
440. C. T. Clark, H. Weissbach, and S. Udenfriend, *J. Biol. Chem.* **210**, 139 (1954).
441. S. Udenfriend, H. Weissbach, and C. T. Clark, *J. Biol. Chem.* **215**, 337 (1955).
442. J. A. Buzard and P. D. Nytch, *J. Biol. Chem.* **237**, 225 (1957).
443. E. Werle and D. Acures, *Z. Physiol. Chem.* **316**, 45 (1959).
444. A.Yuwiler, E. Geller, and S. Eiduson, *Arch. Biochem. Biophys.* **80**, 162 (1959).
445. E. Rosengren, *Acta Physiol. Scand.* **49**, 364 (1960).
446. W. Lovenberg, H. Weissbach, and S. Udenfriend, *J. Biol. Chem.* **237**, 89 (1962).
447. J. A. Oates and A. Sjoerdsma, *Proc. Soc. Exptl. Biol. Med.* **108**, 264 (1961).
448. J. Arendt, S. F. Contractor, and M. Sandler, *Biochem. Pharmacol.* **16**, 413, 419 (1967).
449. H. Blaschko, *Pharmacol. Rev.* **4**, 415 (1952).
450. H. Weissbach, B. G. Redfield, and S. Udenfriend, *J. Biol. Chem.* **229**, 953 (1957).
451. E. A. Zeller, J. Barsky, and E. R. Berman, *J. Biol. Chem.* **214**, 267 (1955).
452. F. C. Happold, *Advan. Enzymol,* **10**, 51 (1950).
453. W. A. Wood, I. C. Gunsalus, and W. W. Umbreit, *J. Biol. Chem.* **170**, 313 (1947).
454. J. A. Hoch, F. J. Simpson, and R. D. DeMoss, *Biochemistry* **5**, 2229 (1966).
455. J. A. Hoch and R. D. DeMoss, *Biochemistry* **5**, 3137 (1966).
456. W. A. Newton and E. E. Snell, *Proc. Natl. Acad. Sci. U.S.* **51**, 382 (1964).
457. W. A. Newton, Y. Morino, and E. E. Snell, *J. Biol. Chem.* **240**, 1211 (1965).
458. R. O. Burns and R. D. DeMoss, *Biochim. Biophys. Acta* **65**, 233 (1962).
459. K. G. Gopinathan and R. D. DeMoss, *Proc. Natl. Acad. Sci. U.S.* **56**, 223 (1966).
460. Y. Morino and E. E. Snell, *J. Biol. Chem.* **242**, 5591, 5602 (1967).
461. Y. Morino and E. E. Snell, *J. Biol. Chem.* **242**, 2793 (1967).
462. Y. Morino and E. E. Snell, *J. Biol. Chem.* **242**, 2800 (1967).

Carbon Catabolism of Amino Acids

Victor W. Rodwell

I. METABOLISM OF LEUCINE, VALINE, AND ISOLEUCINE

A. Leucine Catabolism

In mammalian tissues, and also in certain other life forms, the initial four reactions are similar for valine, isoleucine, and leucine catabolism (Figs. 1 and 2). Thereafter the catabolic pathways diverge with the ultimate formation of distinct amphibolic intermediates from each

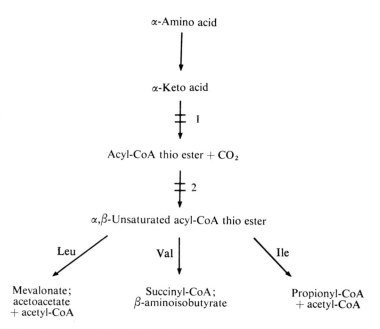

α-Amino acid

α-Keto acid

1

Acyl-CoA thio ester + CO_2

2

α,β-Unsaturated acyl-CoA thio ester

Leu   Val   Ile

| Mevalonate; acetoacetate + acetyl-CoA | Succinyl-CoA; β-aminoisobutyrate | Propionyl-CoA + acetyl-CoA |

FIG. 1. Overall aspects of branched amino acid catabolism. Double lines intersecting arrows mark sites of metabolic blocks in two rare human diseases: 1, maple sugar urine disease, a defect in catabolism of all three amino acids and, 2, isovaleric acid acidemia, a defect solely in leucine catabolism.

amino acid (Fig. 1). It therefore is convenient to discuss certain aspects of catabolism of all three branched amino acids together.

A close metabolic relationship between leucine and isovalerate was inferred early in the present century (1,2). At this time, isovalerate was proposed as a leucine catabolite from results of whole organ perfusion studies (1). Whole-animal studies using deuteroleucine and later leucine-^{14}C and isovalerate-^{14}C in the decade following the Second World War supported this view. An explanation for the long-known ketogenic effect of dietary leucine was provided when isotopic studies showed that leucine and isovalerate formed acetoacetate via an acetate-like intermediate formed from carbon atoms 2 and 3 of leucine (1 and 2 of isovalerate) with carbon atom 3 of leucine (2 of isovalerate) forming the methyl group of acetoacetate (3–5). Subsequently the isopropyl also was found to yield acetoacetate (5,6). All three carbons of the isopropyl group were shown to be incorporated as a unit into acetoacetate (6) with carbon dioxide supplying the fourth carbon atom (6,7). The enzymatic steps in these interconversions and the delineation of the

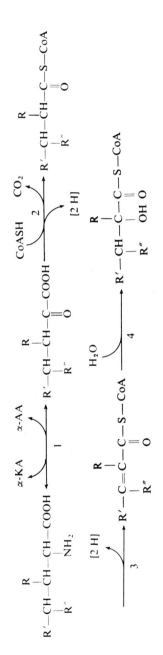

FIG. 2. Homology of the initial four steps in branched amino acid catabolism. For valine, $R = CH_3$; $R' = R'' = H$. For isoleucine, $R = R' = CH_3$; $R'' = H$. For leucine, $R = H$; $R' = R'' = CH_3$. The individual reactions are: 1, transamination; 2, oxidative decarboxylation to an acyl-CoA thio ester; 3, dehydrogenation to an α,β-unsaturated CoA thio ester; and 4, hydration to a β-hydroxyacyl-CoA thio ester.

mechanism of carbon dioxide fixation were later clarified by Coon and by Lynen and their associates (8–11). Intermediates in leucine catabolism are shown in Fig. 3.

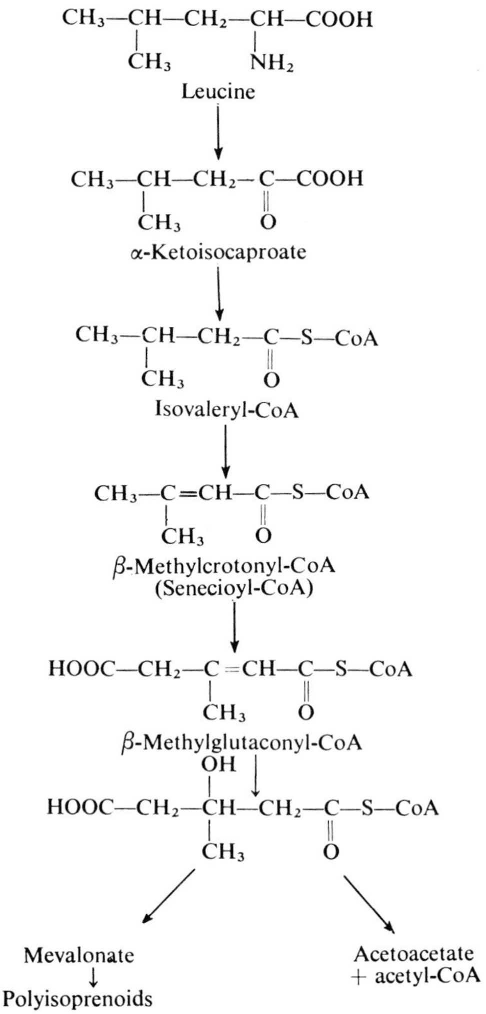

$$CH_3-CH-CH_2-CH-COOH$$
$$| \quad\quad\quad\quad |$$
$$CH_3 \quad\quad\quad NH_2$$

Leucine

↓

$$CH_3-CH-CH_2-C-COOH$$
$$| \quad\quad\quad\quad ||$$
$$CH_3 \quad\quad\quad O$$

α-Ketoisocaproate

↓

$$CH_3-CH-CH_2-C-S-CoA$$
$$| \quad\quad\quad\quad ||$$
$$CH_3 \quad\quad\quad O$$

Isovaleryl-CoA

↓

$$CH_3-C=CH-C-S-CoA$$
$$| \quad\quad\quad\quad ||$$
$$CH_3 \quad\quad O$$

β-Methylcrotonyl-CoA
(Senecioyl-CoA)

↓

$$HOOC-CH_2-C=CH-C-S-CoA$$
$$| \quad\quad\quad\quad ||$$
$$CH_3 \quad\quad O$$

β-Methylglutaconyl-CoA
$$OH \quad |$$
↓

$$HOOC-CH_2-CH-CH_2-C-S-CoA$$
$$| \quad\quad\quad\quad\quad ||$$
$$CH_3 \quad\quad\quad\quad O$$

↙   ↘

Mevalonate   Acetoacetate
↓   + acetyl-CoA
Polyisoprenoids

FIG. 3. Intermediates in leucine catabolism.

1. TRANSAMINATION

Reversible transamination of all three amino acids is catalyzed by cell extracts of most life forms (see Chapter 14). It generally is assumed that the same transaminases can serve both biosynthetic and degradative functions. Recent demonstration of a host of distinct proteins catalyzing the same reaction but fulfilling either anabolic or catabolic roles *in vivo* suggests cautious acceptance of this assumption. For example, *Bacillus subtilis* possesses a dehydrogenase catalyzing reversibly the oxidative deamination of L-leucine, and, to a lesser extent, that of valine and isoleucine (*12*).

2. OXIDATIVE DECARBOXYLATION TO ISOVALERYL-CoA

That isovaleryl-CoA rather than isovalerate was the true intermediate in leucine catabolism became apparent from the work of Coon *et al.*, who also showed that mammalian tissue extracts contained activating enzymes for conversion of isovalerate to isovaleryl-CoA (*9*). In *Proteus vulgaris*, isovaleryl-CoA formation involves intermediate formation of isovaleraldehyde (*13,14*). Although no evidence supports a similar mechanism in mammalian systems, the mammalian enzymes or enzyme catalyzing oxidation of branched α-keto acids do not appear to have been intensively studied. Indirect evidence may be interpreted to indicate either a single, or at least two, oxidative decarboxylases. In maple sugar urine disease (*15–19*), a rare genetic defect in infants, a metabolic block prevents further metabolism of the α-keto acids (Figs. 1–3). These acids accumulate in plasma and urine, imparting to urine a characteristic odor which gave the disease its name. Since all three α-keto acids accumulate, this suggests that a single enzyme catalyzes conversion of all three α-keto acids to acyl-CoA thio esters. By contrast, a recent note reports these reactions to be catalyzed by distinct mammalian enzymes for each branched α-keto acid (*20*). A partially purified enzyme inactive for α-ketoisovalerate but catalyzing oxidative decarboxylation of α-ketoisocaproate and α-keto-β-methylvalerate is known (*20a*). Physical studies (*20b*) suggest that these latter activities reside in a single protein.

Yet another familial error of leucine catabolism is the hypoglycemia induced by leucine or by isovalerate (*21,69*). In rat liver slices, overall oxidation of valine, but not of isoleucine or leucine, is inhibited by hypoglycin and by its α-keto analog (Fig. 4) (*22,23*). That a hypoglycin catabolite rather than hypoglycin itself may be the active inhibitor is suggested by (a) the greater inhibitory effect of the α-keto analog (*23*), (b) the demonstration that rat liver homogenates convert hypoglycin first to the α-keto then to the acetic acid analog and that the latter is the active inhibitor of fatty acid oxidation (*24*), and (c) the observation (*23*) that the site of inhibition of valine oxidation lies beyond isovaleryl-CoA.

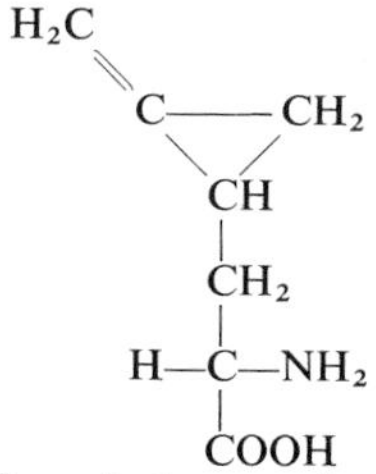

Hypoglycin; α-amino-
β-methylenecyclopropane
propionic acid

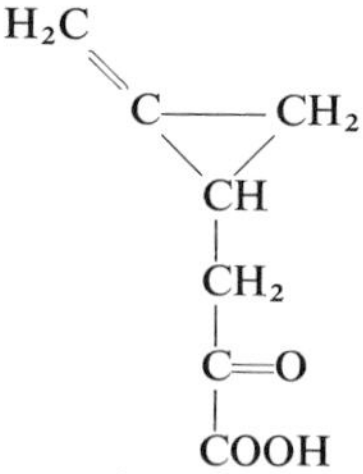

α-Keto analog of hypoglycin;
α-oxo-β-methylencyclopropane
propionic acid

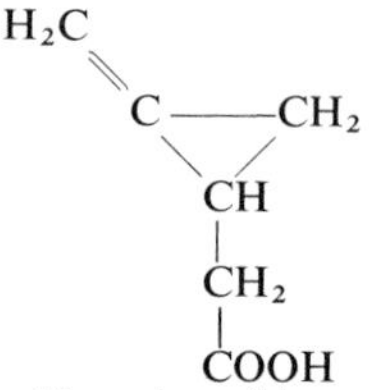

Acetic acid analog of hypoglycin;
β-methylenecyclopropane acetic acid (24)

FIG. 4. Hypoglycins.

3. DEHYDROGENATION TO β-METHYLCROTONYL-CoA

Dehydrogenation of all three acyl-CoA thio esters to α,β-unsaturated acyl-CoA thio esters was originally proposed by Coon et al. (9) by analogy with the dehydrogenation of straight-chain acyl-CoA thio esters. β-Methylcrotonyl-CoA (senecioyl-CoA), tigylyl-CoA and methacrylyl-CoA (the postulated α,β-unsaturated CoA thio ester intermediates in leucine, isoleucine, and valine catabolism, respectively) were converted to β-hydroxyisovaleryl-CoA, α-methyl-β-hydroxybutyryl-CoA, and β-hydroxyisobutyryl-CoA, respectively (9,10). That isovaleryl-CoA is oxidized by a specific oxidoreductase is suggested by the occurrence of yet another genetic defect, isovaleric acid acidemia. Particularly after injection of protein-rich food, isovalerate and N-isovalerylglycine accumulate (24a). Notably, this occurs without simultaneous accumulation of α-methylbutyrate, isobutyrate, or straight-chain fatty acids. Isovalerate presumably arises via deacylation of isovaleryl-CoA which accumulates owing to the absence of a functional, specific oxidoreductase (25).

4. CONVERSION OF β-METHYLCROTONYL-CoA TO β-HYDROXY-β-METHYLGLUTARYL-CoA

Despite the ability of tissue extracts and of crystalline crotonase (L-3-hydroxyacyl-CoA hydro-lyase, EC 4.2.1.17) (26) to convert

β-methylcrotonyl-CoA to β-hydroxyisovaleryl-CoA and the demonstrated formation of β-hydroxyisobutyrate, these reactions do not account for the ketogenic effect of leucine. Conversion of the isopropyl group of leucine to acetoacetate is accompanied by fixation of stoichiometric quantities of CO_2 (6,7). These observations became understandable with the demonstration by Bachhawat, Robinson, and Coon (10) of the ability of heart or liver extracts, and by Knappe and Lynen (27) of mycobacterial extracts, to convert β-methylcrotonyl-CoA plus CO_2 to β-hydroxy-β-methylglutaryl-CoA (HMG-CoA) and to cleave HMG-CoA to acetoacetate and acetyl-CoA (8). Formation of β-methylglutaconyl-CoA as a free intermediate between β-methylcrotonyl-CoA and HMG-CoA, proposed by Lynen et al. (27a), was demonstrated by Himes et al. (11) using purified Achromobacter β-methylcrotonyl-CoA carboxylase [3-methylcrotonyl-CoA : carbon dioxide ligase (ADP), EC 6.4.1.4]. Synthesis of β-methyl-glutaconyl-CoA occurs in two steps, formation of enzyme-biotinyl-CO_2 and transfer of CO_2 to β-methylcrotonyl-CoA (Fig. 5). These two steps were demonstrated independently

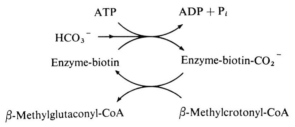

FIG. 5. Stepwise mechanism for carboxylation of β-methylcrotonyl-CoA to β-methylglutaconyl-CoA (11).

by synthesis of carboxylated enzyme by incubation with ATP, Mg^{2+}, and bicarbonate. The "active CO_2" thus formed could then be transferred to β-methylcrotonyl-CoA in the absence of ATP and Mg^{2+} (11). β-Methylglutaconyl-CoA also is an intermediate in liver where, in the absence of crotonase, β-methyl-crotonyl-CoA is carboxylated to β-methylglutaconyl-CoA, then reduced to HMG-CoA (28).

5. CLEAVAGE OF HMG-CoA TO ACETOACETATE AND ACETYL-CoA

As noted above, liver, kidney, and heart tissue cleave HMG-CoA to acetoacetate and acetyl-CoA. The partially purified pig heart cleavage enzyme (3-hydroxy-3-methylglutaryl-CoA acetoacetate-lyase, EC 4.1.3.4) (8) catalyzes HMG-CoA cleavage essentially irreversibly. A comparable activity has been detected in Pseudomonas (29,30). That HMG-CoA is a key intermediate in acetoacetate formation, not only from leucine, but also from acetoacetyl-CoA and from acetyl-CoA, was shown by Lynen *et al.* (*31*).

B. Valine Catabolism

The glycogenic nature of valine, i.e., conversion to amphibolic intermediates readily convertible into glucose and glycogen, became apparent about 25 years ago from metabolic balance studies in whole animals (*32,33*). These conclusions were later confirmed using valine-[13]C (*34,35*). These early isotope studies implicated a three-carbon acid, possibly propionate, as an intermediate in gluconeogenesis from valine (*34*). Subsequent studies utilizing valine-^{14}C implicated isobutyrate and

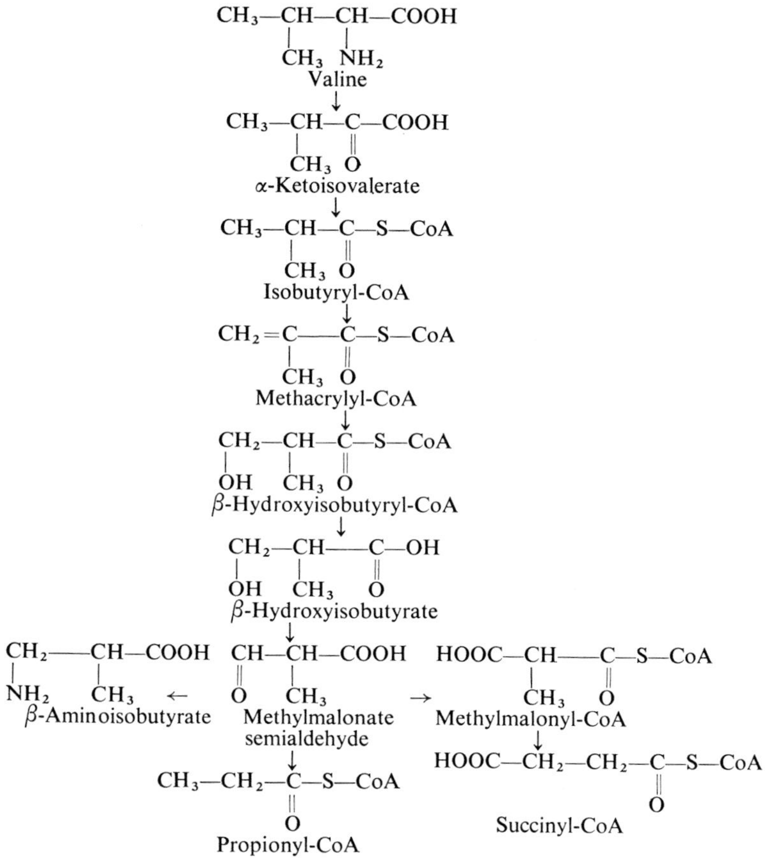

FIG. 6. Intermediates in valine catabolism.

β-hydroxyisobutyrate as intermediates (*36*). Participation of acyl-CoA thio ester intermediates was next proposed by Coon *et al.* (*9*), who soon provided direct evidence for the pathway for valine catabolism to β-aminoisobutyrate (*37,38*) depicted in Fig. 6. Comparable intermediates also appear to be involved in *Pseudomonas* (*39,40*). In *Streptomyces antibioticus*, *N*-succinyl-D-valine is formed from D-valine (*41*).

1. Transamination and (2.) Conversion to Isobutyryl-CoA

See corresponding sections for leucine (above) and Figs. 1 and 2. Initial transamination to α-ketoisovalerate and subsequent decarboxylation to isobutyrate were originally proposed from studies of isobutyrate oxidation (*42,43*) and of valine-^{14}C catabolism (*36*) by kidney and liver homogenates, respectively.

3. Dehydrogenation to Methacrylyl-CoA

Conversion of isobutyryl-CoA to methacrylyl-CoA (Figs. 2 and 6) by dialyzed liver extracts (*37*) presumably is catalyzed by the general acyl-CoA dehydrogenase [acyl-CoA: (acceptor) oxidoreductase, EC 1.3.99.3] (*44*).

4. Hydration of Methacrylyl-CoA to β-Hydroxyisobutyryl-CoA

Although hydration of methacrylyl-CoA to β-hydroxyisobutyryl-CoA can occur nonenzymically (*9,37*), the reaction is catalyzed by crystalline crotonase (*26*), a hydro-lyase of broad specificity for L-β-hydroxyacyl-CoA or -pantothenate thio esters of 4–9 carbons (*45*).

5. Deacylation of β-Hydroxyisobutyryl-CoA

Deacylation of β-hydroxyisobutyryl-CoA, originally postulated (*37*) because the CoA thio ester itself is not a substrate for the subsequent dehydrogenation reaction (*46*), is catalyzed by a pig heart deacylase (3-hydroxyisobutyryl-CoA hydrolase, EC 3.1.2.4) whose only other known substrate is β-hydroxypropionyl-CoA (*47*). The deacylase is widely distributed in animal tissues and in microorganisms (*47*).

6. Oxidation to Methylmalonate Semialdehyde

In contrast to straight-chain fatty acid catabolism (Vol. II, Chapter 8), in valine catabolism deacylation precedes oxidation (*37*). On incubation of β-hydroxyisobutyrate with an ammonium sulfate fraction from pig heart, NAD was reduced and equimolar quantities of a compound whose properties matched those of synthetic methylmalonate accumulated. The reaction was freely reversible (37). β-Hydroxyisobutyrate dehydrogenase (3-hydroxyisobutyrate : NAD oxidoreductase, EC 1.1.1.31), widely distributed in animal and plant tissues and microorganisms, is distinct from β-hydroxypropionate dehydrogenase (48). The 200-fold purified hog kidney dehydrogenase exhibits a high order of substrate and coenzyme specificity ($K_m = 1.2 \times 10^{-4}$ and 5.4×10^{-5} M for DL-β-hydroxyisobutyrate and for NAD, respectively). NADP cannot replace NAD, and neither β-hydroxyisobutyryl-CoA, β-hydroxybutyrate, nor its CoA thio ester, nor a variety of related compounds are oxidized. The reaction is freely reversible, the equilibrium constant favoring methylmalonate semialdehyde reduction (46). Although the absolute configuration of the β-hydroxyisobutyrate formed is not known, racemic β-hydroxyisobutyrate has been resolved (49) and the naturally occurring (+)-isomer shown to have the S configuration (50–52).

7. FATE OF METHYLMALONATE SEMIALDEHYDE

a. Transamination to β-Aminoisobutyrate. Transamination of methylmalonate semialdehyde, forming β-aminoisobutyrate, a normal constituent of urine (53), is catalyzed by various animal, plant, and microbial extracts (38). The crude transaminase from pig kidney is specific for α-ketoglutarate, although other non-α-amino acids (β-alanine, γ-aminobutyrate) appear to function as amino donors (38). Alternatively, methylmalonate semialdehyde may be converted to methylmalonyl-CoA, an intermediate in propionate metabolism (54) and hence to succinyl-CoA (Fig. 6).

b. Conversion to Succinyl-CoA. A second major pathway for methylmalonate semialdehyde involves oxidation to methylmalonate and conversion to methylmalonyl-CoA. Cobamide coenzyme-dependent isomerization of methylmalonyl-CoA to succinyl-CoA, catalyzed by methylmalonyl-CoA mutase (methylmalonyl-CoA CoA-carbonyl muttase, EC 5.4.99.2) is important not only for valine catabolism but also for that of propionate (54) and hence of isoleucine (Fig. 7). In cobalt deficiency, mutase activity is impaired (55,56), producing a dietary metabolic defect in ruminants which utilize large quantities of propionate as an energy source (56). The reaction resembles the isomerization of β-methylaspartate to glutamate (54,57) and favors succinyl-CoA formation (58). The 7000-fold purified sheep liver mutase (MW = 165,000; $K_m = 2.4 \times 10^{-4}$ and 6.2×10^{-5} M for L-methylmalonyl-CoA and for succinyl-CoA, respectively) contains about 2 moles of 5'-deoxyadenosyl-B_{12} per mole. The enzyme and coenzyme are readily dissociable (58). That from *Propionibacterium shermanii* (MW = 56,000; $K_m = 8.5 \times 10^{-5}$ and 3.5×10^{-5} M for L-methylmalonyl-CoA and succinyl-CoA, respectively) (58,59) could not be resolved into apoenzyme and coenzyme (59). Rearrangement occurs via an intramolecular shift of the CoA-carboxyl group (60,61).

Despite similarities in structure, coenzyme requirement, and, superficially, in reaction mechanism, isomerization of *threo*-β-methylaspartate occurs with inversion and that of L-(R)-methylmalonyl-CoA with retention of configuration about carbon 2 (50–52). Sprinson *et al.* favor the interpretation that both substrates are first decarboxylated forming an enzyme-bound enol with subsequent protonation from either side (52). Resolution of the sheep liver enzyme exposes essential SH groups which may be involved in coenzyme binding (58).

c. *Conversion to Propionyl-CoA*. *Pseudomonas aeruginosa* grown on valine as a sole source of carbon contains an enzyme that catalyzes oxidation of methylmalonate semialdehyde, in the presence of NAD and CoA, to propionyl-CoA plus CO_2 (61a).

C. Isoleucine Catabolism

As for leucine and valine, the earliest data concerning isoleucine catabolism derived from dietary studies in intact animals. These studies revealed that, unlike leucine, isoleucine was glycogenic and that it and its presumed catabolite α-methylbutyrate were only weakly ketogenic (62,63). Glycogen synthesis from isoleucine was later confirmed indirectly, using D_2O (64). The availability of specifically labeled methylbutyrate-α-^{14}C revealed that the isoleucine skeleton was cleaved to a two-carbon fragment prior to conversion to acetoacetate. This two-carbon fragment, later identified as acetyl-CoA (65), derived from carbons 3 and 4 of α-methylbutyrate (the β- and γ-carbons of isoleucine), while carbon 1 (the α-carbon of isoleucine) formed CO_2 (66). Further studies with liver slices showed formation of a propionate derivative, later identified as propionyl-CoA (65), which contained the methyl carbon of α-methylbutyrate and hence, by inference, of isoleucine also (67). The current scheme for isoleucine catabolism via CoA-thio ester intermediates (Fig. 7) was put forth by Coon *et al.* (9) coincidentally with pathways for catabolism of the other branched amino acids.

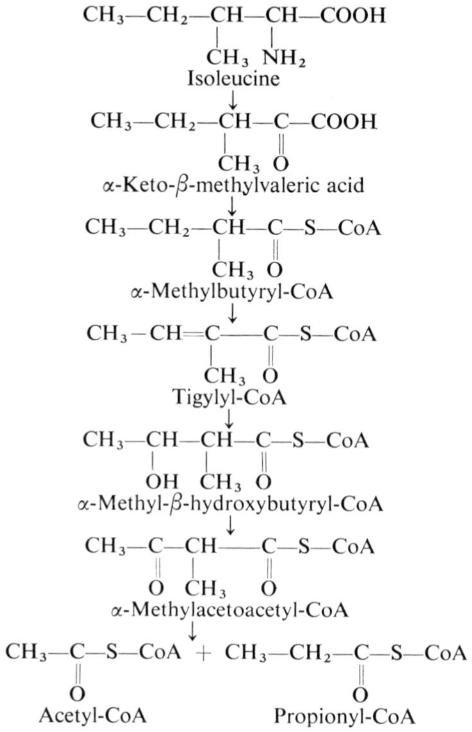

$$CH_3-CH_2-CH-CH-COOH$$
$$\underset{CH_3}{|} \quad \underset{NH_2}{|}$$

Isoleucine

↓

$$CH_3-CH_2-CH-C-COOH$$
$$\underset{CH_3}{|} \quad \underset{O}{||}$$

α-Keto-β-methylvaleric acid

↓

$$CH_3-CH_2-CH-C-S-CoA$$
$$\underset{CH_3}{|} \quad \underset{O}{||}$$

α-Methylbutyryl-CoA

↓

$$CH_3-CH=C-\!\!-\!\!-C-S-CoA$$
$$\underset{CH_3}{|} \quad \underset{O}{||}$$

Tigylyl-CoA

↓

$$CH_3-CH-CH-C-S-CoA$$
$$\underset{OH}{|} \quad \underset{CH_3}{|} \quad \underset{O}{||}$$

α-Methyl-β-hydroxybutyryl-CoA

↓

$$CH_3-C-CH-\!\!-\!\!-C-S-CoA$$
$$\underset{O}{||} \quad \underset{CH_3}{|} \quad \underset{O}{||}$$

α-Methylacetoacetyl-CoA

↓

$$CH_3-C-S-CoA \;+\; CH_3-CH_2-C-S-CoA$$
$$\underset{O}{||} \qquad\qquad\qquad \underset{O}{||}$$

Acetyl-CoA   Propionyl-CoA

FIG. 7. Intermediates in isoleucine catabolism.

1. TRANSAMINATION AND (2.) CONVERSION TO α-METHYLBUTYRYL-CoA

See corresponding sections on leucine (above). Reversible transamination of L-isoleucine and of L-alloisoleucine to α-keto-β-methylvalerate by hog heart tissue and by *Escherichia coli* and lactic acid bacteria was shown by Meister (68).

3. DEHYDROGENATION TO TIGYLYL-CoA

Dehydrogenation of α-methylbutyryl-CoA to tigylyl-CoA (the CoA-thioester of *cis*-2-methyl-2-butenoate) was predicted by analogy to straight-chain fatty acid catabolism (9). When synthetic α-methyl-butyryl-CoA is added to dialyzed liver extracts under anaerobic conditions, artificial electron acceptors are reduced. When crotonase is added, the hydroxamate of α-methyl-β-hydroxybutyrate may also be isolated (65) (see below).

4. Hydration of Tigylyl-CoA to α-Methyl-β-Hydroxybutyryl-CoA

Hydration of tigylyl-CoA, observed spectrophotometrically, is catalyzed either by dialyzed pig heart preparations or by crystalline crotonase (9,65). Since reaction mixtures treated with hydroxylamine contained α-methyl-β-hydroxybutyryl hydroxamate, α-methyl-β-hydroxybutyryl-CoA is presumably the product formed by hydration of tigylyl-CoA (65).

5. Oxidation to α-Methylacetoacetyl-CoA

Overall conversion of tigylyl-CoA to acetyl-CoA was shown by measuring citrate synthesis from tigylyl-CoA in the presence of heart or liver extract, oxaloacetate, and citrate condensing enzyme [citrate oxaloacetate-lyase (CoA-acetylating), EC 4.1.3.7]. NAD was an obligatory cofactor for citrate synthesis, and hence for acetyl-CoA synthesis. α-Methylacetoacetyl-CoA formation from tigylyl-CoA was shown indirectly by adding succinate to convert it to the free acid, decarboxylating the α-keto acids present, and chromatographing the 2,4-dinitrophenylhydrazones of the resulting ketones. Formation of methyl ethyl ketone was dependent on the presence of tigylyl-CoA (65).

6. Thiolysis of α-Methylacetoacetyl-CoA to Acetyl-CoA and Propionyl-CoA

Conversion of tigylyl-CoA to propionyl-CoA was shown by isolation both of propionyl-CoA and of its hydroxamate. Approximately stoichiometric quantities of acetyl-CoA and propionyl-CoA were produced (65). The further catabolism of propionyl-CoA has recently been reviewed (54).

II. METABOLISM OF ARGININE, ORNITHINE, AND PROLINE

Not only the biosynthesis (see Chapter 16), but also the metabolism, of arginine, ornithine, and proline are closely interrelated (Fig. 8). Some aspects of the interconversions of ornithine, citrulline, and arginine were discussed in Chapter 14. In microorganisms, distinct enzymes catalyzing these interconversions are produced when cultures are grown under conditions requiring arginine or ornithine biosynthesis from amphibolic intermediates (restrictive conditions) or whether these amino acids are present in excess (nonrestrictive conditions) as, for

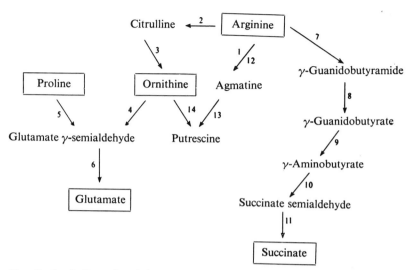

FIG. 8. Catabolism of arginine, ornithine, and proline to amphibolic intermediates. Enzymes catalyzing the numbered reactions are: (1) L-arginine amidinohydrolase (EC 3.5.3.1) or L-arginine : glycine amidinotransferase (EC 2.6.2.1); (2) L-arginine imino-hydrolase (EC 3.5.3.6); (3) carbamoylphosphate:L-ornithine carbamoyl transferase (EC 2.1.3.3); (4) L-ornithine : 2-oxoacid aminotransferase (EC 2.6.1.13); (5) L-amino acid : O_2 oxidoreductase (EC 1.4.3.2); (6) L-glutamate γ-semialdehyde : NAD oxidoreductase; (7) L-arginine oxidase (decarboxylating); (8) γ-guanidobutyramide amidohydrolase; (9) γ-guanidobutyrate amidinohydrolase; (10) γ-aminobutyrate : 2-oxoglutarate trans-aminase; (11) succinate semialdehyde : NAD oxidoreductase; (12) L-arginine carboxy-lyase (EC 4.1.1.19); (13) agmatine amidinohydrolase; and (14) L-ornithine carboxy-lyase (EC 4.1.1.17).

example, when they serve as a major source of energy (*70–72*). Although the concept of distinct " biosynthetic " and " degradative " enzymes is not new, the generality of this phenomenon is perhaps not widely recognized. It is necessary to exercise considerable caution when com-paring enzymes with identical catalytic functions obtained not only from different sources, but even from the same organisms cultured in differing nutritional environments. Certain aspects of the interconversion of arginine, ornithine, and citrulline are here reexamined with specific reference to the properties of the enzymes formed when organisms are cultured under nonrestrictive conditions for these amino acids.

A. Direct Conversion of Arginine to Ornithine

Removal of the formamidine group of arginine to form ornithine can result from transfer of the formamidine group either to water or to a suitable organic acceptor such as glycine. Group transfer to water, catalyzed by arginase (L-arginine amidinohydrolase, EC 3.5.3.1) was discussed in Chapter 14. Two examples of group transfer to an organic acceptor, creatine biosynthesis in animals and streptidine synthesis in microorganisms, are discussed briefly below.

1. CREATINE SYNTHESIS

Creatine biosynthesis in mammalian systems (73) requires two enzymes, an amidinotransferase or transamidinase (L-arginine : glycine amidinotransferase, EC 2.6.2.1) and an N-methyltransferase (S-adenosylmethionine : guanidoacetate N-methyltransferase, EC 2.1.1.2) (Fig. 9). Although primate liver, kidney, and pancreatic tissue are rich sources

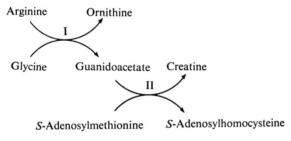

FIG. 9. Arginine metabolism by mammals: creatine biosynthesis. The catalysts are: I, L-arginine : glycine amidinotransferase (EC 2.6.2.1), and II, S-adenosylmethionine : guanidoacetate N-methyltransferase (EC 2.1.1.2).

of transamidinase, this activity is absent from liver tissue of rats, rabbits and dogs (74). For the primate transamidinases, arginine analogs (guanidoacetate, 4-guanidobutyrate, 3-guanidopropionate) serve, albeit less effectively, as formamidine donors (74). Various amines (hydroxylamine, glycine, β-alanine, α-aminobutyrate, canaline, ornithine or δ-aminovalerate) function as formamidine acceptors. A formamidinated-enzyme intermediate has been proposed by Walker (74–78). Repression of transamidinase synthesis, but not of methyltransferase synthesis, occurs in developing chick embryos injected with creatine (79,80).

2. STREPTIDINE SYNTHESIS

The streptidine portion of streptomycin (and of dihydro-, hydroxy-, and mannosidostreptomycins) contains two guanido groups (81) (Fig. 10). These appear to arise via formamidine transfer to an as yet unidentified streptomycin precursor in a reaction catalyzed by an arginine: X amidinotransferase. Synthesis of this enzyme parallels antibiotic production by certain *Streptomyces* strains (*82*).

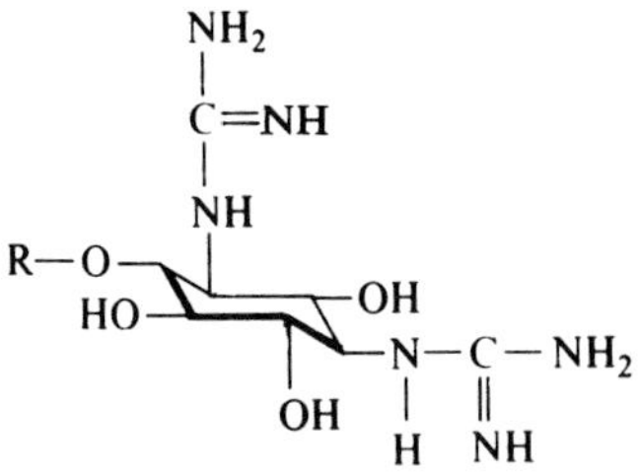

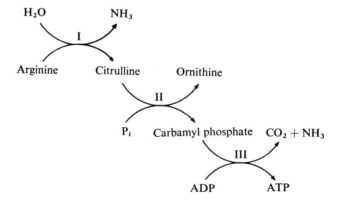

FIG. 10. The streptidine portion of the streptomycins (*81*).

B. Indirect Conversion of Arginine to Ornithine: The Arginine Dihydrolase Pathway

Various yeasts and both aerobic and anaerobic bacteria dissimilate arginine via the so-called arginine dihydrolase pathway (*71,72,83-89*). The pathway, which utilizes three enzymes, can produce one mole of ATP per mole of arginine without further catabolism of the resulting ornithine (Fig. 11). The initial step, conversion to citrulline and ammonia, is catalyzed by arginine deiminase (L-arginine iminohydrolase,

Fig. 11. Arginine catabolism by bacteria and yeast: the arginine dihydrolase pathway. Coupled formation of ATP has been demonstrated for *Mycoplasma hominis* (*88*). The enzymes of the pathway are widespread in *Mycoplasma* (*87*). I, L-arginine iminohydrolase (EC 3.5.3.6); II, carbamoylphosphate:L-ornithine carbamoyltransferase (EC 2.1.3.3); III, ATP:carbamate phosphotransferase (EC 2.7.2.2).

EC 3.5.3.6) (*90*), which appears also to catalyze conversion of canava-nine to *O*-ureidohomoserine (*91*). Purified deiminase from *Mycoplasma hominis* (MW = 73,000) (*88*) and that from *Streptococcus faecalis* (*92*) share similar kinetic properties. Substantially purified ornithine trans-carbamylase (carbamoylphosphate : L-ornithine carbamoyltransferase, EC 2.1.3.3) (see Chapter 14) has also been obtained from these organisms (*88,93*), as well as from *E. coli* (*94*). The estimated molecular weight of the *Mycoplasma* transcarbamylase (136,000) is about double that from *E. coli* (approximately 60,000). This difference may reflect distinct physiological functions, for while the transcarbamylase from *Myco-plasma* performs primarily (probably exclusively) a catabolic role, that from derepressed cells of *E. coli* serves solely a biosynthetic function. Differential regulatory properties, perhaps accompanied by novel tertiary or quaternary structures, might therefore be anticipated. Direct evidence (*71,72*) exists for two distinct ornithine transcarbamylases in *Pseudomonas*. The two izoenzymes are separable by ammonium sulfate fractionation, exhibit different pH optima, and are differentially regula-ted. The enzyme most active at alkaline pH is arginine-repressible, and hence presumably serves a biosynthetic function. The other, which appears to be arginine-inducible, presumably serves a degradative function. *In vivo*, both enzymes catalyze the reaction unidirectionally, but in opposite directions. Comparable "duplication" of this enzyme activity has also been observed in *Bacillus licheniformis* (*95*). The final reaction in the pathway, catalyzed by carbamate kinase (ATP : car-bamate phosphotransferase, EC 2.7.2.2) results in ATP synthesis. This reaction can itself account for a substantial proportion of the energy required for growth of *M. hominis* (*88*).

C. Putrescine Biosynthesis

Putrescine (1,4-diaminobutane), a precursor of the ubiquitous poly-amines spermidine and spermine (*96,97*), is synthesized from ornithine by two distinct, but metabolically related, routes in *E. coli*. Ornithine may either be directly decarboxylated to putrescine, or first converted to arginine which then is converted to putrescine via agmatine (Fig. 12). Which pathway predominates depends in part on the culture conditions employed. Morris and Pardee (*70*) have demonstrated the presence in cell-free extracts of *E. coli* K12 grown under nonspecialized conditions (i.e., minimal media) of ornithine decarboxylase (L-ornithine carboxy-lyase, EC 4.1.1.17), of arginine decarboxylase (L-arginine carboxy-lyase EC 4.1.1.19), and of agmatine amidinohydrolase. That agmatine, and not ornithine, was a free intermediate in putrescine synthesis from

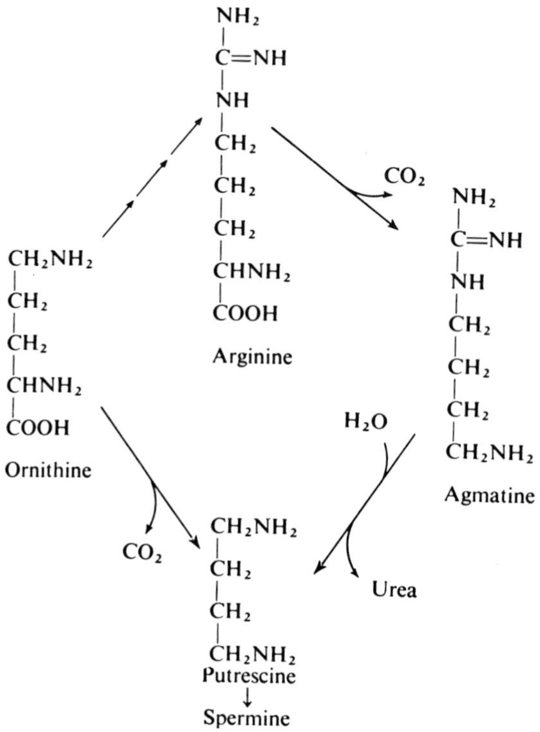

FIG. 12. Putrescine biosynthesis in *E. coli* (70).

arginine was shown by the incorporation of isotope from arginine-^{14}C into a trapping pool of agmatine, while a pool of unlabeled ornithine caused no lag in $^{14}CO_2$ production from arginine-1-^{14}C (77). The above enzymes are described as true "biosynthetic" enzymes, serving a presumably essential function in polyamine synthesis by cells grown under conditions of limiting arginine. When *E. coli* is grown under quite different and specialized conditions (rich media; acid pH), inducible "degradative" ornithine and arginine decarboxylases are formed (98). That these are distinct from their "biosynthetic" counterparts is suggested by their differential heat stability and pH optima (77,99). Morris and Pardee rationalize the existence of these two pathways for putrescine biosynthesis as follows:

"The existence of the two pathways of putrescine biosynthesis appears to provide a unique answer to the regulatory problems encountered in branched biosynthetic pathways. As can be seen from Fig. 12, the pathway of arginine biosynthesis in *E. coli* K-12 is indeed branched, the branch point occurring at ornithine. The most economical route of putrescine synthesis, from the standpoint of energy conservation, is the direct decarboxylation of ornithine. However, when repressible strains of *E. coli* are grown in the presence of arginine, the synthesis of the arginine-biosynthetic enzymes is inhibited. Since the formation of ornithine is blocked during growth on arginine, the organism can no longer synthesize putrescine from this source, but uses the now energetically favorable route from preformed arginine. Consequently, under these conditions, the conversion of arginine to putrescine is the only route of polyamine biosynthesis. This provides a reasonable explanation for the existence of the two pathways of putrescine biosynthesis."

D. Arginine Catabolism via γ-Guanidobutyramide

The initial steps in arginine catabolism in *Streptomyces griseus* (Fig. 13) closely resemble those of lysine catabolism in *Pseudomonas* (Fig. 20). Exposure of mycelia to L-arginine induces formation of the enzymes characteristic of the γ-guanidobutyrate pathway: L-arginine oxidase (decarboxylating), γ-guanidobutyramide amidohydrolase, γ-guanido-butyrate amidinohydrolase, and γ-aminobutyrate transaminase (*100*). Partially purified L-arginine oxidase (decarboxylating) is specific for

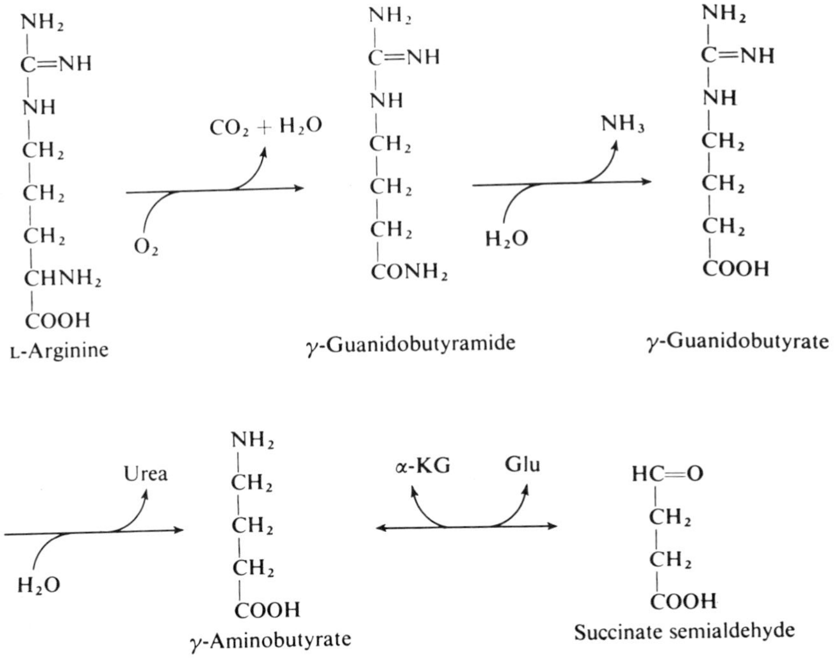

Fig. 13. Arginine catabolism in *Streptomyces griseus*: catabolism via γ-guanidobutyrate (*100–102,107–110*).

arginine ($K_m = 2.3 \times 10^{-3}$ M), and for its next higher and lower homologs, homoarginine and canavanine (*101,102*). A true oxygenase, it requires molecular oxygen as cosubstrate. γ-Guanidobutyrate amidinohydrolase is highly specific for its substrate and is distinct from arginase (*100*).

E. Other Reactions of Arginine

Other important reactions of arginine include phosphorylation to arginine-phosphate, catalyzed by arginine phosphokinase (ATP : L-arginine phosphotransferase, EC 2.7.3.3) (*103*), and conversion to the corresponding α-keto acid, catalyzed by L-amino acid oxidase [L-amino acid : oxygen oxidoreductase (deaminating), EC 1.4.3.2] (*104–106*).

F. Proline Metabolism

Liver homogenates readily oxidize both D- and L-proline (*111–114*). Oxidation catalyzed by liver or kidney (*115*) L- or D-amino acid oxidase produces an equilibrium mixture of Δ^1-pyrroline-5-carboxylate and 2-oxo-5-aminovalerate (Fig. 14). Initial oxidative attack at carbon-5,

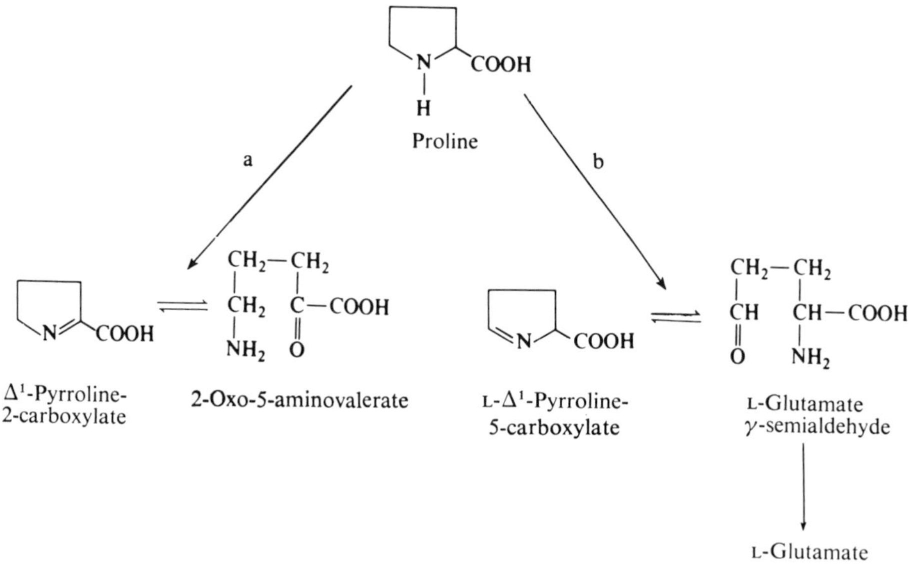

Fig. 14. Proline oxidation. Initial oxidative attack is catalyzed either by L- or D-amino acid oxidase (a) or by L-proline oxidase (b).

catalyzed by L-proline oxidase of liver (*113,114*) or of *E. coli* (*116,117*) yields an equilibrium mixture of L-Δ^1-pyrroline-5-carboxylate and L-glutamate-γ-semialdehyde (*117*) (Fig. 14). Both liver and *E. coli* L-proline oxidases are particulate and are closely associated with components of the electron transport chain. That from rat liver, which is thought to be linked to cytochrome c, also catalyzes oxidation of hydroxy-L-proline (*114*). Proline oxidase appears to fulfill a purely degradative function both in liver and in microorganisms. Synthesis of the *E. coli* oxidase is induced rather than repressed by proline, and reduction of L-Δ^1-pyrroline-5-carboxylate to proline is catalyzed by a soluble, nicotinamide nucleotide-dependent oxidoreductase (see Chapter 16). In liver (*119,120*), *E. coli* (117), and *B. subtilis* (*118*) the equilibrium mixture formed by the proline oxidase reaction is oxidized to L-glutamate by a soluble, NAD(P)-dependent oxidoreductase (*117a*). Whether the open or cyclic form represents the true substrate is not known, although Strecker (*120*), from substrate-analog inhibitor studies, suggests it to be the cyclic form and names the ox liver enzyme Δ^1-pyrroline-5-carboxylate dehydrogenase. In a partially purified state, this enzyme utilizes NAD preferentially and also catalyzes oxidation of Δ^1-pyrroline-3-hydroxy-5-carboxylate and of D-glyceraldehyde (*120*).

In anaerobic bacteria, both L- and D-proline can serve as oxidant for a Stickland reaction* whereby proline is reduced to δ-aminovalerate (*121,122*). Two enzymes, proline racemase (proline racemase, EC 5.1.1.4) and D-proline reductase [5-aminovalerate : NAD oxidoreductase (cyclizing), EC 1.4.1.6] are involved (*123*). L-Proline is first racemized to D-proline, and this then is irreversibly reduced by dithiols to δ-aminovalerate (Fig. 15). D-Proline reductase contains covalently-bound

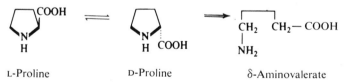

| L-Proline | D-Proline | δ-Aminovalerate |

FIG. 15. Proline catabolism in anaerobic bacteria.

pyruvate, which is essential for activity. It is postulated that the nitrogen atom of proline adds to the carbonyl group of pyruvaté, facilitating reductive cleavage (*123a*). Neither Δ^1-pyrroline-2-carboxylate nor 2-oxo-5-aminovalerate is a free intermediate. The reductase, which has been purified 200-fold (*123a*), does not attack the next higher homolog, DL-pipecolate (*123*).

* The coupled oxidation-reduction of a pair of amino acids (*121*).

III.   METABOLISM OF 4-HYDROXYPROLINE

Of the four stereoisomers of 4-hydroxyproline (Fig. 16), three occur naturally either in the free state or in peptides. All are catabolized by some life form. In mammals, D- and L-hydroxyproline are catabolized

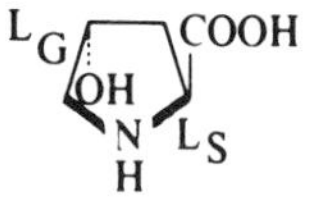

(I) 4-Hydroxy-L-proline 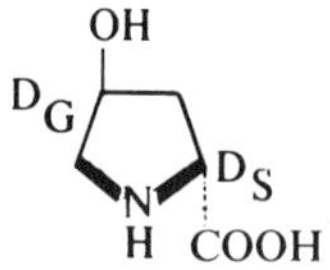   (II) 4-Hydroxy-D-proline

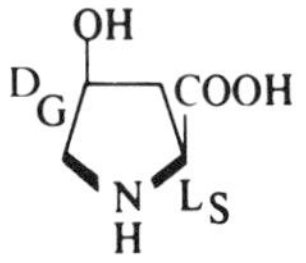

(III) *allo*-4-Hydroxy-L-proline    (IV) *allo*-4-Hydroxy-D-proline

FIG. 16. Stereoisomers of 4-hydroxyproline (*166,167*). (I) occurs in proteins; (III) occurs in the free state in the sandal tree *Santalum album* (*168*) and in the toxic polypeptide phalloidin from *Amanita phalloides* (*169*); (IV) occurs in the polypeptide antibiotic etamycin elaborated by *Streptomyces griseus* (*170*).

by independent routes. Furthermore, catabolism of L-hydroxyproline by *Pseudomonas* proceeds via intermediates different from those found in animals.

A. Hydroxyproline Catabolism in *Pseudomonas*

Both the enzymes and intermediates (Fig. 17) in 4-hydroxyproline catabolism in *Pseudomonas striata* were investigated by Adams and co-workers (*124–133*). Enzymes catalyzing these interconversions are induced in *P. striata* grown on hydroxy-L-proline as the principal source of carbon and nitrogen. Epimerization of hydroxy-L-proline (I, Fig. 16) to *allo*-hydroxy-D-proline (IV, Fig. 16) (*124,125*) is catalyzed by hydroxyproline 2-epimerase (*127*). The purified epimerase is essentially homogeneous by physical criteria, has a molecular weight of about 18,000, and contains no detectable cofactors. It is highly substrate-specific, but catalyzes equilibration of both sets of epimeric pairs, I⇌IV or II⇌III (Fig. 16). The equilibrium constant is about 1. Although experiments using T_2O show a single hydrogen atom, presumably that at carbon 2, to be labilized during epimerization, little else is known of the mechanism (127). Oxidation of *allo*-hydroxy-D-proline to a compound thought to be Δ^1-pyrroline-4-hydroxy-2-carboxylate is catalyzed by a membrane-bound, cytochrome-linked *allo*-hydroxy-D-proline oxidase of *P. striata* (125,126,133). Hydroxy-D-proline is oxidized at about one-eighth the

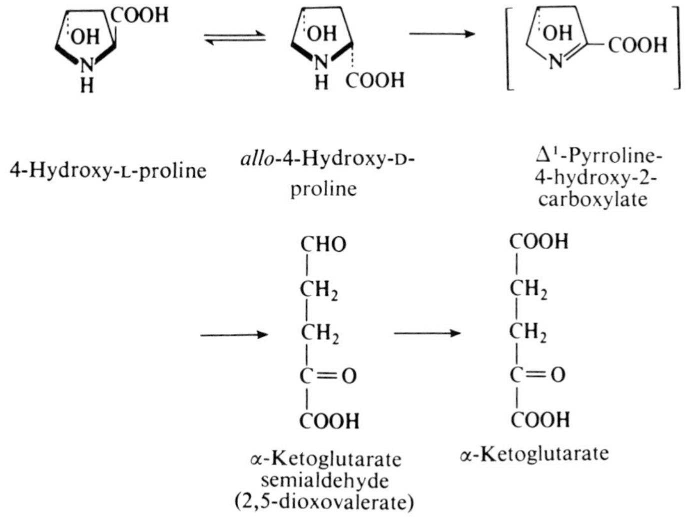

| 4-Hydroxy-L-proline | *allo*-4-Hydroxy-D-proline | Δ^1-Pyrroline-4-hydroxy-2-carboxylate |

α-Ketoglutarate semialdehyde (2,5-dioxovalerate) α-Ketoglutarate

FIG. 17. Intermediates in hydroxyproline catabolism by *Pseudomonas striata*.

rate of *allo*-hydroxy-D-proline, but the other two stereoisomers are not substrates. As with mammalian D-amino acid oxidase (134), the initial reaction product rapidly decomposes to pyrrole-2-carboxylate. Solubilization and some purification of this oxidase have been achieved by the use of EDTA (126). Deamination of Δ^1-pyrroline-4-hydroxy-2-carboxylate to α-ketoglutarate semialdehyde (2,5-dioxovalerate) is catalyzed by a highly substrate-specific deiminase. The equilibrium point of this reaction strongly favors α-ketoglutarate semialdehyde formation (130–132).

Each of the above enzymes is specifically induced in *P. striata* by growth on hydroxyproline. By contrast, cell-free extracts contain two forms of α-ketoglutarate semialdehyde dehydrogenase, the enzyme catalyzing conversion of α-ketoglutarate semialdehyde to α-ketoglutarate. They share similar kinetic properties but are distinguishable by their behavior on electrophoresis or on density gradient centrifugation.

One is induced by growth either on hydroxy-L-proline or on glutarate. The other is present in cultures grown on several other carbon sources (*128a*). The authors suggest that the latter enzyme may represent an activity of lysine rather than of hydroxyproline catabolism (*128,128a*).

B. Hydroxyproline Catabolism in Mammals

1. D-HYDROXYPROLINE

The intermediates in D-hydroxyproline degradation were investigated by Radhakrishnan and Meister (*134*) using hog and sheep kidney preparations. In the presence of catalase, hog kidney D-amino acid oxidase oxidizes both hydroxy-D-proline and *allo*-hydroxy-D-proline to an intermediate with the properties of an equilibrium mixture of Δ^1-pyrroline-4-hydroxy-2-carboxylate and 2-oxo-4-hydroxy-5-amino-valerate (Fig. 18). These intermediates rapidly decompose nonenzymic-

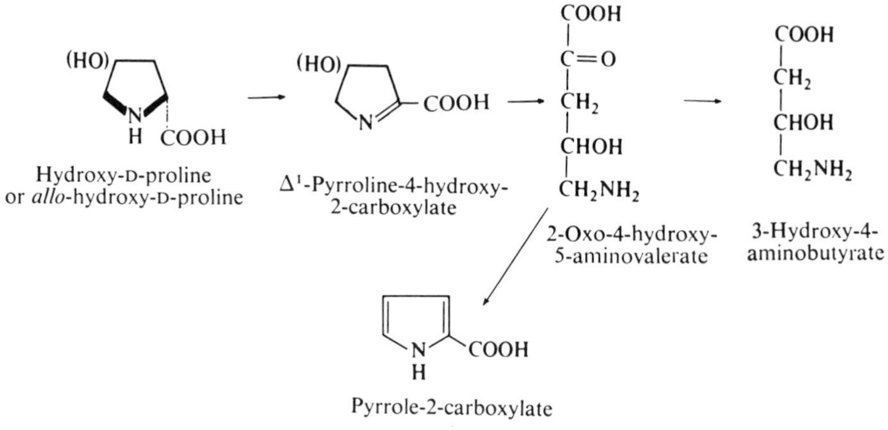

FIG. 18. Intermediates in D-hydroxyproline catabolism in animals.

ally with formation of pyrrole-2-carboxylate. In the absence of catalase, 3-hydroxy-4-aminobutyrate accumulated. Peroxide oxidation of hydroxyproline formed the same products. Pyrrole-2-carboxylate has also been identified as a urinary catabolite in rats fed DL-hydroxyproline-2-^{14}C (*135*).

2. L-HYDROXYPROLINE

Early nutritional and metabolic studies indicated that hydroxyproline was glycogenic (*136,137*) and was converted to glutamate and to alanine

(138–142). The partial oxidation of hydroxy-L-proline by kidney mito-chondria was reported by Taggart and Krakaur *(112)*. Using comparable preparations, Lang and Mayer *(143)* and later Adams *et al.* *(144,145)* demonstrated first the accumulation of γ-hydroxyglutamate semi-aldehyde, trapped as its dinitrophenylhydrazone *(143)*, and subsequently of an equilibrium mixture of L-Δ¹-pyrroline-3-hydroxy-5-carboxylate and γ-hydroxy-L-glutamate-γ-semialdehyde *(144,145)* (Fig. 19). A par-tially purified beef liver NAD(P)H-dependent oxidoreductase reduces

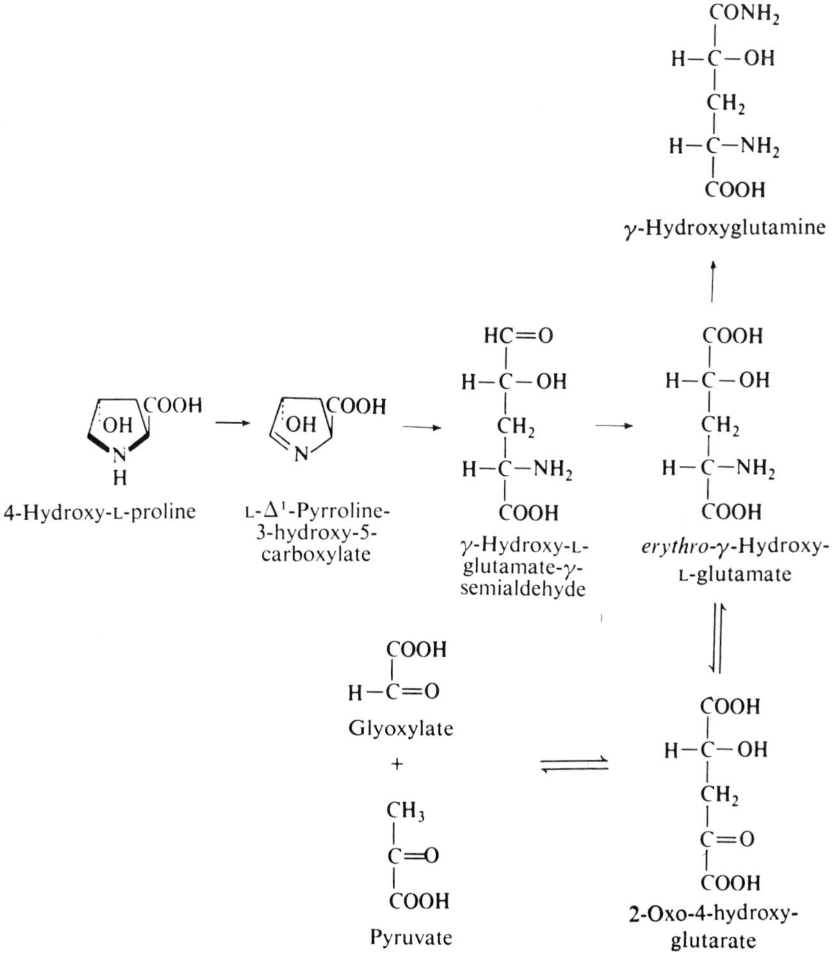

FIG. 19. Intermediates in L-hydroxyproline catabolism in animals.

this equilibrium mixture to hydroxy-L-proline (*145a*). A second partially purified beef liver oxidoreductase catalyzes NAD-dependent oxidation of the equilibrium mixture to *erythro*-γ-hydroxy-L-glutamate (*146*). The latter oxidoreductase appears to be the same enzyme that catalyzes oxidation of the equilibrium mixture of Δ^1-pyrroline-5-carboxylate and glutamate-γ-semialdehyde to glutamate (*146a*).

The further catabolism of γ-hydroxyglutamate, an amino acid first isolated from plants by Virtanen (*147*), has received the attention of several investigators (Fig. 19). Overall conversion of γ-hydroxyglutamate to alanine and glyoxylate by rat liver preparations was first shown by Dekker (*148*). γ-Hydroxyglutamate was next shown to form alanine and glycine in intact rats (*149*). The earlier observation that whole animals converted hydroxyproline to alanine was thus explained. That the three-carbon fragment resulting from γ-hydroxyglutamate cleavage might be pyruvate rather than alanine was soon suggested by Kuratomi and Fukunaga (*150–152*). These investigators showed that rat liver catalyzed (a) reversible transamination of γ-hydroxyglutamate to α-keto-γ-hydroxyglutarate, a reaction first shown in plants (*153*), and (b) reversible cleavage of α-keto-γ-hydroxyglutarate to pyruvate and glyoxylate. These observations were confirmed by Dekker *et al.* (*154–156*). The enzyme, 4-hydroxy-2-ketoglutarate aldolase, has been purified some 500-fold from rat liver, and its kinetic properties have been studied in detail (*156a*). A similar aldolase cleavage occurs also in *Pseudomonas* (*129*). γ-Hydroxyglutamate is a substrate both for glutamate dehydrogenase (*155*) and for glutamine synthetase. The latter reaction forms γ-hydroxyglutamine (*157*). Transamination of *erythro*- or *threo*-γ-hydroxy-L-glutamate is catalyzed by purified rat liver glutamate : aspartate transaminase (*158,159*). Extensively purified α-keto-γ-hydroxyglutarate aldolase from rat liver (*160,161*) shows no optical specificity for the two isomeric forms of its substrate.

C. Hydroxyproline Catabolism in Plants

Despite the widespread occurrence of hydroxyproline in plants, particularly in their cell walls (*162,163*), little is known of its metabolism in plants. Conversion to proline, a minor catabolic route in animals, appears to account for over one-half of the hydroxyproline administered to plant seedlings (*164*). Transamination of γ-hydroxyglutamate (*153*) and cleavage of γ-hydroxyglutamate to pyruvate and glyoxylate also have been demonstrated in whole leaves (*165*).

IV. LYSINE METABOLISM

Not only the biosynthesis (Chapter 16) but also the catabolism of lysine shows considerable variation from one to another life form. No fewer than four distinct catabolic routes for lysine have been proposed for animals, yeast, aerobic and anaerobic bacteria, respectively. At the time of writing, many uncertainties remain, and much work remains to be done before even the intermediates in these various pathways are established beyond all doubt.

A. Lysine Catabolism by Aerobic Bacteria

Decarboxylation of L-lysine by partially purified preparations of L-lysine decarboxylase (L-lysine carboxy-lyase, EC 4.1.1.18) was first described by Gale and co-workers (171–173). Catabolism of L-lysine in *Pseudomonas*, studied intensively in Japan, proceeds via the intermediates shown in Fig. 20. Interconversion of D- and L-lysine is catalyzed

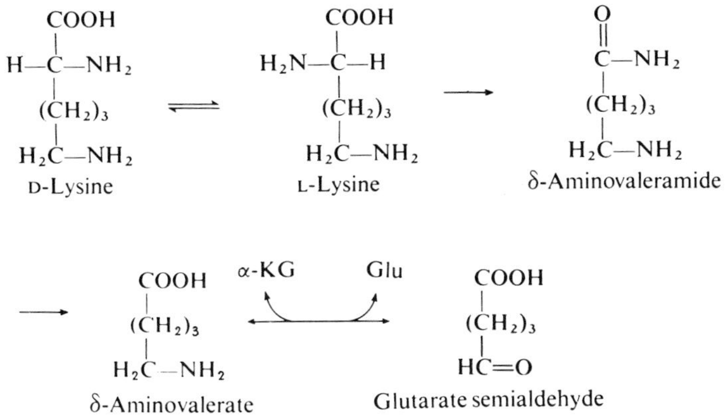

Fig. 20. Lysine catabolism via five-carbon intermediates. The δ-aminovalerate pathway of *Pseudomonas*.

by a partially purified, pyridoxal-P-dependent racemase (174). Conversion of L-lysine to δ-aminovalerate, initially proposed to proceed via α-keto-ε-aminocaproate as intermediate (175), now is known to involve prior formation of δ-aminovaleramide. Oxidation of L-lysine to δ-aminovaleramide is catalyzed by crystalline L-lysine oxygenase of *P. fluorescens* (*176,177*). This monooxygenase (*178*), which contains no heavy metal ions, has a molecular weight of about 191,000 and contains 2 moles of bound FAD per mole. FAD is required for activity and cannot be replaced by FMN. Oxidation, deamination, and decarboxylation apparently all are catalyzed by this enzyme (*176*). From data based on spectral changes observed when lysine is used as reductant under anaerobic conditions, Hayaishi (*178*) concludes that the α-imino analog of lysine is an intermediate. When oxygen is added, δ-aminovaleramide appears. If, however, the enzyme is first reduced, then inactivated, α-keto-ε-aminocaproate rather than δ-aminovaleramide is formed. Hayaishi therefore proposes the mechanism shown in Fig. 21. Lysine is presumed to reduce FAD, which then activates molecular oxygen. This then attacks the α-imino acid forming δ-aminovaleramide.

Formation of δ-aminovalerate, originally regarded as the product of the L-lysine oxygenase reaction (*177,179*), is catalyzed by a deamidase present in less highly purified oxidase preparations (*176*). δ-Aminovaleramide has recently been synthesized (*179a*). A highly substrate-specific δ-aminovaleramide deamidase has been purified 50-fold from cells of *P. putida* grown on L-lysine as sole source of carbon and nitrogen. Other amides, such as valeramide, butyramide, asparagine, glutamine, or glycine, are neither substrates nor inhibitors. δ-Aminovaleramide deamidase exhibits no metal ion requirements, but it is sensitive to inhibition by a variety of cations and by γ-hydroxymercuribenzoate, but not by iodoacetate, iodoacetamide, or *N*-ethylmaleimide (*179b*). Transamination of δ-aminovalerate to glutarate semialdehyde (*180*) is catalyzed by a specific δ-aminovalerate : α-ketoglutarate aminotransferase present in *Pseudomonas* acetone powder extracts (*181*). Both the oxidase and transaminase activities also are present in *Pseudomonas* acetone powder extracts (*181*) and in *P. putida* grown on L-lysine (*182*). Further metabolism of glutarate semialdehyde, while not investigated in detail, may proceed via conversion to glutaryl-CoA. Metabolism of glutaryl-CoA in *P. fluorescens* involves several coenzyme-A thio esters as intermediates (Fig. 22). Oxidative decarboxylation of glutaryl-CoA via glutaconyl-CoA to crotonyl-CoA is catalyzed by a partially purified FAD-flavoprotein which thus far has resisted resolution into dehydrogenase and decarboxylase activities. The further catabolism of crotonyl-CoA to acetyl-CoA appears to proceed via the familiar pathways of fatty acid oxidation (*183*). It is of interest that δ-aminovalerate has also been shown to transaminate (*180*) and to form glutarate in rats (*184*) and that glutarate is accumulated by a lysine-requiring yeast (*185*).

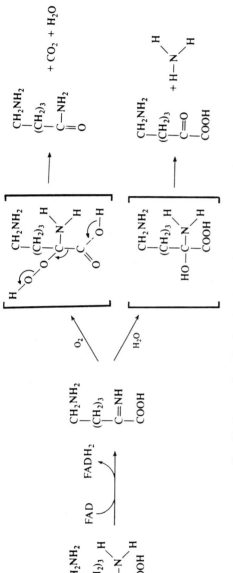

Fig. 21. Proposed mechanism for the L-lysine oxygenase reaction.

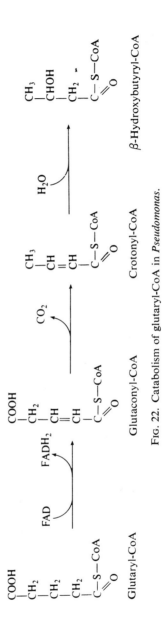

Fig. 22. Catabolism of glutaryl-CoA in *Pseudomonas*.

B. Lysine Catabolism by Anaerobic Bacteria

L-Lysine is fermented by certain Clostridia (*186–191*), by *Fusobacterium nucleatum* (*192*), and by mixed cultures of two strains of *E. coli* (*193*). In Clostridia and *E. coli*, the overall fermentation, which involves no net oxidation or reduction, is represented by:

$$\text{L-Lysine} + P_i + \text{ADP} + 2\,H_2O = \text{acetate} + \text{butyrate} + 2\,NH_3 + \text{ATP}$$

Comparable stoichiometry probably holds also for *F. nucleatum* (*194*). The isotope-labeling pattern of isolated butyrate and acetate derived from DL-lysine-2-^{14}C- and 6-^{14}C suggests two different routes for cleavage of lysine: (a) between carbons 2 and 3, and (b) between carbons 4 and 5 (*187*). Whether this indicates two catabolic routes for L-lysine or independent routes for L- and D-lysine is not known. Both isomers are fermented at equal rates by crude Clostridia extracts (*189*). Cofactor requirements for lysine fermentation include NAD, Fe^{2+}, a cobamide coenzyme such as dimethylbenzimidazolylcobamide coenzyme, ADP, and either acetyl-CoA or coenzyme A plus either acetyl or carbamyl phosphate. One mole of ATP is synthesized per mole of lysine fermented in crude cell extracts, possibly via flavin-, NAD, and disulfide-linked catalysts (*188,189*).

Elucidation of the degradative pathway has been hampered by difficulty in isolating intermediates (*189*). Quite recently, however, L-3,6-diaminohexanoate (β-lysine) was shown to be formed from L-lysine in a readily reversible reaction (*191a*). Cofactor requirements for β-L-lysine formation by Clostridia extracts include coenzyme A, a thiol, pyridoxal phosphate, NAD, and ADP. β-Lysine is fermented more rapidly than is L-lysine itself (*191a*). The next isolable intermediate is L-3,5-diaminohexanoate (*195a*). Formation of 3,5- from 3,6-diaminohexanoate

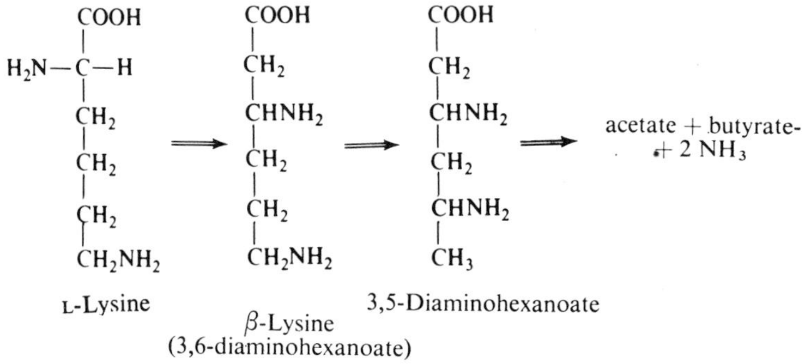

FIG. 23. Intermediates in L-lysine fermentation by Clostridia.

(β-lysine) involves migration of an amino group from C-5 to C-6 and requires a cobamide coenzyme, ATP, Mg^{2+}, pyruvate, a mercaptan, FAD or FMN, and two enzymes, one an orange protein presumed to contain a cobamide coenzyme (195,195a) (Fig. 23).

C. Lysine Catabolism by Yeast

The degradation of L-lysine-6-^{14}C by whole cells and by cell-free extracts of *Hansenula saturnis* has been studied by Rothstein (196,197). This yeast, which utilizes the nitrogen, but not the carbon of lysine, accumulates ϵ-N-acetyllysine, δ-acetamido-α-hydroxycaproate, δ-acetamidovalerate, glutarate, and several as yet unidentified compounds. These observations led Rothstein to propose the following scheme for lysine degradation in yeast via acylated intermediates (Fig. 24). No evidence for pipecolate formation was obtained, and pipecolate could not replace lysine as a nitrogen source for growth. ϵ-Acetylation of lysine would prevent cyclization of the proposed intermediate ϵ-acetamido-α-ketocaproate to Δ^1-piperideine-6-carboxylate and its subsequent reduction to pipecolate. Formation of ϵ-acetoamido-α-hydroxycaproate and of δ-acetamidovalerate might occur, respectively, via reduction or via oxidative decarboxylation of the proposed intermediate ϵ-acetamido-α-ketocaproate. Since the next isolable intermediate was glutarate, the remaining intermediates all are speculative, although analogous or identical reactions occur in bacteria. Other yeasts, unlike *H. saturnis*, can utilize the carbon skeleton of lysine for growth and might degrade glutaryl-CoA via pathways similar to those of *Pseudomonas*.

D. Lysine Catabolism in Mammals

This subject has recently been reviewed by Meister (198). Mammalian catabolism of lysine has been studied primarily in rats, guinea pigs, and man. Early nutritional studies revealed that, although certain ϵ-substituted lysine derivatives can replace lysine in the diet, this is not true for any α-substituted derivative. Thus neither D-lysine (199), α-N-methyllysine, α-N-dimethyllysine (200), α-N-acetyllysine (201,202) nor α-hydroxy-ϵ-aminocaproate can replace L-lysine for growth of young rats. By contrast, either ϵ-N-acetyllysine or ϵ-N-methyllysine (202), and to a limited extent α-amino-ϵ-guanido-n-caproate (homoarginine) (203) substitute for L-lysine, although α-amino-ϵ-ureido-n-caproate (homocitrulline) (204), α-aminoadipate (204,205), pipecolate (204), or α-amino-ϵ-hydroxycaproate (206) cannot. Mammalian tissues are known to contain ϵ-lysine acylase (207) and ϵ-demethylase (202) activity which

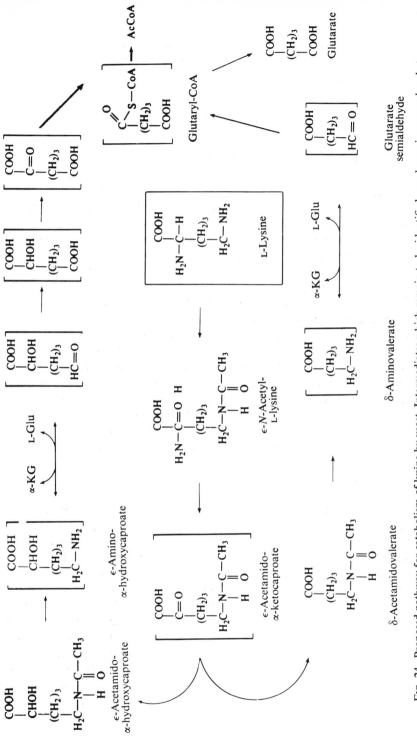

FIG. 24. Proposed pathway for catabolism of lysine by yeast. Intermediates which remain to be identified are shown in square brackets.

can account for utilization of these compounds. As noted by Meister (*198*), the inability of α-*N*-methyllysine to replace dietary lysine presumably results from its slow rate of demethylation, for some ^{15}N from ε-acetyllysine-α-^{15}N is found in the lysine of carcass protein (*208*).

Lysine appears not to participate significantly in reversible trans-amination-deamination reactions in mammalian tissues, as shown by (a) the inability of D-lysine to replace dietary L-lysine, (b) the incorporation of lysine doubly labeled with deuterium and with ^{15}N into protein without appreciable change in the D : ^{15}N ratio (*209,210*), and (c) the nonincorporation of ^{15}N from ^{15}NH$_3$ or from ^{15}N-amino acids into lysine of proteins (*211*).

The first lysine catabolite identified was α-aminoadipate. Borsook and co-workers synthesized lysine-^{14}C and demonstrated its conversion to α-aminoadipate by guinea pig liver homogenates (*212*) and by rat liver slices (*213*). Somewhat later, both α-aminoadipate (*214*) and α-ketoadipate (*215*) were identified in the urine of guinea pigs and rats fed lysine. These observations suggested the following partial metabolic sequence (Fig. 25). Shortly thereafter a new imino acid, pipecolic acid (Fig. 26),

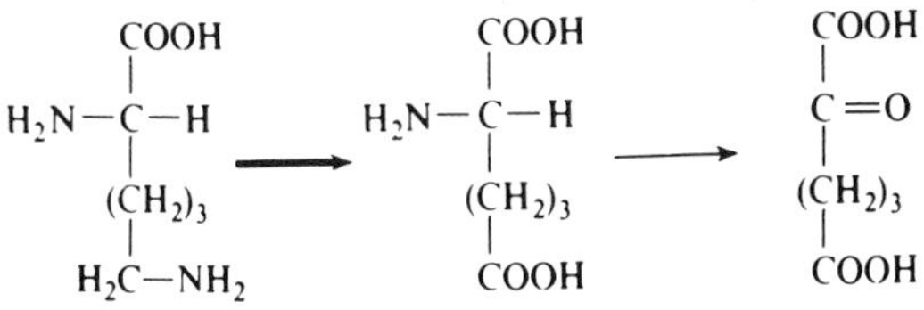

FIG. 25. Conversion of lysine to α-amino- and to α-ketoadipate in mammals.

was discovered in plants (*216–220*) and elsewhere (*221*). Almost immediately, several investigators demonstrated that pipecolate was formed from lysine in higher plants (*222,223*), *Neurospora* (*224*), and rats (*225–228*). Representative data are those of Rothstein and Miller (*225,228*), who isolated pipecolate-^{14}C from the urine of rats administered L-lysine-^{14}C and carrier pipecolate. In a key experiment (*228*)

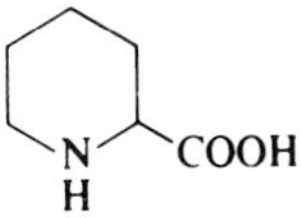

FIG. 26. Pipecolic acid.

it was shown that lysine-ε-^{15}N formed pipecolate with retention of much of the ^{15}N, indicating the ε- and not the α-carbon of lysine forms the imino nitrogen of pipecolate. This in turn suggested α-keto-ε-aminocaproate as an intermediate between lysine and pipecolate (*224,226,229–231*), a postulate supported by the observation that *Neurospora crassa* L-amino acid oxidase converts lysine to a product (presumably Δ^1-piperideine-2-carboxylate) which yields DL-pipecolate on catalytic hydrogenation (*230,231*). Supporting data included the isolation of a conjugate of α-hydroxy-ε-aminocaproate from cultures of *Neurospora* grown with L-lysine-1-^{14}C (*230*). When *N. crassa* mycelia were exposed lysine-^{14}C in the presence of a trapping pool of DL-pipecolate, only L-pipecolate became labeled with ^{14}C (*230*). These authors therefore proposed, in *Neurospora*, a stereospecific reduction of Δ^1-piperideine-2-carboxylate, the cyclic form of α-keto-ε-aminocaproate (*229,231*), to L-pipecolate. Evidence for stereospecific NAD(P)H-dependent reduction of synthetic Δ^1-piperideine-6-carboxylate to L-pipecolate by a partially purified oxidoreductase present in rat and rabbit liver was shortly provided by Meister *et al.* (*232,233*). This reductase activity is widespread in mammalian and plant tissues (*233*). A reasonable interpretation of these observations is that conversion of lysine to pipecolate in mammals proceeds as shown in Fig. 27. It must be noted, however,

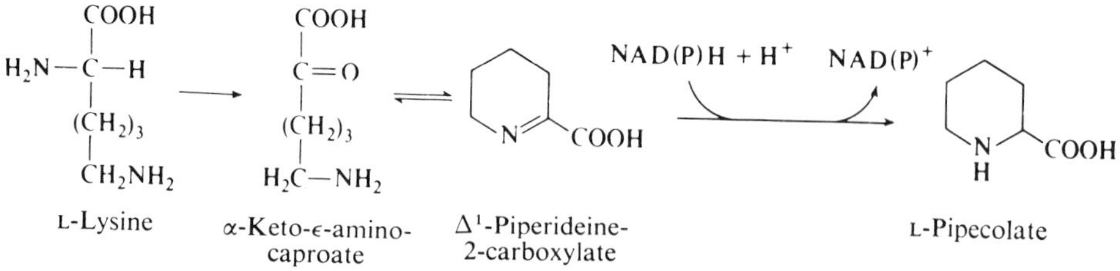

FIG. 27. Generally accepted pathway for conversion of lysine via α-keto-ε-aminocaproate to pipecolate.

that this scheme is based largely on inference, and several disturbing features exist. This was apparent to Greenberg (*234*) who noted that: "...there is substantial evidence that the α-amino group of lysine is removed by some still unknown reaction." Although mammalian tissues readily reduce Δ^1-piperideine-6-carboxylate to pipecolate, the partially purified rat liver oxidoreductase catalyzing this reaction is at present indistinguishable from L-proline : NAD(P) 2-oxidoreductase (EC 1.5.1.1). Furthermore, although formation of α-keto-ε-aminocaproate (in equilibrium with Δ^1-piperideine-6-carboxylate) from lysine has been demonstrated for *Neurospora* (*230,231*), comparable data do not, to this writer's knowledge, exist for a mammalian system. Naturally, the mechanism of Δ^1-piperideine-6-carboxylate formation also remains unclear. Although reports of lysine transamination in microorganisms exist (*235*), Δ^1-piperideine-2-carboxylate formation in mammals by a reversible process such as transamination or oxidative deamination appears to have been effectively excluded. One possibility, suggested by Paik (*236*), is that lysine first is converted to an ε-acyl derivative, then to an ε-acylated form of α-keto-ε-aminocaproate, and finally is deacylated (*236–238*) to α-keto-ε-aminocaproate. Another possibility is suggested by the recent work of Hayaishi *et al.* (*176–178*) on the mechanism of bacterial L-lysine monooxygenase (see above). In the absence of oxygen, this enzyme formed the α-imino acid of lysine, which decomposed nonenzymically to α-keto-ε-aminocaproate. Yet another possibility is suggested by Higashino *et al.* (*239,240*), who identified saccharopine as a major catabolite of lysine oxidation by rat liver mitochondria. This led these authors to suggest that the proposed scheme (Fig. 28)

Lysine → [α-Keto-ε-aminocaproate] → [Δ^1-Piperideine-2-carboxylate]
↓
Pipecolate
↓
α-Aminoadipate ← [α-Aminoadipate-δ- ← [Δ^1-Piperideine-6-carboxylate]
↓   semialdehyde]
α-Ketoadipate

FIG. 28. Proposed pathway for lysine catabolism. Intermediates not actually isolated as catabolites in mammalian systems are shown in square brackets.

for lysine degradation via pipecolate in mammalian liver may be incorrect. They propose that lysine catabolism may proceed via the same intermediates as those of lysine biosynthesis via the α-aminoadipate pathway, i.e.,

Lysine → saccharopine → α-aminoadipate-δ-semialdehyde →

α-aminoadipate → α-ketoadipate

Since Broquist (*241*) has suggested the presence of saccharopine dehydrogenase as diagnostic of the α-aminoadipate biosynthetic pathway, it would be interesting to know whether this activity is detectable in rat liver mitochondria. Such a scheme would not seem to be consistent with the observation that the ε- not the α-nitrogen of lysine forms the ring nitrogen of pipecolate. Possibly saccharopine is formed from

α-aminoadipate semialdehyde rather than from lysine. If so, its formation in liver tissue is not inconsistent with the sequence:

$$\text{Lysine} \Rightarrow \text{pipecolate} \Rightarrow \text{α-aminoadipate-δ-semialdehyde}$$
$$\text{saccharopine} \qquad \text{α-aminoadipate}$$

Fuller exposition of this intriguing observation may well clarify our
knowledge of this still uncertain area of metabolism.

The further catabolism of pipecolate in mammalian systems appears
to proceed in a manner analogous to proline catabolism (Fig. 14).
Parallel observations in rat tissues and in *Pseudomonas* adapted to
growth on pipecolate as a sole source of carbon and nitrogen suggest
that pipecolate is oxidized to Δ^1-piperideine-6-carboxylate, which in
turn is oxidized to α-aminoadipate. Pipecolate is readily oxidized to
α-aminoadipate, glutarate, and CO_2 by rat and beef liver mitochondria
(*242,243*). Presumably it is the L-isomer of pipecolate which forms
α-aminoadipate, for D-pipecolate forms Δ^1-piperideine-2-carboxylate
(*244*). Indirect support for participation of Δ^1-piperideine-6-carboxylate
(*245,246*) as an intermediate between pipecolate and α-aminoadipate is
provided by the work of Rodwell *et al.* on pipecolate catabolism in
P. putida (*247–251*) (Fig. 29). This aerobe is inducible for L-pipecolate
dehydrogenase, a membrane-bound, FAD flavoprotein associated with
the electron transport particle of this organism (*250,251*) which converts

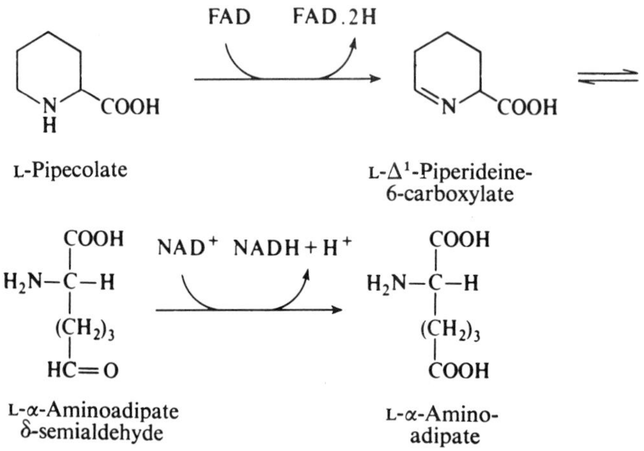

FIG. 29. Pipecolate catabolism in *Pseudomonas*.

L-pipecolate to an equilibrium mixture of Δ^1-piperideine-6-carboxylate and α-aminoadipate δ-semialdehyde (*248*). This mixture is oxidized, apparently irreversibly, to L-α-aminoadipate by a soluble, partially purified NAD(P)-dependent oxidoreductase (*249*). Cell-free extracts also catalyze transamination of α-aminoadipate to α-ketoadipate (*182*). Whether mammalian systems parallel those in *Pseudomonas* is not known. It is interesting to note, however, that the enzymes noted above are inducible in *Pseudomonas putida* as well by growth on L-lysine as on pipecolate.

As noted by Meister (*198*), a surprisingly large number of investigations have been carried out with D-lysine in mammals. In general, however, these reveal that D-lysine is not catabolized but excreted unchanged in the urine (*252*).

E. 5-Hydroxylysine

Early studies showed that both rat and turkey liver homogenates convert 5-hydroxylysine to 5-hydroxypipecolate (*253*). Turkey liver L-amino acid oxidase also converts the L-isomer to a product which on reduction yields 5-hydroxypipecolate (*254*). Difficulties in preparation of the two L-isomers as homogeneous, ^{14}C-labeled materials have in part been overcome by difference studies utilizing either 5-hydroxy-DL-lysine and 5-hydroxy-D-lysine or *allo*-5-hydroxy-DL-lysine and *allo*-5-hydroxy-D-lysine. Both normal and germ-free intact rats catabolize all four isomers (*255*). The asymmetric center at carbon 2 rather than that at carbon 5 appears of primary importance in determining overall metabolism. Thus both L (but not D) isomers are readily absorbed by perfused rat liver and are largely converted to respiratory $^{14}CO_2$ by intact rats (*255*). Intermediates implicated during oxidation of 5-hydroxy-L-lysine (and possibly also of *allo*-5-hydroxy-L-lysine) include α-ketoadipate, glutarate, and crotonate (*256*). This was shown both using a mixture of all four isomers as 6-^{14}C derivatives and using tritiated 5-hydroxy-L-lysine (*256*). Ultimate conversion of the L-isomers to glutarate was further shown by the isolation of labeled glutarate from rats injected with carrier glutarate plus either 5-hydroxy-DL-lysine-6-^{14}C or *allo*-5-hydroxy-DL-lysine-6-^{14}C (*255*). The D-isomers, by contrast, are far less readily catabolized, are not taken up to a significant extent by perfused livers and are less readily converted to CO_2 by intact rats. While the majority of injected 5-hydroxy-D-lysine or *allo*-5-hydroxy-D-lysine is excreted unchanged, significant quantities of 5-hydroxypipecolate are excreted in urine. Although the configuration of isolated 5-hydroxypipecolate about carbon 2 remains to be established, that about carbon 5 is *normal* in both cases. This suggests a metabolic transformation involving carbon 5. Hammerstedt *et al.* (*255*), however, have proposed an alternate scheme for catabolism of the hydroxy-D-lysines which involves loss of asymmetry at carbon 2 followed by stereospecific reduction to either D- or L-pipecolate derivatives (Fig. 30). This must

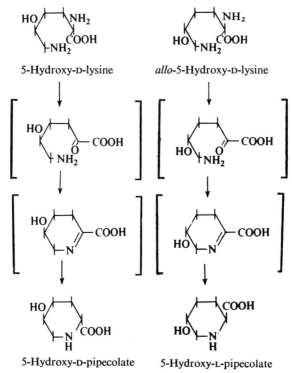

5-Hydroxy-D-pipecolate   5-Hydroxy-L-pipecolate

FIG. 30. Proposed pathway for conversion of the D-isomers of 5-hydroxylysine to 5-hydroxypipecolate (*255*). Bracketed intermediates are hypothetical. D-Isomers with either the *normal* or *allo* configuration at carbon 5 form 5-hydroxypipecolate with unknown configuration at carbon 2, but with the *normal* configuration about carbon 5.

for the present be viewed as hypothesis due to lack of information concerning the configuration about carbon 2 in isolated 5-hydroxypipecolate, the hypothetical nature of the intermediates involved, and the necessity for postulating two reductases, the one reducing a 2,6-double bond to the D-, the other to the L-isomer of 5-hydroxypipecolate.

By contrast to rats, lyophilized cells of *Achromobacter* convert all four isomers of 5-hydroxylysine to 5-hydroxypipecolate. 5-Hydroxypipecolate of unknown configuration prepared biosynthetically from 5-hydroxy-DL-lysine-6-^{14}C is slowly converted to 2-amino-5-hydroxyadipate both by *Achromobacter* and by rat liver mitochondria (*257*). Further metabolism of 2-amino-5-hydroxyadipate to the corresponding α-keto acid and to 2-hydroxyglutarate is reported (*258*).

REFERENCES

1. G. Embden, M. Salmon, and F. Schmidt, *Bei. Chem. Physiol. Pathol.* **8**, 129 (1906).
2. A. I. Ringer, E. M. Fraenkel, and L. Jonas, *J. Biol. Chem.* **14**, 525 (1913).
3. K. Bloch, *J. Biol. Chem.* **155**, 255 (1944).
4. M. J. Coon and S. Gurin, *J. Biol. Chem.* **180**, 1159 (1949).
5. I. Zabin and K. Bloch, *J. Biol. Chem.* **185**, 117 (1950).
6. M. J. Coon, *J. Biol. Chem.* **187**, 71 (1950).
7. G. W. E. Plaut and H. A. Lardy, *J. Biol. Chem.* **192**, 435 (1951).
8. B. K. Bachhawat, W. G. Robinson, and M. J. Coon, *J. Biol. Chem.* **216**, 727 (1955).
9. M. J. Coon, W. G. Robinson, and B. K. Bachhawat, *A Symp. Amino Acid Metab.*, *Baltimore, 1954, Johns Hopkins Univ.*, *McCollum-Pratt Inst. Contrib.* **105**; 431 (1955).
10. B. K. Bachhawat, W. G. Robinson, and M. J. Coon, *J. Biol. Chem.* **219**, 539 (1956).
11. R. H. Himes, D. L. Young, E. Ringelmann, and F. Lynen, *Biochem. Z.* **337**, 48 (1963).
12. B. D. Sanwal and M. W. Zink, *Arch. Biochem. Biophys.* **94**, 430 (1961).
13. S. Sasaki, *Nature* **189**, 400 (1961).
14. S. Sasaki, *J. Biochem.* (*Tokyo*) **51**, 335 (1962).
15. J. H. Menkes, P. L. Hurst, and J. M. Craig, *Pediatrics* **14**, 462 (1954).
16. J. Dancis, J. Hutzler, and M. Levitz, *Biochim. Biophys. Acta* **43**, 342 (1960).
17. J. Dancis, V. Jansen, J. Hutzler, and M. Levitz, *Biochim. Biophys. Acta* **77**, 523 (1963).
18. J. Dancis and M. Levitz, *in The Metabolic Basis of Inherited Disease*, J. B. Stanbury, D. S. Frederickson, and J. B. Wyngaarden, eds., 2nd ed., p. 353. McGraw-Hill, New York, 1966.
19. J. Dancis, J. Hutzler, and M. Levitz, *Biochim. Biophys. Acta* **78**, 85 (1963).
20. H. W. Goedde, M. Hüfner, F. Möhlenbeck, and K. G. Blume, *Biochim. Biophys. Acta* **132**, 524 (1967).
20a. J. L. Connelly, D. J. Danner, and J. A. Bowden, *J. Biol. Chem.* **243**, 1198 (1968).
20b. J. A. Bowden and J. L. Connelly, *J. Biol. Chem.* **243**, 3526 (1968).
21. W. A. Cochrane, W. W. Payne, M. J. Simpkiss, and L. I. Woolf, *J. Clin. Invest.* **35**, 411 (1956).
22. S. J. Patrick and L. C. Stewart, *Can. J. Biochem. Physiol.* **42**, 139 (1964).
23. B. I. Posner and M. S. Raben, *Biochim. Biophys. Acta* **136**, 179 (1967).
24. C. von Holt, J. Chang, M. von Holt, and H. Böhm, *Biochim. Biophys. Acta* **90**, 611 (1964).
24a. K. Tanaka and K. J. Isselbacher, *J. Biol. Chem.* **242**, 2966 (1967).
25. K. Tanaka, M. A. Budd, M. L. Efron, and K. J. Isselbacher, *Proc. Natl. Acad. Sci. U.S.* **56**, 236 (1966).
26. J. R. Stern, A. del Campillo and I. Raw, *J. Biol. Chem.* **218**, 971 (1956).
27. J. Knappe and F. Lynen, *Proc. 4th Intern. Congr. Biochem., Abstr. Communs., Vienna, Sept. 1958*, 49 (1959–1960).
27a. F. Lynen, J. Knappe, E. Lorch, G. Jütting, E. Ringelmann, and J.-P. La Chance, *Biochem. Z.* **335**, 123 (1961).
28. A. del Campillo-Campbell, E. E. Dekker, and M. J. Coon, *Biochim. Biophys. Acta* **31**, 290 (1959).

29. G. M. Fimognari and V. W. Rodwell, *Biochemistry* **4**, 2086 (1965).
30. M. A. Siddiqi and V. W. Rodwell, *J. Bacteriol.* **93**, 207 (1967).
31. F. Lynen, U. Henning, C. Bublitz, B. Sorbo, and L. Kroplin-Rueff, *Biochem. Z.* **330**, 269 (1958).
32. J. S. Butts and R. O. Sinnhuber, *J. Biol. Chem.* **139**, 963 (1941).
33. W. C. Rose, J. E. Johnson, and W. J. Haines, *J. Biol. Chem.* **145**, 679 (1942).
34. W. S. Fones, T. P. Waalkes, and J. White, *Arch. Biochem. Biophys.* **32**, 89 (1951).
35. E. A. Peterson, W. S. Fones, and J. White, *Arch. Biochem. Biophys.* **36**, 323 (1952).
36. D. S. Kinnory, Y. Takeda, and D. M. Greenberg, *J. Biol. Chem.* **212**, 385 (1955).
37. W. G. Robinson, R. Nagle, B. K. Bachhawat, F. P. Kupiecki, and M. J. Coon, *J. Biol. Chem.* **224**, 1 (1957).
38. F. P. Kupiecki and M. J. Coon, *J. Biol. Chem.* **229**, 743 (1957).
39. J. R. Sokatch, *J. Bacteriol.* **92**, 72 (1966).
40. J. E. Norton and J. R. Sokatch, *J. Bacteriol.* **92**, 116 (1966).
41. J. Barchas, E. Katz, V. Beaven, H. Weissbach, and H. Fales, *Biochemistry* **3**, 1684 (1964).
42. W. A. Atchley, *J. Biol. Chem.* **176**, 123 (1948).
43. I. Gray, P. Adams, and H. Hauptmann, *Experientia* **6**, 430 (1950).
44. F. L. Crane, S. Mii, J. G. Hauge, D. E. Green, and H. Beinert, *J. Biol. Chem.* **218**, 701 (1956).
45. J. R. Stern and A. del Campillo, *J. Biol. Chem.* **218**, 985 (1956).
46. W. G. Robinson and M. J. Coon, *J. Biol. Chem.* **225**, 511 (1957).
47. G. Rendina and M. J. Coon, *J. Biol. Chem.* **225**, 523 (1957).
48. H. Den, W. G. Robinson, and M. J. Coon, *J. Biol. Chem.* **234**, 1666 (1959).
49. J. Retey and F. Lynen, *Biochem. Biophys. Res. Commun.* **16**, 358 (1964).
50. M. Sprecher, M. J. Clark, and D. B. Sprinson, *Biochem. Biophys. Res. Commun.* **15**, 581 (1964).
51. M. Sprecher, R. L. Switzer, and D. B. Sprinson, *J. Biol. Chem.* **241**, 864 (1966).
52. M. Sprecher, M. J. Clark, and D. B. Sprinson, *J. Biol. Chem.* **241**, 872 (1966).
53. H. C. Crumpler, C. E. Dent, M. Harris, and R. G. Westall, *Nature* **167**, 307 (1951).
54. Y. Kaziro and S. Ochoa, *Advan. Enzymol.* **26**, 283 (1964).
55. R. M. Smith and K. J. Monty, *Biochem. Biophys. Res. Commun.* **1**, 105 (1959).
56. H. R. Marston, S. H. Allen, and R. M. Smith, *Nature* **190**, 1085 (1961).
57. H. A. Barker, H. Weissbach, and R. D. Smyth, *Proc. Natl. Acad. Sci. U.S.* **44**, 1093 (1958).
58. J. J. B. Cannata, A. Focesi, Jr., R. Mazumder, R. C. Warner, and S. Ochoa, *J. Biol. Chem.* **239**, 3249 (1965).
59. R. W. Kellermeyer, S. H. G. Allen, R. Stjernholm, and H. G. Wood, *J. Biol. Chem.* **239**, 2562 (1964).
60. H. Eggerer, P. Overath, F. Lynen, and E. R. Stadtman, *J. Am. Chem. Soc.* **82**, 2643 (1960).
61. R. W. Kellermeyer and H. G. Wood, *Biochemistry* **1**, 1124 (1962).
61a. J. R. Sokatch, L. E. Sanders, and V. P. Marshall, *J. Biol. Chem.* **243**, 2500 (1968).
62. J. S. Butts, H. Blunden, and M. S. Dunn, *J. Biol. Chem.* **120**, 289 (1937).
63. A. N. Wick, *J. Biol. Chem.* **141**, 897 (1941).
64. L. C. Terriere and J. S. Butts, *J. Biol. Chem.* **190**, 1 (1951).
65. W. G. Robinson, B. K. Bacchhawat, and M. J. Coon, *J. Biol. Chem.* **218**, 391 (1956).
66. M. J. Coon and N. S. B. Abrahamsen, *J. Biol. Chem.* **195**, 805 (1952).
67. M. J. Coon, N. S. B. Abrahamsen, and G. S. Greene, *J. Biol. Chem.* **199**, 75 (1952).
68. A. Meister, *J. Biol. Chem.* **195**, 813 (1952).

69. R. S. de Ropp, J. C. Van Meter, E. C. de Renzo, K. W. McKerns, C. Pidacks, P. H. Bell, E. F. Ullman, S. R. Safir, W. J. Fanshawe, and S. B. Davis, *J. Am. Chem. Soc.* **80**, 1004 (1958).
70. D. R. Morris and A. B. Pardee, *J. Biol. Chem.* **241**, 3129 (1966).
71. V. Stalon, F. Ramos, A. Pierard, and J. M. Waime, *Biochim. Biophys. Acta* **139**, 91 (1967).
72. F. Ramos, V. Stalon, A. Pierard, and J. M. Waime, *Biochim. Biophys. Acta* **139**, 98 (1967).
73. H. Borsook and J. W. Dubnoff, *J. Biol. Chem.* **138**, 389 (1941).
74. J. B. Walker, *Biochim. Biophys. Acta* **73**, 241 (1963).
75. J. B. Walker, *J. Biol. Chem.* **218**, 549 (1956).
76. J. B. Walker, *J. Biol. Chem.* **221**, 771 (1956).
77. J. B. Walker, *J. Biol. Chem.* **224**, 57 (1957).
78. J. B. Walker, *J. Biol. Chem.* **231**, 1 (1958).
79. M. S. Walker and J. B. Walker, *J. Biol. Chem.* **237**, 473 (1962).
80. J. B. Walker, *Proc. Soc. Exptl. Biol. Med.* **112**, 245 (1963).
81. J. R. Dyer and A. W. Todd, *J. Am. Chem. Soc.* **85**, 3896 (1963).
82. J. B. Walker and V. S. Hnilica, *Biochim. Biophys. Acta* **89**, 473 (1964).
83. G. M. Hills, *Biochem. J.* **34**, 1057 (1940).
84. G. C. Schmidt, M. A. Logan, and A. A. Tytell, *J. Biol. Chem.* **198**, 771 (1952).
85. E. L. Oginsky and R. F. Gehrig, *J. Biol. Chem.* **198**, 791, 799 (1952).
86. M. Korzenovsky and C. H. Werkman, *Arch. Biochem. Biophys.* **46**, 174 (1953).
87. M. F. Barile and R. T. Schimke, *Proc. Soc. Exptl. Biol. Med.* **114**, 676 (1963).
88. R. T. Schimke, C. M. Berlin, E. W. Sweeney, and W. R. Carroll, *J. Biol. Chem.* **241**, 2228 (1966).
89. B. M. Mitruka and R. N. Costilow, *J. Bacteriol.* **93**, 295 (1967).
90. P. P. Cohen and G. W. Brown, Jr. *in* "Comparative Biochemistry," M. Florkin and H. S. Mason (eds.), Vol. 2, p. 172. Academic Press, New York, 1960.
91. H. Kihara and E. E. Snell, *J. Biol. Chem.* **226**, 485 (1957).
92. B. Petrack, L. Sullivan, and S. Ratner, *Arch. Biochem. Biophys.* **69**, 186 (1957).
93. J. M. Ravel, M. L. Grona, J. S. Humphreys, and W. Shive, *J. Biol. Chem.* **234**, 1452 (1959).
94. P. Rogers and G. D. Novelli, *Arch. Biochem. Biophys.* **96**, 398 (1962).
95. R. W. Bernlohr, *Science* **152**, 87 (1966).
96. R. C. Greene, *J. Am. Chem. Soc.* **79**, 3929 (1957).
97. H. Tabor, S. M. Rosenthal, and C. W. Tabor, *J. Biol. Chem.* **233**, 907 (1958).
98. E. F. Gale, *Biochem. J.* **39**, 46 (1945).
99. D. R. Morris and A. B. Pardee, *Biochem. Biophys. Res. Commun.* **20**, 697 (1965).
100. N. van Thoai, F. Thome-Beau, and A. Olomucki, *Biochim. Biophys. Acta* **115**, 73 (1966).
101. N. van Thoai and A. Olomucki, *Biochim. Biophys. Acta* **59**, 533 (1962).
102. N. van Thoai and A. Olomucki, *Biochim. Biophys. Acta* **59**, 545 (1962).
103. S. L. Blethan and N. O. Kaplan, *Biochemistry* **6**, 1413 (1967).
104. W. K. Paik and S. Kim, *Biochim. Biophys. Acta* **139**, 49 (1967).
105. M. Nakano and T. S. Danowski, *J. Biol. Chem.* **241**, 2075 (1966).
106. M. Nakano, Y. Tsutsumi, and T. S. Danowski, *Biochim. Biophys. Acta* **139**, 40 (1967).
107. N. van Thoai, J. L. Hatt, and T. T. An, *Biochim. Biophys. Acta* **18**, 589 (1955).
108. N. van Thoai, J. L. Hatt, and T. T. An, *Biochim. Biophys. Acta* **22**, 116 (1956).
109. N. van Thoai, J. L. Hatt, T. T. An, and J. Roche, *Biochim. Biophys. Acta* **22**, 337 (1956).

110. N. van Thoai, F. Thome-Beau, and D. B. Pho, *Biochim. Biophys. Acta* **63**, 128 (1962).
111. H. Weil-Malherbe and H. A. Krebs, *Biochem. J.* **29**, 2077 (1935).
112. J. V. Taggart and R. B. Krakaur, *J. Biol. Chem.* **177**, 641 (1949).
113. K. Lang and G. Schmid, *Biochem. Z.* **322**, 1 (1951).
114. A. B. Johnson and H. J. Strecker, *J. Biol. Chem.* **237**, 1876 (1962).
115. D. Wellner and H. Scannone, *Biochemistry* **3**, 1746 (1964).
116. L. Frank and P. Rybicki, *Arch. Biochem. Biophys.* **95**, 441 (1961).
117. H. J. Strecker, *J. Biol. Chem.* **235**, 2045 (1960).
117a. L. Frank and B. Ranhand, *Arch Biochem. Biophys.* **107**, 325 (1964).
118. G. deHauwer, R. Lavalle, and J. M Waime, *Biochim. Biophys. Acta* **81**, 257 (1964).
119. H. J. Strecker and P. Mela, *Biochim. Biophys. Acta* **17**, 580 (1955).
120. H. J. Strecker, *J. Biol. Chem.* **235**, 3218 (1960).
121. L. H. Stickland, *Biochem. J.* **29**, 288 (1935).
122. T. C. Stadtman, *Biochem. J.* **62**, 614 (1956).
123. T. C. Stadtman and P. Elliott, *J. Biol. Chem.* **228**, 983 (1957).
123a. D. Hodgins and R. H. Abeles, *J. Biol. Chem.* **242**, 5158 (1967).
124. E. Adams, *J. Am. Chem. Soc.* **79**, 6338 (1957).
125. E. Adams, *J. Biol. Chem.* **234**, 2073 (1959).
126. E. Adams and S. L. Newberry, *Biochem. Biophys. Res. Commun.* **6**, 1 (1961).
127. E. Adams and I. L. Norton, *J. Biol. Chem.* **239**, 1525 (1964).
128. E. Adams and G. Rosso, *Biochem. Biophys. Res. Commun.* **23**, 633 (1966).
128a. E. Adams and G. Rosso, *J. Biol. Chem.* **242**, 1802 (1967).
129. L. D. Aronson, R. G. Rosso, and E. Adams, *Biochim. Biophys. Acta* **132**, 200 (1967).
130. R. M. M. Singh and E. Adams, *Science* **144**, 67 (1964).
131. R. M. M. Singh and E. Adams, *J. Biol. Chem.* **240**, 4344 (1965).
132. R. M. M. Singh and E. Adams, *J. Biol. Chem.* **240**, 4352 (1965).
133. T. Yoneya and E. Adams, *J. Biol. Chem.* **236**, 3272 (1961).
134. A. N. Radhakrishnan and A. Meister, *J. Biol. Chem.* **226**, 559 (1957).
135. G. Wolf and C. R. A. Berger, *J. Biol. Chem.* **230**, 231 (1958).
136. J. Kapfhammer and C. Bischoff, *Z. Physiol. Chem.* **172**, 251 (1927).
137. W. C. Hess and I. P. Shaffran, *J. Am. Chem. Soc.* **73**, 474 (1951).
138. H. Weil-Malherbe and H. A. Krebs, *Biochem. J.* **29**, 2077 (1935).
139. R. Gianetto and L. P. Bouthillier, *Can. J. Biochem. Physiol.* **32**, 154 (1954).
140. M. R. Stetten, *J. Biol. Chem.* **181**, 31 (1949).
141. M. R. Stetten, *Symp. Amino Acid Metab., Baltimore, 1955, Johns Hopkins Univ., McCollum-Pratt Inst. Contrib.* **105**, 277 (1955).
142. G. Wolf, W. W. Heck, and J. C. Leak, *J. Biol. Chem.* **223**, 95 (1956).
143. K. Lang and U. Mayer, *Biochem. Z.* **324**, 237 (1953).
144. E. Adams, R. Friedman, and A. Goldstone, *Biochim. Biophys. Acta* **30**, 212 (1958).
145. E. Adams and A. Goldstone, *J. Biol. Chem.* **235**, 3492 (1960).
145a. E. Adams and A. Goldstone, *J. Biol. Chem.* **235**, 3499 (1960).
146. E. Adams and A. Goldstone, *Biochim. Biophys. Acta* **77**, 133 (1963).
146a. E. Adams and A. Goldstone, *J. Biol. Chem.* **235**, 3504 (1960).
147. A. I. Virtanen and P. K. Hietala, *Acta Chem. Scand.* **9**, 175 (1955).
148. E. E. Dekker, *Biochim. Biophys. Acta* **40**, 174 (1960).
149. L. P. Bouthillier, Y. Binette, and G. Pouliot, *Can. J. Biochem. Physiol.* **39**, 1595 (1961).
150. K. Kuratomi and K. Fukunaga, *Biochim. Biophys. Acta* **43**, 562 (1960).
151. K. Kuratomi and K. Fukunaga, *Biochim. Biophys. Acta* **78**, 617 (1963).
152. K. Kuratomi, K. Fukunaga, and Y. Kobyashi, *Biochim. Biophys. Acta* **78**, 629 (1963).
153. A. I. Virtanen and P. K. Hietala, *Acta Chem. Scand.* **9**, 549 (1955).

154. U. Maitra and E. E. Dekker, *Biochim. Biophys. Acta* **51**, 416 (1961).
155. U. Maitra and E. E. Dekker, *J. Biol. Chem.* **238**, 3660 (1963).
156. E. E. Dekker and U. Maitra, *J. Biol. Chem.* **237**, 2218 (1962).
156a. R. G. Rosso and E. Adams, *J. Biol. Chem.* **242**, 5524 (1967).
157. A. Goldstone and E. Adams, *J. Biol. Chem.* **240**, 2077 (1965).
158. A. Goldstone and E. Adams, *J. Biol. Chem.* **237**, 3476 (1962).
159. U. Maitra and E. E. Dekker, *Biochim. Biophys. Acta* **81**, 517 (1964).
160. R. G. Rosso and E. Adams, *Biochem. Biophys. Res. Commun.* **23**, 842 (1966).
161. R. D. Kobes and E. E. Dekker, *Biochem. Biophys. Res. Commun.* **25**, 329 (1966).
162. D. T. A. Lamport, *Advan. Botan. Res.* **2**, 151 (1965).
163. C. H. Van Etten, R. W. Miller, I. A. Wolff, and Q. Jones, *J. Agr. Food Chem.* **11**, 399 (1963).
164. R. Cleland and A. C. Olson, *Biochemistry* **6**, 32 (1967).
165. A. D. Homola and E. E. Dekker, *Biochim. Biophys. Acta* **82**, 207 (1964).
166. C. S. Hudson and A. Neuberger, *J. Org. Chem.* **15**, 24 (1950).
167. B. Witkop and T. Beiler, *J. Am. Chem. Soc.* **78**, 2882 (1956).
168. A. N. Radhakrishnan and K. V. Giri, *Biochem. J.* **58**, 57 (1954).
169. H. Wieland and B. Witkop, *Ann.* **543**, 171 (1940).
170. J. C. Sheehan, H. G. Zachau, and W. B. Lawson, *J. Am. Chem. Soc.* **80**, 3349, (1958).
171. E. F. Gale and H. M. R. Epps, *Nature* **152**, 327 (1943).
172. E. F. Gale and H. M. R. Epps, *Biochem. J.* **38**, 232 (1944).
173. V. A. Najjar, *in Methods in Enzymology* (S. P. Colowick and N. O. Kaplan, eds.), Vol. 2, p. 188. Academic Press, New York, 1955.
174. A. Ichihara, S. Furiya, and M. Suda, *J. Biochem. (Tokyo)* **48**, 277 (1960).
175. M. Suda, T. Kamahora, and H. Hagihara, *Med. J. Osaka Univ.* **5**, 119 (1954).
176. H. Takeda and O. Hayaishi, *J. Biol. Chem.* **241**, 2733 (1966).
177. N. Itada, A. Ichihara, T. Makita, O. Hayaishi, M. Suda, and N. Sasaki, *J. Biochem. (Tokyo)* **50**, 118 (1961).
178. O. Hayaishi, *Bacteriol. Rev.* **30**, 720 (1966).
179. H. Hagihara, H. Hayashi, A. Ichihara, and M. Suda, *J. Biochem. (Tokyo)* **48**, 267 (1960).
179a. M. S. Reitz, D. L. Miller, and V. W. Rodwell, *Anal. Biochem.* **28**, 269 (1969).
179b. M. S. Reitz and V. W. Rodwell, unpublished data.
180. E. Roberts, *Arch. Biochem. Biophys.* **48**, 395 (1954).
181. M. Suda, T. Kamahora, and H. Hagihara, *Med. J. Osaka Univ.* **5**, 119 (1954).
182. R. A. Hartline, Ph.D. Thesis, Univ. of California, San Francisco, Calif., 1965.
183. S. Numa, Y. Ishimura, T. Nakazawa, T. Okazaki, and O. Hayaishi, *J. Biol. Chem.* **239**, 3915 (1964).
184. M. Rothstein and L. L. Miller, *Arch. Biochem. Biophys.* **54**, 1 (1955).
185. J. R. Mattoon and R. D. Haight, *J. Biol. Chem.* **237**, 3486 (1962).
186. T. C. Stadtman, *J. Bacteriol.* **67**, 314 (1954).
187. T. C. Stadtman, *Symp. Amino Acid Metab., Baltimore, 1954, Johns Hopkins Univ., McCollum-Pratt Inst. Contrib.* **105**, 493 (1955).
188. T. C. Stadtman, *J. Biol. Chem.* **237**, 2409 (1962).
189. T. C. Stadtman, *J. Biol. Chem.* **238**, 2766 (1963).
190. T. C. Stadtman and F. H. White, *J. Bacteriol.* **67**, 651 (1954).
191. R. N. Costilow, O. M. Rochovansky, and H. A. Barker, *J. Biol. Chem.* **241**, 1573 (1966).
191a. R. C. Bray and T. C. Stadtman, *J. Biol. Chem.* **243**, 381 (1968).
192. H. C. Jackins and H. A. Barker, *J. Bacteriol.* **61**, 101 (1951).

193. P. M. Dohner and B. P. Cardon, *J. Bacteriol.* **67**, 608 (1954).
194. H. A. Barker, *in The Bacteria* (I. C. Gunsalus and R. Y. Stainer ed.), Vol. 2, p. 173. Academic Press, New York, 1961.
195. T. C. Stadtman and P. Renz, *Federation Proc.* **26**, 343 (1967).
195a. E. E. Dekker and H. A. Barker, *J. Biol. Chem.* **243**, 3232 (1968).
196. M. Rothstein and J. L. Hart, *Biochim. Biophys. Acta* **93**, 439 (1964).
197. M. Rothstein, *Arch. Biochem. Biophys.* **111**, 467 (1965).
198. A. Meister, "Biochemistry of the Amino Acids," 2nd ed. Academic Press, New York, 1965.
199. C. P. Berg, *J. Nutr.* **12**, 671 (1936).
200. W. G. Gordon, *J. Biol. Chem.* **127**, 487 (1939).
201. A. Neuberger and F. Sanger, *Biochem. J.* **37**, 515 (1943).
202. A. Neuberger and F. Sanger, *Biochem. J.* **38**, 125 (1944).
203. C. M. Stevens and J. A. Bush, *J. Biol. Chem.* **183**, 139 (1950).
204. C. M. Stevens and P. B. Ellman, *J. Biol. Chem.* **182**, 75 (1950).
205. E. Geiger and H. J. Dunn, *J. Biol. Chem.* **178**, 877 (1949).
206. R. Gingras, E. Pagé and R. Gaudry, *Science* **105**, 621 (1947).
207. W. K. Paik, L. Bloch-Frankenthal, S. M. Birnbaum, M. Winitz, and J. P. Greenstein, *Arch. Biochem. Biophys.* **69**, 56 (1957).
208. I. Clark and D. Rittenberg, *J. Biol. Chem.* **189**, 529 (1951).
209. N. Weissman and R. Schoenheimer, *J. Biol. Chem.* **140**, 779 (1941).
210. I. Clark and D. Rittenberg, *J. Biol. Chem.* **189**, 521 (1951).
211. G. L. Foster, R. Schoenheimer, and D. Rittenberg, *J. Biol. Chem.* **127**, 319 (1939).
212. H. Borsook, C. L. Deasy, A. J. Haagen-Smit, G. Keighley, and P. H. Lowy *J. Biol. Chem.* **173**, 423 (1948); **176**, 1383 (1948).
213. J. W. Dubnoff and H. Borsook, *J. Biol. Chem.* **173**, 425 (1948).
214. P. Boulanger and G. Biserte, *Compt. Rend.* **232**, 1451 (1951) [*C.A.* **45**, 8110 (1951)].
215. D. Cavallini and B. Mondovi, *Arch. Sci. Biol.* (*Bologna*) **36**, 468 (1952) [*C.A.* **47**, 3436 (1952)].
216. R. M. Zacharius, J. F. Thompson, and F. C. Steward, *J. Am. Chem. Soc.* **74**, 2949 (1952).
217. R. M. Zacharius, J. F. Thompson, and F. C. Steward, *J. Am. Chem. Soc.* **76**, 2908 (1954).
218. A. C. Hulme and W. Arthington, *Nature* **170**, 659 (1952).
219. N. Grobbelaar, R. M. Zacharius, and F. C. Steward, *J. Am. Chem. Soc.* **76**, 2912 (1954).
220. R. I. Morrison, *Biochem. J.* **53**, 474 (1953).
221. J. H. Martin and W. K. Hausmann, *J. Am. Chem. Soc.* **82**, 2079 (1960).
222. P. H. Lowy, *Arch. Biochem. Biophys.* **47**, 228 (1953).
223. N. Grobbelaar and F. C. Steward, *J. Am. Chem. Soc.* **75**, 4341 (1953).
224. R. S. Schweet, J. T. Holden, and P. Lowy, *Federation Proc.* **13**, 293 (1954).
225. M. Rothstein and L. L. Miller, *J. Am. Chem. Soc.* **75**, 4371 (1953).
226. M. Rothstein and L. L. Miller, *J. Am. Chem. Soc.* **76**, 1459 (1954).
227. M. Rothstein and L. L. Miller, *J. Biol. Chem.* **206**, 243 (1954).
228. M. Rothstein and L. L. Miller, *J. Biol. Chem.* **211**, 851 (1954).
229. A. Meister, *J. Biol. Chem.* **206**, 577 (1954).
230. R. S. Schweet, J. T. Holden, and P. H. Lowy, *J. Biol. Chem.* **211**, 517 (1954).
231. R. S. Schweet, J. T. Holden, and P. H. Lowy, *Symp. Amino Acid Metab., Baltimore, 1954, Johns Hopkins Univ., McCollum-Pratt, Inst. Contrib.* **105**, 496 (1955).
232. A. Meister and S. D. Buckley, *Biochim. Biophys. Acta* **23**, 202 (1957).

233. A. Meister, A. N. Radhakrishnan, and S. D. Buckley, *J. Biol. Chem.* **229**, 789 (1957).
234. D. M. Greenberg, *in* "Metabolic Pathways," (D. M. Greenberg, ed.), 2nd ed., Vol. II, p. 113. Academic Press, New York, 1961.
235. K. Soda, T. Tochikura, and H. Katagiri, *Agr. Biol. Chem. (Tokyo)* **25**, 811 (1961).
236. W. K. Paik, *Biochim. Biophys. Acta* **65**, 518 (1962).
237. I. Chibata, T. Ishikawa and T. Tosa, *Nature* **195**, 80 (1962).
238. S. Kim, L. Benoiton, and W. K. Paik, *J. Biol. Chem.* **239**, 3790 (1964).
239. K. Higashino, K. Tsukada, and I. Lieberman, *Biochem. Biophys. Res. Commun.* **20**, 285 (1965).
240. K. Higashino and I. Lieberman, *Biochim. Biophys. Acta* **111**, 346 (1965).
241. S. T. Vaughan and H. P. Broquist, *Federation Proc.* **24**, 218 (1965).
242. M. Rothstein and D. M. Greenberg, *J. Biol. Chem.* **235**, 714 (1960).
243. M. Rothstein, K. E. Cooksey, and D. M. Greenberg, *J. Biol. Chem.* **237**, 2828 (1962).
244. K. E. Cooksey and D. M. Greenberg, *Arch. Biochem. Biophys.* **112**, 238 (1965).
245. A. J. Aspen and A. Meister, *Biochemistry* **1**, 600 (1962).
246. A. J. Aspen and A. Meister, *Biochemistry* **1**, 606 (1962).
247. D. R. Rao and V. W. Rodwell, *J. Biol. Chem.* **237**, 2232 (1962).
248. L. V. Basso, D. R. Rao, and V. W. Rodwell, *J. Biol. Chem.* **237**, 2239 (1962).
249. A. F. Calvert and V. W. Rodwell, *J. Biol. Chem.* **241**, 409 (1966).
250. M. L. Baginsky and V. W. Rodwell, *J. Bacteriol.* **92**, 424 (1966).
251. M. L. Baginsky and V. W. Rodwell, *J. Bacteriol.* **94**, 1034 (1967).
252. M. Rothstein, C. G. Bly, and L. L. Miller, *Arch. Biochem. Biophys.* **50**, 252 (1954).
253. S. Lindstedt and G. Lindstedt, *Arch. Biochem. Biophys.* **85**, 565 (1959).
254. P. Boulanger, R. Osteux, and J. Bertrand, *Biochim. Biophys. Acta* **29**, 534 (1958).
255. R. H. Hammerstedt, P. B. Swan, and L. M. Henderson, *Arch. Biochem. Biophys.* **128**, 243 (1968).
256. C. E. Polan, W. G. Smith, C. Y. Ng, R. H. Hammerstedt, and L. M. Henderson, *J. Nutr.* **91**, 143 (1967).
257. S. Lindstedt, G. Lindstedt, and C. Mitoma, *Arch. Biochem. Biophys.* **119**, 336 (1967).
258. G. Lindahl, G. Lindstedt, and S. Lindstedt, *Arch. Biochem. Biophys.* **119**, 347 (1967).

CHAPTER 16 (Part I)

Biosynthesis of Amino Acids and Related Compounds

David M. Greenberg

I. INTRODUCTION

Synthesis of the amino acids in the body of the vertebrates is limited to those that are not essential in the diet. The nutritionally essential amino acids are the product of organisms with a more autotrophic metabolism, such as various plants and microorganisms. The mechanisms of the synthesis of all the amino acids found in proteins will be included in this chapter. A large number of amino acids not present in proteins have been discovered over the years. Their biosynthesis will not be discussed in this chapter.

In addition to the amino acids themselves the formation of certain biologically important compounds which are derived from amino acids is also included in this chapter. Formation of other important compounds which also involve amino acids are contained in other chapters of this work.

II. BIOSYNTHESIS OF SERINE AND PHOSPHOSERINE

Two pathways are known for the neogenesis of serine from carbohydrate precursors (Fig. 1). Abundant evidence is available for the occurrence of the schemes shown in Fig. 1. The initial evidence for the formation of serine via D-glyceric acid was published by Sallach (1), and the synthesis via 3-phospho-D-glyceric acid by Ichihara and Greenberg (2).

The two pathways are found to coexist in liver and kidneys of a variety of vertebrate species (*3*) and in *Neurospora* (*3a*). This is shown by the presence of the two chief enzymes of each pathway, D-glycerate dehydrogenase and hydroxypyruvate: L-alanine aminotransferase for the nonphosphorylated pathway, and D-glycerate-3-P dehydrogenase

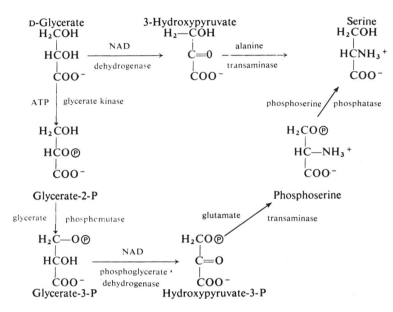

FIG. 1. Pathways of phosphoserine and serine biosynthesis.

and 3-phospho-hydroxypyruvate: L-glutamate transaminase for the phosphorylated pathway (Table I) (*3*). According to Table I the non-phosphorylated pathway is the major route of serine synthesis in the livers of the dog, rat, and frog; the phosphorylated pathway appears to predominate in the liver of the other vertebrate species and in the kidney and brain of all vertebrate species. Subsequently it has been shown that the activity of the enzymes in liver for the phosphorylated pathway is regulated by diet. On a low-protein diet (2% casein) their activities in the rat liver are greatly enhanced as compared to a diet containing 25% casein (*4*). Bridgers (*5*) has shown that the phosphorylated pathway is apparently the only route of serine biosynthesis in mouse brain. This appears to be also true of human KB cells propagated by cell culture (*6*).

In a study of the changes in the levels of D-glycerate dehydrogenase and D-glycerate-3-P dehydrogenase in the rat during development and growth, Johnson *et al.* (*7*) determined that the former enzyme increased

TABLE I

DISTRIBUTION OF ENZYMES OF SERINE BIOSYNTHESIS IN ANIMAL TISSUES[a]

Tissue and animal	Hydroxy-pyruvate-P:L-glutamate transaminase (units/mg protein)	Hydroxy-pyruvate:L-alanine transaminase (units/mg protein)	D-Glycerate-3-P dehydrogenase (units/mg protein)	D-Glycerate dehydrogenase (units/mg protein)
Liver				
Rabbit	210	590	93	117
Dog	<60	550	9	293
Pig	310	260	278	432
Beef	720	185	140	70
Frog			48	1106
Rat (adult)	<20	<60	4	163
Human	120	50	60	85
Chicken	120	<20	580	26
Pigeon			396	37
Kidney				
Dog	80	150	195	152
Rabbit	180	30	254	59
Pig	300	80	302	292
Beef	250	140	377	144
Calf			254	46
Rat	80	30	34	145
Human	<20	<20		
Brain				
Dog	70	<20	64	115
Beef	190	<20		
Heart				
Rat	0	<20		
Beef	0	<20		
Pig	<60	<20		

[a] From Walsh and Sallach (3).

10-fold between the fetal and 13- to 17-day postnatal period (and the latter enzyme decreased 3-fold). Beyond the 13- to 17-day postnatal period, both enzymes decreased.

The phosphorylated pathway has been reported to occur predominantly in a variety of microorganisms (8–10) and in plants (11).

The enzymes of the nonphosphorylated pathway have been shown to occur in a Pseudomonad (12), and L-serine aminotransferase has been partially purified and separated from L-serine-P aminotransferase in a strain of E. coli (13). Separation was made possible by the observation that the L-serine aminotransferase resisted inactivation on being heated to 65°C, whereas the L-serine-P aminotransferase was inactivated.

A variety of evidence has been gathered in support of the phosphorylated pathway of serine biosynthesis in *Salmonella typhimurium* and *Escherichia coli*. Formation of phosphoserine by *E. coli* extracts was shown from D-glycerate-^{14}C. No phosphoserine was produced by a serine-glycine auxotroph of *E. coli* (9). The presence of the three enzymes required for serine biosynthesis by the phosphorylated pathway—D-glycerate-3-P dehydrogenase, 3-P-hydroxypyruvate: glutamate transaminase, and a phosphoserine-specific phosphatase—has been demonstrated in *E. coli* (9), in *S. typhimurium* (8), and in *Peptostreptococcus elsdenii* (10). In two serine-glycine auxotrophs of *S. typhimurium*, one was shown to be lacking in P-3-glycerate dehydrogenase and the other in P-serine phosphatase. The phosphoserine pathway in microörganisms is regulated by the concentration of serine. This functions by inhibiting formation of P-3-glycerate dehydrogenase, the first enzyme of the pathway.

A. Enzymes of the Nonphosphorylated Pathway

1. D-GLYCERATE: NAD OXIDOREDUCTASE (EC 1.1.1.29)

Glycerate dehydrogenase has been partially purified (24-fold) from beef liver (13, 14) by $(NH_4)_2SO_4$ precipitation and calcium phosphate gel chromatography. Oxidation was induced by either NAD or NADPH.

The enzyme was separated almost completely from L-lactate dehydrogenase. The latter is present in much higher concentration in liver than is the D-glycerate dehydrogenase. Hydroxypyruvate, the product of D-glycerate oxidation, is a substrate for L-lactate dehydrogenase and is reduced to L-glycerate. Proof of the formation of hydroxypyruvate from D-glycerate in the enzymic reaction was obtained by forming the 2,4-dinitrophenylhydrazone. This was identified by paper chromatography. Further proof of this was that hydrogenation of the 2,4-dinitrophenylhydrazone yielded serine and alanine.

When the reaction was carried out in the reverse direction with hydroxypyruvate as substrate, D-glycerate was identified as the product formed, by reacting it with specific plant hydroxypyruvate reductase.

2. L-SERINE: PYRUVATE AMINOTRANSFERASE

Serine aminotransferase has been purified about 50-fold from dog liver (1). A requirement for pyridoxal-P was indicated by its stimulating effect on the reaction. Alanine was the only amino acid that was a significant amino group donor in the transamination reaction. The reaction was shown to be reversible, and the stoichiometry was established from the equivalence in product increase to substrate decrease.

Glutamine was found to an excellent amino group donor for the mammalian L-serine aminotransferase. The *E. coli* L-serine aminotransferase differs from the mammalian one in that glutamate is the most active amino group donor, being about 50% better than alanine. Another difference is that the bacterial enzyme is inhibited by P_i, the mammalian enzyme is not.

B. Enyzmes of the Phosphorylated Pathway

In addition to the D-glycerate-3-P dehydrogenase and phosphoserine aminotransferase, this pathway includes a specific L-serine-P phosphatase. The latter is required to complete conversion of phosphoserine to serine.

1. D-GLYCERATE-3-P: NAD OXIDOREDUCTASE

D-Glycerate-3-P dehydrogenase has been prepared in virtually pure state from chicken liver (*14–17*). Highest levels of this enzyme are found in bird liver (*14,15*). It is also high in pork and rabbit liver. The oxidation product of D-glycerate-3-P was identified as hydroxypyruvate-3-P by converting it to the 2,4-dinitrophenylhydrazone. The product of enzymic reduction of hydroxypyruvate-3-P was shown to be D-glycerate-3-P.

The highly purified liver enzyme was indicated to be an essentially pure protein by ultracentrifugation, Biogel P-200 filtration, and starch gel electrophoresis. Multiple active zones of enzyme were found on polyacrylamide gel electrophoresis. The molecular weight of the gylcerate-P dehydrogenase was estimated to be 200,000. The highly purified enzyme preparation was shown to be free of lactate and malate dehydrogenases.

The dehydrogenase has optimum activity at the high alkaline value of pH 9.3 (*15*). The isoelectric point was found to be at pH > 8.6. This indicates a high amide content of aspartate and glutamate (*17*).

The enzyme is activated by decimolar NaCl or KCl and it is stabilized by dithiothreitol (*16*). This suggests that the dehydrogenase is an SH enzyme. This is supported by other evidence.

The Michaelis constants for the substrates were estimated to be: NAD^+, 6×10^{-5}; D-glycerate-3-P, 2.5×10^{-4} M; hydroxypyruvate-3-P, 1×10^{-5} M; NADH, 5×10^{-6} M. The equilibrium constant for the reaction in the direction of glycerate-P oxidation was calculated to be 1×10^{-12}. The low value of this constant shows that the equilibrium is greatly in favor of D-glycerate-3-P formation. Dissociation of NAD^+ from the enzyme also was measured optically and the dissociation constant measured to be 7×10^{-4} M (16).

D-Phosphoglycerate dehydrogenase has also been highly purified and crystallized from *E. coli* B (18). The molecular weight was estimated to be 163,000 and the $s_{20,w}$ 7.7. Optimum activity is at about pH 8.5. The velocity of the reduction of 3-phosphohydroxypyruvate was found to be about 40 times more rapid than the oxidation of the D-phosphoglycerate. An estimate of the equilibrium constant of the reaction in the direction of phosphoglycerate oxidation yielded the value 6×10^{-11} at pH 7.5 and 25°.

The above authors (18,18a) made a careful study of the mechanism of the specific inhibition of this enzyme by L-serine. The data obtained indicates that serine binds at two interacting sites, and that the binding produces conformational changes in the enzyme protein. L-Serine was shown to be a noncompetitive inhibitor with respect to hydroxypyruvate-P and uncompetitive with respect to phosphoglycerate and NADH.

2. L-PHOSPHOSERINE: 2-OXOGLUTARATE AMINOTRANSFERASE

L-Phosphoserine aminotransferase has been shown to occur in all tissues and organisms that possess the phosphorylated pathway of serine biosynthesis (3,4,8,9). Highly purified enzyme has been prepared only from sheep brain (19) with a 500-fold enrichment of enzyme activity. Ultracentrifugation of the aminotransferase gave only a single migrating peak with an $s_{20,w}$ of 4.2S. The molecular weight was calculated to be 96,000. Optimum activity is at pH 8.15, and K_m values were estimated to be 0.25 mM for hydroxypyruvate-3-P and 0.7 mM for glutamate. The transamination reaction was demonstrated to be stoichiometric. Measurement of the equilibrium for serine-P transamination yielded the value of about 0.124 for the apparent equilibrium constant.

Proof that pyridoxal-P was the required coenzyme was obtained by inactivating the enzyme by dialysis against cysteine and then reactivating the aminotransferase with pyridoxal-P. Other evidence in favor of this conclusion is that the enzyme has an absorption peak at 415 mμ, which suggests the presence of bound pyridoxal-P. Reduction of the enzyme with sodium borohydride and subsequent acid hydrolysis yielded an amino acid derivative that was identified to be ε-pyridoxyllysine.

The enzyme was shown to be highly specific for serine-P and gluta-mate. L-Alanine had about 10% of the activity of L-glutamate as the amino group donor.

3. Phosphoserine Phosphohydrolase (EC 3.1.3.3)

Phosphatase with a high specificity for serine-P has been studied in chicken liver (20), rat liver (21), mouse brain (22), and yeast (23). A purification of 20- to 40-fold has been achieved for the enzymes from the above sources. A heating step in the purification procedure elemin-ates many interfering enzymes.

In addition to hydrolysis of the phosphate group of serine-P, the enzyme catalyzes an exchange reaction with serine. Optimum activity of the enzyme from liver and brain for hydrolysis is at pH 6.2–6.3; optimum activity for exchange is at the higher pH of 7.2–7.5. The phosphatase requires a divalent metal ion for activity. The activation potency is in the order $Mg^{2+} > Co^{2+}$, $Ni^{2+} > Zn^{2+} = Fe^{2+}$.

The stereospecificity of phosphoserine is not greatly important in determining the enzyme activity; D-serine-P is nearly as good a sub-strate. K_m values found were 5.8×10^{-5} M for L-serine-P and 4.2 mM for D-serine-P, which shows that the L-amino acid has a greater affinity for the enzyme.

L-Serine-P phosphatase activity is strongly inhibited by both D- and L-serine. The K_i of L-serine, however, is 40-fold less than that of D-serine. This inhibition has been proposed to function as a regulatory step in serine biosynthesis in vertebrates, just as inhibition of D-glycerate-3-P dehydrogenase by serine serves similarly in bacteria.

The exchange reaction has been the subject of considerable investiga-tion (20–30). It is studied by incubating serine-P with serine-^{14}C and enzyme. This leads to the formation of ^{14}C-serine-P. It is interesting that, whereas the enzyme catalyzes hydrolysis of D-serine-P, it does not catalyze the exchange reaction with D-serine.

Bridgers (22) has found that the two enzyme activities of serine-P phosphatase can be varied independently. Thus urea (3 M) was found to inhibit the hydrolysis by 50% and to enhance the phosphotransferase reaction 200%. Sucrose (0.9 M) caused an 80% decrease in phospho-transferase activity, with only a slight loss of hydrolysis. These findings suggest that the binding site on the enzyme for water differs from the binding site for serine required for transferase activity. The reaction mechanism for hydrolysis and exchange proposed by Bridgers is shown in the accompanying scheme, where E, PS, EPS, and S* represent enzyme, serine-P, enzyme-serine-P complex, and labeled or unlabeled serine, respectively.

$$E + PS \rightleftharpoons^{1} EPS \qquad E \cdot PS \cdot S^* \rightleftharpoons^{3} EPS^* + S$$

$$S^* \diagup 2$$

$$\diagdown H_2O$$

$$4 \searrow E + P + S$$

Inhibition of hydrolysis by serine is presumed to result from the binding of the latter to the enzyme-serine-P complex (reaction 2). This is consistent with the uncompetitive nature of the inhibition. Inhibition of hydrolysis by urea is explained by interference with reaction 4. Inhibition of the transferase reaction by sucrose is presumed to be due to interference with reaction 3.

C. Mechanism of Interchange between the Phosphorylated and Nonphosphorylated Pathways

This interchange in the mode of synthesis of serine can be achieved through the activities of the three enzymes, D-glycerate kinase, D-glycerate-2,3-phosphomutase, and D-glycerate-2-P phosphatase.

1. D-GLYCERATE KINASE (ATP: D-GLYCERATE-3-PHOSPHOTRANSFERASE, EC 2.7.1.31)

This enzyme catalyzes the formation of D-glycerate-2-P as shown in

$$\text{D-glycerate} + \text{ATP} \xrightarrow[\text{kinase}]{\text{Mg}^{2+}} \text{D-glycerate-2-P} + \text{ADP} \tag{1}$$

Eq. (1). D-Glycerate kinase has been demonstrated and partially purified from liver (24–26). This enzyme has been shown to be specific for D-glycerate and to require ATP and Mg^{2+} for activity. Lamprecht et al. (26) obtained evidence that it is a mitrochondrial enzyme. Ichihara and Greenberg (24) and Holzer and Holldorf (25) believed that the product formed was D-glycerate-3-P, but Lamprecht et al. (26) established that it was the D-glycerate-2-P. This was done by several enzymatic coupling reactions. Paper and column chromatography do not distinguish between the two isomeric phosphoglycerates. The optimum activity of this enzyme is about pH 7.4, and K_m values of 2.4 and 9.7 mM were obtained for D-glycerate and ATP, respectively.

2. GLYCERATE PHOSPHOMUTASE (2,3-DIPHOSPHO-D-GLYCERATE: 2-PHOSPHO-D-GLYCERATE PHOSPHOTRANSFERASE, EC 2.7.5.3)

This enzyme catalyzes the interconversion of 2-phospho-D-glycerate and 3-phospho-D-glycerate (27–29). It is widely distributed in natural materials. The enzyme in animal tissues requires 2,3-diphospho-D-glycerate as an intermediary cofactor (29).

The combined actions of D-glycerate kinase and glycerate phosphomutase places D-glycerate on the phosphorylated pathway of serine biosynthesis.

The converse, conversion of D-glycerate-3-P to D-glycerate, is completed through the activity of a D-glycerate-2-P phosphohydrolase. An enzyme that catalyzes dephosphorylation of D-glycerate-2-P was demonstrated in beef liver by Fallon and Byrne (30). D-Glycerate-2-P is the most active substrate for this enzyme.

III. BIOSYNTHESIS OF GLYCINE

The major route of glycine formation in most organisms appears to be via the decomposition of serine to glycine and methylene-FH_4. This reaction is catalyzed by the enzyme serine hydroxymethyltransferase (EC 2.1.2.1). The historical development of this reaction is discussed in the Second Edition of this work (Vol. II, p. 177). For discussion of the properties of highly purified serine hydroxymethyltransferase, the reader is referred to Chapter 15.

Evidence for the above conclusion is the observation that liver extracts incubated anaerobically with glycine-^{14}C exhibited slight incorporation of the label into serine (31).

A. Alternative Reactions for Glycine Biosynthesis

1. AMINATIONS OF GLYOXYLIC ACID

The equilibrium of the reaction for transamination of glycine, Eq. (2), strongly favors glycine synthesis.

$$O{=}CH{-}COOH + NH_2{-}CHR{-}COOH \rightarrow NH_2{-}CH_2COOH + O{=}CR{-}COOH$$

$$\text{(2)}$$

Two specific aminotransferases that catalyze this reaction have been found in liver. One of these preferentially utilizes glutamate as the amino group donor (32,33) and the other, alanine (33). With both enzymes, glycine formation is essentially irreversible.

Nakada (32) considerably purified a glycine aminotransferase from rat liver (160-fold) that specifically required L-glutamate as the amino group donor. The pH optimum of this enzyme was at 7.2. A similar enzyme has been purified 200-fold from human liver (33). This enzyme also was able to utilize glutamine, alanine, arginine, and methionine, in the order given, as amino group donors.

Thompson and Richardson (*34*) prepared an aminotransferase from human liver (purified 900-fold) that showed preference for L-alanine as the amino group donor. Serine (84%) and arginine (13%) could also serve as alternative amino group donors. Optimum activity of this enzyme was at pH 8.4. This enzyme was inhibited by the substrate glyoxylate.

2. THE THREONINE ALDOLASE REACTION

Cleavage of L-threonine and L-allothreonine to acetaldehyde and glycine, catalyzed by the above enzyme, has a low degree of activity in mammalian liver and would not be expected to contribute greatly to the formation of glycine. In the strict anaerobe *Clostridium pasteurianum*, this may be a very important route of glycine formation according to Dainty and Peel (*35*). This claim is supported by the observation that this bacterium lacks certain enzymes of the phosphorylated pathway of serine biosynthesis. The threonine aldolase was demonstrated to be present and was partially purified from ultrasonic extracts of *C. pasteurianum*. Incubation with isotopically labeled threonine resulted in the formation of labeled glycine and serine. Threonine could be derived from glucose by a metabolic pathway through aspartate.

3. SYNTHESIS FROM SERINE AND CO_2

A recent proposal for a new route of glycine biosynthesis proceeds according to Eq. (3) (*36,37*).

$$CO_2 + NH_3 + serine + 2 H \underset{PLP}{\overset{FH_4}{\rightleftharpoons}} 2 \text{ glycine} + H_2O \qquad (3)$$

The enzyme that catalyzes this reaction has been solubilized from rat liver mitochondria. This reaction has been demonstrated to occur in the livers and kidneys of mammals and birds. It proceeds better anaerobically than aerobically. Reaction components required to give maximum conversion of $^{14}CO_2$ to glycine are serine, NH_4Cl, $NaH^{14}CO_3$, dithiothreitol, folate.H_4, and pyridoxal-P. Labeled glycine and serine are both formed, the latter by an exchange reaction. The magnitude of the labeling indicates that glycine is formed first and serine is formed from the latter.

Evidence was obtained that the β-carbon of serine is combined with the CO_2 to form the glycine. Methylenefolate·H_4 can replace serine in the reaction, and NADH can substitute for the dithiothreitol.

The reaction evidently results from the action of a multienzyme system. One of the enzymes involved is serine hydroxymethyltransferase, which was identified in the enzyme preparation. The rest of the system is totally or in part the glycine oxidizing enzyme system of Klein and Sagers (*38*) and Baginsky and Huennekens (*39*). The dithiothreitol serves to keep the SH protein of this enzyme system in the reduced state.

The above enzyme system also catalyzes the oxidation of glycine and the exchange reaction between $^{14}CO_2$ and the carboxyl group of glycine. In the oxidation of glycine only the 1-^{14}C labeled material and not the 2-^{14}C compound formed $^{14}CO_2$.

IV. FORMATION OF PHOSPHATIDE BASES

The best known of the nitrogen constituents of the phosphatides are choline, found in lecithin (phosphatidylcholine); ethanolamine, in cephalin (phosphatidylethanolamine); and serine, in phosphatidylserine. More recently the partially methylated bases, methylaminoethanol and dimethylaminoethanol, have been discovered in various organic materials, e.g., *Neurospora* (*40*) and liver (*41*). The phosphatide bases are interrelated by the metabolic cycle shown in Fig. 2.

The precursor of the phosphatide bases is the amino acid serine. It is decarboxylated to aminoethanol and the latter is methylated in a step-wise manner to choline. Formation of aminoethanol by the decarboxyl-ation of serine was established by isotopic labeling experiments (*42,43*). However, discovery of the enzyme that catalyzes this reaction was not achieved until it became evident that the substrates for the enzymic interconversions of the phosphatide bases were phosphatides. Experi-ments with serine-^{14}C by injection into the intact animal and incubation of liver and brain slices and homogenates (*44*) indicated that the amino acid was incorporated into phospholipid and then decarboxylated to become phosphatidylaminoethanol.

Subsequently Borkenhagen *et al.* (*45*) demonstrated the occurrence of a phosphatidyl serine decarboxylase in the mitochondria of rat liver. Proof that the decarboxylation involved phosphatidyl serine was obtained by preparing synthetic dipalmitoyl-L-(+)-glycerophosphoryl-DL-serine-1-^{14}C and demonstrating the release of $^{14}CO_2$. The mito-chondrial enzyme was found to be strongly activated by certain organic solvents such as toluene.

The mechanism of the formation of choline was clarified by the discoveries of Bremer and Greenberg (*46,47*) and of Wilson, Gibson, and Udenfriend (*44,48,49*) that the methylation occurred at the phos-phatide level, namely, that phosphatidylaminoethanol is methylated to form lecithin. S-Adenosylmethionine has been shown to be the methyl group donor, and the enzyme system is present in the microsomal fraction of the cell in all organisms studied.

In experiments with isolated microsomes (47) it was found that, if methionine-$^{14}CH_3$ was used as a substrate, there was a requirement for ATP and Mg^{2+} to secure methylation. Upon substituting S-adenosylmethionine, ATP and Mg^{2+} were no longer essential.

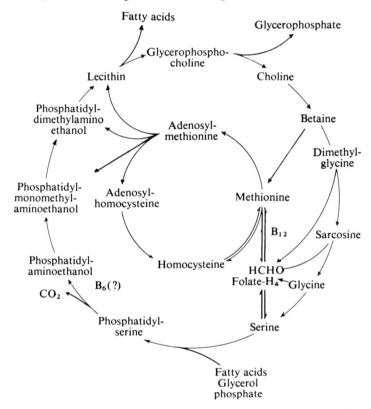

FIG. 2. Metabolic pathways of methylation and demethylation of phosphatide bases.

The initial studies were complicated by the presence of large amounts of endogenous substrates in the cell extracts which obscured the results obtained with added compounds. Subsequently procedures were developed to purify the enzyme preparations and to eliminate the endogenous substrate material. Another obstacle in the work was the difficulty of preparing properly dispersed substrates, both synthetic and those isolated from natural sources. This was true in particular with ethanolamine phosphatide. The consequence of this difficulty was that the results obtained were quite variable.

Early attempts to fractionate the particulate phospholipid methyl-transferase system into individual enzymes failed, but more recently a degree of success has been achieved by the fortunate circumstances that soluble enzyme components can be prepared from *Agrobacterium tumefaciens* and *Neurospora crassa*.

Rehbinder and Greenberg (*50*) obtained a highly dispersed liver microsome preparation free of endogenous substrates by treatment with deoxycholate and sonic oscillation. This enzyme preparation showed an absolute dependence on added monomethyl- or dimethylaminoethanol phosphatides for methyl transfer to form phosphatidylcholine. No re-action could be detected with ethanolamine phosphatide. The results suggested that a single enzyme catalyzed the two-step methylation from monomethyl-ethanolamine to choline.

Kaneshiro and Law (*51*) were successful in obtaining a soluble aminoethanol phosphatide *N*-methyltransferase from *A. tumefaciens* that catalyzed the first methylation step. The enzyme was purified 40-fold. The enzyme was found to be quite unstable, losing half its activity on heating at 40° for 15 minutes. Optimal activity was found at pH 8.4 and 9.6. The K_m of adenosylmethionine was estimated to be $2 \times 10^{-4} M$. *S*-Adenosylhomocysteine was inhibitory. The subsequent methylation steps were catalyzed by the bacterial particles resistant to fractionation.

Scarborough and Nyc (*52*), on the other hand, obtained a soluble monomethylethanolamine phosphatide *N*-methyltransferase from *N. crassa*, while the first methyltransferase activity was particle-bound. The soluble enzyme catalyzed the two-step methylation to phosphatidyl-choline, and evidence was obtained that this was the property of a single enzyme. The optimum activity of the phosphatidylethanolamine *N*-methyltransferase was at pH 8.0. The K_m for adenosylmethionine was determined to be 13 μM for the first methylation step and 18 μM for the second.

Further evidence of the occurrence of two enzymes to methylate ethanolamine to choline was obtained with mutant strains of *N. crassa* (*53*). One of these could only carry out the first methylation step with the accumulation of monomethylaminoethanolamine phosphatide. The second mutant had a subnormal amount of the first enzyme, but readily methylated the mono- and dimethylaminoethanol phosphatides.

A. Demethylation of Choline

Although this is not an anabolic process, discussion of the reactions involved are included to complete the story of the total oxidative metabolism of choline. The reactions of the several steps involved are given in Fig. 3.

The enzymes that catalyze the individual reactions shown in Fig. 3 have been separated from each other and at least partially characterized. Most of these enzymes are part of the mitochondrial complex and thus difficult to solubilize and study.

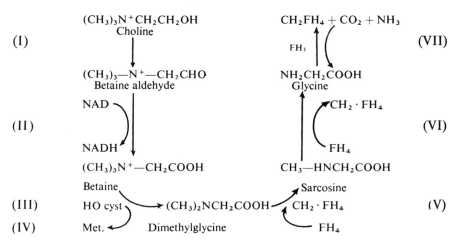

FIG. 3. Pathway of metabolic decomposition of choline. FH_4, tetrahydrofolate.

The particulate enzymes are components of electron transfer units and function in oxidative phosphorylation.

The evidence that oxidation of choline is a two-step process requiring separate enzymes was first obtained by Klein and Handler (54). The two enzymes required to catalyze the oxidation of choline to betaine are choline dehydrogenase (EC 1.1.99.1) (Fig. 3, I) and betaine aldehyde dehydrogenase (EC 1.2.1.8) (Fig. 3, II).

Choline dehydrogenase is associated with the cytochrome electron transport system of mitochondria. It can be solubilized and separated from the electron transport particle by a number of procedures (54–57). The most recent improvement involves treatment with snake venom lipase A (57). The unfractionated mitochondrial particles can utilize 2,6-dichlorophenolindophenol, cytochrome c, and O_2 as electron acceptors in the oxidation of choline. The isolated choline dehydrogenase can be assayed satisfactorily only with phenazine methosulfate.

The partially purified choline dehydrogenase contains a flavin coenzyme and nonheme iron. The optimum activity is at pH 7.6–8.2 at 30–38°. The K_m for choline is 6.5 mM. The dehydrogenase is sensitive to sulfhydryl reagents.

Undamaged mitochondria do not readily oxidize choline, owing to its impermeability. Reagents that produce mitochondrial swelling promote the oxidation of choline (*58–61*). There have been a number of studies on the factors regulating oxidation of choline. It has been concluded that in isolated mitochondria this control is exerted by the intramitochondrial concentration of adenine nucleotides (ADP, ATP) (*60*). These compounds are presumed to regulate the permeability of the mitochondrion to choline, thereby determining the oxidation rate (*61*).

1. BETAINE ALDEHYDE DEHYDROGENASE (Fig. 3, III)

The enzyme that catalyzes the oxidation of betaine aldehyde to betaine is not present in mitochondria, but only in the soluble cytoplasm (*62*). Isolation and partial purification from rat liver of this enzyme was first reported by Rothschild and Barron (*63*). Since then the dehydrogenase has been largely ignored. A recent paper reports a 4- to 8-fold purification from rat liver (*63a*). The betaine aldehyde dehydrogenase was found to utilize NAD specifically in the presence of Mg^{2+}. Its activity was increased by cysteine and it was inhibited by SH reagents, suggesting that it is an SH enzyme. The pH optimum was at 8.8–9.2, and a K_m value for the betaine aldehyde of 0.11–0.36 mM was obtained. Jellinck *et al.* (*64*) prepared crystalline betaine aldehyde and demonstrated its conversion to betaine by a preparation of the dehydrogenase.

In the demethylation of betaine, only one of the methyl groups is removed by methyl transfer, with homocysteine sering as the acceptor.

A relatively unspecific enzyme that can catalyze the transfer of one methyl group from betaine to homocysteine has been purified to an essentially homogeneous state from horse (*65,66*) and rat liver (*67*). This enzyme utilizes preferentially various dimethylthetins. More recently evidence has been obtained and a partial separation achieved of a distinct enzyme that has greater transmethylation activity with betaine than with dimethylthetins (*68*). This enzyme, betaine homocysteine *S*-methyltransferase (EC 2.1.1.5) (Fig. 3, IV), has not been further characterized because of its marked instability.

Further demethylation of dimethylglycine and sarcosine occurs by oxidation of their methyl groups to a product that can be trapped as formaldehyde in the presence of semicarbazide. Investigation of the demethylation of these compounds has mainly been carried out in the laboratory of Mackenzie. A review of the earlier work is contained in (*69*).

Separate enzymes catalyze the demethylation of dimethylglycine and sarcosine (*70,71*). They are recognized and fractionated by the following differences in properties. Sarcosine dehydrogenase is inhibited by methoxyacetic acid, dimethylglycine dehydrogenase is not. Dimethylglycine dehydrogenase is stable to heating at 50° for 5 minutes and is eluted from a DEAE-cellulose column by 0.075 M KHPO$_4$ buffer, sarcosine dehydrogenase is inactivated by heating and requires 0.5 M KCl—0.0375 M KHPO$_4$ for elution from DEAE-cellulose.

Both dehydrogenases are flavoproteins and contain nonheme iron. Dimethylglycine dehydrogenase has been enriched about 14-fold. It increases in activity on heating, and higher specific activities are obtained by batch purification with calcium phosphate gel than by column chromatography on DEAE-cellulose. Optimum activity of dimethylglycine dehydrogenase is at pH 8.5–9.

Sarcosine dehydrogenase is reported to have its optimum activity at pH 7.4–7.5 and a K_m for sarcosine of 1.4 mM (72).

Fractionation of solubilized sarcosine and dimethylglycine dehydrogenases separates them into two components, the enzymes acting on the substrates at 40–60% (NH$_4$)$_2$SO$_4$ saturation and a flavoprotein obtained at 60–80% saturation (73). The purified enzymes can be assayed by reduction of 2,6-dichlorophenolindophenol only when phenazine methosulfate is used as a prior electron acceptor. The latter can be replaced by the soluble electron transfer flavoprotein. The electron transfer flavoprotein is activated by the addition of FAD. Dichlorophenolindophenol can be replaced by submitochondrial particles containing cytochrome c and cytochrome oxidase. The complete O$_2$-utilizing system is termed sarcosine oxidase (EC 1.5.3.1).

The electron transfer flavoprotein has been purified 50-fold. It contains FAD, as shown by release of this from the protein by treatment with acid (74). It has also been found that the sarcosine dehydrogenase electron transfer flavoprotein and the fatty acyl coenzyme A dehydrogenase can substitute for each other in their function (75). The primary dehydrogenases are specific for their particular substrates.

Oxidation of glycine as shown (Fig. 3, VII) is catalyzed by the oxidizing enzyme system discussed in connection with the reverse reaction of glycine biosynthesis (Section III).

V. BIOSYNTHESIS OF HOMOSERINE AND THREONINE

The pathway of biosynthesis of homoserine and threonine has been established by isotopic and enzymic studies. The reactions involved are shown in Fig. 4. All the enzymes and cofactors of the several reactions have been isolated and purified to varying degrees. In recent years major attention has been devoted to study of the regulatory mechanisms controlling synthesis of these amino acids and certain others derived from them.

The pathway shown in Fig. 4 has been established for yeast (*76,77*), *Neurospora* (*78*), *E. coli* and other bacteria (*79–82*), and higher plants (*83*).

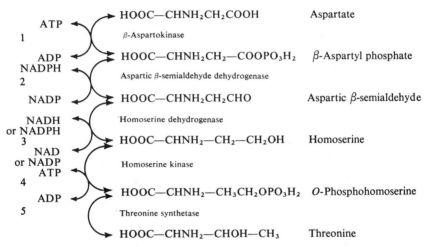

FIG. 4. Pathway of biosynthesis of homoserine and threonine from aspartate.

A. Phosphorylation of Aspartate

The first reaction in the biosynthetic conversion of aspartate to threonine is its phosphorylation to β-aspartyl phosphate (Fig. 4, reaction 1). The enzymic product of the reaction between aspartate and ATP catalyzed by β-aspartate kinase yields β-aspartohydroxyamic acid on treatment with hydroxylamine, and it is quantitatively reduced enzymically to L-aspartic semialdehyde (*77*).

β-Aspartate kinase (EC 2.7.2.4) has been purified from brewer's yeast (*77*), *Rhodopseudomonas spheroides* (*82*), and *E. coli* (*83*), the preparation of highest purity being from *E. coli*. The specific activity of this preparation was 94 μmoles/min/mg protein. The molecular weight was estimated to be about 100,000. The optimum activity of the enzyme covers a wide range of pH from 7.9–9.0. The yeast enzyme exhibited an even greater activity plateau of pH 5-9. Michaelis constants for the *E. coli.* aspartate kinase (*83*) were found to be 4.7 and 4.8 mM for aspartate and ATP, respectively. With the *R. spheroides* enzyme preparation a biphasic substrate concentration-activity curve was obtained, thus the Lineweaver-Burk plots were not linear.

As might be anticipated, β-aspartate kinase required Mg^{2+} for activity. This can be replaced less effectively by Mn^{2+} and Fe^{2+}. Datta and Prakash (82) reported stimulation of the enzyme activity by K^+, Co^+, or Rb^+. The K^+ increased the maximum velocity and the binding of ATP to the enzyme.

The kinase reaction is reversible. An equilibrium constant of 3.5×10^{-4} at pH 8 and 15° has been calculated for the reaction using the yeast enzyme preparation.

Feedback inhibition of homoserine and threonine biosynthesis has been extensively studied. Wormser and Pardee (80) observed that L-threonine competitively inhibited the β-aspartyl phosphate reaction of E. coli. Subsequently it has been shown that regulation of the biosynthesis of the amino acids derived from aspartate is of much greater complexity and varies in different organisms. Many examples of concerted repression have been reported. In most instances the reported repressors are L-threonine and L-lysine (84–87). The repression in this case is counteracted by L-methionine. With enzymes from both Salmonella typhimurium and E. coli., repression by L-threonine and isoleucine has been reported (88).

Two aspartate kinases are contained in E. coli (89), one specifically and noncompetitively inhibited by L-lysine, the second one competitively inhibited by L-threonine. The first one is relatively stable to heat, the second one quite unstable. Each enzyme is protected by its specific inhibitory amino acid; separation can be effected by $(NH_2)_2SO_4$.

Robichon-Szulmajster and Corrivaux (90) deduced the existence of three distinct aspartate kinases in yeast, each with different inhibition sites and sensitive to different inhibitors.

B. Reduction of Aspartic β-Semialdehyde

This reaction (Fig. 4, reaction 2) is catalyzed by aspartate β-semialdehyde dehydrogenase (EC 1.2.1.11) (76,77) with NADPH serving as the hydrogen donor. The optimal pH region of the reaction was found to be at 8.0 in the forward direction and at pH 9.0 in the reverse direction. Estimation of the Michaelis constants for the different substrates yielded values of 0.16 mM for β-aspartyl phosphate and 8.3×10^{-5} M for NADPH in the forward reaction. For the reverse reaction, calculated K_m values were 2.6 mM for aspartate β-semialdehyde, 3.6×10^{-5} M for NADP, and 1.4 mM for phosphate. Participation of phosphate in the reverse reaction was demonstrated by the dependence of the equilibrium level of NADPH on the concentration of phosphate. The product of aspartyl phosphate reduction was established to be L-aspartate β-semialdehyde by its subsequent reaction with NADH, catalyzed by homoserine dehydrogenase to form homoserine. β-Aspartylphosphate formed by aspartate semialdehyde oxidation was identified by its reaction with ADP in the β-aspartate kinase system.

Attempts to estimate the equilibrium constant of the aspartyl phosphate-aspartate semialdehyde reaction yielded the approximate values of 2.5×10^6 at pH 8.7 derived from the forward reaction, and 3.4×10^6 at pH 7.4 from the reverse reaction.

The reaction catalyzed by aspartate β-semialdehyde dehydrogenase is analogous to the one catalyzed by 3-phosphoglyceraldehyde dehydrogenase. This similarity is shown by the parallel inhibition of both enzymes by iodoacetate and by the catalysis of an arsenolysis of its own acyl phosphate substrate by each enzyme.

C. Formation of Homoserine

Reduction of L-aspartate β-semialdehyde to L-homoserine is catalyzed by L-homoserine dehydrogenase (EC 1.1.1.3) (Fig. 4, reaction 3). The enzyme from yeast utilizes NADH preferentially (77); the bacterial enzymes utilize NADPH preferentially (91–93).

L-Homoserine dehydrogenase has been purified about 100-fold from yeast (77) and 1800-fold from *Rhodospirillum rubrum* (92). The enzyme is quite specific for L-aspartate β-semialdehyde and L-homoserine

With only the substrates present, optimum activity of the *R. rubrum* enzyme covers a range of pH 8–9.5. Modifiers alter the character of the pH activity curve. Thus isoleucine, which activates the enzyme, changes the optimum activity curve to give a sharp peak at pH 9.0 (92). L-Homoserine dehydrogenase of *E. coli* is reported to have optimum activity in the forward reaction at pH 11, and at pH 9.0 in the reverse reaction (93).

The reported kinetic constants are of the same order of magnitude for the enzyme from all sources with an occasional exception. The K_m for aspartate β-semialdehyde is 0.12–0.14 mM; for the reduced nicotinamide nucleotide it is 2.4–8.0×10^{-5} M. In the reverse reaction, K_m for homoserine is 1–2 mM, and in the range of $10^{-5} M$ for the oxidized nicotinamide nucleotide. The equilibrium constant of the reaction has been reported as 0.9×10^{11} for the yeast and 1.6×10^{11} for the *E. coli* enzyme.

The enzyme requires K^+ for activity and this is counteracted by Na^+ (92,93).

Homoserine dehydrogenase is also subject to end-product inhibition and repression. In most instances the predominant repressor is L-threonine (*91–93*). Other amino acids that may be repressing are methionine, diaminopimelic acid, and lysine.

Careful studies of inhibition and activation of homoserine dehydrogenase have been carried out by Patti *et al.* (*93*) and Datta and Gest (*92*). The former determined that, with the *E. coli* enzyme, L-threonine exerts both a negative feedback repression and end-product inhibition. Heating destroys the inhibition site for L-threonine on the enzyme.

Both activation and inhibition by threonine required the presence of K^+. The kinetics of L-threonine dehydrogenase were found to be biphasic with respect to homoserine and NADP. A linear relation was obtained by plotting $1/v$ versus $1/(\text{homoserine})^2$ or $1/(\text{NADP})^2$. The authors concluded that there were two binding sites each for homoserine and NADP. These were represented by K_m values of 0.2 and 1 mM for homoserine.

In a further study of the mechanism of L-threonine inhibition of L-homoserine dehydrogenase, Datta and Gest (*94*) determined that the L-threonine induces an aggregation of the enzyme to a dimeric inactive form. L-Homoserine, L-isoleucine, and L-methionine reverse the aggregation and counteract the inhibition by L-threonine.

D. Phosphorylation of Homoserine

The final enzymic steps in the conversion of homoserine to threonine were established by Watanabe *et al.* (*95–97*), who showed that at least two separable enzyme fractions are required (Fig. 4, reactions 4 and 5). These investigators separated homoserine kinase from a second enzyme and confirmed the formation of *O*-homoserine phosphate. Watanabe and co-workers also determined the conversion of *O*-homoserine phosphate to threonine.

Homoserine kinase has been enriched 15- to 25-fold from brewer's yeast extract (*97,98*). ATP is essential for the reaction, as would be anticipated, and evidence was obtained of activation by Mg^{2+}, Mn^{2+}, or Zn^{2+}. Enzyme activity was found to be inhibited by EDTA and PCMB (*97*).

E. Formation of Threonine

The enzyme that catalyzes the conversion of *O*-homoserine phosphate to threonine, threonine synthetase, has been purified 500-fold from *Neurospora* by Flavin and Slaughter (*98*). The purification procedure

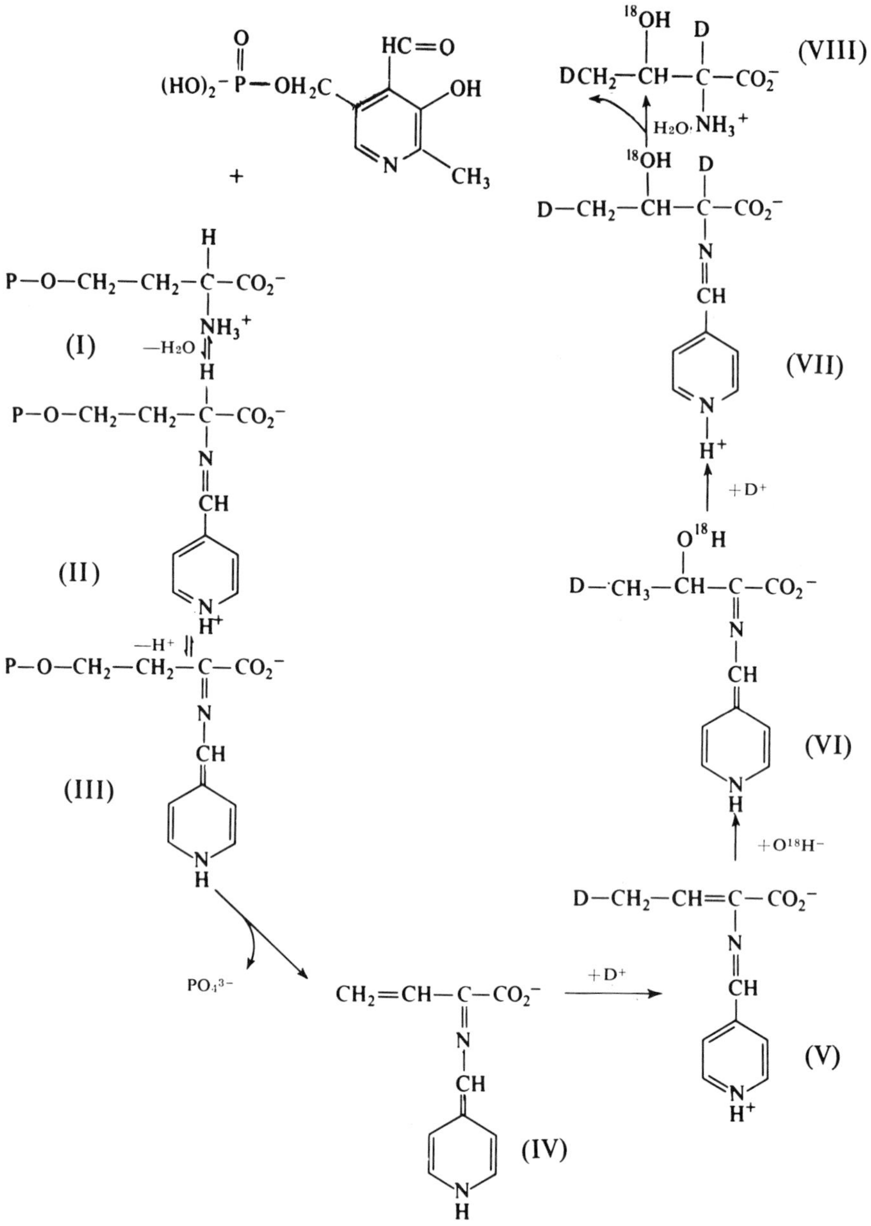

FIG. 5. Probable mechanism of threonine synthetase in formation of threonine. Modified from Flavin and Slaughter (*100*). Experiments with $H_2^{18}O$ and D_2O were performed independently. Pyridine substituents have been omitted from intermediates for convenience.

utilized acetone and ammonium sulfate precipitation and chromatography on a DEAE-cellulose column.

Threonine synthetase has been shown to require pyridoxal-P (98).

In a study of the mechanism of the synthetase reaction Flavin and co-workers determined the incorporation of ^{18}O (99) and of deuterium from the water (100) of the incubation medium into the products of the reaction. It was found that one atom of solvent oxygen was incorporated into threonine, and none into phosphate. This shows that the latter is removed by nonhydrolytic elimination, rather than hydrolysis. Incubation of O-phosphohomoserine with purified threonine synthetase in 100% D_2O resulted in the incorporation of two atoms of D in the newly formed threonine, one in the α-position, the second in the γ-position.

The probable mechanism of the reaction deduced by Flavin and Slaughter (100) from these isotope experiments is illustrated by Fig. 5. The reasoning of Flavin and Slaughter is that the reaction is initiated by Schiff-base formation between enzyme-bound pyridoxal-P and phosphohomoserine, followed by a reversible conversion to the quinoid tautomer in which the α-hydrogen of phosphohomoserine is lost as a proton (Fig. 5, I → II → III). In the quinoic tautomer (III) a β-hydrogen of phosphohomoserine, labilized by being adjacent to a series of conjugated double bonds, is eliminated as a proton, together with phosphate, with the formation of the corresponding derivative of vinylglycine (Fig. 5 III → IV). A solvent proton is now added to the position, coupled with the withdrawal of electrons from the pyridine ring (Fig. 5, VI → V). This leads to the formation of the α,β-unsaturated intermediate (IV). Addition of water in the predicted manner to the α,β-double bond will now introduce a second solvent hydrogen atom, in the α-position, and solvent oxygen in the hydroxyl, to yield the Schiff base of threonine (Fig. 5, V → VI → VII).

VI. BIOSYNTHESIS OF METHIONINE

A. Formation of Homocysteine

Homocysteine may be formed directly by reaction of homoserine of an O-alkylhomoserine with H_2S or by a more circuitous route through the synthesis of cystathionine and cleavage of the latter to homocysteine, pyruvate, and NH_3. The two reaction types can be represented by Eqs. (4–6). Which of these biosynthetic routes represents the major

$$\begin{array}{ccc}
\begin{array}{c} H_2C\!-\!O\!-\!R \\ | \\ CH_2 \\ | \\ CHNH_2 \\ | \\ COOH \\ \textit{O-Alkylhomoserine} \end{array} + H_2S \rightarrow &
\begin{array}{c} H_2C\!-\!SH \\ | \\ CH_2 \\ | \\ CHNH_2 \\ | \\ COOH \\ \text{Homocysteine} \end{array} + HOR & (4)
\end{array}$$

$$\begin{array}{cccc}
\begin{array}{c} H_2\!-\!C\!-\!OR \\ | \\ CH_2 \\ | \\ CHNH_2 \\ | \\ COOH \end{array} +
\begin{array}{c} HS\!-\!CH_2 \\ | \\ HCNH_2 \\ | \\ COOH \end{array} \rightarrow &
\begin{array}{c} H_2\!-\!C\!-\!S\!-\!\!\!-\!CH_2 \\ | \quad\quad | \\ CH_2 \quad HCNH_2 \\ | \quad\quad | \\ CHNH_2 \quad COOH \\ | \\ COOH \\ \text{Cystathionine} \end{array} + HOR & (5)
\end{array}$$

$$\begin{array}{cc}
\begin{array}{c} H_2C\!-\!S\!-\!CH_2 \\ | \quad\quad | \\ CH_2 \quad CHNH_2 \\ | \quad\quad | \\ CHNH_2\,COOH \\ | \\ COOH \end{array} \xrightarrow{+H_2O} &
\begin{array}{c} H_2C\!-\!SH \\ | \\ CH_2 \\ | \\ CHNH_2 \\ | \\ COOH \end{array} +
\begin{array}{c} CH_3 \\ | \\ CO \\ | \\ COOH \end{array} + NH_3 \quad (6)
\end{array}$$

physiological route for homocysteine formation is at present uncertain.

Homoserine has been shown to react directly with H_2S to form homocysteine, catalyzed by extracts of *Neurospora* (*101*). The same authors showed that *O*-substituted homoserine reacts more readily with H_2S than the free homoserine (*102*). They also observed that the enzyme from fungi (*Neurospora*, yeast) preferentially utilizes *O*-acetyl-L-homoserine; the enzyme from bacteria (*E. coli*, *Salmonella*) preferentially utilizes the succinyl derivative.

The magnitude of the difference in rates is shown by the kinetic data given in Table II.

TABLE II

KINETIC DATA OF FORMATION OF HOMOCYSTEINE[a]

Organism	Substrate	K_m (mM)	V_{max} (mμmoles/mg protein/hr) ($\times 10^{-2}$)
Neurospora	Homoserine	0.55	0.2
Neurospora	Acetylhomoserine	4.5	11.1
Yeast	Homoserine	0.06	0.08
Yeast	Acetylhomoserine	5.5	18
E. coli	Homoserine	—	—
E. coli	*O*-Succinylhomoserine	3.9	7.1

[a] From Weibers and Garner (*102*).

Formation of homocysteine by reaction of O-succinylhomoserine with H_2S in the presence of bacterial extracts has been confirmed by Flavin and Slaughter (*103*).

1. FORMATION OF O-SUCCINYL-L-HOMOSERINE

The intermediate formation of O-succinyl-L-homoserine in the biosynthesis of cystathionine catalyzed by bacterial extracts was discovered by Rowbury and Woods (*104*) and by Flavin *et al.* (*105,106*). Rowbury and Woods observed the occurrence of a new amino acid in culture fluids of *E. coli* and *S. typhimurium* that gave a yellow-brown color with ninhydrin. This intermediate was shown to be formed from succinyl-CoA and homoserine in homocysteine or methionine requiring auxotrophs of *E. coli* and *Salmonella*.

By comparison with the synthetic product the new amino acid was identified to be O-succinylhomoserine. Both synthetic and isolated succinylhomoserine yielded cystathionine on incubation with cysteine in the presence of extracts of the above bacteria. Flavin and Slaughter (*106*) synthesized O-succinylhomoserine and showed that it was rapidly spontaneously converted to N-succinylhomoserine.

Incubation of O-succinylhomoserine, but not N-succinylhomoserine, with cysteine in the presence of a *Salmonella* enzyme yielded L-cystathionine. This was proved by isolation of the enzymically formed compound and comparison with synthetic L-cystathionine (*107*).

The enzyme catalyzing formation of O-succinylhomoserine has been tentatively named homoserine O-transsuccinylase. No reports have appeared on its purification and properties aside from the facts that the succinylation reaction proceeds optimally at about pH 7.5 and the enzyme is strongly inhibited by methionine. This is presumed to be a regulatory process controlling L-cystathionine synthesis in *E. coli* (*108*).

Table II shows that *Neurospora* and yeast virtually utilize only O-acetylhomoserine for the synthesis of cystathionine. This has also been shown to be the case for plants (*109*).

2. FORMATION OF CYSTATHIONINE VIA O-ALKYL-L-HOMOSERINES

In bacteria, formation of cystathionine has been established to occur by condensation of O-succinyl-L-homoserine with cysteine according to Eqs. (5) and (6) (*110,111*).

The enzyme catalyzing formation of cystathionine has been obtained in an essentially pure state from a *Salmonella* mutant blocked in a subsequent step of methionine biosynthesis (*110,111*). Isolation of the enzyme cystathionine γ-synthetase was made easier by an assay based on the decomposition of O-succinyl-L-homoserine catalyzed by the same enzyme in the absence of cysteine according to Eq. (7). The α-ketobu-

$$O\text{-Succinyl-L-homoserine} + H_2O \rightarrow \text{succinate} + NH_4^+ + \alpha\text{-ketobutyrate} \qquad (7)$$

tyrate formed in this reaction is determined spectrophotometrically by incubation with NADH and lactate dehydrogenase. The mutant strains selected were me-A and me-B shown in Fig. 6 for the controlling genetic steps in the biosynthesis of methionine.

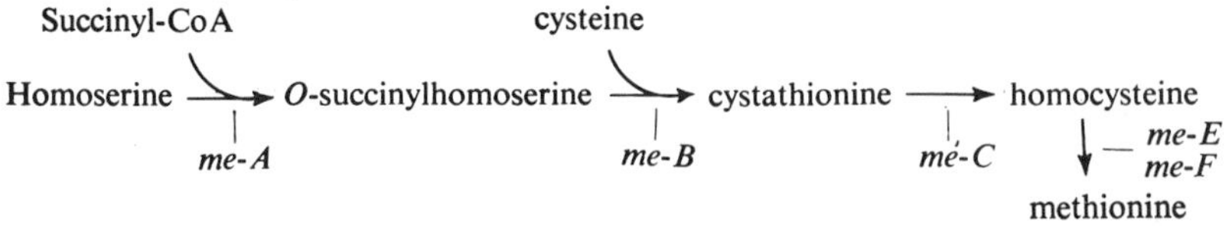

FIG. 6. Genetic pattern of homocysteine and methionine synthesis in *Salmonella typhimurium* (*110*).

The cystathionine γ-synthetase content of these auxotrophs was several-fold higher than in the wild-type *Salmonella* and could be further increased by growing them on certain sulfur compounds that increased the doubling time.

Optimum activity for cystathionine formation is at pH 7.5 and K_m values of the substrates are 7×10^{-5} M for L-cysteine and 4×10^{-3} for O-succinyl-L-homoserine.

About a 225-fold purification of the bacterial extract yielded the essentially pure enzyme. This was a yellow protein of molecular weight 160,000. The color is characteristic of a number of pyridoxal-P enzymes. The enzyme had an absorption maximum at 422 mμ that was independent of pH between 5.5 and 9.2. In 6 M guanidine the enzyme was completely dissociated into 4 subunits of MW 40,000. Each subunit contained 2 reactive cysteine residues, at least one of which was required for catalytic activity. The treatment with guanidine exposed 2 additional SH groups.

Turnover numbers were determined to be 15,000 moles per mole of enzyme per minute for synthesis of cystathionine and 3000 for the decomposition of O-succinylhomoserine.

O-Acetyl-L-homoserine serves as a substrate for the reaction at about 10% of the rate of the succinyl compound. Preliminary evidence has been obtained that, in *Neurospora*, O-acetyl-L-homoserine takes the place of the succinyl derivative in the synthesis of cystathionine (*112*). This evidence consists of the isolation of the acetyl derivative from culture media and the observation that a certain mutant will grow on this compound in place of methionine.

3. FORMATION OF HOMOCYSTEINE FROM CYSTATHIONINE

Bacteria, fungi, and higher plants cleave cystathionine to homocysteine, ammonia, and pyruvate as shown by Eq. (6). In vertebrates the cleavage is to cysteine. This reaction also occurs in *Neurospora*.

Flavin and Slaughter have separated two cystathionine cleavage enzymes from *Neurospora*, one of which catalyzes the β-cleavage to homocysteine, Eq. (6), and the other catalyzes the γ-cleavage to cysteine (*113,114*). Bacteria contain but a single cleavage enzyme catalyzing the reaction represented by Eq. (6) (*113,116*).

The β-cleavage cystathionase was found to be less stable than the γ-cleavage enzyme and to be inhibited by sulfhydryl reagents. The γ-elimination cystathionase that yields cysteine has been purified 400-fold from *Neurospora* (*115*). Its properties resemble closely those of the similar enzyme crystallized from rat liver (*117*). The β-cleavage cystathionase has not been purified appreciably.

Considerable doubt has arisen that cystathionine is an important physiological intermediate in the formation of homocysteine, and from it the biosynthesis of methionine. The view has been expressed that cystathionine may only be a storage form for sulfur in either bacteria or fungi.

Thus Wiebers and Garner (*118*) found virtually no incorporation of the four-carbon chain of cystathionine when labeled with ^{14}C into methionine in *Neurospora*, whereas ^{14}C-homoserine was a significant precursor. As has been previously mentioned, homocysteine can be formed from an *O*-alkyl-L-homoserine by an enzymic direct reaction with H_2S. Dulavier-Klutchko and Flavin (*116*) point out that the pathway of cystathionine synthesis from cysteine has not been established in higher fungi.

B. Formation of Methionine by Methylation of Homocysteine

The route of major physiological importance for the methylation of homocysteine to form methionine appears to be that in which 5-methylfolate-H_4 or its congeners is the methyl group donor.

A number of other methylation reactions have been reported that appear to be of minor physiological significance. Several observers have reported the direct methylation of homocysteine by reaction with methylmercaptan (*113,118*) and an *O*-alkylhomoserine. This explains the ability of certain leaky methione-less mutants to grow sparingly on *S*-methylcysteine.

1. METHYL TRANSFER VIA S-ADENOSYLMETHIONINE

Methylation of homocysteine to methionine has also been reported by transfer of the methyl group from S-adenosylmethionine or methylmethionine (119,120) according to Eqs. (8) and (9).

$$\text{Homocysteine} + \text{adenosylmethionine} \rightarrow \text{methionine} + \text{adenosylhomocysteine} + \text{H}^+ \qquad (8)$$

$$\text{Homocysteine} + \text{methylmethionine} \rightarrow 2 \text{ methionine} + \text{H}^+ \qquad (9)$$

This reaction is catalyzed by an enzyme provisionally named S-adenosylmethionine-homocysteine methyltransferase. It has been demonstrated to occur in cell-free extracts of various microorganisms, rat liver, and seeds of some higher plants. This enzyme has been purified over 500-fold from baker's yeast (119). The weakness of this pathway as a general method of methionine methyl formation is that there is no net synthesis of methionine according to the reactions of Eqs. (8) and (9), and methylmethionine does not appear to be a significant intermediate in the metabolic reactions of methionine.

2. METHIONINE METHYL FORMATION VIA 5-METHYLPTEROYLGLUTAMATES

Progress in clarifying the mechanism of *de novo* methionine methyl synthesis has come from a study of the reaction with extracts of mutant strains of *E. coli*. A long-time study of this problem has been carried out by Woods and co-workers (119,121–130) in which they showed that under certain conditions vitamin B_{12} was required for the synthesis. The subject was taken up later by Buchanan and co-workers (131–136) who discovered that the actual methyl donor was N^5-methylfolate-H_4 and that a flavin nucleotide was involved as a coenzyme in its formation. Recent reviews of this topic are contained in (119–119b).

The first observation in Woods' laboratory was that washed suspensions of *E. coli* could synthesize methionine upon addition of homocysteine with serine as the C_1 donor.

In Buchanan's laboratory the methionine methyl synthesizing system was separated into three partially purified enzyme fractions (133). One of these was serine hydroxymethyltransferase (EC 2.1.2.1) when serine was used as the source of the C_1 moiety. A second fraction of the system was later shown to reduce 5,10-methylenefolate-H_4 to 5'-methylfolate-H_4, and a third was found to contain a vitamin B_{12} cobamide as a prosthetic group.

Woods' group subsequently determined that there are two pathways for methionine methyl formation in *E. coli*, one that has no requirement for a vitamin B_{12} coenzyme but that can utilize only 5-methylpteroyl-triglutamate, and a second pathway that is cobamide-dependent but is indifferent as to the number of glutamyl residues on the methylpteroyl-glutamate. This was confirmed in the laboratory of Buchanan.

5-Methylfolate-H_4 is formed by the reduction of 5,10-methylene-folate-H_4 catalyzed by 5,10-methylenefolate-H_4 reductase (EC 1.1.1.18), a flavoprotein enzyme.

This compound was isolated and identified by Larrabee *et al.* (*131*). After the identification of the new cofactor of the methionine methyl-ating reaction, it became apparent that this was the same compound previously discovered by Donaldson and Keresztesy (*137*) and named by them prefolic A.

Sakami and Ukstins (*138*) showed that 5-methylfolate-H_4 also was the methyl donor for methionine methyl synthesis in mammalian liver. Methylation of homocysteine can proceed in the liver. Methionine is nutritionally essential for vertebrates because they lack the enzyme system that synthesizes the four-carbon chain of methionine.

The vitamin B_{12} methylating system has turned out to be quite complex. The reaction mechanism for the transfer of the methyl group from 5-methylfolate-H_4 to homocysteine has been partially, but not completely, clarified. Required components of the enzyme system are, in addition to 5-methylfolate-H_4, adenosylmethionine, $FADH_2$, NAD, a cobamide coenzyme, and a reducing agent ($FADH_2$ serves in this capacity).

Guest *et al.* (*139*) demonstrated a transfer of methyl from synthetic methyl-B_{12} to homocysteine according to Eq. (10). These authors

$$\text{Methyl-}B_{12} + \text{homocysteine} \rightleftharpoons \text{methionine} + B_{12} \tag{10}$$

postulated that methyl-B_{12} served as an intermediate methyl carrier between 5-methylfolate-H_4 and homocysteine.

In 1963 Mangum and Scrimgeour (*140*) demonstrated a requirement for *S*-adenosylmethionine in the vitamin B_{12} methionine synthesizing system. The function of this compound apparently is to supply a methyl group that is required structurally for activation of the vitamin B_{12} transmethylase (*140a*). Some clarification of the details of the reaction mechanism resulted from experiments performed by Rosenthal *et al.* (*119,141*). These authors found that 5-methylfolate-H_4-homocysteine methyltransferase was not completely specific for either of the above two substrates. *S*-Adenosylhomocysteine exhibited a low degree of activity in place of 5-methylfolate-H_4, and other thiol compounds, such as 2-mercaptoethanol, could serve as methyl group acceptors, although homocysteine was far more active. The reactions with 5-methyl-folate-H$_4$ and S-adenosylmethionine were independent of each other. Other investigators found that methyl iodide and S-adenosyl ethionine could partially replace S-adenosylmethionine (*140a,141a*).

In model experiments Elford *et al.* (see *119*) observed that Factor B, a form of vitamin B$_{12}$ lacking the nucleotide, acquired ^{14}C-methyl in a system containing ^{14}C-methylfolate-H$_4$, adenosinylmethionine, FADH$_2$, Factor B, and the transmethylase from either pig liver or *E. coli*. S-Adenosylmethionine can also serve as a methyl donor to vitamin B$_{12}$ (*142*) or Factor B (*119*). It has been concluded that the fifth and sixth ligands of cobalt in the vitamin B$_{12}$ must be free for it to react as a methyl acceptor.

From the findings discussed above, Elford *et al.* (see *119*) proposed the scheme shown in Fig. 7 to represent the sequence of reactions for the synthesis of methionine methyl.

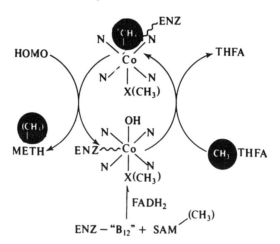

Fig. 7. Proposed mechanism of transmethylation reaction in biosynthesis of methionine (*119*). SAM, CH$_3$THFA, HOMO, and METH refer to S-adenosylmethionine, 5-methyl-tetrahydrofolate, homocysteine, and methionine, respectively.

A great advance in the elucidation of the enzymic mechanism of the methylation reactions catalyzed by the vitamin B$_{12}$ enzyme (N^5-methylfolate-H$_4$-homocysteine transmethylase) has been made by the recent studies of Taylor and Weissbach (*140a,143,144*). These investigators isolated a salmon-colored preparation of 5-methylfolate-H$_4$-homocysteine transmethylase estimated to be about one third pure. The preparation contained a tightly bound vitamin B$_{12}$ derivative which reacts with cyanide. The absorption spectrum of the prosthetic group in the visible region resembles vitamin B_{12r} (a derivative containing Co^+). Estimates made of the sedimentation coefficients and molecular weight of the enzyme are 7.1 S and 140,000, respectively. This enzyme can catalyze the three transmethylation reactions shown in Eqs. (11), (12), and (13).

$$N^5\text{-Methylfolate--}H_4 + \text{homocysteine} \rightarrow \text{methionine} + \text{folate-}H_4 \quad (11)$$

$$S\text{-Adenosyl-L-methionine} + \text{homocysteine} \rightarrow \text{methionine} + S\text{-adenosyl-L-homocysteine} \quad (12)$$

$$\text{Methyl-}B_{12} + \text{homocysteine} \xrightarrow{\text{aerobic}} \text{methionine} + \text{hydroxy-}B_{12} \quad (13)$$

Taylor and Weissbach (143) isolated the cobamide from the 5-methyl-folate-H_4 homocysteine transmethylase and determined that it had properties characteristic of sulfito-B_{12}.

Transmethylation from 5-methylfolate-H_4 or S-adenosylmethionine, according to the reactions of Eqs. (11) and (12), required a reducing system. For a reducing system, 2-mercaptoethanol and cyano-B_{12} can serve but a reducing system that generates reduced flavin mononucleotide is 3–4 times as active. The cyano-B_{12} functions by accelerating the rate of reduction of the 2-mercaptoethanol.

Reaction (13), utilizing 5,6-dimethylbenzimidazolylcobamide methyl, proceeds aerobically.

Propylation of the enzyme inhibits methyl transfer from S-adenosyl-methionine, but not from methyl-B_{12}. The propylated enzyme decomposes on exposure to light. The extent of methyl group transfer from labeled S-adenosylmethionine is not reduced by unlabeled 5-methyl-folate-H_4. This finding suggests that 5-adenosylmethionine and 5-methylfolate-H_4 do not transfer their methyl groups to a common site. As mentioned above, the evidence is suggestive that a methyl group derived from S-adenosylmethionine is required at some site different from the active methylation site to activate the B_{12} enzyme. These results are concordant with the reaction scheme shown in Fig. 7.

It has been long known that methionine served as a repressor of its own synthesis in bacteria. Katzen and Buchanan (145) determined that this resulted from a repression of 5,10-methylenefolate-H_4 reductase. Vitamin B_{12} also is a repressor at high concentrations. This is not a direct action, but is due to an augmentation of methionine biosynthesis.

5,10-Methylenefolate-H_4 reductase was purified 100-fold from de-repressed E. coli (145). Optimum activity of the enzyme is at pH 6.3–6.4. With the purified enzyme it was found that NADH does not directly reduce the 5,10-methylenefolate in the enzyme system. FADH is required for this function. A second enzyme was separated in the enzyme system that is required to catayze the reduction of FAD by NADH.

VII. BIOSYNTHESIS OF HISTIDINE

A. Historical Development

The pattern of histidine biosynthesis began to emerge when certain metabolites accumulated by histidineless mutants of *N. crassa* and *E. coli* were isolated and identified. Other clues were an interrelationship between histidine and purine metabolism shown by a sparing effect by histidine of the purine requirement of *Lactobacillus casei* (*146,147*). A little later it was established that the purines were the source of at least a portion of the carbons and nitrogen of the imidazole ring (*147–149*).

B. Formation of the Alanyl Side Chain

A series of histidineless mutants of *Neurospora* and *Penicillium* were observed to accumulate imidazole-like substances (*150*). Two of these compounds were isolated as the mercury salt and separated on a Dowex 50 column. They were identified to be imidazoleglycerol and imidazoleacetol (*151*). Although these compounds were not active as growth factors for the histidineless mutants, it was considered that they might provide clues to the biosynthesis pathway of histidine.

Previously Vogel *et al.* (*152*) had isolated the compound L-histidinol from an *E. coli* histidineless mutant and determined that this compound was a growth stimulant for another mutant. L-Histidinol was also isolated from mold cultures by Ames and Mitchell. These results indicated a sequence for the formation of the side chain of histidine, and suggested that it was derived from a pentose.

A further advance in developing the scheme of biosynthesis of the histidine side chain was made by the discovery and identification of the phosphate esters of the above-mentioned compounds, namely, imidazoleglycerol phosphate [4-(D-*erythro*-trihydroxyproply-3′-phosphate)-imidazole], imidazoleacetol phosphate [4-(2-keto-3-hydroxypropyl-3′-phosphate)imidazole], and histidinol phosphate [L-4-(2-amino-3-hydroxypropyl-3′-phosphate)imidazole] (*151,153*). Success in obtaining the phosphate esters depended on preventing the hydrolysis of these compounds by phosphatases of the *Neurospora* mycelium. This was accomplished first by freezing the material and then adding the frozen powdered mycelium to boiling water. The phosphate esters were isolated by chromatography on Dowex columns. The products were identified by paper chromatography and by a variety of chemical tests and comparisons with synthetic materials. The phosphates could not be tested as growth factors, since *Neurospora* and mutants of other microorganisms that could be used are impermeable to the phosphate esters. The order of appearance of these compounds in cultures of different histidineless mutants of *Neurospora* is in harmony with the proposed biosynthetic sequence. Definite proof that the above compounds are precursors of histidine was obtained by the isolation of a series of enzymes that could interconvert these phosphate esters and eventually lead to the formation of histidine. The results established the scheme of biosynthesis shown in Fig. 8.

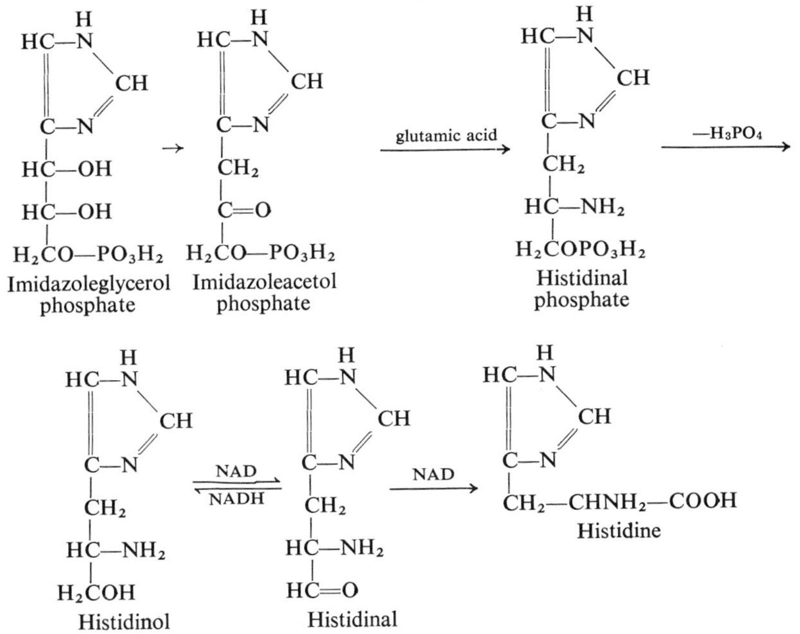

FIG. 8. Pathway of histidine biosynthesis from imidazoleglycerol phosphate.

C. Biosynthesis of the Imidazole Ring

By means of tracer experiments it was determined that various bacteria could form the N-1 and C-2 portion of the imidazole ring of histidine from the C-2 and an attached nitrogen atom of guanine (*147–149*). Later experiments showed that adenine is a more immediate precursor of histidine than is guanine, and also that the N-3 of histidine is derived from the amide nitrogen of glutamine (*153*).

Subsequently Moyed and Magasanik (*153*) obtained the synthesis of D-*erythro*-imidazoleglycerol phosphate ester in cell-free preparations from three enteric bacterial species. This compound is a precursor of histidine. Evidence was also obtained that the imidazoleglycerol phosphate was formed from ribose 5-phosphate, the amide nitrogen of glutamine, and the N-1 and C-2 portion of the adenine ring of ATP. The residue of the ATP appears as 5-amino-l-ribosyl-4-imidazolecarboxamide 5′-phosphate. The latter is a well-known intermediate in purine biosynthesis and can be reconverted to ATP. This provides a cyclic process for the synthesis of the imidazole ring of histidine.

The formation of 5-amino-1-ribosyl-4-imidazolecarboxamide 5′-phosphate by bacterial extracts from ATP, ribose 5-P, and glutamine had been reported previously (*153a*). This reaction was confirmed by Moyed and Magasanik who found, in addition, that a derivative of the above compound accumulated when glutamine was omitted (*153*) that yielded aminoribosylimidazolecarboxamide phosphate on mild acid hydrolysis. These investigators showed that ATP is required not merely to supply energy, but that it is also the source of the aminoribosyl-imidazolecarboxamide phosphate produced.

Since aminoribosylimidazolecarboxamide phosphate is not a known precursor for purine biosynthesis, and various experimental results were obtained which indicated that this compound was not being utilized for purine formation by the bacterial preparations, Moyed and Magasanik were led to suspect that this compound was involved in the biosynthesis of the imidazole ring of histidine. This suggestion was supported by the finding of the accumulation of D-*erythro*-imidazole-glycerol phosphate ester as a product of these reactions.

Incubation of sonic extracts of several *A. aerogenes* bacterial strains with ATP, ribose 5-P, glutamine, and acetyl-P yielded an arylamine and an imidazole. When these incubations were performed with either ribose 5-P-1-^{14}C, AMP-8-^{14}C, or adenine-2-^{14}C as an additional reactant, the arylamine and the imidazole accumulated in sufficient amounts to be isolated and identified. After purification the imidazole was shown to be imidazoleglycerol phosphate and the arylamine, 5-amino-1-ribosyl-4-imidazolecarboxamide 5′-phosphate. The identification was made by paper chromatography, certain chemical properties, and absorption spectra.

The C-1 of ribose 5-P was incorporated into imidazoleglycerol phosphate without dilution. When AMP-8-^{14}C was employed, the amino-ribosylimidazolecarboxamide phosphate had the same radioactivity as the reisolated adenosine 5′-phosphate (AMP); the imidazoleglycerol phosphate had no detectable activity. The C-2 of adenine, on the contrary, was incorporated into imidazoleglycerol phosphate without dilution, but was not incorporated into the aminoribosylimidazolecarboxamide phosphate.

The results support the assumption that the adenine ring is cleaved into two moieties, one containing the C-2 of adenine combined with ribose 5-P to form imidazoleglycerol phosphate, and a residual moiety, containing C-8, that is converted to the arylamine.

D. Genetic and Enzymic Observations

Our present knowledge of the details of the compounds and reactions of histidine biosynthesis is largely derived from genetic and enzyme studies on *Salmonella typhimurium* by Ames, Hartman, Martin, and their co-workers (*154–165*). Hartman *et al.* established that a short region of the chromosome of *Salmonella* contains an operon that stores the genetic information for the specification of the 10 enzymes that catalyze the conversion of PRPP and ATP (see Fig. 9 for meaning of terms) to histidine. The genetic map of the operon has been developed in great detail. The information on the histidine operon is reported in a number of comparatively recent reviews (*154–157*). The intermediates in the pathway have now been isolated and characterized, and assays for all 10 enzymes have been devised.

The reactions of each of the 10 sequential steps of the biosynthetic pathway and the cistrons controlling each enzyme are shown in Fig. 9.

E. Intermediate Compounds

Phosphoribosyl-ATP was isolated from incubation mixtures and isolated by chromatography on Sephadex G-25 and DEAE-cellulose. Its structure was established by chemical analysis and by its conversion to later metabolites in the biosynthesis pathway (*159*). The second intermediate, phosphoribosly-AMP, was isolated and characterized by Smith and Ames (*163*). It was demonstrated that this compound is formed from phosphoribosyl-ATP by one enzyme and is converted to *N*-(5′-phospho-D-ribosylformimino)-5-amino-1-(5′-phosphoribosyl)-4-imidazolecarboxamide [PR-AMP(B.B.M.II)] by another enzyme. PR-AMP was isolated by DEAE-cellulose column chromatography. Its structure was deduced from chemical analysis and from comparative spectral data with respect to PR-ATP and the stoichiometry of its formation.

The two compounds, phosphoribosylformimino AIC-RP and phosphoribulosylformimino AIC-RP (B.B.M. III), were also identified by Smith and Ames before the identification of PR-AMP.

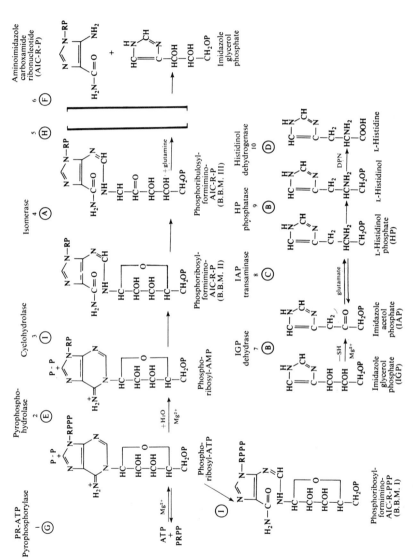

FIG. 9. Sequential steps and cistrons controlling enzymes of histidine biosynthesis (*166*).

By carrying out incubation of PR-ATP in the absence of glutamine with a cell-free extract of A mutants, the compound B.B.M. II accumulated. This is converted to the intermediate by extracts of all except A mutants. Both compounds have absorption spectra different from that of PR-ATP and contain a bound diazotizable amino group. B.B.M. III differs from B.B.M. II in possessing a reducing sugar. In the presence of glutamine, B.B.M. III can be converted to AIC-RP and imidazole-glycerol phosphate (IGP). The two new intermediates were isolated as the previous ones by DEAE-cellulose column chromatography. Deduction of the structures of B.B.M. II and B.B.M. III was made from chemical analysis and the nature of the reactions for their formation and further transformation.

The products from imidazoleglycerol phosphate to histidine had been isolated and identified earlier, as is described in Section VII, A.

F. Enzymes of Histidine Biosynthesis

All of the 10 enzymes in the biosynthesis pathway have been purified to varying degrees. Their approximate molecular weights have been estimated by sucrose gradient centrifugation (*156,162*). The estimated values are recorded in Table III.

ATP-PRPP phosphoribosyltransferase is assayed by the increase in absorbancy at 290 mμ. It was purified from derepressed histidineless

TABLE III

APPROXIMATE MOLECULAR WEIGHTS OF ENZYMES OF HISTIDINE BIOSYNTHESIS

Step	Gene	Enzyme	S	Molecular weight[a] ($\times 10^{-3}$)
1	G	PR-ATP phosphoribosyl transferase	7	215
2	E	PR-ATP pyrophospho-hydrolase	3.5	43
3	I	Cyclohydrolase	3.8	48
4	A	Isomerase	—	29
5	H	Amidotransferase	3.5	44
6	F	Cyclase	3.4	41
7, 9	B	IGP dehydrase-HP phosphatase	79	145
8	C	IAP aminotransferase	—	59
10	D	Histidinol dehydrogenase	—	86

[a] Molecular weight determined by sucrose gradient centrifugation. Data from *156, 162*.

mutants (hisEF-135 and hisE-11). The most recent preparation (*161*) is nearly completely pure. The molecular weight is 215,000 as estimated by equilibrium ultracentrifugation. On treatment with guanidine, polypeptide units of MW 35,000 are obtained, apparently of identical kind. This indicates that the intact anzyme consists of 6 polypeptide units. The amino acid composition has been determined and the aminoterminal amino acid shown to be valine. The enzyme is quite unstable and difficult to preserve.

The enzyme reaction has been shown to be reversible (*159*). This enzyme is the control point in the regulation of histidine biosynthesis. It is strongly inhibited by L-histidine and is coded by a gene immediately adjacent to the operator gene. The enzyme is inhibited by sulfhydryl reagents (*160*). K_m values are 0.2 mM for ATP and 6.7×10^{-5} M for PR-ATP.

The inhibition by histidine has been studied by Martin (*160*). The inhibition is specific and noncompetitive with both substrates. Histidine binding alone does not lead to inhibition. The binding has to be accompanied by a conformational change to produce inhibition. Desensitization to histidine inhibition can be induced by reaction with Hg^{2+} accompanied by a progressive inactivation. Removal of the Hg^{2+} yields a relatively stabilized desensitized enzyme. Sensitivity to histidine can be recovered by incubating the enzyme with ATP, Mg^{2+}, and mercaptoethanol. Enzyme stored at 3° also becomes desensitized to histidine. It is evident that inhibition of the enzyme by histidine is an allosteric effect.

PR-ATP-pyrophosphohydrolase. This is the second enzyme in the sequence. The conversion of PR-ATP to PR-AMP is measured by the evolution of P_i, since the enzyme preparations contain pyrophosphatase. Optimum activity is at pH 9.0 (*164*). K_m for PR-ATP was estimated to be 2.2×10^{-5} M.

Phosphoribosyl-AMP-1,6-cyclohydrolase. This enzyme catalyzes the opening of the purine ring at C-6 to form phosphoribosylformimino-AIC-RP (B.B.M. II). This is a diazotizable amino compound. The K_m for PR-AMP has been estimated to be 2.1×10^{-5} M at pH 8.5. The product PR-AIC has a higher absorption at 290 mμ ($\epsilon = 8 \times 10^3$), than PR-ATP ($\epsilon = 3.6 \times 10^3$), so that enzyme activity can be assayed by increased absorption at the above wavelength. PR-AIC gives the bound Bratten-Marshall reaction, and this can be used as an alternative analytical method.

Gene I, which controls the formation of phosphoribosyl-AMP-1,6-cyclohydrolase consists of 2 cistrons. This indicates that the enzyme consists of at least two dissimilar polypeptide chains, a Ia chain and a

Ib chain (154). The two polypeptides have not as yet been separated and characterized.

Phosphoribosylformiminophosphoribosylaminoimidazolecarboxamide Ketolisomerase. The name of this enzyme is abbreviated to isomerase for convenience. It is the fourth enzyme in the histidine pathway. It is controlled by the A gene of the histidine operon. It has been purified by Margolies and Goldberger (167,168). It catalyzes the conversion of B.B.M. II to B.B.M. III by an Amadori rearrangement (163) involving the isomerization of an aminoaldose to an aminoketose and yielding the product *N*-(5′phospho-D-1′-ribulosylformino)-5-amino-1-5″-phosphoribosyl)-4-imidazolecarboxamide. The molecular weight is estimated to be 29,000 as determined by equilibrium ultracentrifugation. Since the molecular weight is unchanged by reduction or alkylation, the enzyme consists of a single polypeptide chain in agreement with the genetic data. The carboxyl terminal end has been determined to be valine, and the amino acid composition has been analyzed.

The isomerase enzyme was assayed by the reductive properties of B.B.M. III using a tetrazolium dye mix. Measurement was made of the increase in absorption at 520 mμ. The pH optimum of the reaction is at 7.7. The K_m for B.B.M. II is 5.7×10^{-5} M.

Conversion of B.B.M. III to IGP and PR-AIC. There are two steps involved in the conversion of B.B.M. III to the above-mentioned products. The enzyme controlled by the H gene catalyzes addition of the amide group from glutamine. The enzyme of the F gene catalyzes conversion of the compound represented by the brackets (whose structure has not been established) in Fig. 9 to AIC-RP and IGP.

K_m values for glutamine and B.B.M. III have been determined from the overall reaction; these are 1×10^{-2} and 2.4×10^{-5} M, respectively. The F gene enzyme has been purified about 20-fold by chromatography on G-200.

Imidazoleglycerol Phosphate Dehydrase. This, the 7th enzymic reaction, is controlled by gene B. A water molecule is removed from D-*erythro*-imidazoleglycerol phosphate in the reaction, resulting in the formation of imidazoleacetol phosphate (165,170). The ninth reaction is the hydrolysis of L-histidinol phosphate to L-histidinol and P$_i$ (171). In *Neurospora* these two steps are catalyzed by independent proteins (see above and 169), but in *Salmonella* a single protein appears to catalyze both of these nonsequential reactions (171,172). Attempts to separate the two enzyme activities have failed, and there is a considerable body of evidence that they are properties of a single protein.

Imidazoleacetol Phosphate Aminotransferase. This enzyme catalyzes the eighth step of histidine biosynthesis and is controlled by gene C.

The reaction consists of the conversion of imidazoleacetol phosphate and L-glutamate to L-histidinol phosphate and α-ketoglutarate (*165,173*). The purified enzyme has a molecular weight of about 68,000. It dissociates in guanidine hydrochloride into what appears to be two identical subunits.

L-*Histidinol Dehydrogenase.* This is the final enzymic step in histidine biosynthesis. It is controlled by gene D. Gene D is exceedingly complex with numerous subdivisions. The gene is now generally divided into two segments termed Da and Db.

The enzyme is comprised of two different polypeptide subunits elicited by the separate Da and Db cistrons.

Histidinol dehydrogenase catalyzes the two step reactions given in Eq. (14).

$$\text{L-Histidinol} \xrightarrow[NAD^+ \quad NADH+H^+]{} \text{L-Histidinal} \xrightarrow[+H_2O]{NAD^+ \quad NADH+H^+} \text{Histidine} \tag{14}$$

The enzyme has been purified 100-fold from brewer's yeast (*174*). Crystalline histodonal dehydrogenase has been obtained from derepressed cells of *S. typhimurium* (*175,175a*). The amino acid composition has been determined, and evidence has been obtained that the enzyme is a dimer of two probably identical subunits.

Optimum activity was determined as a function of pH with both histidinol and histidinal; the former was at about pH 9.3, the latter at about pH 7.5. The pH peak is sharp for the former and very broad for the latter substrate. Estimated K_m values are 1.15 and 0.78 M for histidinol and histidinal, respectively. The enzyme is specific for NAD^+, which has a K_m of about 1.2 mM. Mn^{2+} increases enzyme activity when histindinol is the substrate, not with histidinal. The enzyme has a molecular weight of 86,000 and is reported to be composed of 4 polypeptide chains, 2 of MW 16,000 and 2 of MW 27,000 (*157*).

Genetic Regulations. Studies of derepression of the enzymes of the histidine operon have indicated that there are two alternative modes of derepression of the first five enzymes of the biosynthetic chain. Mutants

that are incapable of forming 4-amino-5-imidazolecarboxamide ribonucleotide show simultaneous derepression; mutants producing the
above intermediate exhibit a temporal sequence, with about 20 minutes
elapsing between the increase in the first and last enzyme. The presence
of adenine shifts the sequential to simultaneous derepression. From the
above it is concluded that the histidine operon is transcribed into a poly-
cistronic message from one end only in sequential derepression and in
simultaneous derepression translation is initiated at multiple sites (*168*).

VIII.   BIOSYNTHESIS OF AROMATIC AMINO ACIDS

A. Introduction

An extensive body of knowledge has been accumulated on the pathways of biosynthesis of the aromatic amino acids. Much of this has been
derived from the elegant investigations of B. D. Davis and his associates
(*176,177*).

By the use of penicillin to eliminate bacteria capable of growing on
simplified media (glucose, citric acid, and inorganic ions), Davis and
co-workers isolated mutants of *E. coli* and *Aerobacter aerogenes* that
required a mixture of the four aromatic compounds: tyrosine, phenylalanine, tryptophan, and *p*-aminobenzoic acid for growth. A little later
a stimulation by *p*-hydroxybenzoic acid for these mutants was also
discovered. After extensive tests with many compounds it was established that shikimic acid could replace the above aromatic compounds
to secure growth of these mutant bacteria. With this clue to the nature
of the compounds involved in the biosynthesis of the aromatic amino
acids, other intermediates were rapidly discovered and identified. This
was followed by the discovery of enzymes that catalyze individual reactions in the synthetic sequence. Distribution studies on the labeling
of the aromatic products derived from early precursors (various
^{14}C-carbohydrates and their metabolic products) yielded information
necessary to obtain a completed picture of the aromatic biosynthetic
process.

B. Phenylalanine and Tyrosine Biosynthesis

1. INTERMEDIATE CYCLIC COMPOUNDS

Success in the search for the intermediates of aromatic biosynthesis
was greatly favored by the discovery that these compounds could be

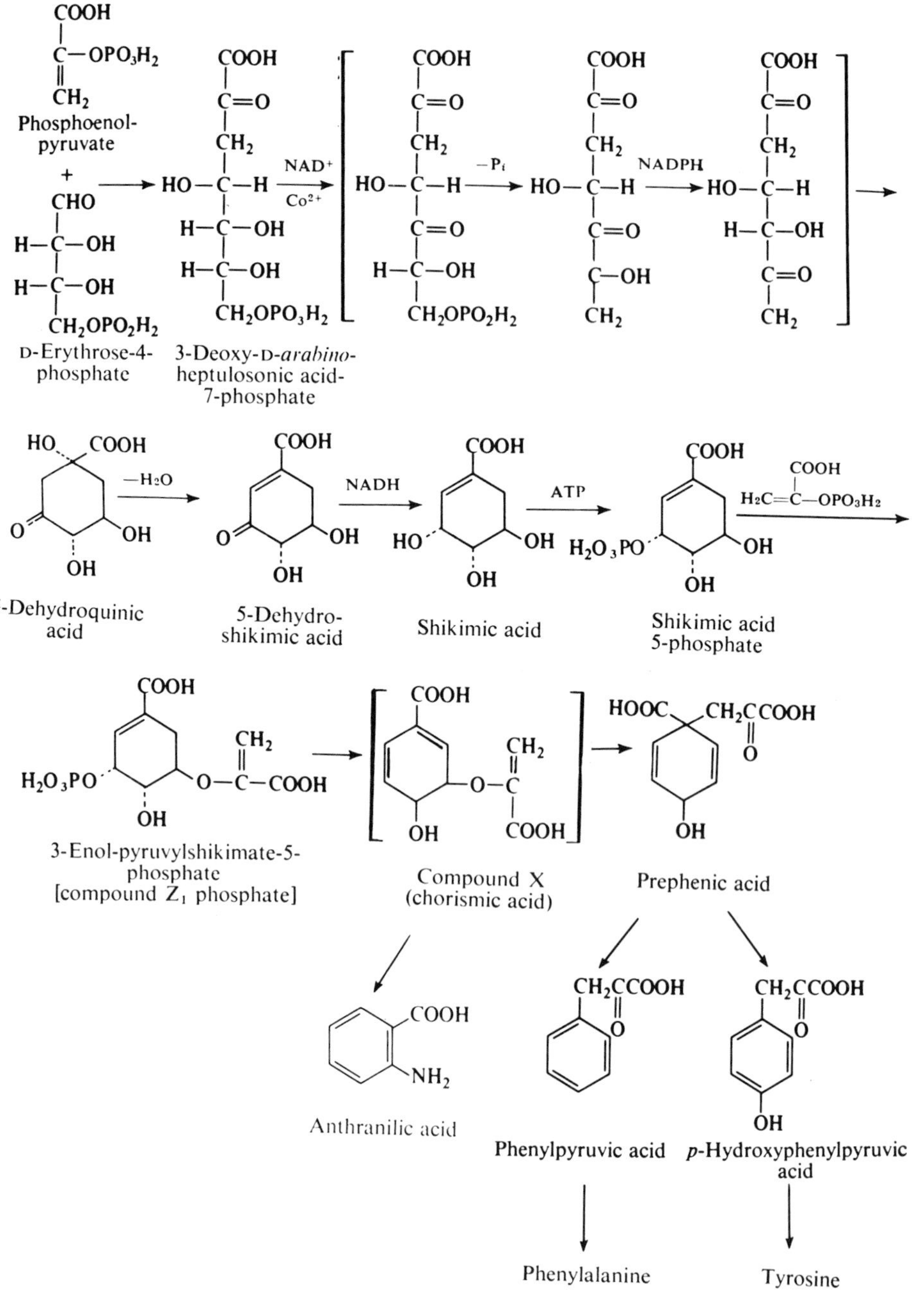

COOH
$C-OPO_3H_2$
CH_2
Phosphoenol-
pyruvate
+
CHO
$H-C-OH$
$H-C-OH$
$CH_2OPO_2H_2$
D-Erythrose-4-
phosphate

$\xrightarrow{NAD^+ \ Co^{2+}}$

COOH
$C=O$
CH_2
$HO-C-H$
$H-C-OH$
$H-C-OH$
$CH_2OPO_3H_2$
3-Deoxy-D-*arabino*-
heptulosonic acid-
7-phosphate

COOH
$C=O$
CH_2
$HO-C-H$
$C=O$
$H-C-OH$
$CH_2OPO_2H_2$

$\xrightarrow{-P_i}$

COOH
$C=O$
CH_2
$HO-C-H$
$C=O$
$C-OH$
CH_2

$\xrightarrow{NADPH}$

COOH
$C=O$
CH_2
$HO-C-H$
$H-C-OH$
$C=O$
CH_2

5-Dehydroquinic
acid

$\xrightarrow{-H_2O}$

5-Dehydro-
shikimic acid

$\xrightarrow{NADH}$

Shikimic acid

$\xrightarrow{ATP}$

Shikimic acid
5-phosphate

$\xrightarrow{H_2C=C-OPO_3H_2 \atop COOH}$

3-Enol-pyruvylshikimate-5-
phosphate
[compound Z_1 phosphate]

Compound X
(chorismic acid)

Prephenic acid

Anthranilic acid

Phenylpyruvic acid

p-Hydroxyphenylpyruvic
acid

Phenylalanine

Tyrosine

separated from most other components of the culture filtrates by chromatography on charcoal and also by the availability of appropriate mutants for bioassay.

The identified intermediates in the aromatic biosynthesis pathway are given in Fig. 10. Shikimic acid ($3\beta,4\alpha,5\alpha$-trihydroxy-$\Delta^{1,6}$-cyclohexene-1-carboxylic acid) was the first one to be isolated and identified, as already mentioned (178). It was found to be the only compound that satisfied the total requirement for growth of the bacterial mutants exhibiting a quintuple aromatic requirement. The amount of shikimic acid producing maximal growth was approximately the same as that of the combined quintuple aromatic supplement.

After this advance it was observed that mutants blocked before shikimic acid accumulated earlier intermediates. One of these was identified to be 5-dehydroshikimic acid (179). The isolated compound was determined to be a seven-carbon monocarboxylic acid (pK 3.2) with two acylable hydroxyl groups. The compound is levorotatory and has an absorption band in the ultraviolet. Besides the chemical evidence, the fact that the compound could be converted to shikimic acid by mutants that accumulate the latter strongly supports the conclusion that this intermediate is dehydroshikimic acid. Purification of dehydroshikimic acid could be followed easily, since it is converted to shikimic acid on autoclaving.

A third intermediate that accumulated in mutants blocked before dehydroshikimic acid was shown to be 5-dehydroquinic acid (180). Dehydroquinic acid was found to be moderately stable in acid and extremely unstable in alkali. On heating in dilute acid (pH 1–5) at 100° it is partly converted to dehydroshikimic acid. The structure of dehydroquinic acid was established from its infrared spectra, its chemical characteristics, analytical composition, and its conversion to 5- dehydroshikimic acid.

Beyond shikimic acid three compounds were detected in culture filtrates that are themselves completely devoid of growth-promoting activity, but which yield growth factors on being autoclaved. One of these was found to yield shikimic acid on acid hydrolysis (181). On isolation this compound was identified as 5-phosphoshikimic acid. The position of the phosphate group was established by exclusion of the other possible positions. The compound is completely hydrolyzed by potato phosphatase to shikimic acid and phosphate. The reason why it is not a growth factor may be its impermeability to the cell. Shikimic acid can be converted to the 5-phosphate by the mutants that accumulate the latter.

The second inactive compound also yields shikimic acid on heating in

FIG. 10. Intermediate compounds of aromatic amino acid biosynthesis (177a).

acid. It is more labile than 5-phosphoshikimic acid. From the pattern of its accumulation Davis concluded that this compound arises at a latter stage in the biosynthetic chain. Subsequently it was established that this unknown compound (Z_1) was 3-enoylpyruvylshikimate 5'-phosphate (*182*).

The most interesting of the three inert compounds is prephenic acid. This compound was found to accumulate in cultures of *E. coli* mutants that require phenylalanine for growth. In acid solution, prephenic acid forms a substance which is an active growth factor for the same phenyl-alanine mutant which accumulates it (*183,184*). This growth-promoting substance has been shown to be phenylpyruvate.

The great instability of prephenic acid, which has a half-life of only 130 hours at pH 7 and 13 hours at pH 6, made it extremely difficult to isolate and characterize. This was accomplished by chromatography on charcoal in alkaline solution (*185*). The purification of prephenic acid could be followed by its acid-catalyzed conversion to phenyl pyruvate. In the conversion of prephenic acid to phenyl pyruvate there was a release of one equivalent of CO_2. A variety of evidence indicated that it has the chemical structure shown in Fig. 10.

Still another intermediate between 3-enoylpyruvylshikimate-5-P and prephenic acid was discovered by Gibson and Gibson (*186,187*) and named chorismic acid (from the Greek: branch point) (2-[2'-hydroxy-5-carboxy-1'-(R)2'(R)-dehydrophenoxy]-acrylic acid). The reason for the name is seen in Fig. 10, namely, that biosynthesis of tryptophan branches off at chorismic acid from the main pathway of the synthesis of pheny-lalanine, tyrosine, and certain other aromatic compounds. Chorismic acid is readily converted by warming into a mixture of prephenate and 4-hydroxybenzoate. In acid it is further converted to phenyl pyruvate.

2. INTERMEDIATE CARBOHYDRATE COMPOUNDS

The bacterial mutant technique failed in establishing noncyclic precursors of the aromatic amino acids. This gap, however, was bridged by isotopic tracer experiments.

The first experimenters studied the distribution of the label in the aromatic amino acids of yeast and other microorganisms grown on such ^{14}C-compounds as acetate, pyruvate, and glucose (*188–191*). This procedure had the defect that the two sides of the aromatic ring could not be distinguished in these amino acids. The results obtained suggested that carbohydrate derivatives were utilized in the formation of the aromatic ring and that the side chain of tyrosine and phenylalanine was almost certainly derived from an intact three-carbon glycolytic fragment, probably pyruvate.

Srinivasan *et al.* (*192*) overcame the difficulty of symmetry by studying the labeling in shikimic acid accumulated by a mutant of *E. coli* with glucose or other carbohydrates as the sole carbon source. The data obtained from the use of variously ¹⁴C-labeled glucose is shown in Fig. 11. The numbers outside the parentheses represent the position

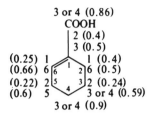

Fig. 11. Incorporation of glucose carbon atoms into shikimic acid. Whole numbers beside each position refer to numbered carbon atom of glucose; adjacent numbers in parentheses represent the fraction of the shikimic atom derived from the numbered glucose atom.

of the label in glucose, and the numbers within the parentheses are the fractions of the carbon atoms incorporated into the indicated carbon atom of shikimic acid. It is to be noted that the results are consistent with the deduction that the carboxyl, C-1, and C-2 portion of shikimic acid is derived from a three-carbon intermediate of glycolysis, while the remaining four-carbon portion might be formed from a tetrose. The findings suggested that synthesis of the cyclic aromatic intermediates might be from a seven-carbon carbohydrate derivative. Sedoheptulose 7-phosphate and the nature of the pentose phosphate cycle was discovered at this time, so that interest shifted to this carbohydrate.

To test this hypothesis, experiments were performed on the conversion of various carbohydrates to 5-dehydroshikimic acid by extracts of *E. coli*. Cells extracts were chosen to eliminate permeability barriers, and dehydroshikimic acid was selected because the equilibrium between dehydroquinic acid and dehydroshikimic acid favored the accumulation of the latter.

Sedoheptulose 7-phosphate was found to be converted to dehydroshikimic acid about as well as various phosphorylated hexoses. When the heptose phosphate was incubated together with fructose 1,6-diphosphate, the degree of conversion was doubled. This led to the suggestion that a reaction product, sedoheptulose 1,7-diphosphate, might be the more direct substrate. Test of sedoheptulose 1,7-diphosphate yielded a very high conversion to dehydroshikimic acid (20%). When NAD was added to this, the degree of conversion was almost quantitative (80%). Since the synthesis of dehydroshikimic acid from sedoheptulose disphosphate involves the removal of two hydrogens as well as two molecules of water, participation of a pyridine nucleotide is understandable.

Further labeling experiments were performed with sedoheptulose 1,7-diphosphate-4,5,6,7-^{14}C. The results showed that carbon atoms 1 to 3 of sedoheptulose diphosphate gave rise to carbon atoms 7, 1, and 2 of shikimic acid, and C-4 to C-7 gave rise to C-3 to C-6 of shikimic acid. The result is consistent with the cyclization of the intact sedoheptulose diphosphate to form shikimic acid. This result, however, is inconsistent with the labeling derived from variously labeled glucose. This is shown in Fig. 12. In the formation of sedoheptulose disphosphate by condensation of a triose phosphate with tetrose phosphate under the influence of aldolase, carbon atoms 1, 2, and 3 of this compound would be derived from the glucose atoms shown in Fig. 12. Cyclization of the

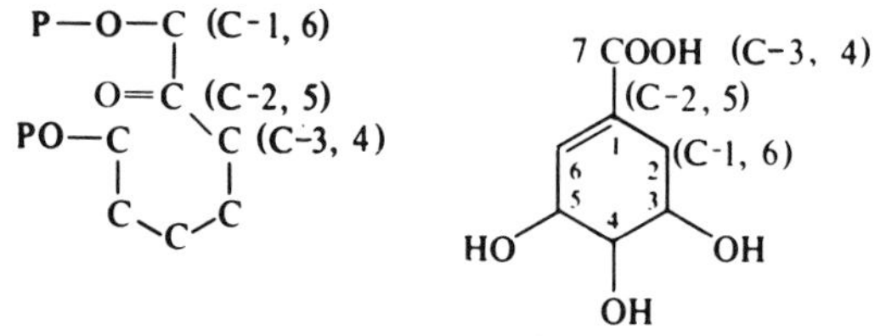

FIG. 12. Distribution of carbon atoms from glucose (C-1, C-2) into sedoheptulose diphosphate shikimic acid.

intact sedoheptulose diphosphate would then yield shikimic acid with the same sequence in positions C-7, C-1, and C-2. The reverse sequence of glucose atoms was actually found in shikimic acid formed from specifically labeled glucose. It would appear that fragment 1, 2, 3 of sedoheptulose 7-P was detached and inverted prior to its incorporation into shikimic acid. This assumption has been corroborated by the demonstration that dehydroshikimic acid is rapidly and quantitatively formed from phosphoenolpyruvate and D-erythose 4-phosphate (*192*). The results indicate that sedoheptulose diphosphate is converted to the above products prior to the synthesis of the cyclic aromatic intermediates. Evidence for this is that, whereas synthesis of dehydroshikimic acid from erythrose phosphate and phosphoenolpyruvate (PEP) is not inhibited by fluoride or iodoacetate, the synthesis from sedoheptulose diphosphate is completely inhibited by these reagents (*193*). The inhibition by fluoride could be reversed by phosphoenolpyruvate, that by iodoacetate by 3-phosphoglycerate. This suggests that glycolysis from triose phosphate or phosphoenolpyruvate is involved in the conversion of sedoheptulose diphosphate to dehydroshikimate.

Srinivasan *et al. (194,195)* speculated that, to initiate the aromatic biosynthesis, D-erythrose 4′-phosphate condenses with phosphoenolpyruvate to form 2-keto-3-deoxy-7-phospho-*d*-glucoheptonic acid in which C-4 and C-5 have the same configuration as C-3 and C-4 of dehydroshikimic acid.

Origin of the side chain of the aromatic amino acids and of prephenic acid from a three-carbon glycolytic intermediate was established by isotopic results, as previously mentioned. Evidence of this is that the β-carbon of tyrosine is derived about equally and almost entirely from C-1 and C-6 of glucose *(189)*, and that the α-carbon was highly labeled while the β-carbon was unlabeled in tyrosine derived from pyruvate-α-^{14}C *(191)*.

C. Enzymes and Reaction Mechanisms

1. 3-Deoxy-D-*arabino*-heptulosonic 7-Phosphate Aldolase
[7-Phospho-2-keto-3-deoxy-D-*arabino*-heptonate: D-erythrose-4-phosphate-lyase (pyruvate-phosphorylating), EC 4.1.2.15]

The reaction which initiates the biosynthesis of the aromatic amino acids is the condensation of pyruvylenol phosphate and D-erythrose 4-phosphate to 3-deoxy-D-*arabino*-heptulosonic 7-phosphate* (DAHP) and P$_i$. The enzyme catalyzing this reaction, DAHP synthetase, has been purified about 60-fold from an *E. coli* mutant *(196)*. The product appeared homogeneous on ultracentrifugation. The reaction product was assayed by reacting it with periodate, which yields β-formylpyruvate. The latter on condensation with thiolbarbituric acid gives an intense pink color. DAHP was isolated from the incubation media by paper chromatography and identified by comparison with the synthetic compound. The isolated DAHP was shown to form 5-dehydroquinate. The reaction was essentially irreversible and proceeded to the quantitative formation of DAHP. Kinetic properties are: maximum activity at pH between 6.4 and 7.4; $K_m = 1.2$ mM for erythrose 4-P and 3.5 mM for PEP. The enzyme is inhibited by a number of sugars and appears to be an SH enzyme.

It is reversibly inhibited by PCMB and attains maximal activity in the presence of thiol compounds, e.g., 2-mercaptoethanol.

---

* This name replaces the previously used, 2-keto-3-deoxy-D-*arabo*-heptonic acid 7-phosphate *(190)*.

Srinivasan and Sprinson suggest that the mechanism of the condensation is initiated by a nucleophilic attack on PEP represented by OH⁻, Eq. (15), resulting in the release of P_i from the open-chain form of DAHP.

$$
\begin{array}{ccc}
\begin{array}{l}
\text{COOH} \quad \text{O} \\
| \quad | \quad | \quad O^{\ominus} \\
\text{C--O--P} \longrightarrow O^{\ominus} \\
| \qquad \qquad \searrow OH^{\ominus} \\
\text{CH}_2 \\
| \\
\text{H}^+\text{O}{=}\text{CH} \\
| \\
\text{HC--OH} \\
| \\
\text{HC--OH} \\
| \\
\text{CH}_2\text{OP}
\end{array}
&
\begin{array}{l}
\text{COOH} \\
| \\
\text{CO} \\
| \\
\text{CH}_2 \\
| \\
\text{HOCH} \\
| \\
\text{HCOH} \\
| \\
\text{HCOH} \\
| \\
\text{CH}_2\text{OP}
\end{array}
&
\begin{array}{l}
\text{COOH} \\
| \\
\text{HOC}---\!\! \\
| \\
\text{CH}_2 \\
| \\
\text{HO--C--H} \quad \text{O} \\
| \\
\text{HC--OH} \\
| \\
\text{HC}---\!\! \\
| \\
\text{CH}_2\text{OP}
\end{array}
\end{array}
\tag{15}
$$

DAHP

The significance of DAHP synthetase for the control of aromatic biosynthesis has been critically reviewed by Doy (*196a*). From a study of the enzyme in *N. crassa* (*196b*) he concluded that DAHP synthetase is an allosteric enzyme activated by its two substrates (erythrose 4-P and PEP) that have a strong cooperative interaction. He also concluded that there are a minimum of two active enzyme sites per substrate species. The reaction kinetics were deduced to be of the ping-pong type with PEP adding first. The enzyme is inhibited by its three amino acid end products phenylalanine, tyrosine, and tryptophan at different enzyme sites in the approximate ratio of 44 : 44 : 10.

2. 5-DEHYDROQUINATE SYNTHETASE

The second reaction step in the aromatic pathway is the synthesis of dehydroquinic acid from DAHP. The enzyme catalyzing this reaction has been purified 100- to 200-fold (*197*). NAD and Co^{2+} are required for enzyme activity. At very low concentrations of Co^{2+} (10^{-6} M), NADH is as active as NAD, but it becomes inhibitory at higher concentrations of the cation (5×10^{-6} M). The optimum activity extends over a plateau region between pH 7.4 and 8.4. A K_m value of 0.1 mM was obtained for NAD.

The conversion of DAHP to 5-dehydroquinate would appear to be a multistep process, because of the great differences in structure. No intermediates could be detected when tests were conducted by a variety of procedures, nor could the enzyme system be resolved into component enzymes. From these findings it was concluded that any intermediates, if they are formed, are enzyme-bound.

Srinivasan *et al.* (cited in *197*) have proposed the hypothetical reaction scheme shown in Fig. 13 to explain possible steps in the overall reaction. This scheme has the advantage of explaining the requirement for catalytic amounts of NAD. The requirement for NAD is invoked to facilitate the elimination of P_i. The essential features of the scheme are the initial oxidation of the hydroxyl group of DAHP with 'NAD as the cofactor. The carbonyl group of C-5 is subsequently reduced with NADH to a hydroxyl group having the original configuration. The resulting 2,6-diketone is finally cyclized to 5-dehydroquinate.

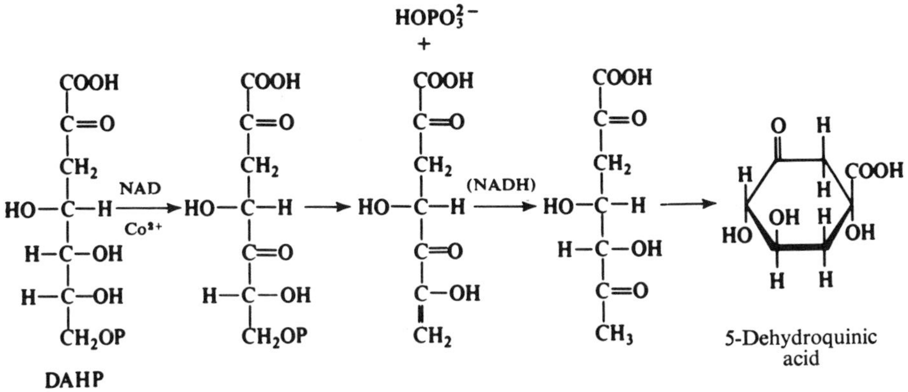

FIG. 13. Hypothetical scheme for the conversion of DAHP to 5-dehydroquinate. Intermediates (II) to (IV) are assumed to be enzyme-bound (*197*).

One of the postulated intermediates between DAHP and 5-dehydroquinate, namely, 3,7-dideoxy-D-*threo*-hepto-2,6-diulosonic acid (II) has been synthesized and tested (*198*). It was not a substrate for DAHP synthase, but is cyclized to 5-dehydroquinate spontaneously at pH 11 by intramolecular aldol condensation or in imidazole buffer at pH 7.

3. 5-DEHYDROQUINATE DEHYDRATASE (5-Dehydroquinate hydro-lyase, EC 4.2.1.10)

An 8-fold-enriched preparation of this enzyme has been obtained from a mutant of *E. coli* (*199*). It is easily assayed by the absorption maximum at 234 mμ shown by 5-dehydroshikimate. The reaction product, 5-dehydroshikimate, was identified by bioassay and on isolation by paper chromatography. The enzyme is apparently specific for dehydroquinic acid. It has a very shallow peak with respect to pH, with a maximum at pH 8.0. The K_m for dehydroquinate was estimated to be 4.4×10^{-5} *M*.

The reaction was demonstrated to be reversible with an equilibrium ratio of (dihydroshikimate/dihydroquinate) of 15 at pH 8.0.

5-Dehydroquinate dehydratase has also been partially purified from cauliflower buds (200). Its properties are very similar to those of the bacterial enzyme.

4. Shikimate Dehydrogenase
   (Shikimate: NADP oxidoreductase, EC 1.1.1.25)

This enzyme has been partially purified from *E. coli* (201) and from etiolated peas (202). It catalyzes the reaction:

$$\text{5-Dehydroshikimate} + \text{NADPH} + \text{H}^+ \rightleftharpoons \text{shikimate} + \text{NADP}^+ \tag{16}$$

The enzyme is specific for NADP. It has a pH optimum at 8.5 and K_m values of 3.1×10^{-5} and 5.5×10^{-5} M for NAD and shikimate, respectively. The equilibrium constants of the reaction as written in Eq. (16) with the omission of H^+ from the reactants were calculated at $30°$ to be 5.7 at pH 7.9 and 27.7 at pH 7.0.

The plant enzyme was purified even more extensively (80-fold). It has a higher pH optimum (pH 10.0). It was shown to be an SH enzyme, being inhibited by PCMB and iodoacetate. Cysteine reverses the PCMB inhibition. From a study of the inhibitory effects on the enzyme of a number of substituted phenolic compounds, Balinsky and Davies (202) concluded that attachment of the two substrates to the dehydrogenase is through the carboxyl and 3- and 4-hydroxyl groups.

Mitsuhashi and Davis (203) discovered a dehydrogenase (EC 1.1.1.24) that catalyzes the oxidation of quinic acid to dehydroquinic acid in certain mutants of *Aerobactor*. NAD is the required pyridine nucleotide coenzyme for this enzyme. The distribution of this enzyme and other evidence led to the conclusion that quinic acid is a side product and is not on the direct path of aromatic biosynthesis.

5. 3-Enoylpyruvylshikimate 5-Phosphate Synthetase

This enzyme has been enriched about 4-fold from a strain of *E. coli* blocked beyond shikimate that accumulates prephenic acid (182). It was demonstrated that 3-enoylpyruvylshikimate 5-phosphate was formed from shikimate 5-P + PEP. The product of the reaction was isolated as the barium salt and identified. Fluoride is required in the incubation medium to prevent the dephosphorylation of the enoylpyruvylshikimate 5-phosphate. The reaction was readily reversible, indicating that the enoyl ether is a high-energy compound. The equilibrium constant for the reaction was estimated to be 15 at pH 6.1 and 37°. Kinetic characteristics of the enzyme are a broad range of optimum activity at pH 5.4–6.2 and K_m values of 0.24 and 0.34 mM for phosphoenoylpyruvate and shikimate 5-P, respectively.

The mechanism of the synthesis is suggested to be very similar to that for the formation of DAHP (Section VIII, C, 1).

6. CHORISMATE SYNTHESIZING ENZYME

This enzyme has been partially separated in a study of the biosynthesis of anthranilate from enoyl SH-5-P, but has not been further characterized (204) (see Section IX).

7. CHORISMATE MUTASE

Study of the enzymic formation of prephenate from chorismate in *A. aerogenes* resulted in the separation of two peaks with chorismate mutase activity on DEAE-cellulose columns (187). One enzyme peak (designated chorismate mutase-P) traveled with prephenate dehydratase and was inhibited by phenylalanine, but not by tyrosine. The second peak (designated chorismate mutase-T) traveled with prephenate dehydrogenase and was inhibited only by tyrosine. The two enzyme activities in each peak were simultaneously affected by mutations. Heating for 4 minutes at 42° caused about a 60% loss of mutase-P and prephenate dehydratase activities, but did not effect the mutase-T and prephenate dehydrogenase activities.

From these observations it is concluded that chorismate mutase-P is intimately and specifically associated with phenylalanine biosynthesis, while mutase-T is correspondingly associated with tyrosine biosynthesis (Fig. 14).

8. PREPHENATE DEHYDRATASE
[Prephenate hydro-lyase (decarboxylating)]

This enzyme catalyzes the conversion of prehenate to phenyl pyruvate. It is assayed by the absorbance of the 320 mμ peak of the latter in alkaline solution. This enzyme has been detected in extracts of *A. aerogenes* as part of the chorismate mutase-P complex but has not been further purified nor characterized (181).

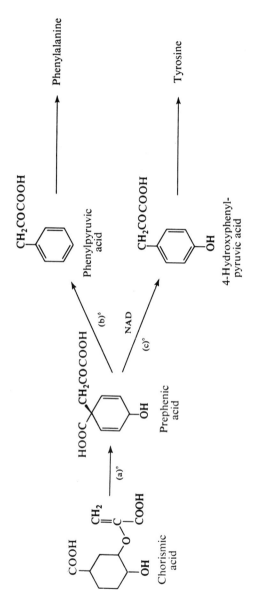

Fig. 14. The conversion of chorismic acid to phenylalanine and tyrosine (187).

9. PREPHENATE DEHYDROGENASE
[Prephenate: NAD oxidoreductase (decarboxylating)]

In *A. aerogenes* this enzyme is part of the chorismate-mutase-T complex (*187*). A soluble preparation of the dehydrogenase has been purified about 8-fold by ammonium sulfate precipitation from sonic extracts of *E. coli* (*205*). The reaction catalyzed is represented by Eq. (17).

$$\text{Prephenate} + \text{NAD}^+ \longrightarrow \text{4-Phenyl pyruvate} + CO_2 + NADH + H^+ \qquad (17)$$

The enzyme was found to have a plateau of activity between pH 6.5 and 9.0. The K_m of prephenate was estimated to be approximately 2 mM at pH 6.5.

Conversion of phenyl pyruvate and 4-hydroxyphenyl pyruvate to phenylalanine and tyrosine, respectively, is accomplished by transamination. These reactions are discussed in Chapter 14, Section III, C.

D. Genetic Regulation

This topic has been investigated by a number of workers in *E. coli* (*206–209*). The major genetic control site apparently is 7-phospho-2-oxo-3-deoxy-D-*arabino*-heptinate aldolase. Smith *et al.* (*206*) reported the occurrence of two aldolases, one strongly inhibited by phenylalanine, the other by tyrosine. Subsequently Doy and Brown (*208*) demonstrated the presence of a third adolase, specifically inhibited by tryptophan. The three aldolases are designated *arom* 1a, 1b, 1c, respectively. The tryptophan-sensitive aldolase is normally strongly associated with the tyrosine one, but can be prepared free from this by suppressing synthesis of the latter by growing the organisms in the presence of tyrosine.

From the levels of enzyme activity and the growth rates on the three aromatic amino acids it has been estimated that the three aldolases control normal amino acid synthesis in the ratio of 47 : 41 : 12 mole % of phenylalanine, tyrosine, and tryptophan, respectively.

Only one DAHP aldolase is found in *B. subtilis* (*209*). It is not subject to feedback inhibition by any of the aromatic amino acids. It is inhibited noncompetitively by prephenic and chorismic acids. An indirect control over DAHP aldolase is suggested by inhibition of the enzymes involved in the utilization of chorismate and prephenate.

IX. BIOSYNTHESIS OF TRYPTOPHAN

The biosynthesis of tryptophan branches off from the main pathway of aromatic biosynthesis at chorismic acid (Fig. 10). It has been known for a number of years that chorismic acid incubated with bacterial extracts in the presence of glutamine, or less effectively with ammonia, is converted to anthranilic acid (*210,211*). It had previously been shown that the amide nitrogen of glutamine was the source of the amino group of anthranilate (*212*). It has also previously been shown that anthranilate could be formed from 3-enolpyruvylshimikate 5-P if fortified with Mg^{2+}, NAD, and a NADH regenerating system (*213*).

From an examination of the structures of chorismate and anthranilate it would appear that the conversion is a very complex operation and probably involves a number of steps. The details of the transformation have not been further resolved to date.*

A. Formation of Indoleglycerophosphate

The further conversion of anthranilate to tryptophan has been analyzed in detail and the intermediate steps are well known. The steps between anthranilate and indoleglycerophosphate are shown in Fig. 15.

The carboxyl carbon is lost in the conversion of anthranilic acid to indole. Consequently two additional carbon atoms must be supplied to complete the pyrrole ring of the indole. The observation that various ribose derivatives could be the source of these two carbon atoms provided the clue that led to the elucidation of the mechanism of indole synthesis in the tryptophan biosynthetic pathway (*214*). Yanofsky determined that sonic extracts of a tryptophan auxotroph of *E. coli* (that also grew on anthranilate or indole) could utilize ribose, ribose 5-P, and 5-phosphoribosylpyrophosphate to form indole from anthranilate. With the two former compounds, ATP was essential for the reaction. With the latter compound, it was not. This result made it appear evident that 5-phosphoribosylpyrophosphate was the more immediate reactant in the condensation with anthranilate.

---

* Anthranilate synthetase has been partially purified from several bacterial sources [I. Baker and I. P. Crawford, *J. Biol. Chem.* **241**, 5571 (1966); A. F. Egan and F. Gilson, *Biochem. Biophys. Acta* **130**, 276 (1966)] and shown to be an aggregate of two protein components [J. Ito and C. Yanofsky, *J. Biol. Chem.* **241**, 4112 (1966)]. H. Zalkin and D. Kling [*Biochemistry* **I**, 3566 (1966)] separated component I from *S. typhimurium* and found that it utilized NH_4 exclusively, in contrast to the native enzyme for which glutamine is the preferred donor. Evidence was obtained that component I is an SH enzyme.

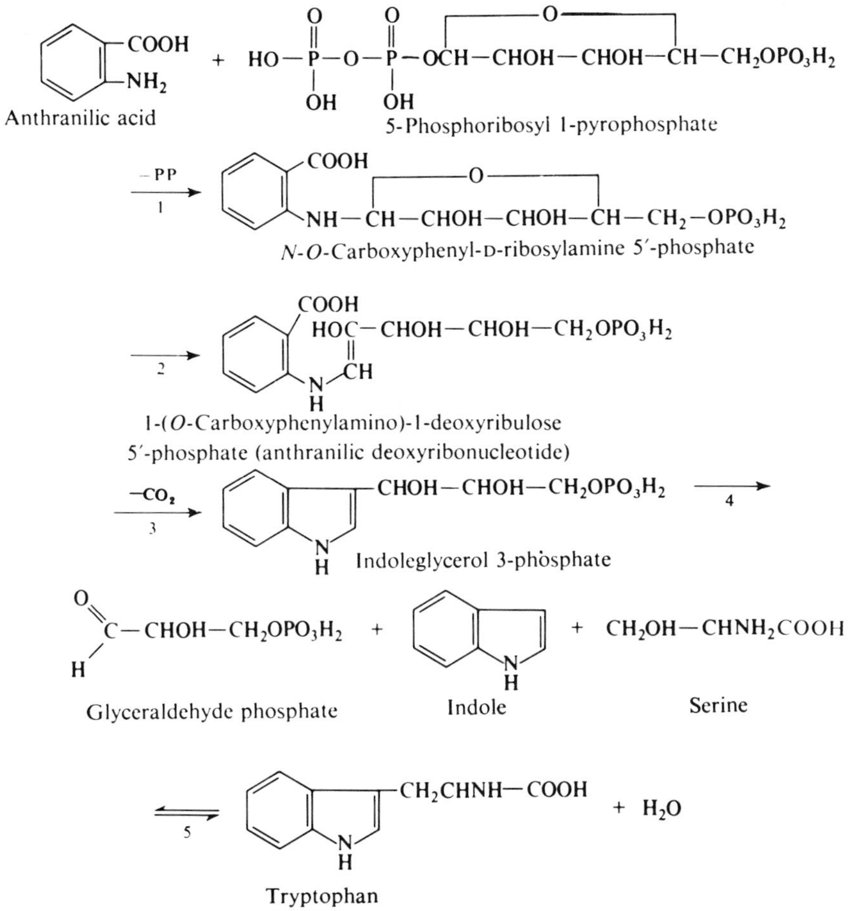

FIG. 15. Proposed scheme for biosynthesis of tryptophan.

Indirect support for this hypothesis was derived from experiments with various ^{14}C-labeled compounds (215). When ribose-1-^{14}C was the carbon donor, the C-2 of indole became exclusively labeled. Use of glucose-1-^{14}C also led to the labeling of the indole C-2, The C-2 of glucose was found to appear in C-2 and C-3 of the pyrrole ring of indole, with most of the label appearing in the C-3. The C-6 of glucose did not appreciably appear in either of the above carbons. The findings showed that utilization of glucose was by a circuitous process; utilization of ribose was by a very direct reaction.

Subsequently Yanofsky (216) separated the enzymes involved in the conversion of anthranilic acid to indole into two fractions and isolated the reaction product, indole-3-glycerole phosphate. One of the enzyme fractions catalyzed the formation of indole-3-glycerol phosphate from anthranilate and 5-phosphoribosylpyrophosphate.

Yanofsky postulated the initial reaction product of the above reaction would be N-O-carboxyphenyl-D-ribosylamine-5-P (reaction 1, Fig. 15). By dehydration and an Amadori-type rearrangement there would be formed anthranilate deoxyribonucleotide (reaction 2, Fig. 15). This would then yield indoleglycerol phosphate by decarboxylation and loss of the hydroxy group on C-2 of the ribulose phosphate side chain.

In support of this hypothesis Doy and Gibson (*217*) isolated and identified the dephosphorylated product 1-(*O*-carboxy-phenylamino)-deoxyribulose from culture filtrates of mutants of *A. aerogenes* and *E. coli*. This compound was prepared synthetically, and the 2,4-diphenyl-hydrazone derivative of it and that of the compound isolated from culture media were found to be identical, completely verifying the proposed structure.

Smith and Yanofsky (*218*) observed the formation of a product formed by extracts of one type of tryptophan-less bacterial mutants on incubation with anthranilate-^{14}C-carboxyl and ^{32}P-ribose-PP that retained both the ^{14}C carbon and the [32]P. While this compound could not be isolated and positively identified, its properties corresponded with those expected of anthranilate deoxyribulonucleotide. These authors also prepared the anthranilate deoxyribulonucleotide synthetically and demonstrated that it was converted to indoleglycerol-P by extracts of bacteria that accumulate the latter.

Wegman and DeMoss (*219*) in experiments with *Neurospora* mutants obtained evidence that N-(5'-phosphoribosyl)-anthranilate was formed by extracts of a mutant blocked in indoleglycero-P formation, not the anthranilate deoxyribulonucleotide.

The name anthranilate-PP-ribose-P phosphoribosyltransferase has been proposed for the enzyme catalyzing the condensation of anthranilate and PP-ribose-P (*219*). It has not been further purified or characterized.

From the steps shown in Fig. 15 it might indeed be assumed that a second enzyme, an isomerase, would be required to catalyze the conversion of anthranilate ribosylnucleotide to anthranilate deoxyribosylnucleotide. Indeed, evidence for such an enzyme was reported for yeast (*220*) and *E. coli* (*221*). It is lacking in *N. crassa* (*219*).

B. Enzymic Formation of Indole 3-Phosphoglycerate

Indole-3-glycerol phosphate synthetase catalyzes the formation of indole 3-phosphoglycerate. It was first highly purified from *E. coli* by Gibson and Yanofsky (*222*) and subsequently crystallized by Creighton and Yanofsky (*223*). By growing a mutant of *E. coli* lacking tryptophan synthetase on a minimal tryptophan medium, the indole glycerophosphate synthetase is increased 60-fold. It was purified another 17-fold to give a total of 100-fold enrichment over the wild-type organism. The enzyme showed optimum activity at pH 8.8. A K_m value 1.75×10^{-5} M was obtained for anthranilate deoxyribulonucleotide.

The crystalline enzyme appeared to be homogeneous by a variety of criteria. Evidence was obtained that it consists of a single polypeptide chain of molecular weight 45,000 with one amino terminal methionine residue. The K_m for anthranilate deoxyribulonucleotide determined with crystalline preparation was lower than that found previously, namely, 5×10^{-6} M. The crystalline enzyme was also found to possess N-5'-phosphoribosylanthranilate isomerase activity. *N. crassa* had previously been shown to contain only two separable enzyme fractions for the conversion of anthranilate to indoleglycerol-P (*219*). One of these was similar in character to the indoleglycerol phosphate synthetase of *E. coli*.

C. Tryptophan Formation from Indole 3-Phosphoglycerate

The early history of the discovery of tryptophan biosynthesis from indole and characterization of the enzyme catalyzing the reaction has been reviewed by Yanofsky (*224*). The enzyme catalyzing the formation of tryptophan, tryptophan synthetase [L-serine hydrolase (adding indole), EC 4.2.1.20], has been shown to consist of two kinds of polypeptide chains in *E. coli* (*225,226*), designated proteins A and B. On the other hand, no separation into subunits has been accomplished with the enzyme from *N. crassa* (*227,228*).

The *N. crassa* tryptophan synthetase has been only partially purified. It is stabilized against inactivation by addition of serine, pyridoxal-P, and EDTA. The molecular weight has been estimated to be about 140,000 (*228*). Altered proteins with tryptophan synthetase characteristics different from that of the wild type have been found in mutants of *N. crassa*.

Both of the polypeptide chains of *E. coli* tryptophan synthetase have now been prepared in crystalline form and have been extensively studied (*229–245*).

Three reactions are catalyzed by the complete enzyme, represented in Eqs. (18), (19), and (20).

$$\text{Indoleglycerol-} \circledP + \text{L-serine} \rightarrow \text{L-tryptophan} + \text{glyceraldehyde 3-} \circledP \qquad (18)$$

$$\text{Indole} + \text{L-serine} \rightarrow \text{L-tryptophan} \qquad (19)$$

$$\text{Indoleglycerol-} \circledP \rightleftharpoons \text{indole} + \text{glyceraldehyde 3-} \circledP \qquad (20)$$

The two proteins of *E. coli* tryptophan synthetase can be separated chromatographically on DEAE-cellulose (*226*), or may be obtained from specific mutants of the bacterium that have lost their capacity to produce one or the other of the subunit proteins (*225*).

The first of the tryptophan synthetase component proteins to be crystallized was designated protein A (*226*). Crystalline protein A was found to have a molecular weight of 29,500. The isoelectric point was found to be at pH 4.9–5.0. Protein A was found to lack tryptophan and to possess 3 cysteinyl residues, one of which reacts as a free SH group. Digestion with trypsin yielded 25 peptides. The amino terminal end of protein A consists of a methionine residue (*232*). The carboxyl terminal residue was identified as serine (*232*).

Purification of protein B to a near homogeneous state was achieved several years later (*234*). It turned out to be much larger than protein A with an estimated molecular weight of 108,000 and a sedimentation coefficient of 5.0 S. Pyridoxal-P was shown to be a cofactor for protein B, approximately 2 moles of the coenzyme being bound per mole of protein. It was also shown that the purified protein B requires a thiol compound and pyridoxal-P for stability and maximal enzyme activity.

While the complete enzyme, composed of proteins A and B, is required to achieve maximal, catalytic activity for all three reactions shown in Eqs. (18)–(20) and also for the performance of reaction (18), the subunit protein B can catalyze reaction (19), at a reduced rate, while protein A can catalyze reaction (20). The catalytic activity of B in reaction (18) is enhanced by monovalent cations, NH_4^+ being the most effective.

Kinetic studies on the reactions of substrates with protein B and the complete enzyme AB have yielded the following K_m values: Indole (with NH_4^+), 2×10^{-4} M and 1.65×10^{-4} M; pyridoxal-P, 9×10^{-6} M and 1.1×10^{-6} M; L-serine, 1×10^{-3} M and 2.5×10^{-2} M for B alone and AB, respectively. In the case of serine the constants are approximate since it did not follow Michaelis kinetics (*230*).

From the estimated molecular weight of the AB complex it has been concluded that it consists of two A units and one B unit. This conclusion is supported by the observation that the AB complex contains equal amounts of A and B enzymic activity. It also seems probable that the B protein is a dimer composed of two identical polypeptide chains (*237*).

Study of the association of the subunits of the A and B proteins of tryptophan synthetase by sucrose gradient centrifugation indicated that the fully associated enzyme consisted of $\alpha 2$ and $\beta 2$ subunits (*237*). Protein A is composed of one α subunit, while protein B is composed of two β subunits. The associated enzyme has a sedimentation coefficient of 6.4 S and is in mobile equilibrium with the free subunits. The α subunits are believed to bind on two identical independent sites of the $\beta2$ subunit. Pyridoxal-P and serine together strongly increase the association of the two subunits. The above conclusions are supported by a study of the combining ratios of α and β to form A2 and B2 and a comparison of the amino acid composition (238).

Hardman and Yanofsky (235) by chemical modification studies concluded that at least 2 of the 3 cysteine residues of protein A are required for enzyme activity. Histidine and methionine also have been concluded to be essential for enzyme activity. This is based on the observation that, in certain mutants with enzymically defective A protein, these amino acids are missing.

Two cysteine residues and a histidine and methionine residue are assumed to be at the binding site of the enzyme for the substrate.

The amino acid sequences surrounding the SH groups of the A protein have been determined by Guest and Yanofsky (236). Each of the 3 cysteines was a component of a sequence of different amino acids. Modification of the A protein with ethyleneimine confirmed that a net of 2 cysteine residues are essential for enzymic activity, previously indicated by studies with iodoacetate and other procedures.

Yanofsky and co-workers (233–235,239–243) have determined the complete amino acid sequence of A protein from analysis of the composition of peptides obtained by tryptic digestion (25 peptides), chymotryptic digestion (48 peptides), and treatment of the protein with cyanogen bromide (242). It was determined that the A protein consists of a single polypeptide chain containing 267 amino acid residues. It was confirmed that the amino terminal residue is methionine and the carboxyl terminal residue, serine.

D. Regulatory Mechanisms of Tryptophan Synthesis

A detailed description of the genetics of the tryptophan operon in *E. coli* and in *N. crassa* is outside the scope of this work. A number of reviews of this have been published (224,244).

Tryptophan synthesis is subject to both feedback inhibition and enzyme repression. The feedback point of control is the enzyme anthranilate synthetase. The enzymes specific for tryptophan biosynthesis are coordinately controlled after the branch point of aromatic biosynthesis at chorismic acid. Tryptophan and compounds that can be converted to tryptophan, e.g., indole, or structural analogs of tryptophan, e.g., 5-methyl tryptophan, repress the formation and inhibit the activity of anthranilate synthetase (246,247). A single locus cistron of anthranilate synthetase is important in the control of tryptophan biosynthesis by regulating repression of anthranilate synthetase. Mutations of this cistron lead to alterations in the capacity of the cell to synthesize other enzymes of the tryptophan pathway. Deletions over the anthranilate synthetase cistron results in nonresponsible transcription of the tryptophan gene (248). Matsushiro et al. concluded that the operation of the tryptophan gene was located at the anthranilate synthetase cistron extremity of the tryptophan operon.

Yanofsky et al. (249) have reported that 30 amino acid changes in mutants of the A protein of tryptophan synthetase are consistent with the in vitro amino acid code and the expectation that single mutational events are limited to single nucleotide substitutions.

X. SYNTHETIC REACTIONS INVOLVING TYROSINE

A. Biosynthesis of Epinephrine Compounds

The major pathway of norepinephrine and epinephrine biosynthesis appears to involve the sequence tyrosine → dopa → dopamine → norepinephrine → epinephrine (Fig. 16). This scheme was proposed by Blaschko (250) and by Holtz (251) in 1939. Some additional minor pathways have since been suggested as shown in Fig. 16 (252).

Isotopic experiments on the conversion of labeled phenylalanine to epinephrine were first performed by Gurin and Delluva (253). In a more detailed study Udenfriend and co-workers (254,255) showed the conversion of labeled phenylalanine and tyrosine to norepinephrine and epinephrine in rat adrenals. Tyrosine yielded products of much higher specific activity than did phenylalanine, showing its closer proximity to the end reactions.

Undenfriend and Wyngaarden (255) also demonstrated the conversion of ^{14}C-dopa to radioactive epinephrine and norepinephrine. With this precursor the labeling was twice as great as with tyrosine. The epinephrine compounds isolated after administration of ^{14}C-labeled tyramine and phenylethylamine were totally nonradioactive, showing that the latter were not on the pathway of formation of the epinephrine compounds, whereas hydroxytyramine was very actively converted to epinephrine (255).

The utilization of the methyl group of methionine for the formation of the methyl group of epinephrine was first shown by Keller et al. (256). The conversion of norepinephrine to epinephrine was shown

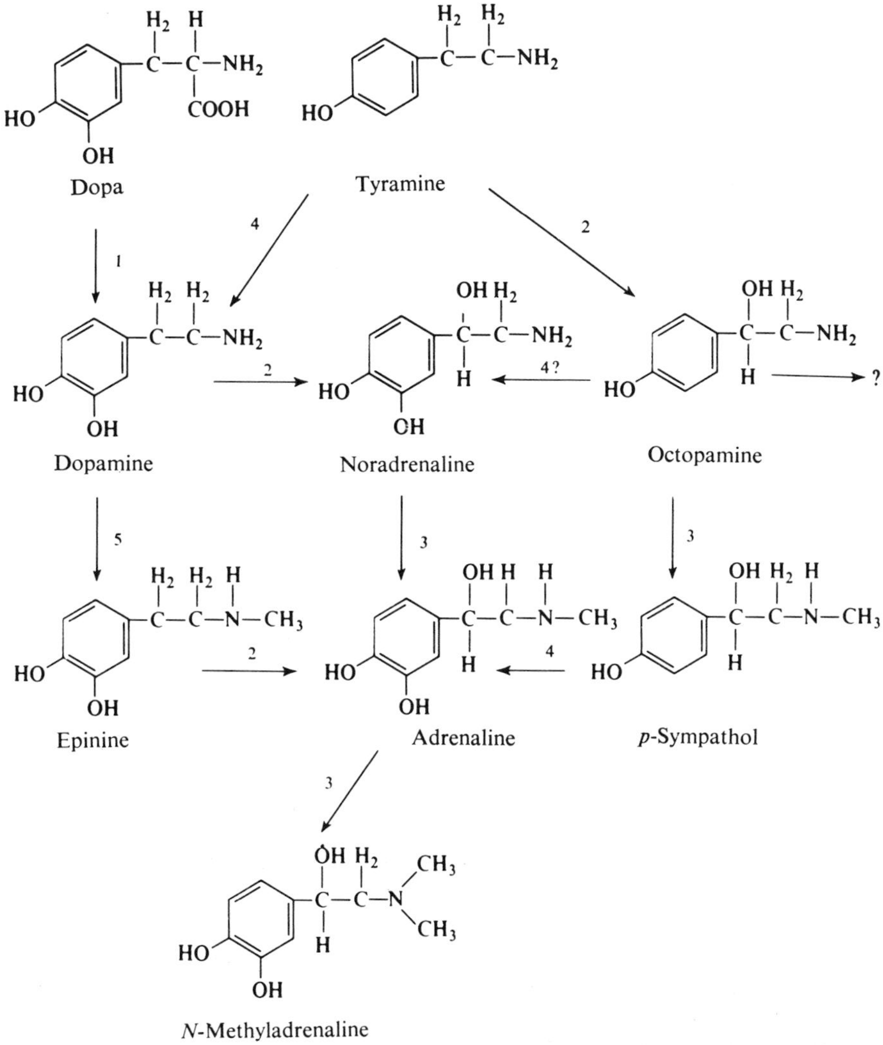

FIG. 16. Pathways in the formation of catecholamines. Heavy arrows show the main routes. 1, Dopa decarboxylase. 2, Dopamine-β-oxidase. 3, Phenylethanolamine-N-methyltransferase. 4, Catechol-forming enzyme. 5, Nonspecific N-methyltransferase (277).

later by the isolation of radioactive epinephrine from rat adrenals after the injection of norepinephrine-^{14}C (257).

The subsequent developments in the details of the individual reactions and modifications in the original scheme came from the isolation and characterization of individual enzymes catalyzing the several reaction steps.

B. Enzymes of the Epinephrine Biosynthetic Pathway

1. HYDROXYLATION OF TYROSINE

The hydroxylation of tyrosine to 3,4-dihydroxyphenylalanine (dopa) was formerly assumed to be solely a function of the enzyme tyrosinase. More recently an enzyme hydroxylating tyrosine has been discovered that is believed to function physiologically in epinephrine biosynthesis (258). This enzyme has been demonstrated in brain, adrenal medulla, and sympathetic nerves. It apparently is specific for the hydroxylation of L-tyrosine. The enzyme has been solubilized and enriched 120-fold from beef adrenals. The enzyme showed a definite requirement for tetrahydropteridines and some evidence for stimulation by Fe^{2+} (258,259). The enzyme has its optimum activity at pH 6.0, and the K_m for L-tyrozine is reported to be $1 \times 10^{-5} M$.

The hydroxylation of tyrosine by tyrosinase has been the subject of recent investigation (260). 6,7-Dimethyltetrahydropteridine at a comparatively high concentration was found to eliminate the induction period, which is characteristic of this enzyme. There is no good evidence that tyrosinase requires a tetrahydropteridine cofactor.

2. DECARBOXYLATION OF 3,4-DIHYDROPHENYLALANINE (DOPA)

Dopa decarboxylase (3,4-dihydrophenylalanine carboxylase, EC 4.1.1.26) was the first enzyme in the pathway of epinephrine biosynthesis to be discovered (251,261). The enzyme from most tissue sources is not highly specific, and it catalyzes decarboxylation of aromatic amino acids in the order dopa > o-tyrosine > m-tyrosine > 5-hydroxytryptophan (262–265). p-Tyrosine is not a substrate for the enzyme. An exception is dopa carboxylase prepared from embryos of the housefly (Calliphora erythrocephala) which catalyzes decarboxylation of only 5-hydroxytryptophan in addition to dopa; the former at a greatly reduced rate (266). As might be anticipated, pyridoxal-P is the coenzyme for dopa decarboxylase. Some instances have been reported of stimulation by Fe^{2+} (262,266). Optimum activity of the enzyme is at pH 7.0–7.2. K_m values are reported to be $4.0 \times 10^{-4} M$ for dopa and $2 \times 10^{-5} M$ for hydroxytryptophan with the guinea pig enzyme (262). Approximately the same K_m values have been obtained with the beef adrenal enzyme (263).

3. HYDROXYLATION OF 3,4-DIHYDROXYPHENYLETHYLAMINE (DOPAMINE)

Norepinephrine is formed by this reaction. Dopa β-hydroxylase [3,4-dihydroxyphenylethylamine, ascorbate: oxygen oxidoreductase (hydroxylating), EC 1.14.2.1] is present in adrenal medulla (267). It is a particulate enzyme, but can be solubilized by nonionic detergents (268). The reaction catalyzed by the enzyme is shown in Eq. (21).

$$+ \text{ dehydroascorbic acid} + H_2O$$

The enzyme has been partially purified and various aspects of the reaction studied (268–273). The enzyme is quite unspecific. It has been shown that it can catalyze the β-hydroxylation of a great variety of phenylethylamines and N-substituted derivatives thereof. Phenylpropylamines are also substrates, but much poorer ones. The primary phenylethylamines were better substrates than the corresponding secondary amines (271–274).

It has been shown that epinene (N-methyl-3,4-dihydroxyphenethylamine) is hydroxylated directly to epinephrine (272).

As shown in Eq. (21), ascorbic acid is required specifically as the hydrogen acceptor for the hydroxylation reaction (268,269). It has also been established that the O_2 molecule, and not oxygen from water, is the source of the hydroxyl oxygen (270). Fumarate or, less effectively, certain other dicarboxylic acids enhance the reactivity of the enzyme. The mechanism of how fumarate acts has not been explained. The enzyme is inhibited by PCMB. This is reversed by glutathionine. Presence of the substrate prevents the inactivation by PCMB. The results suggest that an SH group is part of the active enzyme center (275).

4. METHYLATION OF NOREPINEPHRINE

The enzyme catalyzing this reaction has been partially purified from monkey adrenal glands. The name suggested for it is phenylethanolamine N-methyltransferase (276). S-Adenosylmethionine serves as the methyl group donor. As indicated by its name, the enzyme can catalyze the *N*-methylation of a variety of phenylethanolamine derivatives, naturally occurring and synthetic. The β-hydroxy group appears to be a requirement for activity. The optimum activity of the enzyme is at pH 7.5–8.2 in phosphate buffer, and pH 8.0–9.0 in Tris buffer. The K_m for normetanephrine, which was used as the test substrate in studying the enzyme, was estimated to be $5 \times 10^{-5} M$. The enzyme was inhibited by PCMB.

5. ALTERNATIVE PATHWAYS OF FORMATION OF EPINEPHRINE COMPOUNDS

As shown in Fig. 16, the epinephrine compounds can be formed by hydroxylation of a number of monophenolic derivatives (*277*). Considerable evidence exists that traces of the epinephrine compounds appear on administration of such compounds. Axelrod (*277*) has discovered an enzyme in rabbit liver microsomes that catalyzes hydroxylation of *p*- and *m*-sympathol to epinephrine and *p*- and *m*-tyramine to dopamine. The enzyme also catalyzes formation of catechols from a variety of phenols. NADP is a required electron acceptor for the reaction.

C. Biosynthesis of Thyroid-Active Amino Acids

Two thyroid-active iodinated amino acids have been discovered in thyroid tissue. The first of these, thyroxine, was isolated in 1915 by Kendall, and the structure was established in 1927 by Harrington and Barger (*278*). The second one, L-3,5,3′-triiodothyronine, was isolated and identified in 1952 (*279,280*).

Study of the formation of iodinated derivatives by thyroid tissue has revealed that monoiodotyrosine is formed first, followed by diodotyrosine (*281*). Later thyroxine and triiodothyronine are found (*282*) (see Fig. 17). Iodide concentration and organic iodine binding by the thyroid gland are separate processes. They can be distinguished by the fact that concentration of iodide is inhibited by thiocyanate, while thiourea permits iodide concentration and inhibits the organic binding of the iodine.

After a long period of inconclusive results it has become definitely established that the iodination of the thyroid-active amino acids is catalyzed by a peroxidase-type enzyme present in the mitochondria of thyroid tissue (*283–287*).

The thyroid peroxidase has been solubilized by treatment with either deoxycholate or trypsin. It has been purified about 400-fold. The purified preparation shows an absorption maximum at 410 mμ, characteristic of heme proteins. This, and the fact that enzyme activity is inhibited by CN$^-$ or azide, is proof of its typical peroxidase character. The molecular weight has been estimated by gel filtration to be 64,000 by one group of investigators (287) and 104,000 by another (284).

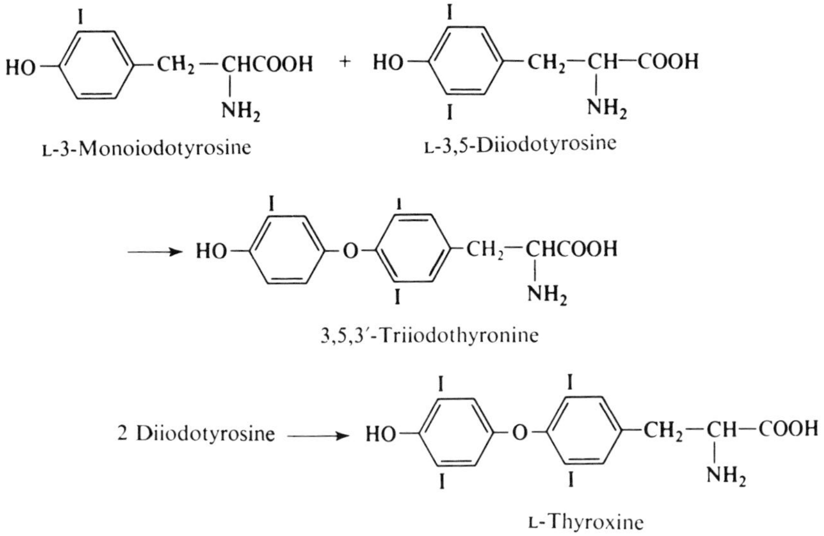

FIG. 17. Precursors and synthesis of thyroid-active amino acids.

The enzyme catalyzes iodination of tyrosine, thryroglobulin, and also of nonthyroid-active proteins. The reaction with tyrosine yields initially a higher rate of formation of 3-iodotyrosine. This is overtaken by the formation of 3,5-diiodotyrosine on continued incubation. Evidence also was obtained of the formation of 3,5-diiodo-4-hydroxy-phenyl pyruvate during iodination of tyrosine. The latter has been proposed as an intermediate in thyroxine synthesis.

With thryoglobulin as the substrate there is obtained, on hydrolysis of the protein, diiodotyrosine, monoiodotyrosine, and thyroxine, decreasing in amount in the above order. Coval and Taurog (287) conclude that 3,5-diiodo-4-hydroxyphenyl pyruvate is not an intermediate in thyroxin formation in thyroglobulin or other proteins. The reaction is presumed to take place by condensation of juxtapositioned diiodotyrosine residues in the protein. Hydrogen peroxide is required in the in vitro iodination of tyrosine residues. The reaction may be assumed to proceed according to Eqs. (22) and (23).

$$4I^- + 4H^+ + 2H_2O_2 \longrightarrow 2I_2 + 4H_2O \tag{22}$$

$$2I_2 + HO-\!\!\!\left\langle\begin{array}{c}\\\end{array}\right\rangle\!\!\!-CH_2CHNH_2COOH \longrightarrow HO-\!\!\!\left\langle\begin{array}{c}I\\\\I\end{array}\right\rangle\!\!\!-CH_2CHNH_2COOH + \quad (23)$$
$$2\,HI$$

In the *in vitro* iodinating system, and H_2O_2 generating system is more effective than added H_2O_2, although the latter will serve. Monoamine oxidase (EC 1.4.3.4) has been hypothesized to provide such an H_2O_2 generating system (*287a*).

The purified thyroid peroxidase catalyzes iodination over a wide range of iodide concentrations and tyrosine concentration. The K_m of I^- for iodination of thyroglobulin was estimated to be $2.5 \times 10^{-4} M$. High I^- concentrations ($> 1 \times 10^{-4} M$) inhibit the enzyme reaction. This may be important for the antithyroid action of excess I^-. Iodination also is inhibited by antithyroid compounds. The results of the above investigations preclude the necessity for a tyrosine iodinase to explain iodination of tyrosine residues. The findings suggest that peroxidase may be involved in the coupling reaction.

XI. GLUTAMINE AND ASPARAGINE BIOSYNTHESIS

Glutamine participates in numerous biosynthetic reactions involving the formation of important biological compounds. These include the purines, aminosugars, histidine, *p*-aminobenzoic acid, and anthranilic acid. In addition, both glutamine and asparagine are utilized in the synthesis of proteins.

The subject of glutamine biosynthesis has been extensively reviewed by Meister and co-workers (*288–290*). On the other hand, little is known about asparagine biosynthesis. Interest in the enzyme catalyzing its synthesis has been generated recently as a result of the observation that treatment with the enzyme asparaginase is beneficial in certain cases of leukemia (*291*).

A. Glutamine Biosynthesis

The reaction of major physiological importance catalyzed by glutamine synthetase [L-glutamate ammonia ligase (ADP), EC 6.3.1.2] is represented by Eq. (24).

$$
\begin{array}{l}
\text{COO}^- \\
|\\
\text{CH}_2 \\
|\\
\text{CH}_2 \quad + \text{NH}_4{}^+ + \text{ATP} \xrightarrow{\text{Mg}^{2+}(\text{Mn}^{2+})} \\
|\\
\text{HC—NH}_3{}^+ \\
|\\
\text{COO}^- \\
\text{Glutamate}
\end{array}
\qquad
\begin{array}{l}
\text{CONH}_2 \\
|\\
\text{CH}_2 \\
|\\
\text{CH}_2 \quad + \text{ADP} + \text{P}_i + \text{H}_2\text{O} \quad (24) \\
|\\
\text{HC—NH}_3{}^+ \\
|\\
\text{COO}^- \\
\text{Glutamine}
\end{array}
$$

An analogous reaction, Eq. (25), is the formation of glutamate hydroxamate. Formation of the hydroxamate has widely been used in

$$
\text{Glutamate} + \text{NH}_2\text{OH} + \text{ATP} \xrightarrow{\text{Mg}^{2+}} \gamma\text{-glutamate hydroxamate} + \text{ADP} + \text{P}_i \quad (25)
$$

the study of glutamine synthetase because of the ease of its spectro-photometric determination by reaction with Fe^{3+}.

A third reaction, Eq. (26), that has been important in obtaining information on the mechanism of the synthesis of glutamine is the formation of pyrrolidine carboxylate in the absence of ammonia.

$$
\begin{array}{l}
\text{CH}_2\text{—COOH} \\
\quad\;\; | \quad\;\, \text{NH}_2 \quad + \text{ATP} \rightarrow \\
\text{CH}_2\text{—CCOOH}
\end{array}
\qquad
\begin{array}{l}
\text{CH}_2\text{—C}{=}\text{O} \\
\quad\;\; | \qquad\;\; \text{NH} \quad + \text{H}_2\text{O} + \text{ADP} + \text{P}_i \quad (26) \\
\text{CH}_2\text{—C—COOH}
\end{array}
$$

Glutamine synthetase has been found to utilize a large variety of dicarboxylic amino acids as substrates. These include L- and D-glutamic acid (291), α-methylglutamic acid (292), β-aminoglutaric acid (293), and α-aminoadipic acid (294).

Glutamine synthetase also catalyzes a γ-glutamyl transfer reaction in which the amide group of glutamine is replaced by hydroxylamine and an arsenolysis reaction that results in the hydrolysis of glutamine to glutamate and NH_3 (288).

The glutamine synthetase from peas and brain catalyzes formation of L-glutamine and L-glutamate hydroxamate at about the same rates, while with D-glutamate the reaction with hydroxylamine is considerably faster.

A significant observation in the stereo specificity of glutamine synthetase is that, with β-glutamate as a substrate, D-β-glutamine is formed.

The behavior of α-aminoadipic acid parallels that of glutamate in many respects. L- and D-Homoglutamine and the analogous ω-hydroxy-amic acids are formed. In the absence of NH_3 or NH_2OH the product of the enzymic reaction is 6-piperidone-2-carboxylic acid. Formation of the hydroxamates of α-aminoadipic acid is much faster (over 30-fold) than of the homoglutamines.

B. Properties of Glutamine Synthetase

The most highly purified mammalian preparation has been prepared in a homogeneous state from sheep brain (*295*). It has a molecular weight of about 525,000 and is reported to consist of 8 subunit polypeptides. The steps in the purification consist of extraction of acetone powder of brain, isoelectric precipitation, differential heat inactivation, adsorption on calcium phosphate gel, precipitation by ammonium sulfate, and electrophoresis. The differential heat inactivation was made possible by the observation that ATP and Mg^{2+} protected the enzyme against heat inactivation. Optimum activity of the enzyme is at pH 7.0–7.4 (*288*). Estimates of the K_m values of L-glutamate, NH_4^+, and NH_2OH are 2.5×10^{-3}, 1.8×10^{-4}, and 1.5×10^{-4} M, respectively. With D-glutamate the constants are about one order of magnitude greater.

It has been observed that the synthesis of glutamine does not proceed to completion. On the basis of measurements of both the forward and reverse reactions, an equilibrium constant for the synthesis of L-glutamine was estimated to be 1.2×10^{-3} at pH 7.0 and 37°.

The glutamine synthetase from *E. coli* differs markedly from that of the mammalian species. It exhibits a cumulative feedback inhibition by eight different products of glutamine metabolism (*288a,288b*). The eight inhibitory compounds consist of alanine, glycine, histidine, tryptophan, CTP, AMP, carbamyl phosphate, and glucosamine-6-P. Each compound alone causes only partial inhibition. The enzyme is also regulated by repression (*288a*).

The bacterial glutamine synthetase has been prepared in homogeneous crystalline form from derepressed *E. coli* cells. The molecular weight has been estimated to be about 600,000 and shown to consist of 12 identical subunit polypeptide chains of 50,000 molecular weight. In the synthesis of glutamine, Mn^{2+} is the most active divalent cation. Synthesis of the hydroxamate analog proceeds better with Mn^{2+}. The bacterial glutamine synthetase has its optimum activity at pH 7.7 with Mg^{2+} as activator and at pH 7.0 with Mn^{2+}. Some values of the Michaelis constants in the presence of Mg^{2+} are 2.4×10^{-3} M for glutamate, 6.8×10^{-4} M for ATP, and 1.8×10^{-3} M for NH_3 (*288a*).

Study of the inhibition reaction has indicated that there are separate binding sites for each of the eight inhibitors and each acts independently (*288b*). Three of the inhibitors are noncompetitive with the substrate, three are partially competitive against glutamate, and two are partially competitive against NH_3.

The enzyme is completely dissociated into its subunits in a medium containing 0.01 M EDTA and 1.0 M urea at pH 8.5. It is also disaggregated by guanidine HCl.

Removal of the dissociating chemicals causes an aggregation, but not back to the native enzyme.

C. Reaction Mechanism

According to the investigations of Meister and co-workers, the synthesis of glutamine involves a two-step mechanism. The first step is the activation of glutamate, presumably by formation of γ-glutamyl phosphate attached to the enzyme. This reaction is of relatively low specificity. The second step is the reaction of the activated glutamate with ammonia or hydroxylamine, which is optically more specific.

The formation of the intermediate activated complex of glutamate is supported by the following evidence: (a) The formation of pyrrolidone carboxylic acid in the absence of ammonia proceeds at an extremely rapid rate compared to the formation of this compound from glutamate. This is consistent with the formation of a highly reactive enzyme-bound γ-glutamyl derivative that readily undergoes cyclization to pyrrolidone carboxylate (288). (b) It has been observed that the synthesis of each mole of glutamine is accompanied by a transfer of one ^{18}O atom from glutamate to P_i (296). (c) Employing large amounts of purified enzyme, it has been observed that in the presence of glutamate-^{14}C, ATP, and Mg^{2+} the radioactivity is sedimented along with the enzyme upon ultracentrifugation. (d) As mentioned above, ATP and Mg^{2+} protect the glutamine synthetase against heat inactivation, and in the absence of ATP or Mg^{2+} there is no binding of glutamate to the enzyme. (e) Perhaps the most convincing experiment is the demonstration that β-aminoglutaryl phosphate, which is comparatively stable, can react with ADP in the presence of the enzyme to synthesize ATP (297). The inability to demonstrate the existence of an acyl-glutamate is attributed to an extreme degree of instability and reactivity.

The greater stereospecificity of the second step in the overall reaction sequence is shown by the fact that only α-methyl-L-glutamate is a substrate for the enzyme and that in the reaction of β-glutamic acid only the D-isomer of the product is formed. This is illustrated by Fig. 18.

Meister (288) concludes that studies on the enzyme mechanism support the conclusion that: "The amino acid substrate of glutamine synthetase is in an extended conformation when attached to the active site of the enzyme and that the side of the substrate molecule bearing the amino group and the undersurface of the molecule are in close contact with the enzyme and the nucleotide. The studies have made it

Glutamic acid	α-Methylglutamic acid	β-Glutamic acid
COOH	COOH	COOH
$\mid$	$\mid$	$\mid$
CH_2	CH_2	CH_2
$\mid$	$\mid$	$\mid$
CH_2	CH_2	$H—C—NH_2$
$\mid$	$\mid$	$\mid$
$H—C—NH_2$	$CH_3—C—NH_2$	CH_2
$\mid$	$\mid$	$\mid$
COOH	COOH	COOH
L- and D-isomers are active	only L-isomer is active	only the D-isomer of the product is formed

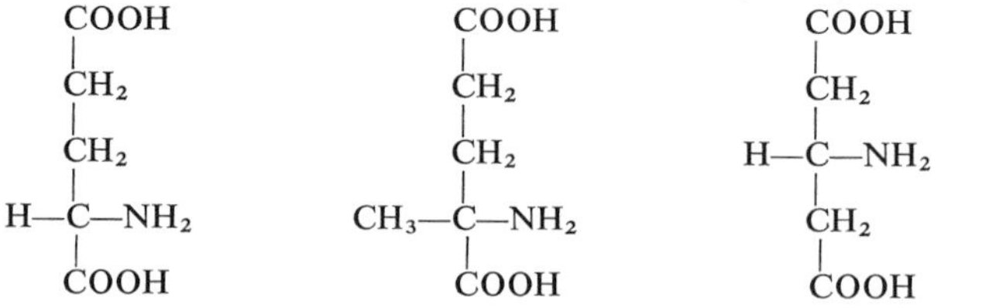

FIG. 18. Stereospecific reactions of glutamine synthetase (290).

possible to understand the curious optical specificity of glutamine synthetase in terms of the conformation of the amino acid substrates on the active site."

D. Asparagine Biosynthesis

In contrast to glutamine, knowledge of asparagine biosynthesis is quite meager. Different reactions appear to be involved for the synthesis of asparagine in bacteria and in the mammal.

The reaction in bacteria is represented by Eq. (27).

$$\text{L-Aspartate} + NH_3 + ATP \xrightarrow{Mn^{2+}} \text{L-asparagine} + AMP + P_i \qquad (27)$$

Partially purified preparations of bacterial asparagine synthetase have been obtained from *Lactobacillus casei* (297) and from *Streptococcus bovis* (298). The degree of purification was only 10- to 20-fold. The enzymes from both organisms have very similar properties. Mn^{2+} is the most active metal-ion cofactor, although Mg^{2+} is also effective. The enzyme is specific for L-aspartate and is inhibited by the reaction product, L-asparagine. In addition to forming asparagine, L-aspartyl-hydroxamate is formed with hydroxylamine.

It is postulated that the synthesis of asparagine in bacteria proceeds through the formation of an enzyme-bound β-aspartyladenylate (297).

Studies on asparagine synthetase have been reported for several rodent cancers (*299–302*) and for embryonic chick liver (*303*). The most advanced results were those of Patterson and Orr with the Novikoff hepatoma (*299*). These investigators purified the enzyme over 100-fold. The most important reaction appears to be represented by Eqs. (28) and (29).

$$\text{L-Aspartate} + \text{ATP} + \text{Enz} \xrightarrow{\text{Mg}^{2+}} \text{Enz-}\beta\text{-aspartyladenylate} + \text{PP} \qquad (28)$$

$$\text{Enz-}\beta\text{-aspartyladenylate} + \text{L-glutamine} \rightarrow \text{Enz} + \text{asparagine} + \text{glutamate} + \text{AMP} \qquad (29)$$

The synthesis can occur with ammonia, but this requires a much higher concentration than of glutamine. This is shown by the Michaelis constants which were estimated to be 1.1 mM for glutamine and 120 mM for ammonia. The reaction with ammonia proceeds in the same manner as found with the bacterial enzyme. The pH-activity curve with glutamine as a substrate exhibited a plateau over the range pH 6.6–8.0. With ammonia there was a linear increase with increasing pH from 6.0 to 8.6. This would be expected if the true substrate was ammonia, not ammonium ion.

Other properties of the Novikoff hepatoma L-asparagine synthetase are: K_m of L-aspartate = 0.58 mM, and of ATP = 0.11 mM. L-Asparagine was found to inhibit the enzyme at higher concentrations. Mg²⁺ was the most active metal ion; Mn²⁺ was about 20% as active. With hydroxylamine the enzyme catalyzes formation of β-aspartylhydroxamate.

Horowitz *et al.* (*300*) found that most normal tissue of the mouse contained very little asparagine synthetase. Certain tumors insensitive to asparaginase contained very high activity, and tumors dependent on asparagine contained little or none (*300–302*).

REFERENCES

1. H. J. Sallach, *in* "Amino Acid Metabolism" (W. D. McElroy and B. Glass, eds.), p. 782, Johns Hopkins Press, Baltimore, 1955; *J. Biol. Chem.* **223**, 1101 (1956).
2. A. Ichihara and D. M. Greenberg, *Proc. Natl. Acad. Sci. U.S.* **41**, 605 (1955); *J. Biol. Chem.*, **224**, 331 (1957).
3. D. A. Walsh and H. J. Sallach, *J. Biol. Chem.* **241**, 4068 (1966).
3a. G. A. Sojka and H. R. Garner, *Biochim. Biophys. Acta* **148**, 42 (1967).
4. H. J. Fallon, E. J. Hackney, and W. L. Byrne, *J. Biol. Chem.* **241**, 4157 (1966).
5. W. F. Bridgers, *J. Biol. Chem.* **240**, 4591 (1965).
6. L. I. Pizer, *J. Biol. Chem.* **239**, 4219 (1964).
7. B. E. Johnson, D. A. Walsh, and H. J. Sallach, *Biochim. Biphys. Acta* **85**, 202 (1964).
8. H. E. Umbarger and M. A. Umbarger, *Biochim. Biophys. Acta* **62**, 193 (1962); H. E. Umbarger, M. A. Umbarger, and J. Siu, *J. Bacteriol.* **85**, 1431 (1963).
9. L. I. Pizer, *J. Biol. Chem.* **238**, 3934 (1963).
10. J. Somerville, *Biochem. J.* **96**, 50P (1965).

11. J. Hanford and D. D. Davies, *Nature* **182**, 532 (1958).
12. P. J. Large and J. R. Quale, *Biochem. J.* **87**, 386 (1963).
13. L. Blatt, F. E. Dorer, and H. J. Sallach, *J. Bacteriol.* **92**, 668 (1966).
14. J. E. Willis and H. J. Sallach, *J. Biol. Chem.* **237**, 910 (1962).
15. J. E. Willis and H. J. Sallach, *Biochim. Biophys. Acta* **81**, 39 (1964).
16. D. A. Walsh and H. J. Sallach, *Biochemistry* **4**, 1076 (1965).
17. D. A. Walsh and H. J. Sallach, *Biochim: Biophys. Acta* **146**, 126 (1967).
18. E. Sugimoto and L. I. Pizer, *J. Biol. Chem.* **243**, 2081, 2090 (1968).
18a. J. Rosenbloom, E. Sugimoto, and L. I. Pizer, *J. Biol. Chem.* **243**, 2099 (1968).
19. H. Hirsch and D. M. Greenberg, *J. Biol. Chem.* **242**, 2283 (1967).
20. F. C. Neuhaus and W. L. Byrne, *Biochim. Biophys. Acta* **28**, 223 (1958); *J. Biol. Chem.* **234**, 113 (1959); **235**, 2019 (1960).
21. F. Borkenhagen and E. P. Kennedy, *Biochim. Biophys. Acta* **28**, 222 (1958); *J. Biol. Chem.* **234**, 849 (1959).
22. W. F. Bridgers, *J. Biol. Chem.* **242**, 2080 (1967).
23. M. Schram, *J. Biol. Chem.* **232**, 1169 (1958).
24. A. Ichihara and D. M. Greenberg, *J. Biol. Chem.* **225**, 949 (1957).
25. H. Holzer and A. Holldorf, *Biochem. Z.* **329**, 283 (1957).
26. W. Lamprecht, T. Diamanstein, F. Heinz, and P. Balde, *Z. Physiol. Chem.* **318**, 97 (1959).
27. R. W. Cowgill and L. I. Pizer, *J. Biol. Chem.* **223**, 885 (1956).
28. B. K. Joyce and S. Grisolia, *J. Biol. Chem.* **234**, 1330 (1959).
29. S. Grisolia and B. K. Joyce, *J. Biol. Chem.* **234**, 1335 (1959).
30. H. J. Fallon and W. L. Byrne, *Biochim. Biophys. Acta* **105**, 43 (1965).
31. D. A. Richert, R. Amberg, and M. Wilson, *J. Biol. Chem.* **237**, 99 (1962).
32. H. I. Nakada, *J. Biol. Chem.* **239**, 468 (1964).
33. J. S. Thompson and K. E. Richardson, *Arch. Biochem. Biophys.* **117**, 599 (1966).
34. J. S. Thompson and K. E. Richardson, *J. Biol. Chem.* **242**, 3614 (1967).
35. R. H. Dainty and J. L. Peel, *Biochem. J.* **100**, 81P (1966).
36. H. Kawasaki, T. Sato, and G. Kikuchi, *Biochem. Biophys. Res. Commun.* **23**, 227 (1966).
37. T. Sato, G. Motokawe, H. Kochi, and G. Kikuchi, *Biochem. Biophys. Res. Commun.* **28**, 495 (1967).
38. S. M. Klein and R. D. Sagers, *J. Biol. Chem.* **241**, 197, 206 (1966).
39. M. Baginsky and F. M. Huennekens, *Arch. Biochem. Biophys,* **120**. 703 (1967).
40. J. F. Nyc, *J. Biol. Chem.* **223**, 811 (1956).
41. J. Bremer and D. M. Greenberg, *Biochim. Biophys. Acta* **35**, 287 (1959).
42. M. Levine and H. Tarver, *J. Biol. Chem.* **184**, 427 (1950).
43. D. M. Greenberg and S. C. Harris, *Proc. Soc. Exptl. Biol. Med.* **75**, 683 (1950).
44. J. D. Wilson, K. D. Gibson, and S. Udenfriend, *J. Biol. Chem.* **235**, 3530 (1960).
45. L. F. Borkenhagen, E. P. Kennedy, and J. Fielding, *J. Biol. Chem.* **236**, PC28 (1961).
46. J. Bremer, P. H. Figard and D. M. Greenberg, *Biochim. Biophys. Acta* **43**, 477 (1960).
47. J. Bremer and D. M. Greenberg, *Biochim. Biophys. Acta* **37**, 173 (1960); **46**, 205 (1961).
48. K. D. Gibson, J. D. Wilson, and S. Udenfriend, *J. Biol. Chem.* **236**, 673 (1961).
49. J. D. Wilson, K. D. Gibson, and S. Udenfriend, *J. Biol. Chem.* **235**, 3213 (1960).
50. D. Rehbinder and D. M. Greenberg, *Arch. Biochem. Biophys.* **108**, 110 (1965).
51. T. Kaneshiro and J. H. Law, *J. Biol. Chem.* **239**, 1705 (1964).
52. G. A. Scarborough and J. F. Nyc, *Biochim. Biophys. Acta* **146**, 111 (1967).
53. G. A. Scarborough and J. F. Nyce, *J. Biol. Chem.* **242**, 238 (1967).
54. J. R. Klein and P. Handler, *J. Biol. Chem.* **140**, 537 (1942).

55. K. Ebisuzaki and J. N. Williams, *Biochem. J.* **60**, 644 (1955).
56. M. Korzenovsky and B. V. Anda, *Biochim. Biophys. Acta* **29**, 463 (1958).
57. T. Kimura and T. P. Singer, *in* "Methods in Enzymology" (S. P. Colowick and N. O. Kaplan, eds.), Vol. 5, p. 562. Academic Press, New York, 1962.
58. G. R. Williams, *J. Biol. Chem.* **235**, 1192 (1960).
59. G. Bianchi and G. F. Azone, *J. Biol. Chem.* **239**, 3947 (1964).
60. T. Kagawa, D. R. Wilken, and H. A. Lardy, *J. Biol. Chem.* **240**, 1836 (1965).
61. D. R. Wilken, T. Kagawa, and H. A. Lardy, *J. Biol. Chem.* **240**, 1843 (1965).
62. K. T. N. Yue, P. J. Russell, and D. J. Mulford, *Biochim. Biophys. Acta* **118**, 191 (1966).
63. H. A. Rothschild and E. S. G. Barron, *J. Biol. Chem.* **209**, 511 (1954).
63a. A. M. Goldberg and R. E. McCaman, *Biochim. Biophys. Acta* **167**, 186 (1968).
64. M. Jellinck, D. R. Strength, and S. A. Thorpe, *J. Biol. Chem.* **234**, 1171 (1959).
65. J. Durell, D. G. Anderson, and G. L. Cantoni, *Biochim. Biophys. Acta* **26**, 270 (1957).
66. J. Durell and G. L. Cantoni, *Biochim. Biophys. Acta* **35**, 515 (1959).
67. H. J. Fromm and R. C. Nordlie, *Arch. Biochem. Biophys.* **81**, 363 (1959).
68. W. A. Klee, H. H. Richards, and G. L. Cantoni, *Biochim. Biophys. Acta* **54**, 157 (1961).
69. C. G. Mackenzie, "Symposium on Amino Acid Metabolism, Baltimore, 1954," *Johns Hopkins Univ. McCollum-Pratt Inst.* 684 (1955).
70. D. D. Hoskins and C. G. Mackenzie, *J. Biol. Chem.* **236**, 177 (1961).
71. W. R. Frisell and C. G. Mackenzie, *J. Biol. Chem.* **237**, 94 (1962).
72. C. G. Mackenzie and D. D. Hoskins, *in* "Methods in Enzymology" (S. P. Colowick and N. O. Kaplan, eds.), Vol. 5, p. 738. Academic Press, New York, 1962.
73. D. D. Hoskins and C. G. MacKenzie, *J. Biol. Chem.* **236**, 177 (1961).
74. W. R. Frisell, J. R. Cronin, and C. G. Mackenzie, *J. Biol. Chem.* **237**, 2975 (1962).
75. H. Beinert and W. R. Frisell, *J. Biol. Chem.* **237**, 2988 (1962).
76. S. Black and N. G. Wright, "Symposium on Amino Acid Metabolism, Baltimore, 1954," *Johns Hopkins Univ. McCollum-Pratt Inst.*, 591 (1955).
77. S. Black and N. G. Wright, *J. Biol. Chem.* **213**, 27, 39, 51 (1955).
78. M. Flavin and C. Slaughter, *J. Biol. Chem.* **235**, 1103 (1960).
79. G. N. Cohen and M. L. Hirsch, *J. Bacteriol.* **67**, 182 (1954).
80. E. H. Wormser and A. B. Pardee, *Arch. Biochem. Biophys.* **78**, 416 (1958).
81. E. Bilinski and W. B. McConnell, *Can. J. Biochem. Physiol.* **35**, 305 (1957).
82. P. Datta and L. Prakash, *J. Biol. Chem.* **241**, 5827 (1966).
83. P. Truffa-Bachi and G. N. Cohen, *Biochim. Biophys. Acta* **113**, 531 (1966).
84. L. Burlant, P. Datta, and H. Gest, *Science* **148**, 1351 (1965).
85. P. Datta and H. Gest, *Nature* **203**, 1259 (1964).
86. P. Datta and H. Gest, *Proc. Natl. Acad. Sci. U.S.* **52**, 1004 (1964).
87. H. Paulus and E. Gray, *J. Biol. Chem.* **239**, PC4008 (1964).
88. M. Freundlich, *Biochem. Biophys. Res. Commun.* **10**, 277 (1963).
89. E. R. Stadtman, G. N. Cohen, G. LeBras, and H. Robichon-Szulmajster, *J. Biol. Chem.* **236**, 2033 (1961).
90. H. Robichon-Szulmajster and D. Corrivaux, *Biochim. Biophys. Acta* **73**, 248 (1967).
91. K. D. Gibson, A. Neuberger, and G. H. Tait, *Biochem. J.* **84**, 483 (1962).
92. P. Datta and H. Gest, *J. Biol. Chem.* **240**, 3023 (1965).
93. J.-C. Patte, G. LeBras, T. Loviny, and G. N. Cohen, *Biochim. Biophys. Acta* **67**, 16 (1963).
94. P. Datta and H. Gest, *Proc. Natl. Acad. Sci. U.S.* **52**, 1004 (1964).
95. Y. Watanabe, S. Konishi, and K. Shimura, *J. Biochem. (Tokyo)* **42**, 837 (1955).
96. Y. Watanabe and K. Shimura, *J. Biochem. (Tokyo)* **43**, 283 (1956).

97. Y. Watanabe, S. Kinoshi, and K. Shimura, *J. Biochem. (Tokyo)* **44**, 299 (1957).
98. M. Flavin and C. Slaughter, *J. Biol. Chem.* **235**, 1103 (1960).
99. M. Flavin and T. Kono, *J. Biol. Chem.* **235**, 1109 (1960).
100. M. Flavin and C. Slaughter, *J. Biol. Chem.* **235**, 1112 (1960).
101. J. L. Wiebers and H. R. Garner, *J. Biol. Chem.* **242**, 12 (1967).
102. J. L. Wiebers and H. R. Garner, *J. Biol. Chem.* **242**, 5644 (1967).
103. M. Flavin and C. Slaughter, *Biochim. Biophys. Acta* **132**, 400 (1967).
104. R. J. Rowbury and D. D. Woods, *J. Gen. Microbiol.* **36**, 341 (1964).
105. M. Flavin, C. Delavier-Klutchko, and C. Slaughter, *Science* **143**, 50 (1964).
106. M. Flavin and C. Slaughter, *Biochemistry* **4**, 1370 (1965).
107. M. M. Kaplan and M. Flavin, *Biochim. Biophys. Acta* **104**, 390 (1965).
108. R. J. Rowbury and D. D. Woods, *J. Gen. Microbiol.* **42**, 155 (1966).
109. J. Giovanelli and S. Harvey Mudd, *Biochem. Biophys. Res. Commun.* **27**, 150 (1967).
110. M. M. Kaplan and M. Flavin, *J. Biol. Chem.* **241**, 4463 (1966).
111. M. M. Kaplan and M. Flavin, *J. Biol. Chem.* **241**, 5781 (1966).
112. S. Nagai and M. Flavin, *J. Biol. Chem.* **241**, 3861 (1966).
113. M. Flavin and C. Slaughter, *Biochim. Biophys. Acta* **132**, 400, 406 (1967).
114. M. Flavin and C. Slaughter, *J. Biol. Chem.* **239**, 2212 (1964).
115. M. Flavin and A. Segal, *J. Biol. Chem.* **239**, 2220 (1964).
116. C. Delavier-Klutchko and M. Flavin, *J. Biol. Chem.* **240**, 2537 (1965).
117. Y. Matsuo and D. M. Greenberg, *J. Biol. Chem.* **234**, 516 (1959).
118. J. L. Weibers and H. R. Garner, *J. Bacteriol.* **88**, 1718 (1964).
119. "Transmethylation and Methionine Biosynthesis" (S. K. Shapiro and F. Schlenk, eds.), p. 200. Univ. of Chicago Press, 1965.
119a. H. Weissbach and H. Dickerman, *Physiol. Rev.* **45**, 80 (1965).
119b. H. Weissbach and R. Taylor, *Federation Proc.* **25**, 1649 (1966).
120. L. Abrahamson and S. K. Shapiro, *Arch. Biochem. Biophys.* **109**, 376 (1965).
121. M. A. Foster, K. M. Jones, and D. D. Woods, *Biochem. J.* **80**, 519 (1961).
122. F. Gibson and D. D. Woods, *Biochem. J.* **74**, 160 (1960).
123. J. R. Guest, C. W. Helleiner, M. J. Cross, and D. D. Woods, *Biochem. J.* **76**, 396 (1960).
124. J. R. Guest and D. D. Woods, *Biochem. J.* **77**, 422 (1960).
124a. J. R. Guest, S. Friedman, M. J. Delworth, and D. D. Woods, *Ann. N.Y. Acad. Sci.* **112**, 774 (1964).
125. K. M. Jones, J. R. Guest, and D. D. Woods, *Biochem. J.* **79**, 566 (1961).
126. R. L. Kisliuck and D. D. Woods, *Biochem. J.* **75**, 467 (1960).
127. J. Szulmajster and D. D. Woods, *Biochem. J.* **75**, 3 (1960).
128. J. R. Guest, M. A. Foster, and D. D. Woods, *Biochem. J.* **92**, 488 (1964).
129. M. A. Foster, G. Tejerina, J. R. Guest, and D. D. Woods, *Biochem. J.* **92**, 476 (1964).
130. J. R. Guest, S. Friedman, M. A. Foster, G. Tejerina, and D. D. Woods, *Biochem. J.* **92**, 497 (1964).
131. A. R. Larrabee, S. Rosenthal, R. E. Cathou, and J. M. Buchanan, *J. Am. Chem. Soc.* **83**, 4094 (1961).
132. J. M. Buchanan, H. L. Elford, R. E. Loughlin, B. M. McDougall, and S. Rosenthal, *Ann. N.Y. Acad. Sci.* **112**, 756 (1964).
132a. F. T. Hatch, S. Takeyama, R. E. Cathou, A. R. Larrabee, and J. M. Buchanan, *J. Am. Chem. Soc.* **81**, 6525 (1959).
133. F. T. Hatch, A. R. Larrabee, R. E. Cathou, and J. M. Buchanan, *J. Biol. Chem.* **236**, 1095 (1961); S. Takeyama, F. T. Hatch, and J. M. Buchanan, *J. Biol. Chem.* **236**, 1102 (1961).
134. S. Rosenthal and J. M. Buchanan, *Acta Chem. Scand.* **17**, S288 (1963).

135. R. E. Cathou and J. M. Buchanan, *J. Biol. Chem.* **238**, 1746 (1963).
136. R. E. Laughlin, H. L. Elford, and J. M. Buchanan, *J. Biol. Chem.* **239**, 2888 (1964).
137. K. O. Donaldson and J. C. Keresztesy. *J. Biol. Chem.* **234**, 3235 (1959); *Biochem. Biophys. Res. Commun.* **5**, 289 (1961).
138. W. Sakami and I. Ukstins, *J. Biol. Chem.* **236**, PC50 (1961).
139. J. R. Guest, S. Friedman, and D. D. Woods, *Nature* **195**, 340 (1962).
140. J. H. Mangum and K. G. Scrimgeour, *Federation Proc.* **21**, 242 (1962).
140a. R. T. Taylor and H. Weissbach, *J. Biol. Chem.* **242**, 1517 (1967).
141. S. Rosenthal, L. C. Smith, and J. M: Buchanan, *J. Biol. Chem.* **240**, 836 (1965).
141a. S. S. Kerwar, J. H. Mangum, K. G. Scrimgeour, J. D. Brodie, and F. M. Huennekens, *Arch. Biochem. Biophys.* **116**, 305 (1966).
142. H. L. Elford, R. E. Loughlin, and J. M. Buchanan, *Federation Proc.* **23**, 480 (1964).
143. R. T. Taylor and H. Weissbach, *J. Biol. Chem.* **242**, 1502, 1509 (1967); *Arch. Biochem. Biophys.* **123**, 109 (1968).
144. R. Ertel, N. Brot, R. Taylor, and H. Weissbach, *Arch. Biochem. Biophys.* **126**, 353 (1968).
145. H. M. Katzen and J. M. Buchanan, *J. Biol. Chem.* **240**, 825 (1965).
146. H. P. Broquist and E. E. Snell, *J. Biol. Chem.* **180**, 59 (1949).
147. B. Magasanik, H. S. Moyed, and D. Karibian, *J. Am. Chem. Soc.* **78**, 1510 (1956); B. Magasanik, *J. Am. Chem. Soc.*, **78**, 5449 (1956).
148. C. Mitoma and E. E. Snell, *Proc. Natl. Acad. U.S.* **41**, 891 (1955).
149. A. Neidle and H. Waelsch, *Federation Proc.* **16**, 255 (1957).
150. B. N. Ames and H. K. Mitchell, *J. Am. Chem. Soc.* **74**, 252 (1952).
151. B. N. Ames, H. K. Mitchell, and M. B. Mitchell, *J. Am. Chem. Soc.* **75**, 1015 (1953).
152. H. J. Vogel, B. D. Davis, and E. S. Mingioli, *J. Am Chem. Soc.* **73**, 1897 (1951).
153. H. S. Moyed and B. Magasanik, *J. Biol. Chem.* **235**, 149 (1960).
153a. S. H. Love, *J. Bacteriol.* **72**, 628 (1956).
154. B. N. Ames and P. E. Hartman, *in* Molecular " Basis of Neoplasia," Univ. of Texas, M. D. Anderson Hospital and Tumor Inst., p. 322, 1961; *Cold Spring Harbor Symp. Quant. Biol.* **28**, 349 (1963).
155. B. N. Ames and R. G. Martin, *Ann. Rev. Biochem.* **33**, 235 (1964).
156. J. C. Loper, M. Grabnar, R. C. Stahl, Z. Hartman, and P. E. Hartman, *Brookhaven Symp. Biol.* **17**, 15 (1964).
157. B. N. Ames, R. F. Goldberger, P. E. Hartman, R. G. Martin, and J. R. Roth, *in* " Regulation of Nucleic Acid and Protein Biosynthesis " (E. C. Slater, ed.), Vol. 10, BBA Library, Elsevier Publ. Co., New York.
158. R. G. Martin and B. N. Ames, *J. Biol. Chem.* **236**, 1372 (1961).
159. B. N. Ames, R. G. Martin, and B. J. Garry, *J. Biol. Chem.* **236**, 2019 (1961).
160. R. G. Martin, *J. Biol. Chem.* **238**, 257 (1963).
161. M. J. Voll, E. Appella, and R. G. Martin, *J. Biol. Chem.* **242**, 1760 (1967).
162. H. J. Whitfield, D. W. E. Smith, and R. G. Martin, *J. Biol. Chem.* **239**, 3288 (1964).
163. D. W. E. Smith and B. N. Ames, *J. Biol. Chem.* **239**, 1848 (1964).
164. D. W. E. Smith and B. N. Ames, *J. Biol. Chem.* **240**, 3056 (1965).
165. B. N. Ames, B. J. Garry, and L. A. Herzenberg, *J. Gen. Microbiol.* **22**, 369 (1960).
166. H. P. Broquist and J. S. Trupin, *Ann. Rev. Biochem.* **35**, 251 (1966).
167. M. N. Margolies and R. F. Goldberger, *J. Biol. Chem.* **241**, 3262 (1966).
168. M. A. Berberich, P. Venetiamer, and R. F. Goldberger, *J. Biol. Chem.* **241**, 4426 (1966).
169. M. N. Margolies and R. F. Goldberger, *J. Biol. Chem.* **242**, 256 (1967).
170. B. N. Ames, *J. Biol. Chem.* **228**, 131 (1957).

171. B. N. Ames, *J. Biol. Chem.* **226**, 583 (1957).
172. J. C. Loper, *Proc. Natl. Acad. Sci. U.S.* **47**, 1440 (1961).
173. B. N. Ames and B. L. Horecker, *J. Biol. Chem.* **220**, 113 (1956).
174. E. Adams, *J. Biol. Chem.* **217**, 325 (1955).
175. J. C. Loper and E. Adams, *J. Biol. Chem.* **240**, 788 (1965); J. C. Loper, *J. Biol. Chem.* **243**, 3234 (1968).
175a. J. Yourno and I. Ino, *J. Biol. Chem.* **243**, 3273 (1968); J. Yourno, *J. Biol. Chem.* **243**, 3273 (1968).
176. B. D. Davis, *Harvey Lectures Ser.* **50**, 230 (1956).
177. D. B. Sprinson, *Advan. Carbohydrate Chem.* **15**, 235 (1960).
177a. A. Meister, "Biochemistry of the Amino Acids," 2nd ed., vol. 2 pp. 890–891, Academic Press, New York, 1965.
178. B. D. Davis, *J. Biol. Chem.*, **191**, 315 (1951).
179. I. I. Salamin and B. D. Davis, *J. Am. Chem. Soc.* **75**, 5567 (1953).
180. V. Weiss, B. D. Davis, and E. S. Mingioli, *J. Am. Chem. Soc.* **75**, 5572 (1953).
181. B. D. Davis and E. S. Mingioli, *J. Bacteriol.* **66**, 129 (1953).
182. J. G. Levin and D. B. Sprinson, *J. Biol. Chem.* **239**, 1142 (1964).
183. B. D. Davis, *Science* **118**, 251 (1953).
184. M. Katagiri and R. Sato, *Science* **118**, 250 (1953).
185. V. Weiss, C. Gilvarg, E. S. Mingioli, and B. D. Davis, *Science* **119**, 774 (1954).
186. M. I. Gibson and F. Gibson, *Biochim. Biophys. Acta* **65**, 160 (1962); *Biochem. J.* **90**, 248 (1964).
187. R. G. H. Cotton and F. Gibson, *Biochim. Biophys. Acta* **100**, 76 (1965); *Biochim. Biophys. Acta* **147**, 222 (1967).
188. J. Baddiley, G. Ehrensvärd, E. Klein, L. Reio, and E. Saluste, *J. Biol. Chem.* **183**, 177 (1950).
189. C. Gilvarg and K. Bloch, *J. Biol. Chem.* **193**, 339 (1951); **199**, 680 (1952).
190. C. Cutinelli, G. E. Ehrensvärd, L. Reio, E. Saluste, and R. Stjernholm, *Acta Chem. Scand.* **5**, 353 (1951).
191. R. C. Thomas, V. H. Cheldelin, B. E. Christensen, and C. H. Wang, *J. Am. Chem. Soc.* **75**, 5554 (1953).
192. P. R. Srinivasan, H. T. Shigeura, M. Sprecher, D. B. Sprinson, and B. D. Davis, *J. Biol. Chem.* **220**, 447 (1956).
193. E. B. Kalan, B. D. Davis, P. R. Srinivasan, and D. B. Sprinson, *J. Biol. Chem.* **223**, 907 (1956).
194. P. R. Srinivasan, D. B. Sprinson, E. B. Kalan, and B. D. Davis, *J. Biol. Chem.* **223**, 913 (1956).
195. P. R. Srinivasan, M. Katagiri, and D. B. Sprinson, *J. Am. Chem. Soc.* **77**, 4943 (1955); *J. Biol. Chem.* **234**, 713 (1959).
196. P. R. Srinivasan and D. B. Sprinson, *J. Biol. Chem.* **234**, 716 (1959).
196a. C. H. Doy, *Rev. Pure Appl. Chem.* **18**, 41 (1968).
196b. C. H. Doy, *Biochim. Biophys. Acta* **159**, 352 (1968).
197. D. B. Sprinson, J. Rothschild, and M. Sprecher, *J. Biol. Chem.* **238**, 3170, 3176 (1963).
198. M. Adlesberg and D. B. Sprinson, *Biochemistry* **3**, 1855 (1964).
199. S. Mitsuhashi and B. D. Davis, *Biochim. Biophys. Acta* **15**, 54 (1954).
200. D. Balinsky and D. D. Davies, *Biochem. J.* **80**, 300 (1961).
201. H. Yanin and G. Gilvarg, *J. Biol. Chem.* **213**, 787 (1955).
202. D. Balinsky and D. D. Davies, *Biochem. J.* **80**, 292, 296 (1961).
203. S. Mitsuhashi and B. D. Davis, *Biochim. Biophys. Acta* **15**, 268 (1954).
204. A. Rivera, Jr., and P. R. Srinivasan, *Biochemistry* **2**, 1063 (1963).

205. I. Schwink and E. Adams, *Biochim. Biophys. Acta* **36**, 102 (1959).
206, L. C. Smith, J. M. Ravel, S. R. Lax, and W. Shive, *J. Biol. Chem.* **237**, 3506 (1962).
207. K. D. Brown and C. H. Doy, *Biochim. Biophys. Acta* **77**, 170 (1963).
208. C. H. Doy and K. D. Brown, *Biochim. Biophys. Acta* **104**, 377 (1965).
209. R. A. Jensen and E. W. Nester, *J. Mol. Biol.* **12**, 468 (1965).
210. M. I. Gibson and F. Gibson, *Biochem. J.* **90**, 248 (1964).
211. J. M. Edwards, F. Gibson, J. M. Jackman, and J. S. Shannon, *Biochem. Biophys. Acta* **93**, 78 (1964).
212. P. R. Srinivasan and A. Rivera, Jr., *Biochemistry* **2**, 1059 (1963).
213. A. Rivera, Jr., and P. R. Srinivasan, *Biochemistry* **2**, 1063 (1963).
214. C. Yanofsky, *Biochim. Biophys. Acta* **16**, 594 (1955).
215. C. Yanofsky, *J. Biol. Chem.* **217**, 345 (1955).
216. C. Yanofsky, *Biochim. Biophys. Acta* **20**, 438 (1956); *J. Biol. Chem.* **223**, 171 (1956).
217. C. H. Doy and F. Gibson, *Biochem. J.* **72**, 586 (1959).
218. O. H. Smith and C. Yanofsky, *J. Biol. Chem.* **235**, 2051 (1960).
219. J. Wegman and J. A. DeMoss, *J. Biol. Chem.* **240**, 3781 (1965).
220. O. H. Smith, *Bacteriol. Proc.* 29 (1965).
221. J. A. DeMoss, *Biochem. Biophys. Res. Commun.* **18**, 850 (1965).
222. F. Gibson and C. Yanofsky, *Biochim. Biophys. Acta* **43**, 489 (1960).
223. T. E. Creighton and C. Yanofsky, *J. Biol. Chem.* **241**, 4616 (1966).
224. C. Yanofsky, *Bact. Rev.* **24**, 221 (1960).
225. C. Yanofsky, and J. Stadles, *Proc. Natl. Acad. Sci. U.S.* **44**, 245 (1958).
226. I. P. Crawford and C. Yanofsky, *Proc. Natl. Acad. Sci. U.S.* **44**, 461 (1958).
227. J. A. DeMoss and D. M. Bonner, *Proc. Natl. Acad. Sci. U.S.* **45**, 1405 (1959).
228. W. C. Mohler and S. R. Suskind, *Biochim. Biophys. Acta* **43**, 228 (1960).
229. D. A. Wilson and I. P. Crawford, *J. Biol. Chem.* **240**, 4801 (1965).
230. M. Hatanaka, E. A. White, K. Honbata, and I. P. Crawford, *Arch. Biochem. Biophys.* **97**, 596 (1962).
231. U. Henning, D. R. Helinski, F. C. Chao, and C. Yanofsky, *J. Biol. Chem.* **237**, 1523 (1962).
232. B. C. Carlton and C. Yanofsky, *J. Biol. Chem.* **237**, 1531 (1962); **238**, 636 (1963).
233. M. Carsiotis and S. R. Suskind, *J. Biol. Chem.* **239**, 4227 (1964); *Biochem. Biophys. Res. Commun.* **18**, 877 (1965).
234. D. A. Wilson and I. P. Crawford, *J. Biol. Chem.* **240**, 4801 (1965).
235. J. K. Hardman and C. Yanofsky, *J. Biol. Chem.* **240**, 725 (1965).
236. J. R. Guest and C. Yanofsky, *J. Biol. Chem.* **241**, 1 (1966).
237. T. E. Creighton and C. Yanofsky, *J. Biol. Chem.* **241**, 980 (1966).
238. M. E. Goldberg, T. E. Creighton, R. L. Baldwin, and C. Yanofsky, *J. Mol. Biol.* **21**, 71 (1966).
239. J. R. Guest, B. C. Carlton, and C. Yanofsky, *J. Biol. Chem.* **242**, 5397 (1967).
240. B. C. Carlton, J. R. Guest, and C. Yanofsky, *J. Biol. Chem.* **242**, 5422 (1967).
241. G. R. Drapeau and C. Yanofsky, *J. Biol. Chem.* **242**, 5413 (1967).
242. G. R. Drapeau and C. Yanofsky, *J. Biol. Chem.* **242**, 5434 (1967).
243. J. R. Guest, G. R. Drapeau, B. C. Carlton, and C. Yanofsky, *J. Biol. Chem.* **242**, 5442 (1967).
244. D. A. Wilson and I. P. Crawford, *Bacteriol. Proc.* 92 (1964).
245. C. Yanofsky, D. R. Helinski and B. D. Maling, *Cold Spring Harbor Symp. Quant. Biol.* **26**, 11 (1961).
246. H. S. Moyed, *J. Biol. Chem.* **235**, 1098 (1960).
247. R. L. Somerville and C. Yanofsky, *J. Mol. Biol.* **11**, 747 (1965).

248. A. Matsushiro, K. Sato, J. Ito, S. Kida, and F. Imamoto, *J. Mol. Biol.* **11**, 54 (1965).
249. C. Yanofsky, J. Ito, and V. Horn, *Cold Spring Harbor Symp. Quant. Biol.* **31**, 151 (1966).
250. H. Blaschko, *J. Physiol. (London)* **96**, 50 (1939).
251. P. Holtz, *Naturwiss,* **27**, 724 (1939).
252. J. Axelrod, *Science* **140**, 499 (1963).
253. S. Gurin and A. M. Delluva, *J. Biol. Chem.* **170**, 545 (1947).
254. S. Udenfriend, J. R. Cooper, C. T. Clark, and J. E. Baer, *Science* **117**, 663 (1953).
255. S. Udenfriend and J. B. Wyngaarden, *Biochim. Biophys. Acta* **20**, 48 (1956).
256. E. B. Keller, R. A. Boissonnas, and V. deVigneaud, *J. Biol. Chem.* **183**, 627 (1950).
257. D. T. Masuoka, H. F. Schott, R. I. Akawie and W. G. Clark, *Proc. Soc. Exptl. Biol. Med.* **93**, 5 (1956).
258. T. Nagatsu, M. Levitt, and S. Udenfriend, *J. Biol. Chem.* **239**, 2910 (1964).
259. A. R. Brenneman and S. Kaufman, *Biochem. Biophys. Res. Commun.* **17**, 177 (1964).
260. S. H. Pomerantz, *J. Biol. Chem.* **238** 2351 (1963); **241**, 161 (1966).
261. H. Blaschko, *Biochem. J.* **36**, 571 (1942).
262. W. Lovenberg, H. Weissbach, and S. Udenfriend, *J. Biol. Chem.* **237**, 89 (1962).
263. J. H. Fellman, *Enzymologia* **20**, 366 (1959).
264. R. Ferrini and A. Glasser, *Biochem. Pharmacol.* **13**, 798 (1964).
265. J. Awapara, R. P. Sandman, and C. Hanly, *Arch. Biochem. Biophys.* **98**, 520 (1962).
266. G. E. Schires, *Z. Physiol. Chem.* **332**, 70 (1963).
267. N. Kirshner, *J. Biol. Chem.* **226**, 821 (1957).
268. E. Y. Levin, B. Levenberg, and S. Kaufman, *J. Biol. Chem.* **235**, 2080 (1960).
269. E. Y. Levin and S. Kaufman, *J. Biol. Chem.* **236**, 2043 (1961).
270. W. J. Smith and N. J. Kirshner, *J. Biol. Chem.* **237**, 1890 (1962).
271. M. Goldstein and J. F. Contsera, *J. Biol. Chem.* **237**, 1898 (1962).
272. W. F. Bridgers and S. J. Kaufman, *J. Biol. Chem.* **237**, 526 (1962).
273. C. R. Creveling, J. B. Van der Schoot, and S. Udenfriend, *Biochem. Biophys. Res. Commun.* **8**, 215 (1962).
274. C. R. Creveling, J. W. Daly, B. Witkop, and S. Udenfriend, *Biochim. Biophys. Acta* **64**, 125 (1962).
275. M. Goldstein, M. R. McKereghan, and E. Lauber, *Biochim. Biophys. Acta* **77**, 161 (1963).
276. J. Axelrod, *Science* **140**, 499 (1963).
277. J. Axelrod, *J. Biol. Chem.* **237**, 1657 (1962).
278. C. R. Harrington, G. Barger, and B. Weiss, *Biochem. J.* **21**, 169 (1927).
279. J. Gross and R. Pitt-Rivers, *Biochem. J.* **63**, 645 (1952).
280. J. Roche, S. Lersitzky, and R. Michel, *Biochim. Biophys. Acta* **11**, 220 (1953).
281. J. B. Trunnell and P. Wade, *J. Clin. Endocrinol. Metab.* **15**, 107 (1955).
282. R. Pitt-Rivers, *Brit. Med. Bull.* **16**, 118 (1960).
283. N. M. Alexander and B. J. Corcoran, *J. Biol. Chem.* **237**, 243 (1962).
284. T. Hosaya and M. Morrison, *J. Biol. Chem.* **242**, 2828 (1967).
285. C. P. Mahoney and R. P. Igo, *Biochim. Biophys. Acta* **113**, 507 (1966).
286. A. Taurog and E. M. Howells, *J. Biol. Chem.* **241**, 1329 (1966).
287. M. L. Coval and A. Taurog, *J. Biol. Chem.* **242**, 5510 (1967).
287a. A. G. Fischer, A. R. Schulz, and L. Oliner, *Biochim. Biophys. Acta* **159**, 460 (1968).
288. A. Meister *in* "The Enzymes," 2nd ed. (P. D. Boyer, H. Lardy, and K. Myrback, eds.), Vol. 6, p. 443, Academic Press, New York, 1962.
288a. C. A. Woolfolk, B. Shapiro, and E. R. Stadtman, *Arch. Biochem. Biophys.* **116**, 177 (1966).

288b. C. A. Woolfolk and E. R. Stadtman, *Arch. Biochem. Biophys.* **118**, 736 (1967); *ibid* **122**, 174 (1967).

288c. R. C. Valentine, B. M. Shapiro, and E. R. Stadtman, *Biochemistry* **7**, 2143 (1968).

289. A. Meister, P. R. Krishnaswamy, and V. Pamiljans, *Federation Proc.* **21**, 1013 (1962).

290. A. Meister, *Federation Proc.* **27**, 100 (1968).

291. L. Levintow and A. Meister, *J. Am. Chem. Soc.* **75**, 3039 (1953).

292. N. Lichtenstein, H. E. Ross, and P. P. Cohen, *J. Biol. Chem.* **201**, 117 (1953)

293. E. Khedouri, V. P. Wellner, and A. Meister, *Biochemistry* **3**, 824 (1964).

294. V. P. Wellner, M. Zoukis, and A. Meister, *Biochemistry* **5**, 3509 (1966).

295. V. Pamiljans, P. R. Krishnaswamy, G. Dumville, and A. Meister, *Biochemistry* **1**, 153 (1962).

296. J. E. Varner, D. H. Slocum, and G. C. Webster, *Arch. Biochem. Biophys.* **73**, 508 (1958).

297. J. M. Ravel, S. J. Norton, J. S. Humphreys, and W. Shive, *J. Biol. Chem.* **237**, 2845 (1962).

298. J. J. Burchall, E. C. Reichelt, and M. J. Wolin, *J. Biol. Chem.* **239**, 1794 (1964).

299. M. K. Patterson, Jr., and G. R. Orr, *J. Biol. Chem.* **243**, 376 (1968).

300. B. Horowitz, B. Madsas, A. Meister, L. J. Old, and E. A. Boyse, *Proc. Am. Assoc. Cancer Res.* **9**, 33 (1968).

301. M. K. Patterson, Jr., and G. R. Orr, *Biochem. Biophys. Res. Commun.* **26**, 228 (1967).

302. M. D. Prager and N. Bachinsky, *Biochem. Biophys. Res. Commun.* **31**, 43 (1968).

303. S. M. Arfin, *Biochim. Biophys. Acta* **136**, 233 (1967).

CHAPTER 16 (Part II)

Biosynthesis of Amino Acids and Related Compounds

Victor W. Rodwell

I. THE GLUTAMATE FAMILY OF AMINO ACIDS

Suggestions of a close metabolic relationship between glutamate, ornithine, arginine, and proline appeared as long ago as the turn of the century. Two short reviews (*1,2*) summarize the implications of these earlier investigations and describe both the techniques and the state of

317

knowledge in this field in 1955. The availability of ^{14}C-labeled compounds in the early 1950s permitted application of the isotope competition technique to amino acid biosynthesis in micro-organisms. These studies (3–6) indicated that:

1. Proline, ornithine, and arginine constitute the "glutamate family" whose biosynthesis originates with glutamate.
2. Proline biosynthesis involves glutamate γ-semialdehyde and Δ^1-pyrroline-5-carboxylate as intermediates.
3. Ornithine biosynthesis involves acylated intermediates.
4. Microorganisms utilize preformed amino acids preferentially. This observation stimulated study of intracellular regulatory mechanisms.

II. THE PROLINES

A. Introduction

The weight of evidence presently available supports the view that proline biosynthesis occurs as shown in Fig. 1.

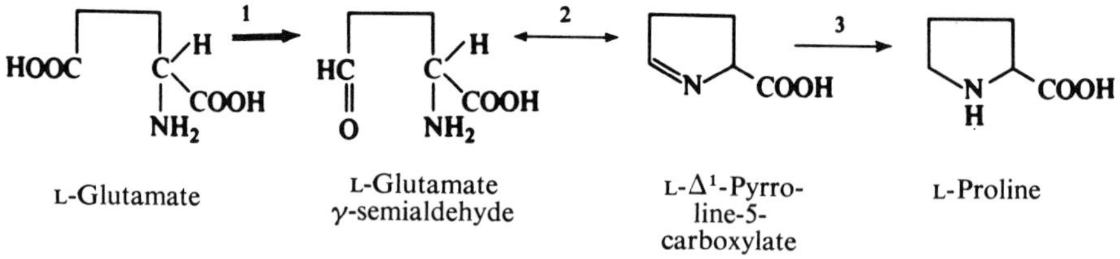

FIG. 1. Intermediates in proline biosynthesis. The double arrow indicates uncertainty as to the possible intermediates between glutamate and its semialdehyde. In Section II, B, subheadings correspond to reaction numbers of this figure.

Although whole cells of *Escherichia coli* form proline from either ornithine or *N*-acetylglutamate, an intermediate in ornithine biosynthesis (Fig. 4), the results of isotope dilution experiments argue against this being a major route for proline biosynthesis (6). A possible role for *N*-acetylglutamate in proline biosynthesis was nevertheless recently reinvestigated by Reed and Lukens (7). Cell-free extracts catalyze proline synthesis from *N*-acetylglutamate, probably via the reactions shown in Fig. 2. Reed and Lukens (7) nevertheless concur with the earlier conclusion (6) that this pathway is probably not a major route for proline biosynthesis in growing wild-type cells, since a mutant, *E. coli*

W 55-25, although it appears to possess all the enzymes for catalysis of these reactions (Fig. 2), is nevertheless auxotrophic for proline.

Pentahomoserine (α-amino-δ-hydroxycaproate) can replace proline or ornithine for growth of certain *Neurospora* auxotrophs. Its biological activity no doubt results from oxidation to glutamate γ-semialdehyde, a reaction catalyzed by an ω-hydroxy-α-amino acid dehydrogenase from a *Neurospora* mutant lacking Δ¹-pyrroline-5-carboxylate reductase (*8,9*).

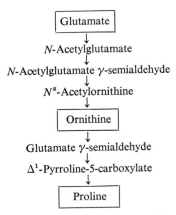

FIG. 2. Biosynthesis of proline via *N*-acetylated intermediates demonstrated in cell-free extracts of *E. coli* (*7*).

B. Individual Reactions in L-Proline Biosynthesis

1. CONVERSION OF GLUTAMATE TO GLUTAMATE γ-SEMIALDEHYDE

Although isotopic and nutritional evidence support a role for glutamate γ-semialdehyde as an intermediate in proline biosynthesis (*2*), conversion of glutamate to glutamate γ-semialdehyde has yet to be demonstrated in cell-free extracts (*7*). The reaction may involve intermediate formation of an activated glutamate such as γ-glutamyl phosphate (*10*) or γ-adenyl glutamate. Intermediates of this type are formed during reduction of aspartate (see Fig. 10) and α-aminoadipate (see Section IV, C, 5) to the corresponding ω-semialdehydes.

Glutamate γ-semialdehyde is in rapid nonenzymic equilibrium with the cyclic compound Δ¹-pyrroline-5-carboxylate (see Section II, B, 2). Vogel and Davis (*11*) and later others (*12,13*) observed that certain *E. coli* mutants auxotrophic for proline accumulated Δ¹-pyrroline-5-carboxylate. This replaced proline for growth of a second proline auxotroph. Accumulation of Δ¹-pyrroline-5-carboxylate was enhanced by addition of its precursor, glutamate. Unlabeled glutamate γ-semi-aldehyde added to growing cultures of *Neurospora crassa* substantially decreases incorporation of isotope from glutamate-^{14}C into proline (*14*). This further supports a role for this compound in proline biogenesis in *Neurospora*.

2. Ring Closure of Glutamate γ-Semialdehyde to Δ^1-Pyrroline-5-carboxylate

Aminoaldehydes such as glutamate γ-semialdehyde and α-amino-adipate δ-semialdehyde reversibly dehydrate between the carbonyl and ammonium groups forming cyclic Schiff bases such as Δ^1-pyrroline-5-carboxylate or Δ^1-piperideine-6-carboxylate. The open and ring forms are in rapid, nonenzymic equilibrium, for chemical reduction produces the saturated ring structure, and chemical oxidation the corresponding straight-chain dicarboxylic α-amino acid (*11,15–17*). It is thus not necessary to postulate enzymic catalysis of reaction 2 (Fig. 1). Whether such catalysis nevertheless occurs has not been studied, possibly because of lack of suitable analytical techniques.

3. Reduction of L-Δ^1-Pyrroline-5-carboxylate to Proline

Δ^1-Pyrroline-5-carboxylate is excreted by double mutants of *E. coli* blocked both in regulation of proline synthesis and in conversion of Δ^1-pyrroline-5-carboxylate to proline (*13*). It replaces proline for growth of other proline auxotrophs of *E. coli* (*11*). Reduction of chemically synthesized Δ^1-pyrroline-5-carboxylate (*11,16*) to proline has been observed in cell-free extracts from microorganisms (*18–20*) and mammals (*15, 20–22*). Enzymes catalyzing this reduction have been partially purified from *N. crassa* (*18,19*), rat liver (*21*), and calf liver (*22*).

Pyrroline-5-carboxylate reductase [L-proline : NADP(H) 5-oxido-reductase, EC 1.5.1.2] of calf liver (*22*), which catalyzes irreversibly the reduction of Δ^1-pyrroline-5-carboxylate to proline, is distinct both from the enzyme catalyzing the analogous reduction of Δ^1-3-hydroxy-pyrroline-5-carboxylate to 4-hydroxyproline (*15*) and from proline oxidase of the mitochondrial fraction of liver. $V_{\max}$ for NADH ($K_m = 8.4 \times 10^{-4}$ M) is greater than that for NADPH ($K_m = 2.0 \times 10^{-5}$ M). Studies with substrate analog inhibitors suggest that the cyclic form ($K_m = 3.3 \times 10^{-4}$ M) rather than glutamate γ-semialdehyde, as proposed for the rat liver reductase (*21*), is the actual substrate. Although the calf liver reductase is unstable in the absence of reducing agents, free SH groups do not appear to be involved in the catalytic mechanism.

The rat liver reductase (21) exhibits comparable kinetic constants for substrate ($K_m = 2.1 \times 10^{-4}$ M) and NADH ($K_m = 2.5 \times 10^{-4}$ M), utilizes NADH preferentially as reductant, but is considerably more sensitive to inhibition by thiol reagents. Adenine nucleotides inhibit competitively with NADH.

The reductase from N. crassa exhibits comparable affinity for Δ^1-pyrroline-5-carboxylate ($K_m = 4.5 \times 10^{-4}$ M) but utilizes NADPH far more efficiency than NADH. A role for this enzyme in proline biosynthesis is suggested by its absence from a proline auxotroph of N. crassa. Evidence consistent with repression of reductase synthesis was obtained when Neurospora was grown in proline-rich media (18,19).

Partially purified hog liver Δ^1-pyrroline-2-carboxylate reductase [L-proline : NAD(P) 2-oxidoreductase, EC 1.5.1.1] catalyzes irreversibly the reduction of the isomeric Δ^1-pyrroline-2-carboxylate to proline. Substrate analog inhibitor studies suggest that the open chain structure is the actual substrate. NADPH ($K_m = 1.2 \times 10^{-5}$ M) is utilized preferentially to NADH ($K_m = 1.7 \times 10^{-4}$ M) as reductant (23). Analogous reactions were previously observed in crude extracts of Neurospora and plant tissues (20). Although proline auxotrophs of N. crassa or of Aerobacter lacking Δ^1-pyrroline-5-carboxylate reductase can utilize Δ^1-pyrroline-2-carboxylate for growth, no additional evidence supports a role for Δ^1-pyrroline-2-carboxylate in proline biosynthesis. The physiological significance of this enzyme is thus not clear.

C. Regulation of L-Proline Biosynthesis

Observations suggestive of tight regulatory control of proline biosynthesis include: the preferential utilization by E. coli of preformed proline (2); the failure of growing cultures of wild-type E. coli to excrete either proline or intermediates of proline biosynthesis even when grown in the presence of a large excess of glutamate (13); and the selection of "uncontrolled" mutants of E. coli which excrete either proline or glutamate γ-semialdehyde (13).

To facilitate study of regulatory controls, E. coli mutants lacking control of proline biosynthesis (C⁻-mutants) were selected by their ability to grow in the presence of the proline antagonist 3,4-dehydroproline. From the C⁻-mutants, proline auxotrophs were produced (Fig. 3). Strains 55-1 and WPI-30 both excrete glutamate γ-semialdehyde. Addition of 1 μg proline per ml culture fluid to growing cultures produces immediate cessation of glutamate γ-semialdehyde excretion in 55-1 but is without effect on excretion by WPI-30. This cessation was so rapid as to suggest feedback inhibition. When glutamate γ-semialdehyde was added to cultures of wild-type *E. coli* W, proline was excreted. L-Δ^1-Pyrroline-5-carboxylate reductase activity was not inhibited by proline, nor was its synthesis repressed by proline. Control thus appears to be exercised prior to, but not subsequent to, glutamate γ-semialdehyde.

Name	Description	Presumed genetic constitution
W	Wild type	$C^+Pro_1{}^+Pro_2{}^+$
55-1	W, lacking L-Δ^1-pyrroline-5-carboxylate reductase	$C^+Pro_1{}^+Pro_2{}^-$
W-2	W, lacking ability to synthesize glutamate γ-semialdehyde	$C^+Pro_1{}^-Pro_2{}^+$
WPI	W, deficient control of proline synthesis	$C^-Pro_1{}^+Pro_2{}^+$
WPI-30	WPI, lacking L-Δ^1-pyrroline-5-carboxylate reductase	$C^-Pro_1{}^+Pro_2{}^-$
WPI-5	WPI, lacking ability to synthesize glutamate γ-semialdehyde	$C^-Pro_1{}^-Pro_2{}^+$

FIG. 3. Mutants of *E. coli* used to investigate regulation of proline biosynthesis (13).

Possible regulation via proline transport has also been suggested (24). Intact and sonified membrane preparations from wild-type *E. coli* and from a transport-minus mutant appear to bind proline loosely by an energy-requiring process (24).

D. D-Proline

Synthesis of D-proline may occur via reversal of the D-amino acid oxidase reaction [D-amino acid: oxygen oxidoreductase (deaminating), EC 1.4.3.1] (25).

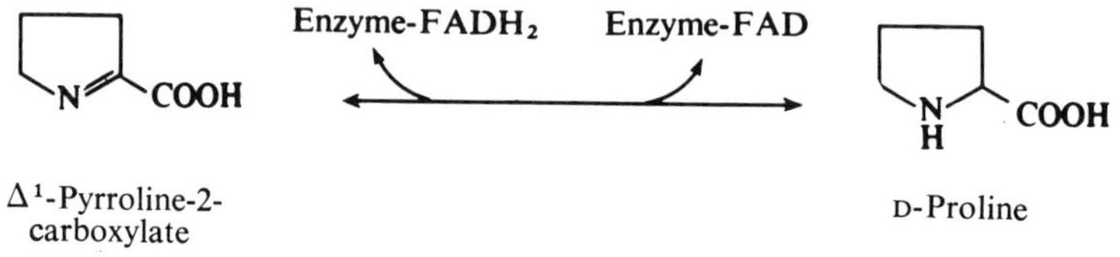

Δ^1-Pyrroline-2-
carboxylate   D-Proline

In *Clostridium sticklandii*, interconversion of D- and L-proline is catalyzed by a proline racemase (proline racemase, EC 5.1.1.4) (26).

E. 4-Hydroxyproline

Essentially all the hydroxyproline and hydroxylysine in animal tissues occurs in collagen (27). Free dietary hydroxyproline is virtually unavailable to growing rats (28,29), and conversion of proline to hydroxyproline of collagen has repeatedly been shown to proceed via intermediates other than free imino acids (30,31). The two possibilities generally considered are that the substrate for hydroxylation is a prolyl-sRNA or that it is a prolyl polypeptide. Studies of collagen synthesis in embryonic tissue provide evidence in support of both these hypotheses.

Using $^{18}O_2$, Fujimoto and Tamiya (32) and Prockop et al. (33) demonstrated that the oxygen of hydroxyproline is derived from air rather than from water. Hydroxylation, catalyzed by the microsomal fraction (34,35), thus presumably involves an oxygenase reaction. 4-Ketoproline and 3,4-dehydroproline were ruled out as intermediates using proline-3,4-^{3}H by the interpretation that tritium from a single hydrogen position was lost during hydroxylation (36). Fujita et al. (37) reinvestigated the hydroxylation mechanism using chemically synthesized cis- and trans-4-^{3}H-prolines. Synthesis of trans-4-hydroxy-L-proline by chick embryos occurred with 94% retention of tritium in cis- and 98% loss of tritium from trans-4-^{3}H-L-proline. Proline hydroxylation apparently involves front-side displacement with retention of configuration at C-4 without intermediate formation of either 3,4-dehydro- or 4-ketoproline.

Collagen synthesis, achieved in cell-free systems from chick embryo tissue (38,39), occurs by mechanisms comparable to those observed for other proteins (40). Its synthesis, but not that of the macromolecular collagen precursor which undergoes hydroxylation, is reported to be inhibited by cortisol (41). Although some uncertainty still remains regarding the macromolecular collagen precursor which undergoes hydroxylation, the weight of evidence favors a polypeptide substrate, possibly one of molecular weight about 15,000 (42). Manner and Gould (43) reported that, although puromycin blocked incorporation of radioactivity from proline-^{14}C into collagen hydroxyproline by chick embryos, it failed to abolish formation of free hydroxyproline. They concluded that free hydroxyproline was formed by reactions not involving synthesis of polypeptide-bound hydroxyproline. These authors (43) also reported the presence of radioactive hydroxyproline in the sRNA fraction after incubation of proline-^{14}C with isolated chick embryo sRNA and 105,000 × g supernatant fraction. Similarly Coronado et al. (44) reported radioactive hydroxyproline and hydroxylysine in the sRNA fraction after incubation of this fraction with proline-^{14}C or lysine-^{14}C and the pH 5 amino acid-activating enzymes from chick embryo. These results have been confirmed in yet a third laboratory for hydroxyproline synthesis by chick embryo and wound granulation tissue (45).

Evidence against prolyl-sRNA and favoring peptide-bound proline as the substrate for hydroxylation is provided by the work of Udenfriend, Prockop, Lukens, and their associates (27,34,39,42,46–49). Peterkofsky and Udenfriend reported a 30-minute lag before initiation of incorporation of radioactivity into hydroxyproline of microsomal-bound protein. If puromycin, which inhibited incorporation when added initially, was added after 30 minutes, little inhibition of subsequent incorporation occurred. Again, incorporation was inhibited by ribonuclease added initially, but not when added after 30 minutes. These observations suggest that radioactive proline was incorporated into a molecule that subsequently could be converted to protein-bound hydroxyproline by reactions insensitive to puromycin or to ribonuclease. Since these reagents are known to inhibit assembly of polypeptide chains from amino acids present as aminoacyl-sRNA compounds (50), it was inferred that the precursor contained proline already bound in a polypeptide linkage. Lukens (49) found that charging of isolated chick embryo sRNA with proline-^{14}C, catalyzed by the pH 5 enzyme fraction of chick embryos, occurred without detectable synthesis of hydroxyprolyl-sRNA. Unlike Manner and Gould (43), Lukens found that puromycin strongly inhibited formation of free hydroxyproline. Since neither charging of sRNA (51) nor release of an amino acid from aminoacyl-sRNA (50) is inhibited by puromycin, Lukens concluded that the precursor which is hydroxylated is a proline polypeptide. A variety of collagen-forming systems (chick embryo, granuloma tissue, fetal rat skin, hen oviduct) accumulate a proline-rich, hydroxyproline-deficient, collagenase-degradable protein ("protocollagenase") on incubation with proline-^{14}C under nitrogen (27,35,47,48).

Both hydroxyproline and 4-ketoproline occur in the polypeptide actinomycin antibiotics. The source of both the hydroxyproline and ketoproline of the actinomycin peptide elaborated by *Streptomyces antibioticus* is known to be L-proline (52). In contrast to mammalian systems, free hydroxyproline formed from proline is incorporated directly into the peptide of actinomycin I (52). Although rabbit kidney contains a ketoproline reductase catalyzing reduction of 4-keto-L-proline to hydroxyproline (53), no analogous reaction was observed in *S. antibioticus* (52).

F. 3-Hydroxyproline

3-Hydroxy-L-proline has been detected in rat tail and sponge collagen (54–56) and in the antibiotic telomycin (57,58). The structure of that derived from collagen has been established as *trans*-3-hydroxy-2-

proline (*56*). Both *cis*- and *trans*-isomers occur in telomycin (*58*). Nothing is known about biosynthesis of 3-hydroxyproline.

III. ORNITHINE AND ARGININE

A. Introduction

Formation of arginine via citrulline, one of the first biosynthetic pathways discovered and studied at an enzymic level, originally was proposed for urea synthesis in liver (*59*). The same reaction sequence was later shown for *Neurospora* (*60*), *Penicillium* (*61*), and certain lactic acid bacteria (*62*). Carbamyl phosphate synthesis, interconversion of ornithine, arginine, and citrulline, and urea formation are discussed in Chapter 14. Certain general aspects of ornithine biosynthesis were noted above (see Section I).

The intermediates and enzymes currently thought to be involved in ornithine and arginine biosynthesis are shown in Fig. 4 and Table I. The best available evidence suggests that acylated intermediates are involved in ornithine biosynthesis in most life forms. Vogel (*2*) proposed that the acylated pathway might have evolved to circumvent the tendency of glutamate γ-semialdehyde (destined for proline synthesis) to undergo nonenzymic cyclization (see Fig. 1 and Section II, B, 2).

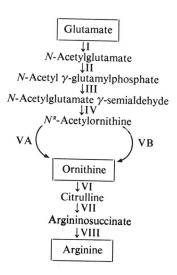

FIG. 4. Biosynthesis of ornithine and arginine via acylated intermediates. The names of the enzymes catalyzing the numbered reactions are given in Table I.

Some uncertainties exist in *Neurospora* and related organisms. Tracer experiments in *Neurospora* (*14*) and in *Torulopsis utilis* (*66*) indicate that, unlike in Enterobacteriaceae, ornithine synthesis does not involve acylated intermediates. It was originally proposed (*2*) that, in *Neurospora*,

TABLE I

ENZYMES OF ARGININE SYNTHESIS[a]

No.	Enzyme	Substrate
I	N-Acetylglutamate synthetase (Acetyl-CoA: L-glutamate N-acetyltransferase, EC 2.3.1.1)	Glutamate
II	N-Acetyl-γ-glutamokinase (ATP: N-acetyl-L-glutamate 5-phosphotransferase[b])	N-Acetylglutamate
III	N-Acetylglutamate γ-semialdehyde dehydrogenase	N-Acetyl-γ-glutamyl phosphate
IV	Acetylornithine δ-transaminase (N^{α}-Acetyl-L-ornithine : 2-oxoglutarate aminotransferase, EC 2.6.1.11)	N-Acetylglutamate γ-semialdehyde
VA	Acetylornithinase	N^{α}-Acetylornithine
VB	Ornithine acetyltransferase (N^{α}-Acetyl-L-ornithine : L-glutamate N-acetyltransferase[c])	N^{α}-Acetylornithine
VI	Ornithine transcarbamylase (Carbamoylphosphate: L-ornithine carbamoyltransferase, EC 2.1.3.3)	Ornithine
VII	Argininosuccinate synthetase [L-Citrulline: L-aspartate ligase (AMP), EC 6.3.4.5]	Citrulline
VIII	Argininosuccinase (L-Argininosuccinate arginine-lyase, EC 4.3.2.1)	Argininosuccinate

[a] Modified from Vogel *et al.* (*63*). In Section III, B, subsection numbers correspond to the Roman numerals for enzymes of this table and Fig. 4. The above pathway was postulated in essentially this form over a decade ago by Vogel (*2*). Reactions are identified in Fig. 4. Reactions VA (catalyzed by acetylornithinase) and VB (catalyzed by ornithine acetyltransferase) are alternative routes for deacylation of N^{α}-acetylornithine.

[b] Proposed name (*64*).

[c] Proposed name (*65*).

ornithine arose via transamination of glutamate γ-semialdehyde. The conversion of glutamate to glutamate γ-semialdehyde, discussed under proline biosynthesis, has yet to be demonstrated in cell-free extracts, and the nature of the intermediates involved remains unsettled (see

Section II, B). A role for ornithine transaminase (L-ornithine: 2-oxoacid aminotransferase*) in ornithine biosynthesis was suggested both by its detection in *Neurospora* (*14,67,68*), heart muscle tissue (*69*), and rat liver mitochondria (*70*) and by its absence from *E. coli* and other bacteria known to utilize acylated intermediates (*68*). Scḫer and Vogel (*68*) report that ornithine transaminase is present in all gram-positive bacteria, yeasts, green algae, higher plants, and animal tissues examined. No activity was detectable in any gram-negative bacteria or blue-green algae tested. The reaction is freely reversible (*67*) and might account for the ability of either proline or ornithine to satisfy the growth requirement for certain *Neurospora* auxotrophs (*67*). Crude *Neurospora* ornithine transaminase is optimally active at pH 8.0 with α-ketoglutarate as cosubstrate (*67*).

As pointed out by Davis (*71*) and by Meister (*10*), this pathway probably is not of major biosynthetic significance. Recent studies with *Saccharomyces cerevisiae* implicate acylated intermediates; thus arginine biosynthesis in yeast may resemble that in Enterobacteriaceae. The arguments for and against a nonacylated pathway are reviewed by Meister (*10*), who notes that a reinvestigation of ornithine biosynthesis in *Neurospora* is in order. In mammals, ornithine transaminase may serve a degradative rather than a biosynthetic function. This is suggested by its induction by high-protein diets (*70*) and its repression by carbohydrate feeding (*72*).

B. Ornithine Synthesis via Acylated Intermediates

1. SYNTHESIS OF *N*-ACETYLGLUTAMATE

Synthesis of *N*-acetylglutamate has been demonstrated in cell-free extracts of wild-type *E. coli* (*73*) and in an *E. coli* mutant strain blocked in reaction II (Fig. 4). Acetylglutamate formation was determined by bioassay using a second mutant lacking enzyme I (Fig. 4) (*74*). In

* Suggested name (*70*).

those organisms utilizing enzyme I, this first enzyme of arginine bio-synthesis is regulated by feedback inhibition by arginine (*74*). As with all other enzymes of this pathway, synthesis of *N*-acetylglutamate syn-thetase is subject to repression by arginine (*74,75*).

In certain microorganisms, *N*-acetyl-L-glutamate formation may be catalyzed by enzyme VB (see Section III, B, 5) (*76,77*). In these micro-organisms, enzyme II is subject to feedback inhibition by arginine (see Section III, C).

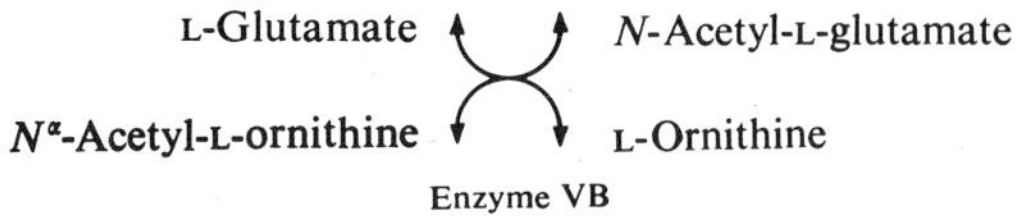

$$\text{L-Glutamate} \quad\text{N-Acetyl-L-glutamate}$$
$$N^{\alpha}\text{-Acetyl-L-ornithine} \quad\text{L-Ornithine}$$
$$\text{Enzyme VB}$$

2. FORMATION OF *N*-ACETYL-γ-GLUTAMYL PHOSPHATE

N-Acetyl-L-glutamate   *N*-Acetyl-γ-glutamyl-phosphate

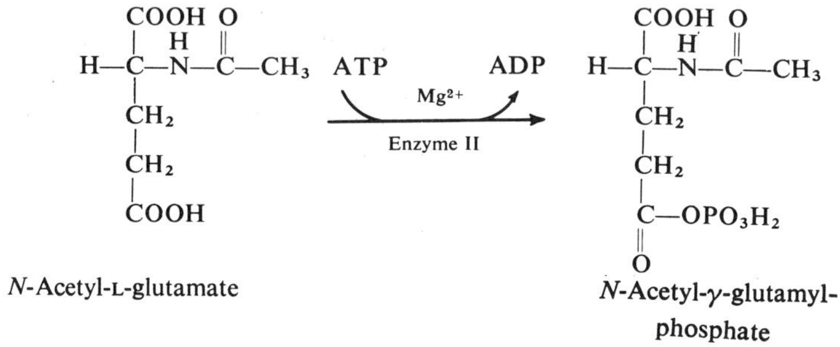

The above reaction is analogous to that catalyzed by asparate kinase (ATP: L-aspartate 4-phosphotransferase, EC 2.7.2.4). *N*-Acetyl-γ-glutamokinase (enzyme II) has been separated from enzyme III in *E. coli* extracts and exhibits an absolute requirement for ATP and for a divalent metal ion (*78*).

The kinase, partially purified (120-fold) from the fresh water alga *Chlamydomonas reinhardti*, requires Mg^{2+} or Co^{2+} for activity and is optimally active at pH 5.5. The kinetics are of the normal Michaelis-Menten type over the pH range 5.5–7.5. Apparent K_m values at pH 5.5 are $1.5 \times 10^{-2}\ M$ for *N*-acetyl-L-glutamate and $1.6 \times 10^{-3}\ M$ for ATP. Kinase activity is inhibited by L-arginine and, less effectively, by arginine analogs such as L-canavanine and L-citrulline. Inhibition by arginine is maximal at pH 7.5 and is competitive with respect to *N*-acetylglutamate but uncompetitive with ATP. Arginine also protects the kinase from both heat and urea denaturation (*64*). In *Chlamy-domonas*, *N*-acetylglutamate kinase is thus a key regulatory enzyme. If, as seems likely, *N*-acetylglutamate synthesis in *Chlamydomonas* occurs via reaction VB rather than reaction I, the kinase (enzyme II) is in a sense the first enzyme of ornithine and arginine biosynthesis, and its inhibition by arginine might therefore be anticipated.

3. REDUCTIVE DEPHOSPHORYLATION OF N-ACETYL-γ-GLUTAMYL PHOSPHATE TO N-ACETYLGLUTAMATE γ-SEMIALDEHYDE

N-Acetyl γ-glutamyl phosphate

N-Acetylglutamate γ-semialdehyde

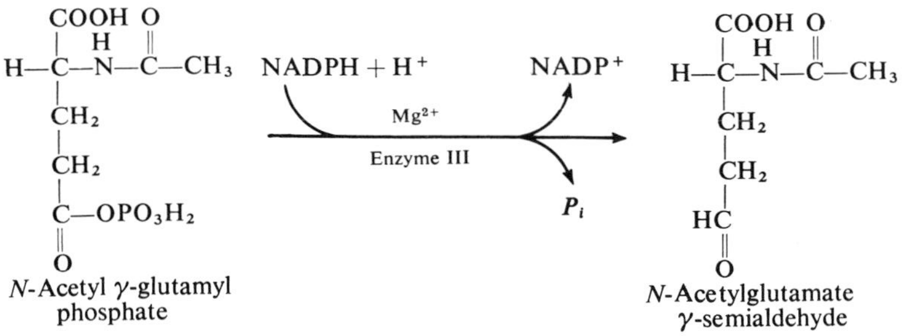

A dehydrogenase catalyzing the above reaction was detected in partially purified extracts of *E. coli* (*78*). Mixtures containing enzymes II and III convert N-acetylglutamate to N-acetylglutamate γ-semi-aldehyde (*78*).

4. TRANSAMINATION OF N-ACETYLGLUTAMATE γ-SEMIALDEHYDE FORMING N^α-ACETYLORNITHINE

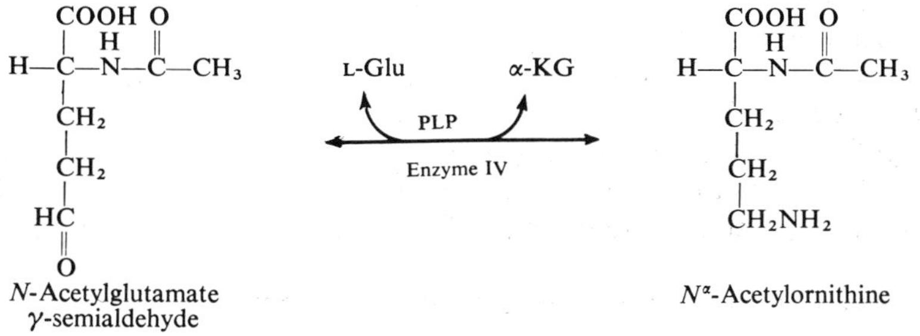

N-Acetylglutamate γ-semialdehyde

N^α-Acetylornithine

Acetylornithine δ-transaminase (N^α-acetylornithine:2-oxoglutarate aminotransferase, EC 2.6.1.11) has been purified some 30-fold from *E. coli* W. The reaction is freely reversible. The transaminase is distinct from transaminase A, is essentially specific both for N^α-acetylornithine ($K_m = 3.4 \times 10^{-4} M$) and for α-ketoglutarate ($K_m = 2.5 \times 10^{-3} M$), requires pyridoxal phosphate ($K_m = 1.7 \times 10^{-6} M$), and is inhibited by PCMB and by Cu^{2+}, but not by EDTA. Transaminase synthesis is repressible by arginine (79).

Acetylornithine δ-transaminase appears to be the site of action of two hydrazone compounds, 2-hydrazino-3(4-imidazolyl)propionate (HIPA) and α-hydrazino-n-caproate, which are potent inhibitors for growth of wild-type S. typhimurium (80). A mutant of Salmonella which lacks the ability to transport histidine or HIPA is insensitive to inhibition by the latter, although it retains sensitivity to α-hydrazino-n-caproate. Despite the structural resemblance to histidine, ornithine

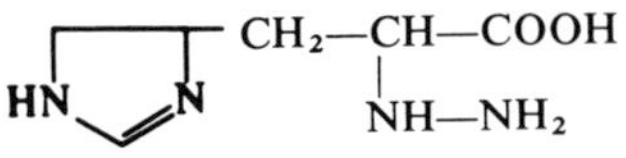

2-Hydrazino-3(4-imidazolyl)propionate
(HIPA)

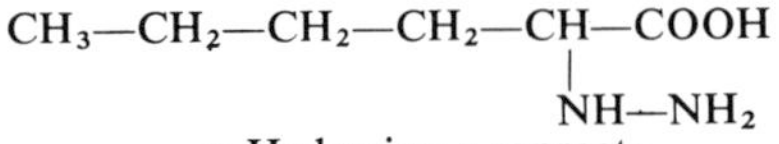

α-Hydrazino-n-caproate

biosynthesis and arginine biosynthesis rather than histidine metabolism are impaired when wild-type Salmonella are grown in the presence of HIPA. This inhibition is partially relieved by addition of aspartate, arginine, ornithine, or citrulline, and almost totally abolished by aspartate plus arginine. Both HIPA and α-hydrazino-n-caproate react rapidly and nonenzymically with pyridoxal-P, forming products of unknown structure. Acetylornithine δ-transaminase partially purified from E. coli or from a crude extract of S. typhimurium were 50% inhibited by $5 \times 10^{-4} M$ HIPA and about 85% inhibited by $1 \times 10^{-4} M$ α-hydrazino-n-caproate. Both the histidine and ornithine biosynthetic pathways involve pyridoxal-P-dependent transaminations. Shifrin et al. (80) speculate that the reason for the greater sensitivity of ornithine biosynthesis is that pyridoxal-P is less tightly bound by N-acetylornithine transaminase (E. Jones, unpublished data) than by imidazolylacetol phosphate transaminase (R. G. Martin, unpublished data).

5. CONVERSION OF N^α-ACETYLORNITHINE TO ORNITHINE

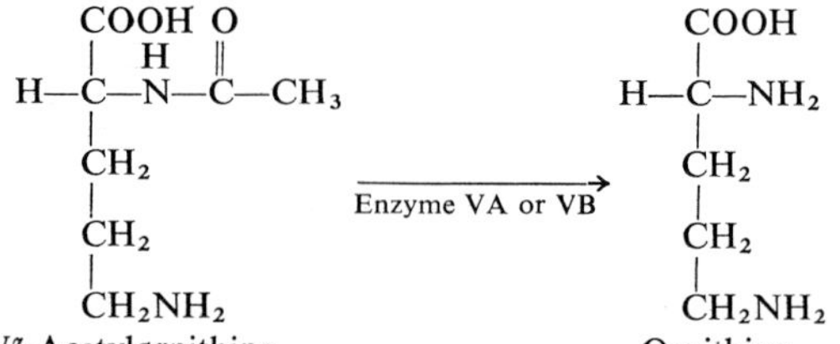

N^α-Acetylornithine

Ornithine

In Enterobacteriaceae the final step in ornithine biosynthesis is a hydrolysis, catalyzed by enzyme VA, acetylornithinase. In *Chlamydomonas reinhardti (81)* and *S. cerevisiae (77)*, acetyl. transfer is to glutamate, catalyzed by enzyme VB, ornithine acetyltransferase (N^α-acetyl-L-ornithine: L-glutamate-N-acetyltransferase*).

A role for acetylornithinase in ornithine biosynthesis is suggested by its general occurrence in Enterobacteriaceae *(76)* and its absence from *Neurospora*, certain ornithine auxotrophs of *E. coli (82)* and *B. subtilis (83)*. The presence of acetylornithinase in *Mycoplasma* and bacterial L-forms was used by Smith *(84)* as a guide to taxonomy. Partially purified *E. coli* acetylornithinase is optimally active at pH 7.0 and exhibits a K_m for N^α-acetylornithine of $2.8 \times 10^{-3} M$ *(82)*. Both *E. coli* and *B. subtilis* acetylornithinases are specifically stimulated by Co^{2+} and by glutathione *(82,83)*.

Although the pathway of ornithine biosynthesis in photosynthetic algae is ill-defined, isotopic data *(81)* suggest that acylated intermediates are involved. Ornithine acetyltransferase purified 60-fold from *Chlamydomonas reinhardti* catalyzed freely reversible acetyl transfer between N^α-acetyl-L-ornithine $(K_m = 5.5 \times 10^{-3} M)$ and L-glutamate $(K_m = 1.3 \times 10^{-2} M)$. The acetyltransferase was optimally active over a broad pH range (pH 7.5–9.0) and showed no detectable requirement for cofactors. It catalyzed hydrolysis of N^α-acetylglutamate at about 1% the rate for acetyl transfer *(65)*.

6. Conversion of Ornithine to Arginine

The enzymes catalyzing conversion of ornithine to arginine are discussed in Chapter 14.

C. Regulation of Ornithine and Arginine Biosynthesis

The literature on regulation of arginine biosynthesis is voluminous and beyond the scope of this chapter. Comparative studies of control mechanisms in microorganisms reveal considerable variations both between *(85,86)* and within *(87)* species. What follows is a summary of certain observations which interested this author.

Of the 8 arginine genes in *E. coli*, 4 (those corresponding to enzymes II, III, VA, and VIII) are closely linked (Fig. 5). The remainder are scattered throughout the genome. Only 2 of the 4 clustered genes (II and III) correspond to sequential enzymes, and the R_{arg} (regulatory) gene is not closely linked to any of the 8 synthetic genes *(63)*. Udaka suggests

* Suggested name *(65)*.

(86) that each strain may exhibit a species-specific pattern for regulation of arginine biosynthesis. The distribution of enzymes II (N-acetyl-glutamokinase), VA (acetylornithinase), and VB (ornithine acetyl-transferase), the sensitivity of enzyme II to inhibition by arginine, and

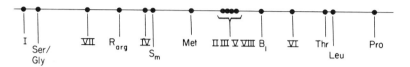

FIG. 5. Genes of arginine synthesis in *E. coli* W are represented schematically by Roman numerals, which correspond to the enzymes listed in Fig. 4. R_{arg} is the regulatory gene of the arginine pathway. No particular sequence is implied for the clustered genes corresponding to enzymes II, III, V, and VIII. Several reference markers are included. From Vogel *et al.* (63).

the relative repressibility of enzyme VI (ornithine transcarbamylase) by arginine were tested for 12 microorganisms (Table II). Two broad groups were recognized, based on the relative activity of enzyme VA.

TABLE II

ENZYMIC LEVELS AND CONTROLS IN THE ARGININE
BIOSYNTHETIC PATHWAYS IN MICROORGANISMS[a]

Strain	Relative specific activity		Percent inhibition of Enz II by 5×10^{-3} M L-arginine	Repression of Enz VI by L-arginine[c]
	VA[b]	VB[b]		
Escherichia coli K-12, *Aerobacter aerogenes*, *Serratia marcescens*	14–33	<0.05	<5	10
Proteus vulgaris	19	<0.05	<5	1–2
Bacillus subtilis, *B. megaterium*, *B. polymyxa*	0.01–0.05	<0.05	<5	4–10
Agrobacterium tumefaciens	<0.005	1.6	>30	1–2
Alcaligenes faecalis	0.5	0.6	70–100	—
Streptomyces griseus	0.1	0.06	70–100	1–2
Pseudomonas fluorescens	0.1–0.3	0.8–2.8	80–100	1–2
Micrococcus glutamicus	0.02	0.05	80–100	1–2

[a] From Udaka (86).

[b] VA = acetylornithinase; VB = ornithine acetyltransferase (Table I).

[c] The approximate ratio of the partially derepressed level (the level in medium minus arginine) to the repressed level.

The enteric bacteria and *Bacillus* species revealed no VB activity comparable to their VA activity, and enzyme II of this group was not inhibited by arginine. In bacteria with significant VB activity, this enzyme was significantly inhibited by arginine. Enzyme VI was significantly repressed by arginine only in enteric bacteria and in *Bacillus* species.

In many wild-type cells and mutants, arginine represses synthesis of all 8 of the enzymes of arginine biosynthesis. For example, in *B. subtilis*, synthesis of enzymes II through VA is repressed by arginine, as demonstrated in *B. subtilis* 8^{a+}, a revertant of the arginine auxotroph, Strain 8, which lacks enzyme VA (acetylornithinase) (*83*). In other strains of mutants (including representatives of *Escherichia*, *Aerobacter*, and *Serratia*), synthesis of enzymes IV (acetylornithine transaminase) (*63*), VA (acetylornithinase) (*88*), or VI (ornithine transcarbamylase) (*89*) appears to be induced rather than repressed by arginine. The arginine-inducible acetylornithine δ-transaminase of a mutant of *E. coli* has been studied by Vogel *et al.* (*88*). Revertants of *E. coli* arginine auxotrophs were screened for repressibility by arginine. One class exhibited relatively low transaminase levels after growth without added arginine, but relatively high levels after growth with exogenous arginine. The mutant transaminase, instead of being repressible, was now inducible, and the inducer was what normally is a repressor. Simultaneously, synthesis of all other enzymes of arginine biosynthesis was repressed. The enzymes from repressible and inducible strains differ in their affinity for N^α-acetylornithine ($K_m = 0.5 \times 10^{-3}$ and $2.6 \times 10^{-3} M$ for the repressible and inducible transaminases, respectively) (*88*).

Sercarz and Gorini (*90*) suggest that, in *E. coli*, exogenous arginine controlling repressor synthesis may comprise a pool separate from endogenously produced arginine destined for general protein synthesis. Although intermixing of these pools is rapid in wild-type organisms, it is impaired in *E. coli* BC 28, a super-derepressed mutant. BC 28 is still sensitive to repression by exogenous arginine, but insensitive to repression by endogenous arginine which, in order to repress, must first leave, then reenter the cell. The authors suggest that permease-bound arginine may represent the exogenous pool. A comparable compartmentalization of endogenous from exogenous ornithine in *N. crassa* was earlier noted by Vogel and Kopac (*14*). While exogenous ornithine was readily converted to proline, endogenous ornithine destined for arginine synthesis was not. These results were originally interpreted to indicate physical separation of the arginine and proline biosynthetic pathways in *Neurospora*.

S. cerevisiae appears to possess a permease (*91*) specific for L-arginine (*92*). Arginine transport is inhibited noncompetitively by histidine and competitively by the L-isomers of canavanine, lysine, and of ornithine, and by D-arginine. That the permease is an arginine rather than a lysine-arginine permease is suggested by detection in arginine permease-less mutants of *S. cerevisiae* of a lysine-specific permease with a higher affinity for lysine (*93*).

IV. LYSINE AND HYDROXYLYSINE

A. Comparative Biochemistry of L-Lysine Biosynthesis

Two distinct biosynthetic pathways, each named for a key intermediate, lead from amphibolic intermediates to lysine. For the α-aminoadipate pathway the amphibolic raw materials are acetyl-CoA and oxalacetate; for the diaminopimelate pathway they are pyruvate and

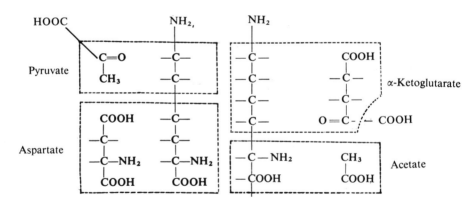

Fig. 6. The diaminopimelate (left) and α-aminoadipate (right) pathways for lysine biosynthesis. From Meister (*10*).

aspartate (Fig. 6). The diaminopimelate pathway also serves for synthesis of *meso*- and of LL-diaminopimelate which occur in the cell walls of many blue-green algae and bacteria (*94*).

The distribution patterns of lysine biosynthesis in different life forms are shown in Table III. Particular lower fungi (phycomycetes) may utilize one or the other pathway. Vogel (*95*) notes that the α-aminoadipate pathway appears confined to phycomycetes producing posteriorly uniflagellate zoospores or aplanospores, and the diaminopimelate pathway to those producing biflagellate or anteriorly uniflagellate zoospores. The phylogenetic implications of these patterns, emphasized by Vogel (*96–100*), predict which pathway should be utilized in essentially any life form. Vogel's predictions have been repeatedly confirmed (*101,102*).

TABLE III

LYSINE BIOSYNTHESIS IN DIFFERENT LIFE FORMS[a]

2,6-Diaminopimelate pathway	α-Aminoadipate pathway
Bacteria	
Pseudomonads	
Eubacteria	
Actinomycetes	
Lower Fungi	
Hyphochytriales	Chytrids
Saprolegniales	Blastocladiales
Leptomitales	Mucorales
Higher Fungi	
	Ascomycetes
	Basidiomycetes
Green Organisms	
Green algae	Euglenids
Ferns	
Flowering plants	

[a] From Vogel (*95*).

B. The Diaminopimelate Pathway

Intermediates in the pathway leading from aspartate and pyruvate to lysine are named in Fig. 7. This discussion begins with dihydropicolinate synthesis. The reactions leading from aspartate to aspartate β-semialdehyde are discussed above (see Chapter 16, Part I, Section V) and under regulation of lysine biosynthesis (Section IV, D).

Aspartate
↓
β-Aspartyl-phosphate
↓
Aspartate β-semialdehyde + Pyruvate
↘   ↙
2,3-Dihydrodipicolinate
↓
Δ¹-Tetrahydrodipicolinate
↓
N-Succinyl-ε-keto-L-α-aminopimelate
↓
N-Succinyl-LL-α,ε-diaminopimelate
↓
L-α,ε-Diaminopimelate
↓
meso-α,ε-Diaminopimelate
↓
Lysine

FIG. 7. Intermediates in the diaminopimelate pathway for lysine biosynthesis.

1. CONDENSATION OF PYRUVATE WITH ASPARTATE β-SEMIALDEHYDE FORMING 2,3-DIHYDRODIPICOLINATE

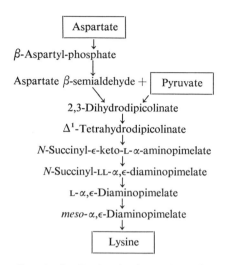

Pyruvate + L-aspartate   2,3-Dihydrodipicolinate
β-semialdehyde

After lysine was assigned to the aspartate family of amino acids for *E. coli* by Abelson and Vogel (*103*), aspartate and pyruvate were reported as precursors of diaminopimelate in *E. coli* (*104*). That aspartate semialdehyde was the starting point for lysine biosynthesis in *E. coli* was shown by Gilvarg (*105*). Dihydropicolinate synthetase, which catalyzes condensation of pyruvate with aspartate β-semialdehyde to form a dihydropicolinic acid, was purified over 1000-fold from *E. coli* (*106*). The reaction product, presumed to be 2,3-dihydropicolinate, was converted to Δ¹-tetrahydrodipicolinate by an *E. coli* oxidoreductase which utilizes 2,3-dihydrodipicolinate as substrate (see Section IV, B, 2) (*106,107*). The synthetase is specific for both substrates. Neither glutamate γ-semialdehyde nor succinate semialdehyde can replace aspartate β-semialdehyde ($K_m = 1.3 \times 10^{-4} M$), and oxalacetate cannot substitute for pyruvate ($K_m = 2.5 \times 10^{-4} M$). The optimum pH is at 8.4, and the equilibrium strongly favors formation of the cyclic product. The condensation reaction is specifically inhibited by L-lysine (see Section IV, D).

The overall reaction may be viewed as an aldol-type condensation between the β-carbon of pyruvate and γ-carbon of L-aspartate β-semialdehyde. Dehydration produces Δ^3-2-oxo-6-aminopimelate, the open-chain form of 2,3-dihydrodipicolinate, which again dehydrates with ring closure, forming 2,3-dihydrodipicolinate. The reaction is somewhat analogous to the condensation of phosphoenolpyruvate with erythrose-4-phosphate forming 2-keto-3-deoxy-D-*arabo*-heptonic acid 7-phosphate plus P_i *(108)* or with arabinose-4-phosphate to form 2-keto-3-deoxy-8-phosphooctanoate plus P_i *(109)*. The driving force, attributable in the above analogies to hydrolysis of phosphopyruvate, may here be due to the accompanying ring closure.

2. REDUCTION OF THE C-C DOUBLE BOND OF 2,3-DIHYDRODIPICOLINATE FORMING Δ^1-PIPERIDEINE-2,6-DICARBOXYLATE

2,3-Dihydrodipicolinate Δ^1-Piperideine-2,6-
                                   dicarboxylate

Dihydrodipicolinate reductase purified some 135-fold from wild-type *E. coli* *(107)* catalyzes reduction of 2,3-dihydrodipicolinate to Δ^1-piperideine-2,6-dicarboxylate (Δ^1-tetrahydrodipicolinate). NADP is about twice as effective an oxidant as is NAD. A role for this enzyme in lysine biosynthesis is suggested by its absence from the lysine-diaminopimelate auxotroph *E. coli* M-203 *(107)*.

3. SUCCINYLATION OF THE AMINO GROUP OF THE OPEN FORM OF Δ^1-PIPERIDEINE-2,6-DICARBOXYLATE FORMING N-SUCCINYL-ε-KETO-L-α-AMINOPIMELATE

Early observations implicating succinylated intermediates in lysine biosynthesis were the isolation of N-succinyl-L-α-diaminopimelate, and later of N-succinyl-ε-keto-α-aminopimelate, from culture media of a lysine-diaminopimelate auxotroph of *E. coli* *(110–112)*. Crude extracts of *E. coli* catalyze conversion of α-amino-ε-ketopimelate to diamino-pimelate in the presence of ATP and succinate. With partially purified preparations, activity is stimulated by addition of succinate thiokinase (*113*). Although assay conditions for Δ^1-tetrahydropicolinate succinylase are given by Farkas and Gilvarg (*107*), an intensive study of this enzyme remains to be carried out.

Δ^1-Piperideine-
2,6-dicarboxylate

N-Succinyl-ε-keto-
L-α-aminopimelate

Kindler and Gilvarg (*114*) have suggested that succinylation performs a function analogous to that of acetylation in ornithine biosynthesis: to prevent cyclization before addition of the second amino group.

4. Transamination of *N*-Succinyl-ε-keto-L-α-aminopimelate Forming *N*-Succinyl-L-α,ε-diaminopimelate

N-Succinyl-ε-keto-
L-α-aminopimelate

N-Succinyl-L-α,ε-
diaminopimelate

The reaction is catalyzed by *N*-succinyldiaminopimelate transaminase which has been purified over 100-fold from a mutant of *E. coli* by Peterkofsky and Gilvarg (*115*). The transaminase is present in the blue-green alga *Nostoc muscorum* and is widely distributed both in gram-positive and in gram-negative bacteria capable of lysine biosynthesis. Significantly, this enzyme is absent both from brewer's yeast, which synthesizes lysine via the α-aminoadipate pathway, and from some, but not all, lysine auxotrophs (absent from *Lactobacillus casei, Euglena gracilis*, and pig heart tissue; present in extracts of the lysine auxotroph *E. coli* M-26-26). The transaminase is distinct from transaminases A and B and from *N*-acetylornithine transaminase. This is indicated by its resolution from transaminase A on DEAE-cellulose columns and by the marked divergence in substrate specificity from the other two transaminases. *N*-Succinyldiaminopimelate transaminase appears absolutely specific both for L-glutamate $(K_m = 5 \times 10^{-3} M)$ and for *N*-succinyl-ϵ-keto-α-aminopimelate $(K_m = 5 \times 10^{-4} M)$. Aged preparations are stimulated by added pyridoxal-P. The reaction is inhibited in either direction by α-ketoglutarate, but at low concentrations of ketoacid is freely reversible. K_{eq} for the reaction as written above is 1.2 at pH 8.0 (115).

5. DEACYLATION OF *N*-SUCCINYL-L-α,ϵ-DIAMINOPIMELATE

N-Succinyl-L-α,ϵ-
diaminopimelate

L-α,ϵ-Diamino-
pimelate

N-Succinyldiaminopimelate deacylase (*N*-succinyl-L-α,ϵ-diamino-pimelate succinyl-hydrolase), which catalyzes the last step in diaminopimelate synthesis and the penultimate reaction in lysine biosynthesis, occurs in organisms (*B. cereus*, *Corynebacterium diphtheriae*) utilizing diaminopimelate as a cell-wall constituent. Its role in lysine biosynthesis is suggested by: its absence both from lysine auxotrophs whose cell walls either do (*E. coli* D-1) or do not contain diaminopimelate (*Lactobacillus casei*; group A and group B streptococci) and from pig heart and ox liver tissue; and its presence in *Micrococcus lysodeikticus*, which lacks diaminopimelate but can synthesize lysine. The reaction is analogous to the deacylation of acetylornithine by N^α-acetylornithinase (see Section III, B, 5). Although diaminopimelate and acetylornithine deacylases are distinct enzymes, they share many common properties, including activation by divalent metals, particularly Co^{2+}. Partially purified *E. coli* *N*-succinyldiaminopimelate deacylase has a broad pH optimum (7–9) and is specific for its substrate $(K_m = 1.3 \times 10^{-3} M)$. The corresponding acetyl derivative is not deacylated (*114*).

6. EPIMERIZATION OF L-α,ε-DIAMINOPIMELATE TO meso-α,ε-DIAMINOPIMELATE

$$
\begin{array}{ccc}
\text{COOH} & & \text{COOH} \\
| & & | \\
\text{H}_2\text{N--C--H} & & \text{H--C--NH}_2 \\
| & \rightleftharpoons & | \\
(\text{CH}_2)_3 & & (\text{CH}_2)_3 \\
| & & | \\
\text{H--C--NH}_2 & & \text{H--C--NH}_2 \\
| & & | \\
\text{COOH} & & \text{COOH} \\
\text{L-α,ε-Diamino-} & & \textit{meso-}\text{α,ε-Di-} \\
\text{pimelate} & & \text{aminopimelate}
\end{array}
$$

Diaminopimelate epimerase (2,6-LL-diaminopimelate 2-epimerase, EC 5.1.1.7) is widely distributed in bacteria (*94*). The reactants and products are diastereoisomers and do not form a racemic mixture. The reaction may be visualized as

$$
\begin{array}{cc}
\text{D} & \rightleftharpoons \text{L} \\
| & | \\
\text{L} & \text{L}
\end{array}
$$

The equilibrium constant is about 1, and the reaction is readily reversible. The epimerase has not been extensively purified owing to its instability and to lack of a simple assay procedure (*116*).

7. DECARBOXYLATION OF meso-α,ε-DIAMINOPIMELATE TO L-LYSINE

$$
\begin{array}{ccc}
\text{COOH} & & \\
| & & \\
\text{H--C--NH}_2 & \xrightarrow[\text{PLP}]{\text{CO}_2} & \text{CH}_2\text{--NH}_2 \\
| & & | \\
(\text{CH}_2)_3 & & (\text{CH}_2)_3 \\
| & & | \\
\text{H--C--NH}_2 & & \text{H--C--NH}_2 \\
| & & | \\
\text{COOH} & & \text{COOH} \\
\textit{meso-}\text{α,ε-Diamino-} & & \text{L-Lysine} \\
\text{pimelate} & &
\end{array}
$$

Diaminopimelate decarboxylase (*meso*-2,6-diaminopimelate carboxy-lyase, EC 4.1.1.20), which catalyzes the terminal step in lysine biosynthesis, is present in bacteria (*117*), *Mycoplasma* (*84*), and in higher plants (*118*). The reaction is a decarboxylation of a D-amino acid, since it involves removal of CO_2 from the asymmetric carbon atom in the D-configuration (*119*). Although originally reported to be constitutive both in *E. coli* and in *A. aerogenes* (*117,120*), diaminopimelate decarboxylase is up to 80% repressed in *E. coli* grown in the presence of lysine (*121*). Both the bacterial and plant decarboxylases require pyridoxal-P and are highly specific for the *meso*-form of the substrate ($K_m = 1.7 \times 10^{-3}$, 2.8×10^{-3}, and 3.5×10^{-4} M for the enzymes from *E. coli*, *A. aerogenes*, and *Lemna perpusilla*, respectively). The enzyme is distinct from L-lysine decarboxylase (*118,122,123*). The decarboxylase from the green plant *L. perpusilla* has been purified about 25-fold (*118*), and that from *E. coli* 200-fold (*123*). Maximal decarboxylase activity is observed during growth of *E. coli* B on media containing complex sources of nitrogen (*124*). The MW of the purified *E. coli* enzyme is estimated as approximately 200,000 (*123*).

C. The α-Aminoadipate Pathway

The evidence for the overall pathway has recently been reviewed by Meister (*10*), by Broquist (*125*), and by Maragoudakis and Strassman (*126*). That synthesis of α-aminoadipate occurs via reactions analogous to those of the citric acid cycle (Fig. 8, reactions 1 to 5) was proposed

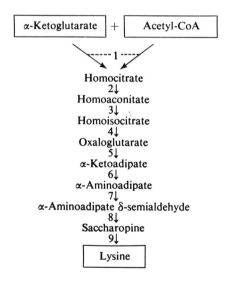

FIG. 8. Intermediates in the α-aminoadipate pathway for lysine biosynthesis. In the discussion the subsection numbers correspond to the numbers of the reactions in this figure.

in 1953 by Strassman and Weinhouse (*127*). Detailed evidence for the individual events is of more recent vintage.

1. CONDENSATION OF ACETYL-CoA WITH α-KETOGLUTARATE
   FORMING HOMOCITRATE

$$CH_2-\overset{\overset{\displaystyle O}{\|}}{C}-S-CoA$$

$$+$$

$$O=\underset{\underset{\displaystyle CH_2-COOH}{|}}{\underset{\displaystyle CH_2}{|}}{C}-COOH$$

α-Ketoglutarate

CoA—SH

$$CH_2-COOH$$
$$HO-\underset{|}{C}-COOH$$
$$\underset{|}{CH_2}$$
$$CH_2-COOH$$

Homocitrate

Certain lysine auxotrophs of yeast (*128*) or of *N. crassa* (*126*) accumulate and excrete homocitrate. The overall conversion of acetate and α-ketoglutarate to α-ketoadipate by cell-free extracts of baker's yeast was recently demonstrated by Lewis *et al.* (*129*). The above condensation reaction is catalyzed by dialyzed crude extracts of a blocked mutant of *S. cerevisiae*. In view of the close similarity to citrate synthesis, it is interesting that the absolute configuration of the homocitrate formed is opposite to that of enzymically synthesized citrate (Fig. 9) (*130*).

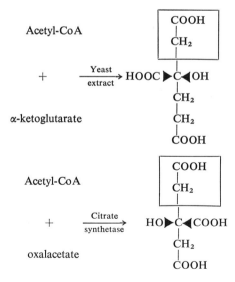

FIG. 9. Absolute configurations of citrate and homocitrate. Atoms derived from acetate are shown boxed (*130*).

Possibly this difference serves the function of avoiding mutual antagonism between citrate and lysine biosynthesis.

2. CONVERSION OF HOMOCITRATE TO HOMOISOCITRATE

$$
\begin{array}{c}
CH_2-COOH \\
| \\
HO-C-COOH \\
| \\
CH_2 \\
| \\
CH_2-COOH
\end{array}
\quad
\xrightarrow[\text{}]{H_2O}
\quad
\left[
\begin{array}{c}
C-COOH \\
\| \\
C-COOH \\
| \\
CH_2 \\
| \\
CH_2-COOH
\end{array}
\right]
\quad
\xrightarrow[\text{}]{H_2O}
\quad
\begin{array}{c}
HO-CH-COOH \\
| \\
CH-COOH \\
| \\
CH_2 \\
| \\
CH_2-COOH
\end{array}
$$

Homocitrate   *cis*-Homoaconitate   Homoisocitrate

A yeast mutant, LY-4, accumulates homocitrate and *cis*-homoaconitate, but not homoisocitrate, in its growth medium, and enzymically prepared homocitrate-^{14}C is converted to lysine by wild-type yeast (*126,131*). Distinct yeast enzymes catalyzing formation either of homoisocitrate or of homocitrate from *cis*-homoaconitate were studied by Strassman and Ceci (*132*). Homoaconitase, the enzyme catalyzing the equilibrium *cis*-homoaconitate ⇌ homoisocitrate, is specific for the *cis*-form of the substrate. That homoaconitase is distinct from *cis*-aconitase [citrate (isocitrate) hydro-lyase, EC 4.2.1.3] appears clear both from partial resolution during purification (*132,133*) and from their independent genetic control (*134*). Synthesis of homoaconitase by yeast is repressed by lysine, although the activity is not lysine-inhibited (*134*).

3. OXIDATIVE DECARBOXYLATION OF HOMOISOCITRATE TO α-KETOADIPATE

$$
\begin{array}{c}
HO-CH-COOH \\
| \\
CH-COOH \\
| \\
CH_2 \\
| \\
CH_2 \\
| \\
COOH
\end{array}
\quad
\xrightarrow[Mg^{2+}]{NAD^+ \quad NADH+H^+}
\quad
\left[
\begin{array}{c}
O=C-COOH \\
| \\
CH-COOH \\
| \\
CH_2 \\
| \\
CH_2 \\
| \\
COOH
\end{array}
\right]
\quad
\xrightarrow[Mg^{2+}]{CO_2}
\quad
\begin{array}{c}
O=C-COOH \\
| \\
CH_2 \\
| \\
CH_2 \\
| \\
CH_2 \\
| \\
COOH
\end{array}
$$

Homoisocitrate   Oxaloglutarate   α-Ketoadipate

Conversion of homoisocitrate to α-ketoadipate by dialyzed crude yeast extracts supplemented with Mg^{2+} and NAD was demonstrated by Strassman *et al.* (*135*). The enzyme, homoisocitrate dehydrogenase, resembles yeast isocitrate dehydrogenase [L$_s$-isocitrate: NAD oxidoreductase (decarboxylating), EC 1.1.1.41] in its requirement for NAD and for Mg^{2+}. The apparent K_m for homoisocitrate is $1.4 \times 10^{-3} M$, and the pH optimum 7.8.

Accumulation of glutarate by a lysine auxotroph of *S. cerevisiae* may reflect a side reaction, oxidative decarboxylation of α-ketoadipate (*136*).

4. TRANSAMINATION OF α-KETOADIPATE

α-Ketoadipate can replace lysine for growth of certain lysine auxotrophs (*101,137*), and transamination of α-aminoadipate was detected as early as 1939 (*138*). Pyridoxal-P-dependent transamination of α-ketoadipate has been demonstrated in yeast and *Neurospora* (*139,140*) and in *Pseudomonas* (*141*), but the transaminases have not been purified or studied in detail.

5. REDUCTION OF α-AMINOADIPATE TO α-AMINOADIPATE δ-SEMIALDEHYDE

α-Aminoadipate was reported as a lysine precursor in *Neurospora* by Mitchell and Houlahan (*142*) and by Windsor (*143*). Evidence for the above reactions is indirect, for α-aminoadipate δ-semialdehyde has yet to be isolated as a product of α-aminoadipate reduction, as the semialdehyde cyclizes nonenzymically to Δ^1-piperideine-6-carboxylate (*17,144*). In the presence of ATP, Mg^{2+}, glutathione, NADPH, and a crude yeast enzyme, Δ^1-piperideine-6-carboxylate formation may be inferred by formation of its *o*-aminobenzaldehyde adduct (*144–147*). Coenzyme A is not required. The overall reaction may be catalyzed by at least 2 enzymes, the first activating α-aminoadipate, possibly to α-aminoadipyl-AMP (*148,149*). The second may catalyze the NADPH- and glutathione-dependent reduction to α-aminoadipate δ-semialdehyde. The nature of the activated intermediate is not clear. Sagisaka and Shimura (*148*) reported an α-aminoadipate-dependent ATP-pyrophosphate exchange reaction catalyzed by their crude yeast enzyme and suggested an acyl-AMP intermediate. A compound with properties consistent with this postulate was later isolated by Mattoon *et al.* (*149*). Although the overall requirements for reduction of α-aminoadipate (ATP, Mg^{2+}, NADPH) parallel those for reduction of aspartate via β-aspartyl phosphate to aspartate β-semialdehyde (see Chapter 16, Part I, Section V, B), ^{32}P exchange with pyrophosphate would appear to exclude α-aminoadipyl δ-phosphate from consideration as an activated intermediate.

An additional mechanism for α-aminoadipate semialdehyde formation is by oxidation of the corresponding alcohol, hexahomoserine, catalyzed by an ω-hydroxy-α-amino acid dehydrogenase of *N. crassa* (*9*).

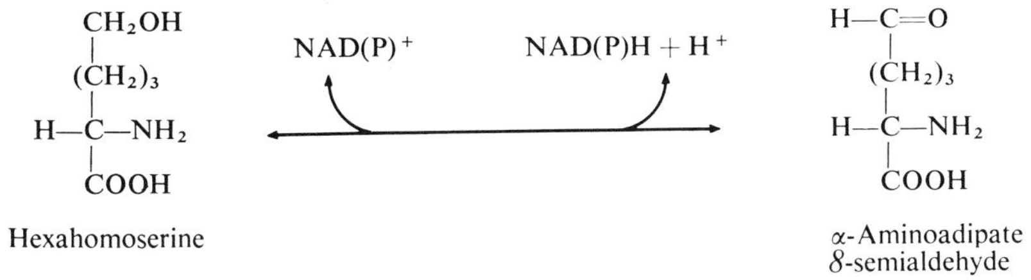

Hexahomoserine   α-Aminoadipate δ-semialdehyde

Although hexahomoserine can substitute for lysine for growth of certain lysine auxotrophs of *N. crassa*, hexahomoserine probably is not normally involved in lysine biosynthesis by the α-aminoadipate pathway.

6. REDUCTIVE CONDENSATION OF GLUTAMATE WITH α-AMINOADIPATE δ-SEMIALDEHYDE FORMING SACCHAROPINE

Saccharopine, a new intermediate in the α-aminoadipate pathway, was first detected in baker's yeast by Darling and Larsen (*150,151*). Its

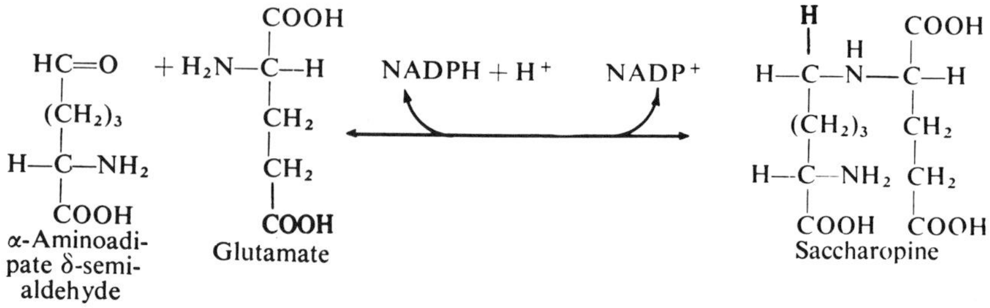

$$
\begin{array}{ccccc}
\text{HC=O} & \text{COOH} & & & \text{H} \quad \text{COOH} \\
| & | & & & | \quad \text{H} \quad | \\
(\text{CH}_2)_3 & \text{H}_2\text{N—C—H} \quad \text{NADPH + H}^+ \quad \text{NADP}^+ & & \text{H—C—N—C—H} \\
| & | & & & | \quad \quad | \\
\text{H—C—NH}_2 & \text{CH}_2 & & & (\text{CH}_2)_3 \quad \text{CH}_2 \\
| & | & & & | \quad \quad | \\
\text{COOH} & \text{COOH} & & & \text{H—C—NH}_2 \quad \text{CH}_2 \\
& & & & | \quad \quad | \\
& & & & \text{COOH} \quad \text{COOH}
\end{array}
$$

α-Aminoadipate δ-semi-aldehyde Glutamate Saccharopine

isolation was later confirmed by Morimoto and Yamano (*152*). Isolation of an unidentified lysine precursor with properties similar to those of saccharopine was reported by Matoon *et al.* (*149*).

The role of saccharopine in lysine biosynthesis has recently been worked out by Broquist and his collaborators (*153–159*) using *Neurospora* and *Saccharomyces cerevisiae* mutants blocked at the following points in lysine biosynthesis (*155,157*).

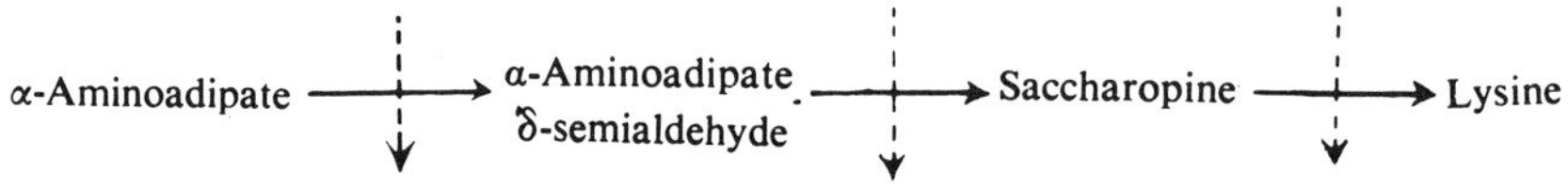

α-Aminoadipate ⟶ α-Aminoadipate δ-semialdehyde ⟶ Saccharopine ⟶ Lysine

In each case the anticipated intermediates accumulated and were conclusively identified.

α-Aminoadipate δ-semialdehyde-glutamate reductase [ε-*N*-(L-glutaryl-2)-L-lysine: NAD(P) oxidoreductase (L-2-aminoadipate semialdehyde-forming)] reversibly catalyzes the above reaction. The reductase, present in yeast, *Neurospora*, and *Euglena*, was purified over 100-fold from baker's yeast and its kinetic and molecular properties studied. In the direction written, NADPH is the obligatory reductant ($K_m = 2.2 \times 10^{-5} M$) and the optimum pH is about 7. For the reverse reaction, NAD can replace NADP, and the pH optimum is 9.8. The K_m for saccharopine at pH 9.8 was $9.2 \times 10^{-4} M$. The reductase is a sulfhydryl enzyme and has a molecular weight of about 73,000 (*157,158*).

7. OXIDATIVE CLEAVAGE OF SACCHAROPINE TO L-LYSINE AND α-KETOGLUTARATE

The terminal reaction in lysine biosynthesis, oxidative cleavage of saccharopine to L-lysine plus α-ketoglutarate, is reversibly catalyzed by saccharopine dehydrogenase [ε-*N*-(L-glutaryl-2)-L-lysine:NAD oxidoreductase (L-lysine-forming)] of yeast, higher and lower ascomycetes, and *Euglena*. Its presence has been suggested to be diagnostic for the

$$
\begin{array}{c}
\text{COOH} \\
\text{H} \quad | \\
\text{H}_2\text{C}—\text{N}—\text{C}—\text{H} \\
| \qquad | \\
(\text{CH}_2)_3 \quad \text{CH}_2 \\
| \qquad | \\
\text{H}—\text{C}—\text{NH}_2 \quad \text{CH}_2 \\
| \qquad | \\
\text{COOH} \quad \text{COOH}
\end{array}
\quad
\underset{\text{NADH} + \text{H}^+}{\overset{\text{NAD}^+}{\rightleftharpoons}}
\quad
\left[
\begin{array}{c}
\qquad\qquad \text{COOH} \\
\text{H}_2\text{C}—\text{N}=\!=\text{C} \\
| \qquad\qquad | \\
(\text{CH}_2)_3 \qquad \text{CH}_2 \\
| \qquad\qquad | \\
\text{H}—\text{C}—\text{NH}_2 \quad \text{CH}_2 \\
| \qquad\qquad | \\
\text{COOH} \qquad \text{COOH}
\end{array}
\right]
\quad
\underset{\alpha - \text{KG}}{\overset{\text{H}_2\text{O}}{\rightleftharpoons}}
\quad
\begin{array}{c}
\text{CH}_2\text{NH}_2 \\
| \\
\text{CH}_2 \\
| \\
\text{CH}_2 \\
| \\
\text{CH}_2 \\
| \\
\text{H}—\text{C}—\text{NH}_2 \\
| \\
\text{COOH}
\end{array}
$$

Saccharopine   Lysine

α-aminoadipate pathway (*156,159,160*). The combined effect of this plus the previous reaction mimics transamination. The resemblance is superficial, for saccharopine is a stable intermediate and nicotinamide nucleotides rather than pyridoxal-P function as coenzymes.

$$
\text{α-Aminoadipate δ-semialdehyde} \xrightarrow[\qquad]{\overset{\text{L-Glu} \qquad\qquad \text{α-KG}}{}} \text{L-lysine}
$$

Yeast saccharopine dehydrogenase, purified over 700-fold, has an apparent molecular weight of 49,000 and a turnover number of about 1130 moles per minute per 49,000 gm enzyme. It is inhibited by sulfhydryl reagents, is specific for L-lysine ($K_m = 1.2 \times 10^{-2} M$) and for α-ketoglutarate ($K_m = 4.4 \times 10^{-4} M$), and preferentially utilizes NAD as oxidant ($K_m = 4.6 \times 10^{-5} M$) (*156,159*).

Experiments with *Aspergillus nidulans* (*161*) and *Euglena gracilis* (*162*) could be interpreted to mean that α-aminoadipate δ-semialdehyde is aminated to lysine via cyclic intermediates such as pipecolate. It is therefore of great interest that Vaughn and Broquist (*156*) have now shown that these organisms possess saccharopine dehydrogenase. Jones and Broquist (*157*) therefore suggest that lysine synthesis in these organisms also, as well as all others utilizing the α-aminoadipate pathway, probably proceeds via the reactions discussed here with saccharopine as an isolable intermediate.

D. Regulation of L-Lysine Biosynthesis

Regulation of lysine biosynthesis via the diaminopimelate pathway is exerted primarily at two points: conversion of aspartate to β-aspartyl phosphate, and condensation of aspartate β-semialdehyde with pyruvate (Fig. 10).

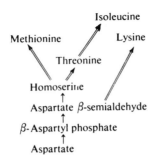

FIG. 10. Aspartate metabolism in *E. coli*. From Datta and Gest (*163*).

1. β-ASPARTOKINASE

Aspartate semialdehyde serves as a precursor, not only of lysine, but also of threonine, methionine, and isoleucine (Fig. 10). Lysine and threonine exert tight regulatory control over their synthesis at the level of conversion of aspartate to β-aspartyl phosphate, the reaction catalyzed by β-aspartokinase (ATP: L-aspartate 4-phosphotransferase, EC 2.7.2.4) (Fig. 10). This represents the first control point in lysine biosynthesis via the diaminopimelate pathway.

The regulation of enzymes catalyzing reactions prior to the divergence point of a branched biosynthetic pathway presents novel features. Repression or feedback inhibition by any one of the ultimate products would also decrease the rate of synthesis of the other end products of the pathway. Repression of synthesis of such enzymes may be effected in two general ways:

1. Independent repression by a specific end product of one of a set of multiple enzymes catalyzing the same biosynthetic reaction (*164*).
2. "Multivalent" repression of a single enzyme by the concerted action of several of the end products of the pathway (*165,166*).

Regulation of feedback inhibition of the activity of such enzymes exhibits similar patterns:

1. Independent inhibition by a specific end product of one of a set of multiple enzymes, catalyzing the same biosynthetic reaction (*164*).
2. "Multivalent," "concerted" (*167*), or "coordinated" (*168*) inhibition of a single enzyme exerted by the simultaneous presence of several end products, any one of which alone is an ineffective inhibitor.

Each of the above modes of regulation applies to β-aspartokinase in one or another organism. In *E. coli*, feedback inhibition by lysine is independent rather than concerted. At least two aspartokinases are formed, one inhibited by L-lysine and another by L-threonine (*164,169*). By contrast, the β-aspartokinases of *B. polymyxa*, *B. subtilis* (*166*), and *Rhodopseudomonas capsulatus* (*167*) are subject to concerted feedback inhibition by L-lysine plus L-threonine. Lysine and threonine are the only aspartate-derived amino acids which affect the β-aspartokinase of *B. polymyxa* or *Rsp. capsulatus*. β-Aspartokinase of *Rsp. capsulatus* is effectively inhibited by a combination of L-lysine and L-threonine, but is essentially uninhibited by either amino acid alone. Concerted inhibition was incomplete even at high concentrations of lysine plus threonine, was reversible, and was noncompetitive with respect to L-aspartate. Either regulator amino acid alone protects against heat inactivation, although both together offer still more protection. L-Aspartate offers no protection. This was interpreted by Datta and Gest (*167*) to indicate that β-aspartokinase of *Rsp. capsulatus* possesses regulatory sites distinct from the aspartate (substrate) site.

2. Condensation of Pyruvate with Aspartate β-Semialdehyde

E. coli 2,3-dihydrodipicolinate synthetase activity is specifically inhibited by low concentrations of L-lysine (*168*). This reaction therefore represents the second control point in lysine biosynthesis postulated by Stadtman *et al.* (*164*).

Lysine transport into *S. cerevisiae* is mediated by at least two permeases (*91*), one transporting both lysine and arginine (*92*). The second is more specific for lysine (*93*).

E. D-Lysine

Lysine racemase (lysine racemase, EC 5.1.1.5), which is widespread in bacteria (*170*), can account for D-lysine synthesis from the L-isomer.

F. Hydroxylysine

5-Hydroxylysine has been repeatedly isolated from acid hydrolysates of gelatin and isinglass (*171–177*), but probably does not occur in other mammalian proteins (*178*). Small quantities of hydroxylysine have been reported in wool (*173,179*), in a phosphatide of *Mycobacterium phlei* (*180*), as a phosphate ester in mammalian muscle (*181*) and, in an as yet unconfirmed report, trypsin (*182*). The structure has been unequivocally established as α,ε-diamino-δ-hydroxycaproate (*183–187*). All four

$$CH_2-CH-CH_2-CH_2-CH-COOH$$

with OH on the second CH and NH₂ groups:

OH on carbon 2, NH_2 on carbon 1 (CH_2) and on the carbon bearing COOH.

5-Hydroxylysine

optical isomers have been prepared (*188*), and the diastereoisomers have been separated by ion-exchange chromatography (*177,189*).

In rats, dietary lysine is an obligatory precursor of collagen hydroxylysine (*190–193*). Fed or injected lysine-^{14}C, but not hydroxylysine, is incorporated into both the lysine and the hydroxylysine of collagen (*194*). The mechanism of hydroxylation appears to differ significantly from that of proline hydroxylation (see Section II, E). Tritium from ^3H_2O and ^{18}O from $H_2{}^{18}O$, but not from $^{18}O_2$, is incorporated into hydroxylysine by chick embryos. Hydroxylation may thus involve either dehydrogenation followed by hydration or an oxygenase reaction with subsequent equilibration of the hydroxyl group with water (*195*). Certain analogies to proline hydroxylation nevertheless exist. During conversion of 4,5-tritiated lysine to hydroxylysine in chick embryo and rat tissue, a single hydrogen, that replaced by the hydroxyl group, was lost. Thus neither a 4,5-unsaturated intermediate nor 5-ketolysine is an intermediate (*178*). Popenoe et al. (*178*) suggest as possible mechanisms for lysine hydroxylation the routes shown in Fig. 11.

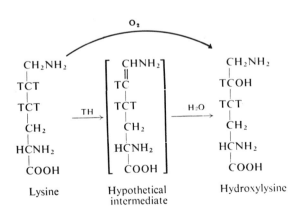

FIG. 11. Possible routes for lysine hydroxylation. These include either direct attack of O_2 on carbon atom 5 or dehydrogenation between carbon atoms 5 and 6 followed by addition of water (*178*).

V. VALINE, ISOLEUCINE, AND LEUCINE

A. Introduction

Biosynthesis of the branched amino acids (valine, isoleucine, leucine) involves several homologous intermediates (Fig. 12). Only 9 enzymes are required, 4 of which (those catalyzing reactions 2–5, Fig. 12) possess

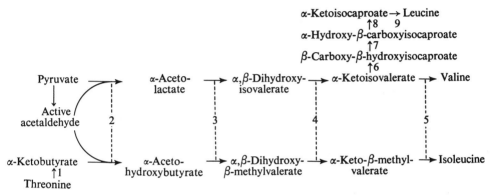

FIG. 12. Pathway of biosynthesis of valine, isoleucine, and leucine. Redrawn from Freundlich *et al.* (*165*). In the discussion the subsection number corresponds to the reaction number on this chart. The dashed lines connecting reactions 2–5 indicate that single enzymes are believed to catalyze analogous reactions in valine-leucine and in isoleucine biosynthesis.

dual substrate specificities for homologous substrates (*165,196*). By 1960 the intermediates (Fig. 12) all were known. Since then, attention has focused on the kinetic and regulatory features of the enzymes involved. Although it now is recognized that these pathways share both metabolic and regulatory features, even prior to 1950 certain observations hinted at these relationships. Gladstone (*197*) observed that an excess of any one branched amino acid inhibited growth of *Bacillus anthracis*. The toxic effect of valine was specifically reversed by leucine (and *vice versa*), but that of isoleucine could only be counteracted by addition of both valine and leucine. He suggested that an excess of one amino acid might prevent synthesis of another. As noted by Umbarger (*198*), Gladstone thus predicted the existence of "false feedback" (*199*) over a generation ago.

Branched-chain amino acid auxotrophs of bacteria and of *Neurospora* have greatly facilitated elucidation of these pathways. Using a valine-isoleucine auxotroph of *N. crassa*, Bonner *et al.* (*200*) discovered that the corresponding α-keto acids could substitute for valine or for isoleucine and proposed the α-keto acids as precursors. Shortly thereafter Bonner (*201*) reported that, in *Neurospora*, exogenous α-keto-β-methylvalerate (the α-keto acid of isoleucine) specifically inhibited amination of α-ketoisovalerate (the α-keto acid of valine). The intermediates in the biosynthetic pathways were proposed by Strassman, Weinhouse, and their associates in the early 1950s (*202–206*) from the labeling patterns of branched amino acids isolated from *Torulopsis utilis* supplied ^{14}C-labeled precursors. Following implication of an α,β-dihydroxy acid in valine and isoleucine biosynthesis by Myers and Adelberg (*207*), an analogous route was proposed for isoleucine biosynthesis. Acetaldehyde was assumed to condense either with a three- or four-carbon precursor forming either α-acetolactate or α-aceto-α-hydroxybutyrate, respectively. Pinacol-type rearrangements of α-acetolactate (Fig. 13)

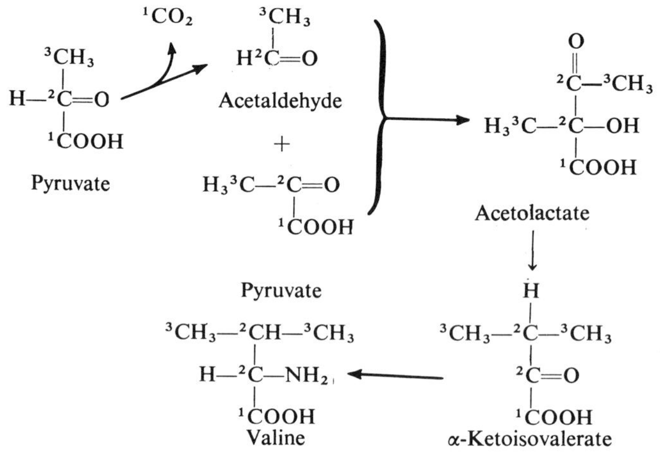

FIG. 13. Pathway for valine biosynthesis proposed by Strassman *et al.* (*202*). The numbered carbon atoms indicate precursor-product relationships.

or of α-acetohydroxybutyrate (Fig. 14) were proposed to provide the carbon skeleton of valine and isoleucine. Simultaneously Adelberg (*208*) provided evidence consistent with this scheme that, in *N. crassa*, threonine provided carbon atoms 1, 2, 4, and 5 of isoleucine. Either α-aminobutyrate or α-ketobutyrate was soon shown to replace iso-

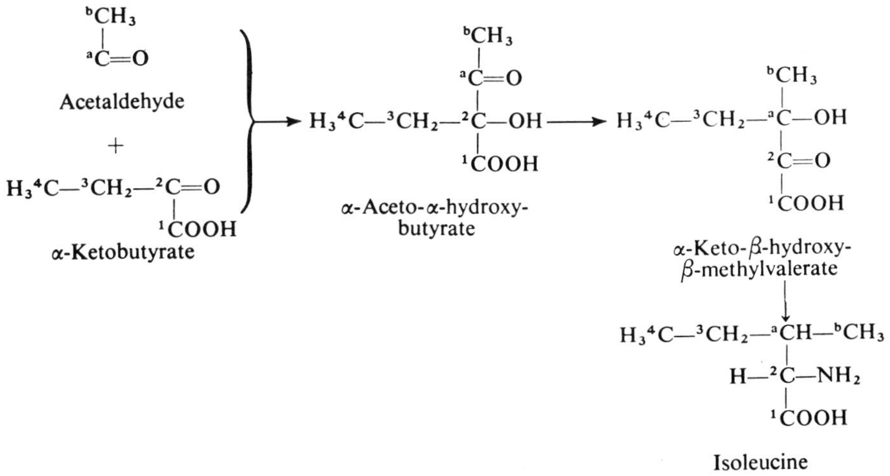

FIG. 14. Pathway for isoleucine biosynthesis proposed by Strassman *et al.* (*204*). Numbered and lettered carbon atoms illustrate precursor-product relationships.

leucine for growth of an isoleucine auxotroph of *E. coli* (*209,210*). Using isotope competition, Abelson (*5*) assigned isoleucine to the aspartate family of amino acids and showed that both threonine and α-ketobutyrate were likely intermediates in isoleucine biosynthesis by *E. coli*.

B. Individual Reactions

1. FORMATION OF α-KETOBUTYRATE

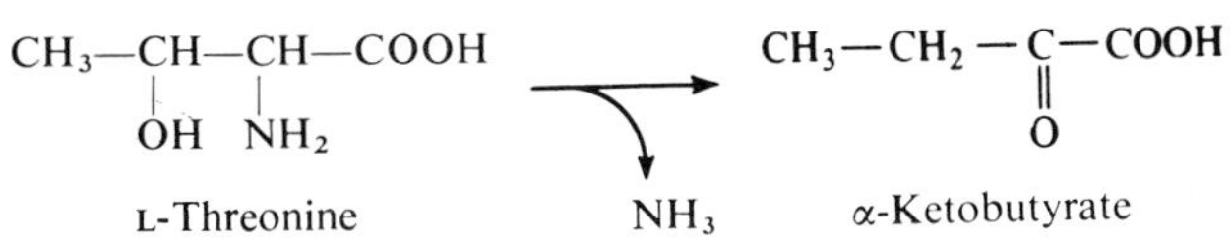

Deamination of L-threonine to α-ketobutyrate, catalyzed by L-threonine dehydratase (L-threonine hydro-lyase deaminating, EC 4.2.1.16), is a nonoxidative deamination involving dehydration of threonine and subsequent rehydration of α-iminobutyrate. The reaction is discussed in detail elsewhere (Chapter 14). *E. coli* possesses two distinct L-threonine dehydratases, one presumed to serve biosynthetic, and the other degradative, functions. Each deaminase exhibits distinctive regulatory features. The "biosynthetic" deaminase is specifically inhibited by isoleucine (211,212). The "degradative" threonine deaminases of E. coli (213) and of Clostridium tetanomorphum (214,215) are regulated by nucleotide diphosphates but are insensitive to isoleucine. In cells possessing both deaminases, only the biosynthetic enzyme is presumed to function in isoleucine biosynthesis.

Biosynthetic L-threonine deaminase, which catalyzes an irreversible reaction unique to isoleucine biosynthesis, was one of the first feedback-regulated enzymes studied (211,216,217). Inhibition of the E. coli biosynthetic deaminase by L-isoleucine was first reported by Umbarger (216). Straight lines were obtained when the ratio of noninhibited to inhibited velocity was plotted versus the square of the isoleucine concentration (216). A regulatory function for this inhibition was suggested by Umbarger, who noted that the threonine-sparing effect of isoleucine in threonine auxotrophs of E. coli (218) could best be comprehended in terms of feedback inhibition. Additional significant properties of the E. coli biosynthetic deaminase include: stimulation by low levels of isoleucine (219); apparent second-order kinetics (211); resistance to thermal inactivation in the presence of isoleucine (219); and either activation or inhibition by valine at low or high substrate concentrations, respectively (220). Both two- and three-site models have been proposed to account for the kinetics of end-product inhibition by isoleucine (219,221,222). From kinetic analysis, Changeaux (221) inferred that 1 mole of deaminase possessed a minimum of 2 binding sites for substrate and 2 for isoleucine. However, the Salmonella typhimurium deaminase exhibited normal substrate kinetics when precautions were taken to correct initial velocity data for enzyme instability at low L-threonine concentrations (223,224). Maeba and Sanwal (224) observed that, in their hands, the E. coli deaminase also exhibited normal substrate kinetics—results at variance with previous reports (219,220). They suggest a model for the S. typhimurium enzyme with separate substrate and regulatory sites, but without interaction between multiple substrate sites, if present.

2. Synthesis of α-Acetolactate and of α-Acetohydroxybutyrate

α-Hydroxyethyl-2-thiamine pyrophosphate ("active acetaldehyde") (I) condenses with an α-keto acid (II) bearing a methyl or an ethyl side chain (R) forming an α-hydroxy-β-keto acid (III). For valine and leucine biosynthesis, $R = -CH_3$, (II) is pyruvate, and (III) is α-acetolactate, For isoleucine biosynthesis, $R = -C_2H_5$, (II) is α-ketobutyrate, and (III) is α-aceto-α-hydroxybutyrate. Acetohydroxy acid synthetase has no systematic name but might be termed a 2-hydroxy-3-oxoacid lyase.

$$
\begin{array}{ccccc}
\underset{\text{(I)}}{\underset{\overset{|}{\text{OH}}}{\overset{\text{CH}_3}{\underset{|}{\text{TPP—CH}}}}} + \underset{\text{(II)}}{\underset{\overset{||}{\text{O}}}{\overset{\text{R}}{\underset{|}{\text{C—COOH}}}}} & \xrightarrow{\text{Mg}^{2+}} & \text{TPP.H} + \underset{\text{(III)}}{\underset{\overset{||}{\text{O}}}{\overset{\text{CH}_3}{\underset{|}{\text{C}}}}} \underset{\overset{|}{\text{OH}}}{\overset{\text{R}}{\underset{|}{\text{C—COOH}}}}
\end{array}
$$

a. Introduction. α-Acetolactate formation from pyruvate, postulated by Watt and Krampitz (*225*), was proposed by Strassman *et al.* (*202*) as the initial reaction in valine biosynthesis. These authors (*202*) suggested an aldol-type condensation of a two- and a three-carbon fragment derived from pyruvate or lactate and proposed that the substrates for acetolactate synthesis were pyruvate and acetaldehyde. Analogous proposals for α-acetohydroxybutyrate synthesis advanced by Adelberg (*208*) and by Strassman *et al.* (*204*) envisaged an initial condensation of acetaldehyde with α-ketobutyrate. Metabolic studies

$$
\left.
\begin{array}{c}
\overset{\text{O}}{\overset{||}{\text{H—C—CH}_3}} \\[2mm]
+ \\[2mm]
\underset{\overset{|}{\text{COOH}}}{\text{R—C=O}}
\end{array}
\right\}
\longrightarrow
\begin{array}{c}
\overset{\text{O}}{\overset{||}{\text{C—CH}_3}} \\[2mm]
\text{R—C—OH} \\[2mm]
\underset{}{\overset{|}{\underset{|}{\text{COOH}}}}
\end{array}
$$

Pyruvate (R = —CH₃)
or α-ketobutyrate (R = —C₂H₅)

α-Acetolactate (R = —CH₃)
or α-acetohydroxybutyrate
(R = —C₂H₅)

soon provided supportive evidence for these proposals. Acetolactate formation from pyruvate was demonstrated in extracts of *A. aerogenes* (*227*), *E. coli* (*228–230*), *S. typhimurium* (*230*), *N. crassa* (*231*), green plants (*232,233*), and pigeon breast muscle (*226,234*). Acetolactate accumulation by a valine auxotroph of *E. coli* which was supplied growth-limiting quantities of valine was inhibited when valine was added (*228,229*). Umbarger *et al.* (*228,229*) therefore suggested acetolactate synthesis as the locus for regulation of valine biosynthesis by feedback inhibition.

b. Active Acetaldehyde. All acetohydroxy acid synthetases require thiamine pyrophosphate (TPP) as cofactor. That the two-carbon moiety undergoing condensation is α-hydroxyethyl-2-thiamine pyrophosphate ("active acetaldehyde") was shown by Holtzer *et al.* (*235,236*) for a variety of systems forming α-acetolactate, α-acetohydroxybutyrate, and other products derived from "active acetaldehyde." Pyruvate condenses with enzyme-bound TPP, forming "active pyruvate" (reaction 1, Fig. 15), which decarboxylates to enzyme-bound "active acetaldehyde" (reaction 2, Fig. 15). This may exchange with free TPP, regenerating enzyme-bound TPP and liberating α-hydroxyethyl-2-thiamine pyrophosphate or, in the presence of an appropriate enzyme, condense with appropriate acceptors forming α-acetolactate or α-acetohydroxybutyrate (reaction 3, Fig. 15).

In *E. coli* a pyruvate oxidase [pyruvate:lipoate oxidoreductase (acceptor-acetylating)] specifically concerned with branched amino acid biosynthesis catalyzes synthesis of acetaldehyde-thiamine pyrophosphate from pyruvate. For this enzyme, valine acts as an inhibitor competitive with pyruvate. Synthesis of different pyruvate oxidases appears to be independently regulated. When a valine auxotroph of *E. coli* was grown under conditions where valine limited growth, levels of the "biosynthetic" oxidase rose 15-fold, while those of the "classical" oxidase were unaffected (*237*).

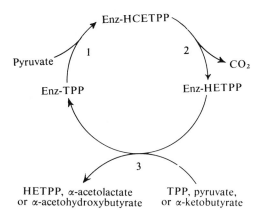

FIG. 15. Proposed mechanism for pyruvate decarboxylation with formation of free α-hydroxyethyl-2-thiamine pyrophosphate (HETPP; "active acetaldehyde"), α-acetolactate or α-acetohydroxybutyrate. TPP = thiamine pyrophosphate; HCETPP = α-hydroxy-α-carboxyethyl-thiamine pyrophosphate ("active pyruvate"); Enz = enzyme. Redrawn from Holtzer *et al.* (*236*).

c. *Acetohydroxy Acid Synthetase.* *E. coli* and *Aerobacter* elaborate two distinct acetohydroxy acid synthetases with widely divergent pH optima. The synthetase active at alkaline pH is thought to function primarily in branched-chain amino acid biosynthesis, while that active at low pH may fulfill a primarily degradative function such as acetoin formation. Biosynthetic acetohydroxy acid synthetases of *E. coli* or

Aerobacter are both valine-repressible and valine-inhibited, exhibit an absolute requirement for Mg^{2+}, are inactive at low pH, and produce optically active α-acetolactate (*227,229*). By contrast, the *N. crassa* synthetase shares none of these properties (*231*). The acetohydroxy acid synthetase of green plants appears to resemble that of *Neurospora* more closely than that of bacteria. Mg^{2+} is required, but the plant synthetase is active at low pH and is not valine-sensitive. The possibility should be considered that the main function of the *Neurospora* and plant aceto-hydroxy acid synthetases studied to date may be unrelated to branched amino acid biosynthesis (*233*). The *E. coli* acetohydroxy acid synthetase also catalyzes formation of α-acetohydroxybutyrate from α-ketobutyrate. Leavitt and Umbarger (*196*) measured simultaneous formation of α,β-dihydroxy-β-methylbutyrate and α,β-dihydroxy-β-methylvalerate (the dihydroxy acid intermediates in leucine-valine and in isoleucine biosynthesis, respectively) by coupling the synthetase reaction with acetohydroxy acid isomeroreductase plus NADPH (see Section V, B, 3). Pyruvate alone formed α,β-dihydroxy-β-methylbutyrate. When α-ketobutyrate was added, α,β-dihydroxy-β-methylbutyrate synthesis decreased with an accompanying increase in α,β-dihydroxy-β-methyl-valerate formation. The valine-insensitive acetohydroxy acid synthetase of *Neurospora* likewise catalyzes α-acetohydroxybutyrate synthesis from α-ketobutyrate (*231*).

3. REDUCTIVE ISOMERIZATION OF THE α-HYDROXY-β-KETO ACIDS TO α,β-DIHYDROXY ACIDS

$$
\underset{\text{(III)}}{\overset{\displaystyle R}{H_3C-\underset{\displaystyle O}{\overset{\displaystyle |}{C}}-\underset{\displaystyle OH}{\overset{\displaystyle |}{C}}-COOH}}
\quad\xrightarrow[\text{Mg}^{2+}]{\text{NADPH}+\text{H}^+ \qquad \text{NADP}^+}\quad
\underset{\text{(IV)}}{\overset{\displaystyle R}{HC_3-\underset{\displaystyle OH}{\overset{\displaystyle |}{C}}-\underset{\displaystyle OH}{\overset{\displaystyle |}{CH}}-COOH}}
$$

The products of the previous reaction next undergo a pinacol-type isomerization with reduction to form the carbon skeletons of valine and leucine. The reaction involves intramolecular migration of the R group from the α- to the β-carbon plus a net reduction. For valine and leucine biosynthesis, $R = -CH_3$ and the product (IV) is α,β-dihydroxy-isovalerate (α,β-dihydroxy-β-methylbutyrate). For isoleucine biosynthesis, $R = -C_2H_5$ and (IV) is α,β-dihydroxy-β-methylvalerate. Conversion of the α-hydroxy-β-keto acids to branched-chain amino acids and/or their known precursors has been demonstrated for *E. coli* (*228,238*), *A. aerogenes* (*228*), *S. cerevisiae* (*238,239*), and *N. crassa* (*240*). Participation of α,β-dihydroxy acids as intermediates in branched amino acid biosynthesis was first recognized by Adelberg *et al.* (*241, 242*). During unsuccessful attempts to isolate the α-keto acid corresponding to isoleucine from culture filtrates of a valine-isoleucine auxotroph of *N. crassa*, α,β-dihydroxy-β-methylvalerate and α,β-dihydroxy-β-methylbutyrate accumulated. These dihydroxy acids satisfied the isoleucine or valine requirements of isoleucine-valine auxotrophs of *E. coli*, respectively. Using isotopes and a double mutant of *N. crassa* obtained by mating a strain unable to synthesize threonine with one accumulating the valine dihydroxy acid, Adelberg (*243*) showed formation of α,β-dihydroxy-β-methylbutyrate from pyruvate and of α,β-dihydroxy-β-methylvalerate from threonine.

Reductive isomerization of either substrate is catalyzed by a reductoisomerase present in yeast (*244,245*), *E. coli* (*246,247*), *Aerobacter* (*246*), *S. typhimurium* (*248*), *N. crassa* (*247,249*), and green plants (*250*). The reaction, which appears to proceed without formation of the corresponding α-keto acids or of other free intermediates, may involve a concerted mechanism (*247*). Although purified reductoisomerase does not catalyze reduction of the corresponding α-keto-β-hydroxy acids (*251*), most sources examined do exhibit a distinct reductase activity catalyzing this reaction. All purified reductoisomerases show an absolute requirement for NADPH and are optimally active at pH 7.5–8.5, but the enzymes from various sources differ in several other respects. Although both the *N. crassa* and *Salmonella* enzymes require Mg^{2+} for activity, and the *N. crassa* enzyme is unstable in the absence of NADPH and Mg^{2+} (*248,249*), there is no detectable metal requirement for the plant enzyme (*250*). The microbial enzymes, but not that

TABLE IV

K_m VALUES FOR REDUCTOISOMERASE FROM VARIOUS SOURCES

Source	$K_m{}^a$	
	α-Acetohydroxybutyrate ($M \times 10^4$)	α-Acetolactate ($M \times 10^4$)
N. crassa	1.6	3.2
S. typhimurium	3.4	1.9
Phaseolus vulgaris	1.2	55

[a] K_m values are for racemic substrates.

from green plants, are sensitive to sulfhydryl reagents. Differences also are apparent in the reported K_m values. While those for the *N. crassa* and *Salmonella* enzymes differ little between substrates, those of the plant enzyme differ substantially (Table IV). The purified *N. crassa* enzyme contains considerable (40–50%) lipid.

4. Dehydration of the Dihydroxy Acids to α-Keto Acids

The final reaction common to biosynthesis of all three branched amino acids is dehydration of an α,β-dihydroxy-β-alkyl acid to the corresponding α-keto-β-alkyl acid. For valine and leucine biosynthesis,

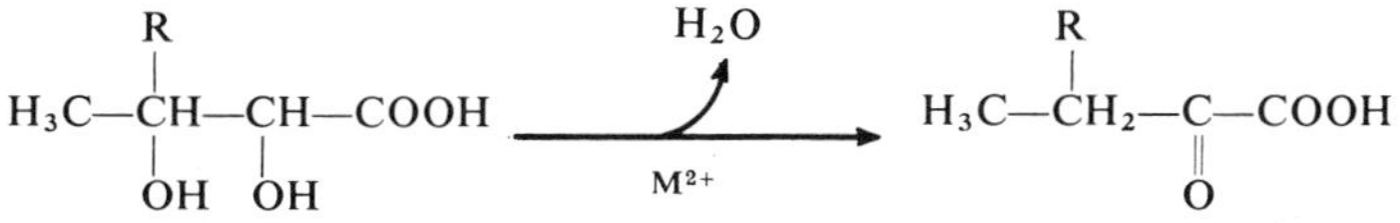

$$
\begin{array}{ccc}
\overset{\displaystyle R}{\underset{\displaystyle OH\ \ OH}{H_3C-CH-CH-COOH}} & \xrightarrow[M^{2+}]{H_2O} & \overset{\displaystyle R}{\underset{\displaystyle O}{H_3C-CH_2-C-COOH}} \\
(IV) & & (V)
\end{array}
$$

$R = -CH_3$ and (V) is α-ketoisovalerate. For isoleucine biosynthesis, $R = -C_2H_5$ and (VI) is α-keto-β-methylvalerate.

Myers and Adelberg reported in 1954 that extracts of wild-type *E. coli* and *N. crassa* catalyzed conversion of either dihydroxy acid to its corresponding α-keto acid (*207*). In the presence of glutamate and pyridoxal-P, valine and isoleucine were formed. Extracts of mutant strains blocked in branched-chain amino acid synthesis were unable to catalyze these reactions.

A single enzyme, dihydroxy acid dehydratase (2,3-dihydroxy acid hydro-lyase, EC 4.2.1.9) catalyzes the above reaction. Dihydroxy acid dehydratase occurs in all life forms independent of an exogenous supply of branched amino acids, but is absent from mammalian tissue, certain lactic acid bacteria, and other life forms auxotrophic for these amino acids (*252,253*). Partially purified preparations of dihydroxy acid dehydratase, from *E. coli* (*254*), *S. cerevisiae* (*255*), spinach (*256*), beans (*257*), and *N. crassa* (*258*), share many common properties, including optimum pH and a requirement for Mg^{2+} or Mn^{2+} for activity, for maintenance of activity or for both. Although the K_m values vary several-fold, the ratios of the K_m's for the two substrates vary little (Table V). During purification the ratio of activities with the two substrates remains essentially constant, both activities are lost at equal rates on heating, and the two substrates compete with one another for enzyme. The *N. crassa* dehydratase, purified some 150-fold to an apparently homogeneous state, contains 45–50% lipid which is not

TABLE V

KINETIC PARAMETERS OF DIHYDROXY ACID DEHYDRATASES FROM VARIOUS SOURCES

Organism	K_m		Optimum pH	Ratio: $\dfrac{V_{max}\,DIV}{V_{max}\,DMV}$
	DIV[a] $(M \times 10^4)$	DMV[a] $(M \times 10^4)$		
E. coli (254)	0.8	1.7	7.8–7.9	2.4
N. crassa (258)	5.8	12	7.7–8.3	1.2
Phaseolus radiatus (257)	9	24	8.0	0.89
Spinacea oleracea (256)	20	63	8.0–8.2	0.45

[a] DIV = α,β-dihydroxyisovalerate; DMV = α,β-dihydroxy-β-methylvalerate.

phospholipid (258). In *N. crassa* the enzymes catalyzing reactions 2, 4, and 5 (Fig. 12) appear to be associated in a complex with the mitochondrial fraction (258,259).

5. (ALSO 9). TRANSAMINATION OF THE α-KETO ACID FORMING AN α-AMINO ACID

$$\underset{(V)}{CH_3\!-\!\underset{\overset{|}{R}}{CH}\!-\!\underset{\overset{\|}{O}}{C}\!-\!COOH} \quad\overset{Glu \qquad \alpha\text{-}KG}{\underset{PLP}{\rightleftarrows}}\quad \underset{(VI)}{CH_3\!-\!\underset{\overset{|}{R}}{CH}\!-\!\underset{\overset{|}{NH_2}}{CH}\!-\!COOH}$$

For valine biosynthesis, $R = -CH_3$ and (VI) is valine. For isoleucine biosynthesis, $R = -C_2H_5$ and (VI) is isoleucine. Ultimately (reaction 9, Fig. 12), an analogous transamination of α-ketoisocaproate results in leucine synthesis. Before this can occur, α-ketoisovalerate must undergo conversion to α-ketoisocaproate via reactions 6, 7, and 8.

The α-keto acid analogs of valine and of isoleucine were shown to be the immediate precursors of valine and isoleucine, respectively, in studies using mutants of *Neurospora* (201) and *E. coli* (209,210,260). The presence in *E. coli* of enzymes catalyzing transamination of all three branched-chain amino acids, recognized by Feldman and Gunsalus (261), was reported by Umbarger and Magasanik (262) to be catalyzed by a distinct valine and isoleucine transaminase. Rudman and Meister (263) later demonstrated the presence in *E. coli* and *A.*

aerogenes of two distinct transaminases. One, transaminase B (L-leucine: 2-oxoglutarate aminotransferase, EC 2.6.1.6), catalyzed reactions between α-ketoglutarate and either valine, leucine, or isoleucine. Similar transaminases are present in *Neurospora* (*264–266*) and beef heart (*267–269*) and can function for synthesis of all three branched amino acids. Although transaminase B activity was absent from a mutant strain auxotrophic only for isoleucine, this is attributed to the presence of limiting quantities of an enzyme catalyzing transamination between ketovaline and either alanine or α-aminobutyrate (*10,263*). The reader is referred to Chapter 14 for a more complete discussion of transamination.

6. Synthesis of α-Isopropylmalate

α-Ketoisovalerate → β-Hydroxy-β-carboxyisocaproate (α-isopropylmalate)

Early studies of leucine biosynthesis by yeast, *N. crassa*, and *E. coli* implicated α-ketoisovalerate and α-ketoisocaproate as intermediates and indicated that carbon atoms 1 and 2 of leucine were derived from acetate (*5,10,270*). On the basis of isotope-labeling data, Strassman *et al.* (*206*) proposed that the initial step in leucine biosynthesis involved condensation of acetyl-CoA with α-ketoisovalerate, producing α-isopropylmalate. Synthetic α-isopropylmalate was next shown to replace leucine for growth of one class of leucine auxotrophs of *Salmonella* (*271*). α-Isopropylmalate, isolated from culture filtrates of a leucine auxotroph of *N. crassa*, was converted to α-ketoisocaproate by cell-free extracts of wild-type *Salmonella* or *Neurospora* (*272*).

Synthesis of α-isopropylmalate from acetyl-CoA and α-ketoisovalerate has now been studied in cell-free extracts of *E. coli* (*273*), yeast (*274*), *N. crassa* (*275–277*), and *Salmonella* (*271,275*, Kohlhaw and Umbarger, unpublished data). An essentially homogeneous α-isopropylmalate synthetase was obtained from *Neurospora* (*276*) and its

TABLE VI

KINETIC PARAMETERS OF α-ISOPROPYLMALATE SYNTHETASE

| Source of enzyme | $(S)_{0.5}(M \times 10^4)$ | | | | | | $(M)_{0.5}$ concentration of L-leucine required to produce 50% inhibition at pH | | |
| | Acetyl-CoA at pH | | | α-Ketoisovalerate at pH | | | | | |
	6.5	7.5	8.5	6.5	7.5	8.5	6.5	7.5	8.5
N. crassa (wild-type) (276)	0.35	0.25			0.1		2×10^{-5}	1.5×10^{-4}	
N. crassa FLR$_{92}$ (leucine-resistant) (276)		0.2			0.1				
S. typhimurium C-19 (regulator negative)[a]	1.0		1.5	1.4		0.75	6×10^{-6}		2×10^{-4}
S. typhimurium FL-241 (leucine-insensitive)[a]	2.0		1.6	4.0		0.60	1×10^{-2}		1×10^{-1}

[a] Kohlhaw, Calvo, and Umbarger, unpublished data.

amino acid composition determined (277). The synthetase (MW = about 143,000) is reported to be a trimer of apparently identical subunits. Both the substrate and the feedback inhibition kinetics of this enzyme have been investigated in detail by Webster and Gross (276). Activity is optimal over the pH range 7–8, and K^+ is required for activity. Either pyruvate or α-ketobutyrate can replace α-ketoisobutyrate as substrate, although their K_m values are more than two orders of magnitude higher. The synthetase from a fluoroleucine-resistant, leucine-insensitive strain, also studied, was some 500-fold less sensitive to inhibition by leucine than was the enzyme from wild-type cells. Feedback inhibition by leucine is exerted in a complex, pH-dependent manner (Table VI). The pH optimum for leucine inhibition does not coincide with that for enzyme activity in the absence of leucine. At pH 7.5, leucine appears to inhibit competitively with acetyl-CoA, and α-ketoiso-butyrate does not reverse the inhibition. At lower pH, both binding of acetyl-CoA and the catalytic mechanism appear to be affected. No dissociation into subunits was observed on addition of leucine at any pH. A variety of hybrid enzymes produced by two mutants also was studied. They all exhibited varying degrees of leucine sensitivity (276,277).

α-Isopropylmalate synthetase (MW = approximately 160,000) has recently been obtained in an essentially homogeneous state from a regulator-negative, leucine-insensitive strain of *S. typhimurium* (Kohlhaw, Calvo, and Umbarger, unpublished data). Overall, its properties closely resemble those of the *Neurospora* enzyme. Either NH_4^+ or K^+ (K_m = approximately $10^{-3} M$) is required for activity, which is optimal at pH 8.5. A similar lack of specificity prevails with respect to the α-keto acid substrate, but again the K_m values for pyruvate and α-ketobutyrate are much higher than those for α-ketoisovalerate. Although the substrate constants for the wild-type and the leucine-resistant enzymes differ little (Table VI), a marked difference is apparent in sensitivity to inhibition by leucine. As with the *Neurospora* enzyme, leucine inhibition is far more effective at low pH. Although in the presence of leucine the substrate kinetics of the wild-type *Salmonella* enzyme are normal at both pH's tested, the inhibition kinetics are sigmoidal at pH 8.5. This is not true for the *Neurospora* synthetase.

7. INTERCONVERSION OF α- AND β-ISOPROPYLMALATE

A role for β-hydroxy-β-carboxyisocaproate (α-isopropylmalate) and for α-hydroxy-β-carboxyisocaproate (β-isopropylmalate) in leucine biosynthesis was originally suggested on the basis of isotope distribution patterns by Strassman *et al.* (*204*). Supporting evidence for these

$$
\begin{array}{ccc}
\text{H}_3\text{C} & \text{COOH} & \\
\diagdown & \mid & \\
\text{CH}-\text{C}-\text{CH}_2-\text{COOH} \rightleftharpoons \\
\diagup & \mid & \\
\text{H}_3\text{C} & \text{OH} &
\end{array}
\qquad
\begin{array}{ccc}
\text{CH}_3 & \text{COOH} & \\
\diagdown & \mid & \\
\text{C}-\text{CH}-\text{CH}-\text{COOH} \\
\diagup & \mid & \\
\text{CH}_3 & \text{OH} &
\end{array}
$$

β-Hydroxy-β-carboxyiso-   α-Hydroxy-β-carboxyisocaproate (α-isopropylmalate)   caproate (β-isopropylmalate)

two compounds as intermediates in leucine biosynthesis in *S. typhimurium*, *E. coli*, and *N. crassa* includes: their excretion by one class of leucine mutants; their utilization for leucine biosynthesis by others; and the presence in cell-free extracts of enzymes catalyzing their synthesis, interconversion, and conversion to leucine (*271–273,278,279*).

Isomerization of β- and α-hydroxy-β-carboxyisocaproate, a readily reversible reaction with an equilibrium constant slightly favoring β-isopropylmalate synthesis, is catalyzed by a soluble α-isopropylmalate isomerase partially purified from *N. crassa*. The isomerase is maximally active at pH 7.0, requires Mg^{2+}, and has K_m values of $6.7 \times 10^{-4} M$ (α-isopropylmalate) and $6.4 \times 10^{-5} M$ (β-isopropylmalate). The isomerase also catalyzes synthesis of dimethylcitraconate from either substrate—a situation reminiscent of the formation of *cis*-aconitate from citrate or isocitrate by aconitase. Whether dimethylcitraconate functions as an intermediate in leucine synthesis is not known, although the authors do not favor its exclusion as an intermediate (*278*).

8. CONVERSION of β-ISOPROPYLMALATE TO α-KETOISOCAPROATE

$$
\begin{array}{ccc}
\text{H}_3\text{C} & \text{COOH} & \\
\diagdown & \mid & \\
\text{CH}-\text{CH}-\text{CH}-\text{COOH} \\
\diagup & \mid & \\
\text{H}_3\text{C} & \text{OH} &
\end{array}
$$

NAD   NADH+H⁺

$$
\begin{array}{ccc}
\text{H}_3\text{C} & & \text{O} \\
\diagdown & & \parallel \\
\text{CH}-\text{CH}_2-\text{C}-\text{COOH} \\
\diagup & & \\
\text{H}_3\text{C} & &
\end{array}
$$

α-Hydroxy-β-carboxy- CO_2 α-Ketoisocaproate
isocaproate

The reaction is catalyzed by α-hydroxy-β-carboxyisocaproate dehydrogenase [*threo*-D_s-2-hydroxy-3-carboxyisocaproate:NAD oxidoreductase (decarboxylating), EC 1.1.1.X] which has been partially purified from *S. typhimurium* (*279*). The absolute configuration of the substrate, established as *threo*-D_s-α-hydroxy-β-carboxyisocaproate (*280*) is the same as that of isocitrate formed in the citric acid cycle. The dehydrogenase is specific for NAD, exhibits a pH optimum of 9.5, and requires both mono- and divalent cations. Genetic analysis supports the view that a single cistron controls its formation (*281*).

9. (SEE SECTION V, B, 5)

C. Regulation of Branched Amino Acid Biosynthesis

The regulation of valine, isoleucine, and leucine biosynthesis in microorganisms has recently been reviewed in detail in a monograph by Umbarger (*198*). Only salient features of this regulation will be alluded to here.

In *Salmonella* the closely linked genes controlling steps 1 through 5 (Fig. 12) form the *ilva* cluster (Fig. 16). The genes controlling reactions 6

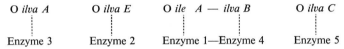

$$O\ ilva\ A \qquad O\ ilva\ E \qquad O\ ile\ \ A - ilva\ B \qquad O\ ilva\ C$$
$$\text{Enzyme 3} \qquad \text{Enzyme 2} \qquad \text{Enzyme 1—Enzyme 4} \qquad \text{Enzyme 5}$$

FIG. 16. The *ilva* cluster. O denotes an operator. Redrawn from Umbarger (*198*).

through 8 constitute the leucine operon (Fig. 17) (*282,283*). For the leucine operon, cistron I, II, and III mutants as well as operator-constitutive and regulator-negative mutants of *S. typhimurium* are known. The cistrons are clustered on the chromosome and all are inactivated by a single point mutation (*281*). Coordinate repression and derepression of all three enzymes, 7–9, are observed when leucine auxotrophs are grown in media containing either excess- or growth-limiting quantities of leucine, respectively (*284*).

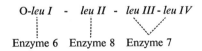

$$O\text{-}leu\ I \quad - \quad leu\ II \quad - \quad leu\ III \text{-} leu\ IV$$
$$\text{Enzyme 6} \quad \text{Enzyme 8} \quad \text{Enzyme 7}$$

FIG. 17. The leucine operon. Redrawn from Umbarger (*198*).

Regulation of leucine synthesis in other microorganisms can differ significantly from that in *Salmonella*. In *B. subtilis* (*284*), but not in *S. typhimurium* (*285*), nor in *E. coli* (*286*), leucine markers are linked to those for valine and isoleucine. Regulation in *Neurospora* appears to depend significantly on feedback inhibition. In *Neurospora* the four leucine genes are not organized as an operon but are randomly distributed over the genome. While synthesis of the first enzyme of the pathway, α-isopropylmalate synthetase, is repressed by leucine, synthesis of subsequent enzymes appears to be regulated by induction by α-iso-propylmalate. Excess leucine therefore both inhibits by feedback and represses synthesis of the first enzyme. It is thought that, as a consequence of lowered levels of α-isopropylmalate (the product of the first enzyme), subsequent enzymes are not induced (*276–287*).

In wild-type cells grown on complex media, levels of enzymes 1 through 9 are strongly repressed, possibly because of high intracellular levels of leucine. This is apparent when cells are grown on minimal media, resulting only in doubling the enzyme levels. Enzymes 2 through 4 are essential not only for synthesis of branched amino acids but also of pantothenate. Repression of these enzymes is "multivalent," that is, repression requires an ample supply of leucine, valine, and isoleucine (*165*). Enzyme 1 also is subject to multivalent control. As was shown using a pantothenate-requiring mutant, when pantothenate limits growth, the *ilva* enzymes are derepressed. Derepression of the *leu* enzymes results only from a lowered amount of leucine, although this may arise indirectly during growth of an *ilva* mutant in media deficient in valine.

Regulation also is achieved at the level of enzyme activity. Enzymes subject to end-product inhibition (Table VII) were discussed above (see Section V, B).

TABLE VII

FEEDBACK-INHIBITED ENZYMES OF
BRANCHED AMINO ACID BIOSYNTHESIS

Enzyme	Inhibitory regulator
Threonine deaminase	Isoleucine
Acetohydroxy acid synthetase	Isoleucine
α-Isopropylmalate synthetase	Leucine

REFERENCES

1. M. R. Stetten, *in* "A Symposium on Amino Acid Metabolism" (W. D. McElroy and H. B. Glass, eds.), p. 277. Johns Hopkins Press, Baltimore, 1955.
2. H. J. Vogel, *in* "A Symposium on Amino Acid Metabolism" (W. D. McElroy and H. B. Glass, eds.), p. 335. Johns Hopkins Press, Baltimore, 1955.
3. P. H. Abelson, E. T. Bolton, and E. Aldous, *J. Biol. Chem.* **198**, 165 (1952).
4. P. H. Abelson, E. T. Bolton, and E. Aldous, *J. Biol. Chem.* **198**, 173 (1952).
5. P. H. Abelson, *J. Biol. Chem.* **206**, 335 (1954).
6. H. J. Vogel, P. H. Abelson, and E. T. Bolton, *Biochim. Biophys. Acta* **11**, 584 (1953)

7. D. E. Reed and L. N. Lukens, *J. Biol. Chem.* **241**, 264 (1966).
8. T. Yura and H. J. Vogel, *Biochim. Biophys. Acta* **24**, 648 (1957).
9. T. Yura and H. J. Vogel, *J. Biol. Chem.* **234**, 339 (1959).
10. A. Meister, "Biochemistry of the Amino Acids" (2 vols.). Academic Press, New York, 1965.
11. H. J. Vogel and B. D. Davis, *J. Am. Chem. Soc.* **74**, 109 (1952).
12. H. J. Strecker, *J. Biol. Chem.* **225**, 825 (1957).
13. A. Baich and D. J. Pierson, *Biochim. Biophys. Acta* **104**, 397 (1965).
14. R. H. Vogel and M. J. Kopac, *Biochim. Biophys. Acta* **36**, 505 (1959).
15. E. Adams and A. Goldstone, *J. Biol. Chem.* **235**, 3492 (1960).
16. H. J. Strecker, *J. Biol. Chem.* **235**, 2045 (1960).
17. L. V. Basso, D. R. Rao, and V. W. Rodwell, *J. Biol. Chem.* **237**, 2239 (1962).
18. T. Yura and H. J. Vogel, *Biochim. Biophys. Acta* **17**, 582 (1955).
19. T. Yura and H. J. Vogel, *J. Biol. Chem.* **234**, 335 (1959).
20. A. Meister, A. N. Radhakrishnan, and S. D. Buckley, *J. Biol. Chem.* **229**, 789 (1957).
21. M. E. Smith and D. M. Greenberg, *J. Biol. Chem.* **226**, 317 (1957).
22. J. Peisach and H. J. Strecker, *J. Biol. Chem.* **237**, 2255 (1962).
23. P. L. Petrakis and D. M. Greenberg, *Biochim. Biophys. Acta* **99**, 78 (1965).
24. H. R. Kaback and T. F. Deuel, *Federation Proc.* **26**, 393 (1967).
25. A. N. Radhakrishnan and A. Meister, *J. Biol. Chem.* **233**, 444 (1958).
26. T. C. Stadtman and P. Elliott, *J. Biol. Chem.* **228**, 983 (1957).
27. D. J. Prockop and K. Juva, *Proc. Natl. Acad. Sci. U.S.* **53**, 661 (1965).
28. S. Pedersen and H. B. Lewis, *J. Biol. Chem.* **154**, 705 (1944).
29. M. Womack and W. C. Rose, *J. Biol. Chem.* **171**, 37 (1947).
30. M. R. Stetten, *J. Biol. Chem.* **181**, 31 (1949).
31. N. M. Green and D. A. Lowther, *Biochem. J.* **71**, 55 (1959).
32. D. Fujimoto and N. Tamiya, *Biochem. J.* **84**, 333 (1962).
33. D. Prockop, A. Kaplan, and S. Udenfriend, *Biochem. Biophys. Res. Commun.* **9** 162 (1962).
34. B. Peterkofsky and S. Udenfriend, *Proc. Natl. Acad. Sci. U.S.* **53**, 335 (1965).
35. D. J. Prockop and K. Juva, *Biochem. Biophys. Res. Commun.* **18**, 54 (1965).
36. D. J. Prockop, P. S. Ebert, and B. M. Shapiro, *Arch. Biochem. Biophys.* **106**, 112 (1964).
37. Y. Fujita, A. Gottlieb, B. Peterkofsky, S. Udenfriend, and B. Witkop, *J. Am. Chem. Soc.* **86**, 4709 (1964).
38. B. Peterkofsky and S. Udenfriend, *Biochem. Biophys. Res. Commun.* **6**, 184 (1961).
39. B. Peterkofsky and S. Udenfriend, *J. Biol. Chem.* **238**, 3966 (1963).
40. D. J. Prockop, B. Peterkofsky, and S. Udenfriend, *J. Biol. Chem.* **237**, 1581 (1962).
41. W. H. Daughaday and I. K. Mariz, *J. Biol. Chem.* **237**, 2831 (1962).
42. K. Juva and D. J. Prockop, *J. Biol. Chem.* **241**, 4419 (1966).
43. G. Manner and B. S. Gould, *Biochim. Biophys. Acta* **72**, 243 (1963).
44. A. Coronado, E. Mardones, and J. E. Allende, *Biochem. Biophys. Res. Commun.* **13**, 75 (1963).
45. D. S. Jackson, D. Watkins, and A. Winkler, *Biochim. Biophys. Acta* **87**, 152 (1964).
46. B. Peterkofsky and S. Udenfriend, *Biochem. Biophys. Res. Commun.* **12**, 257 (1963).
47. A. A. Gottlieb, B. Peterkofsky, and S. Udenfriend, *J. Biol. Chem.* **240**, 3099 (1965).
48. A. A. Gottlieb, A. Kaplan, and S. Udenfriend, *J. Biol. Chem.* **241**, 1551 (1966).
49. L. N. Lukens, *J. Biol. Chem.* **240**, 1661 (1965).
50. D. Nathans and F. Lipmann, *Proc. Natl. Acad. Sci. U.S.* **47**, 497 (1961).
51. M. B. Yarmolinsky and G. L. de la Haba, *Proc. Natl. Acad. Sci. U.S.* **45**, 1721 (1959).

52. E. Katz, D. J. Prockop, and S. Udenfriend, *J. Biol. Chem.* **237**, 1585 (1962).
53. T. E. Smith and C. Mitoma, *J. Biol. Chem.* **237**, 1177 (1962).
54. F. Irreverre, K. Morita, A. V. Robertson, and B. Witkop, *Biochem. Biophys. Res. Commun.* **8**, 453 (1962).
55. K. A. Piez, E. A. Eigner, and M. S. Lewis, *Biochemistry* **2**, 58 (1963).
56. J. S. Wolff, III, J. D. Ogle, and M. A. Logan, *J. Biol. Chem.* **241**, 1300 (1966).
57. J. C. Sheehan and J. G. Whitney, *J. Am. Chem. Soc.* **84**, 3980 (1962).
58. J. C. Sheehan and J. G. Whitney, *J. Am. Chem. Soc.* **85**, 3863 (1963).
59. H. A. Krebs and K. Henseleit, *Z. Physiol. Chem.* **210**, 33 (1932).
60. A. M. Srb and N. H. Horowitz, *J. Biol. Chem.* **154**, 129 (1944).
61. D. Bonner, *Am. J. Botany* **33**, 788 (1946).
62. B. E. Volcani and E. E. Snell, *J. Biol. Chem.* **174**, 893 (1948).
63. H. J. Vogel, D. F. Bacon, and A. Baich, *in* "Informational Macromolecules", (H. J. Vogel, V. Bryson, and J. O. Lampen, eds.), p. 293. Academic Press, New York, 1963.
64. A. Faragó and G. Dénes, *Biochim. Biophys. Acta* **136**, 6 (1967).
65. M. Staub and G. Dénes, *Biochim. Biophys. Acta* **128**, 82 (1966).
66. M. Strassman and S. Weinhouse, *J. Am. Chem. Soc.* **74**, 1726 (1952).
67. J. R. S. Fincham, *Biochem. J.* **53**, 313 (1953).
68. W. I. Scher, Jr., and H. J. Vogel, *Proc. Natl. Acad. Sci. U.S.* **43**, 796 (1957).
69. P. S. Cammarata and P. P. Cohen, *J. Biol. Chem.* **187**, 439 (1950).
70. N. Katunuma, M. Okada, T. Matsuzawa, and Y. Otsuka, *J. Biochem. (Tokyo)* **57**, 445 (1965).
71. B. D. Davis, *Advan. Enzymol.* **16**, 247 (1955).
72. H. C. Pitot and C. Peraino. *J. Biol. Chem.* **238**, PC 1910 (1963).
73. W. K. Maas, G. D. Novelli, and F. Lipmann, *Proc. Natl. Acad. Sci. U.S.* **39**, 1004 (1953).
74. S. Vyas and W. K. Maas, *Arch. Biochem. Biophys.* **100**, 542 (1963).
75. W. K. Maas, *Cold Spring Harbor Symp. Quant. Biol.* **26**, 183 (1961).
76. R. H. De Deken, *Biochem. Biophys. Res. Commun.* **8**, 462 (1962).
77. R. H. DeDeken, *Biochim. Biophys. Acta* **78**, 606 (1963).
78. A. Baich and H. J. Vogel, *Biochem. Biophys. Res. Commun.* **7**, 491 (1962).
79. A. M. Albrecht and H. J. Vogel, *J. Biol. Chem.* **239**, 1872 (1964).
80. S. Shifrin, B. N. Ames, and G. FerroLuzzi-Ames, *J. Biol. Chem.* **241**, 3424 (1966).
81. G. A. Hudock, *Biochem. Biophys. Res. Commun.* **9**, 551 (1962).
82. H. J. Vogel and D. M. Bonner, *J. Biol. Chem.* **218**, 97 (1956).
83. R. H. Vogel and H. J. Vogel, *Biochim. Biophys. Acta* **69**, 174 (1963).
84. P. F. Smith, *J. Bacteriol.* **92**, 164 (1966).
85. A. C. Wilson and A. B. Pardee, *in* "Comparative Biochemistry," Vol. 6 (M. Florkin and H. S. Mason, eds.), p. 73. Academic Press, New York, 1964.
86. S. Udaka, *J. Bacteriol.* **91**, 617 (1966).
87. L. Gorini, W. Gundersen, and M. Burger, *Cold Spring Harbor Symp. Quant. Biol.* **26**, 173 (1961).
88. H. J. Vogel, A. M. Albrecht, and C. Cocito, *Biochem. Biophys. Res. Commun.* **5**, 115 (1961).
89. L. Gorini, *Proc. Natl. Acad. Sci. U.S.* **46**, 682 (1960).
90. E. E. Sercarz and L. Gorini, *J. Mol. Biol.* **8**, 254 (1964).
91. G. N. Cohen and J. Monod, *Bacteriol. Rev.* **21**, 169 (1957).
92. M. Grenson, M. Mousset, J. M. Wiame, and J. Bechet, *Biochim. Biophys. Acta* **127**, 325 (1966).

93. M. Grenson, *Biochim. Biophys. Acta* **127**, 339 (1966).
94. M. Antia, D. S. Hoare, and E. Work, *Biochem. J.* **65**, 448 (1957).
95. H. J. Vogel, *Proc. 5th Intern. Congr. Biochem. Moscow* **3**, 341 (1963).
96. H. J. Vogel, *Proc. Natl. Acad. Sci. U.S.* **45**, 1717 (1959).
97. H. J. Vogel, *Biochim. Biophys. Acta* **34**, 282 (1959).
98. H. J. Vogel, *Biochim. Biophys. Acta* **41**, 172 (1960).
99. H. J. Vogel, *Nature* **189**, 1026 (1961).
100. H. J. Vogel, *Cold Spring Harbor Symp. Quant. Biol.* **26**, 163 (1961).
101. H. P. Broquist, A. V. Stiffey, and A. M. Albrecht, *Appl. Microbiol.* **9**, 1 (1961).
102. S. W. Tanenbaum and K. Kaneko, *Biochemistry* **3**, 1314 (1964).
103. P. H. Abelson and H. J. Vogel, *J. Biol. Chem.* **213**, 355 (1955).
104. C. Gilvarg, *J. Biol. Chem.* **233**, 1501 (1958).
105. C. Gilvarg, *J. Biol. Chem.* **237**, 482 (1962).
106. Y. Yugari and C. Gilvarg, *J. Biol. Chem.* **240**, 4710 (1965).
107. W. Farkas and C. Gilvarg, *J. Biol. Chem.* **240**, 4716 (1965).
108. P. R. Srinivasan and D. B. Sprinson, *J. Biol. Chem.* **234**, 716 (1959).
109. D. H. Levin and E. Racker, *J. Biol. Chem.* **234**, 2532 (1959).
110. C. Gilvarg, *Biochim. Biophys. Acta* **24**, 216 (1957).
111. C. Gilvarg, *J. Biol. Chem.* **234**, 2955 (1959).
112. C. Gilvarg, *J. Biol. Chem.* **236**, 1429 (1961).
113. C. Gilvarg, *Federation Proc.* **21**, 4 (1962).
114. S. H. Kindler and C. Gilvarg, *J. Biol. Chem.* **235**, 3532 (1960).
115. B. Peterkofsky and C. Gilvarg, *J. Biol. Chem.* **236**, 1432 (1960).
116. E. Work *in* "Methods in Enzymology," Vol. 5 (S. P. Colowick and N. O. Kaplan, eds.), p. 858. Academic Press, New York, 1962
117. D. L. Dewey and E. Work, *Nature* **169**, 533 (1952).
118. Y. Shimura and H. J. Vogel, *Biochim. Biophys. Acta* **118**, 396 (1966).
119. E. Work, *in* "Methods in Enzymology," Vol. 5 (S. P. Colowick and N. O. Kaplan, eds.), p. 864. Academic Press, New York, 1962.
120. D. L. Dewey, D. S. Hoare, and E. Work, *Biochem. J.* **58**, 523 (1954).
121. J.-C. Patte, T. Loviny, and G. N. Cohen, *Biochim. Biophys. Acta* **58**, 359 (1962).
122. D. S. Hoare and E. Work, *Biochem. J.* **61**, 562 (1955).
123. P. J. White and B. Kelly, *Biochem. J.* **96**, 75 (1965).
124. A. Maretzki and M. F. Mallette, *J. Bacteriol.* **83**, 720 (1962).
125. H. P. Broquist and J. S. Trupin, *Ann. Rev. Biochem.* **35**, 231 (1966).
126 M E. Maragoudakis and M. Strassman, *J. Biol. Chem.* **241**, 695 (1966).
127. M. Strassman and S. Weinhouse, *J. Am. Chem. Soc.* **75**, 1680 (1953).
128. R. W. Hogg and H. P. Broquist, *Federation Proc.* **25**, 2941 (1966); *J. Biol. Chem.* **243**, 1839 (1968).
129. M. A. Weber, A. N. Hoagland, J. Klein, and K. Lewis, *Arch. Biochem. Biophys.* **104**, 257 (1964).
130. U. Thomas, M. G. Kalyanpur, and C. M. Stevens, *Biochemistry* **5**, 2513 (1966).
131. M. E. Maragoudakis, H. Holmes, L. N. Ceci, and M. Strassman, *Federation Proc.* **25**, 2940 (1966).
132. M. Strassman and L. N. Ceci, *J. Biol. Chem.* **241**, 5401 (1966).
133. M. Strassman, L. N. Ceci, and M. E. Maragoudakis, *Federation Proc.* **24**, 555 (1965).
134. C. Scheifinger, S. Ogura, and M. Ogura, *Federation Proc.* **25**, 2939 (1966).
135. M. Strassman, L. N. Ceci, and B. E. Silverman, *Biochem. Biophys. Res. Commun.* **14**, 268 (1964).
136. J. R. Mattoon and R. D. Haight, *J. Biol. Chem.* **237**, 3486 (1962).

137. S. Bergstrom and M. Rottenberg, *Acta Chem. Scand.* **4**, 553 (1950).
138. A. E. Braunstein, *Advan. Protein Chem.* **3**, 1 (1947).
139. H. P. Broquist and A. V. Stiffey, *Federation Proc.* **18**, 198 (1959).
140. J. G. DeBoever, "Enzymatic Conversion of Ketoadipic Acid to Aminoadipic Acid in Fungi: Relation to Lysine Biosynthesis." M.S. Thesis, Univ. of Illinois, Urbana. Ill., 1963.
141. V. W. Rodwell and R. A. Hartline, *Federation Proc.* **25**, 3440 (1965).
142. H. K. Mitchell and M. B. Houlahan, *J. Biol. Chem.* **174**, 883 (1948).
143. E. Windsor, *J. Biol. Chem.* **192**, 607 (1951).
144. R. L. Larson, W. D. Sandine, and H. P. Broquist, *J. Biol. Chem.* **238**, 275 (1963).
145. S. Sagisaka and K. Shimura, *Nature* **184**, 1709 (1959).
146. S Sagisaka and K Shimura, *J. Biochem.* (*Tokyo*) **51**, 27 (1962).
147. S. Sagisaka and K. Shimura, *J. Biochem.* (*Tokyo*) **51**, 398 (1962).
148. S. Sagisaka and K. Shimura, *Nature* **188**, 1189 (1960).
149. J. R. Mattoon, T. A. Moshier, and T. H. Kreiser, *Biochim. Biophys. Acta* **51**, 615 (1961).
150. S. Darling and P. O. Larsen, *Acta Chem. Scand.* **15**, 743 (1961).
151. A. Kjaer and P. O. Larsen, *Acta Chem. Scand.* **15**, 750 (1961).
152. H. Morimoto and M. Yamano, *Biochem. Z.* **340**, 155 (1964).
153. M. H. Kuo, P. P. Saunders, and H. P. Broquist, *Biochem. Biophys. Res. Commun.* **8**, 227 (1962).
154. M. H. Kuo, P. P. Saunders, and H. P. Broquist, *J. Biol. Chem.* **239**, 508 (1964).
155. J. S. Trupin and H. P. Broquist, *J. Biol. Chem.* **240**, 2524 (1965).
156. S. T. Vaughan and H. P. Broquist, *Federation Proc.* **24**, 497 (1965).
157. E. E. Jones and H. P. Broquist, *J. Biol. Chem.* **240**, 2531 (1965).
158. E. E. Jones and H. P. Broquist, *J. Biol. Chem.* **241**, 3430 (1966).
159. P. P. Saunders and H. P. Broquist, *J. Biol. Chem.* **241**, 3435 (1966).
160. S. T. Vaughn "Saccharopine Dehydrogenase, a Marker of the Aminoadipic Acid Pathway of Lysine Biosynthesis." M.S. Thesis, Univ. of Illinois, Urbana, Ill., 1965.
161. A. J. Aspen and A. Meister, *Biochemistry* **1**, 600 (1962).
162. M. Rothstein and E. M. Saffran, *Arch. Biochem. Biophys.* **101**, 373 (1963).
163. P. Datta and H. Gest, *Nature* **203**, 1259 (1964).
164. E. R. Stadtman, G. N. Cohen, G. LeBras, and H. deRobichon-Szulmajster, *J. Biol. Chem.* **236**, 2033 (1961).
165. M. Freundlich, R. O. Burns, and H. E. Umbarger, *Proc. Natl. Acad. Sci. U.S.* **48**, 1804 (1962).
166. H. E. Umbarger, *Science* **145**, 674 (1964).
167. P. Datta and H. Gest, *Proc. Natl. Acad. Sci. U.S.* **52**, 1004 (1964).
168. Y. Yugari and C. Gilvarg, *Biochim. Biophys. Acta* **62**, 612 (1962).
169. E. R. Stadtman, *Bacteriol. Rev.* **27**, 170 (1963).
170. H. T. Huang and J. W. Davisson, *J. Bacteriol.* **76**, 495 (1958).
171. S. B. Schryver, H. W. Buston, and D. H. Mukherjee, *Proc. Roy. Soc.* (*London*) *Ser. B* **98**, 58 (1925).
172. D. D. Van Slyke, A. Hiller, R. T. Dillon, and D. MacFayden, *Proc. Soc. Exptl. Biol. Med.* **38**, 548 (1938).
173. A. J. P. Martin and R. L. M. Synge, *Biochem. J.* **35**, 294 (1941).
174. J. G. Heathcote, *Biochem. J.* **42**, 305 (1948).
175. J. C. Sheehan and W. A. Bolhofer, *J. Am. Chem. Soc.* **72**, 2466 (1950).
176. L. W. Inskip, *J. Am. Chem. Soc.* **73**, 5463 (1951).
177. P. B. Hamilton and R. A. Anderson, *J. Biol. Chem.* **213**, 249 (1955).

178. E. A. Popenoe, R. B. Aronson, and D. D. Van Slyke, *J. Biol. Chem.* **240**, 3089 (1965).
179. H. Kersten and L. Zürn, *Nature* **184**, 1490 (1959).
180. M. Barbier and E. Lederer, *Biochim. Biophys. Acta* **8**, 590 (1952).
181. A. H. Gordon, *Nature* **162**, 778 (1948).
182. T. Viswanatha and F. Irreverre, *Biochim. Biophys. Acta* **40**, 564 (1960).
183. S. Bergström and S. Lindstedt, *Arch. Biochem.* **26**, 323 (1950).
184. S. Bergström and S. Lindstedt, *Acta Chem. Scand.* **5**, 157 (1951).
185. J. R. Weisiger, *J. Biol. Chem.* **186**, 591 (1950).
186. J. C. Sheehan and W. A. Bolhofer, *J. Am. Chem. Soc.* **72**, 2469 (1950).
187. J. C. Sheehan and W. A. Bolhofer, *J. Am. Chem. Soc.* **72**, 2472 (1950).
188. W. S. Fones, *J. Am. Chem. Soc.* **75**, 4865 (1953).
189. K. A. Piez, *J. Biol. Chem.* **207**, 77 (1954).
190. F. M. Sinex and D. D. Van Slyke, *J. Biol. Chem.* **216**, 245 (1955).
191. K. A. Piez and R. C. Likins, *J. Biol. Chem.* **229**, 101 (1957).
192. D. D. Van Slyke and F. M. Sinex, *J. Biol. Chem.* **232**, 797 (1958).
193. E. A. Popenoe and D. D. Van Slyke, *J. Biol. Chem.* **237**, 3491 (1962).
194. F. M. Sinex, D. D. Van Slyke, and D. R. Christman, *J. Biol. Chem.* **234**, 918 (1959).
195. D. Fujimoto and N. Tamiya, *Biochem. Biophys. Res. Commun.* **10**, 498 (1963).
196. R. I. Leavitt and H. E. Umbarger, *J. Biol. Chem.* **236**, 2486 (1961).
197. G. P. Gladstone, *Brit. J. Exptl. Pathol.* **20**, 189 [*C.A.* **33**, 86714 (1939)].
198. H. E. Umbarger, Lecture on Theoretical and Applied Aspects of Modern Microbiology, Univ. of Maryland, March 31, 1965.
199. H. S. Moyed, *J. Biol. Chem.* **235**, 1098 (1960).
200. D. Bonner, E. L. Tatum, and G. W. Beadle, *Arch. Biochem.* **3**, 71 (1943).
201. D. Bonner, *J. Biol. Chem.* **166**, 545 (1946).
202. M. Strassman, A. J. Thomas, and S. Weinhouse, *J. Am. Chem. Soc.* **75**, 5135 (1953).
203. M. Strassman, A. J. Thomas, and S. Weinhouse, *J. Am. Chem. Soc.* **77**, 1261 (1955).
204. M. Strassman, A. J. Thomas, L. A. Locke, and S. Weinhouse, *J. Am. Chem. Soc.* **76**. 4241 (1954).
205. M. Strassman and S. Weinhouse, *in* "Amino Acid Metabolism" (W. D. McElroy and B. Glass, eds.), p. 452. Johns Hopkins Press, Baltimore, 1955.
206. M. Strassman, L. A. Locke, A. J. Thomas, and S. Weinhouse, *J. Am. Chem. Soc.* **78**, 1599 (1956).
207. J. W. Myers and E. A. Adelberg, *Proc. Natl. Acad. Sci. U.S.* **40**, 493 (1954).
208. E. A. Adelberg, *J. Am. Chem. Soc.* **76**, 4241 (1954).
209. H. E. Umbarger·and J. H. Mueller, *J. Biol. Chem.* **189**, 277 (1951).
210. H. E. Umbarger and E. A. Adelberg, *J. Biol. Chem.* **192**, 883 (1951).
211. H. E. Umbarger and B. Brown, *J. Bacteriol.* **73**, 105 (1957).
212. H. E. Umbarger and B. Brown, *J. Biol. Chem.* **233**, 415 (1958).
213. W. A. Wood and I. C. Gunsalus, *J. Biol. Chem.* **181**, 171 (1949).
214. H. R. Whiteley, *J. Biol. Chem.* **241**, 4890 (1966).
215. H. R. Whiteley and M. Tahara, *J. Biol. Chem.* **241**, 4881 (1966).
216. H. E. Umbarger, *Science* **123**, 848 (1956).
217. H. E. Umbarger and B. Brown, *J. Bacteriol.* **70**, 241 (1955).
218. H. E. Umbarger, *in* "Amino Acid Metabolism" (W. D. McElroy and B. Glass, eds.), p. 442. Johns Hopkins Press, Baltimore, 1955.
219. J.-P. Changeux, *Cold Spring Harbor Symp. Quant. Biol.* **26**, 313 (1961).
220. M. Freundlich and H. E. Umbarger, *Cold Spring Harbor Symp. Quant. Biol.* **28**, 505 (1963).
221. J.-P. Changeux, *Cold Spring Harbor Symp. Quant. Biol.* **28**, 497 (1963).

222. J. Monod, J.-P. Changeux, and F. Jacob, *J. Mol. Biol.* **6**, 306 (1963).
223. C. Cennamo and D. Carretti, *Biochim. Biophys. Acta* **122**, 371 (1966).
224. P. Maeba and B. D. Sanwal, *Biochemistry* **5**, 525 (1966).
225. D. Watt and L. O. Krampitz, *Federation Proc.* **6**, 301 (1947).
226. E. Juni, *J. Biol. Chem.* **195**, 715, 727 (1952).
227. Y. S. Halpern and H. E. Umbarger, *J. Biol. Chem.* **234**, 3067 (1959).
228. H. E. Umbarger, B. Brown. and E. J. Eyring, *J. Am. Chem. Soc.* **79**, 2980 (1957).
229. H. E. Umbarger, and B. Brown, *J. Biol. Chem.* **233**, 1156 (1958).
230. R. H. Bauerle, M. Freundlich, F. C. Stormer, and H. E. Umbarger, *Biochim. Biophys. Acta* **92**, 142 (1964).
231. A. N. Radhakrishnan and E. E. Snell, *J. Biol. Chem.* **235**, 2316 (1960).
232. T. Satyanarayana and A. N. Radhakrishnan, *Biochim. Biophys. Acta* **56**, 197 (1962).
233. T. Satyanarayana and A. N. Radhakrishnan, *Biochim. Biophys. Acta* **77**, 121 (1963).
234. E. Juni and G. A. Heym, *J. Biol. Chem.* **218**, 365 (1956).
235. H. Holzer and G. Kohlhaw, *Biochem. Biophys. Res. Commun.* **5**, 452 (1961).
236. H. Holzer, F. da Fonseca-Wollheim, G. Kohlhaw, and Ch. W. Woenckhaus, *Ann. N.Y. Acad. Sci.* **98**, 453 (1962).
237. R. I. Leavitt, *Bacteriol. Proc.*, 103 (1962).
238. K. F. Lewis and S. Weinhouse, *J. Am. Chem. Soc.* **80**, 4913 (1958).
239. M. Strassman, J. B. Shatton, M. E. Corsey, and S. Weinhouse, *J. Am. Chem. Soc.* **80**, 1771 (1958).
240. R. P. Wagner, A. N. Radhakrishnan, and E. E. Snell, *Proc. Natl. Acad. Sci. U.S.* **44**, 1047 (1958).
241. E. A. Adelberg, D. M. Bonner, and E. L. Tatum, *J. Biol. Chem.* **190**, 837 (1951).
242. E. A. Adelberg and E. L. Tatum, *Arch. Biochem. Biophys.* **29**, 235 (1950).
243. E. A. Adelberg, *in* "Amino Acid Metabolism" (W. D. McElroy and B. Glass, eds.), p. 419. Johns Hopkins Press, Baltimore, 1955.
244. Y. Watanabe, K. Hayashi, and K. Shimura, *Biochim. Biophys. Acta* **31**, 583 (1959).
245. M. Strassman, J. B. Shatton, and S. Weinhouse, *J. Biol. Chem.* **235**, 700 (1960).
246. H. E. Umbarger, B. Brown, and E. J. Eyring, *J. Biol. Chem.* **235**, 1425 (1960).
247. A. N. Radhakrishnan, R. P. Wagner, and E. E. Snell, *J. Biol. Chem.* **235**, 2322 (1960).
248. F. B. Armstrong and R. P. Wagner, *J. Biol. Chem.* **236**, 2027 (1961).
249. K. Kiritani, S. Narise, and R. P. Wagner, *J. Biol. Chem.* **241**, 2047 (1966).
250. T. Satyanarayana and A. N. Radhakrishnan, *Biochim. Biophys. Acta* **110**, 380 (1965).
251. F. B. Armstrong and R. P. Wagner, *J. Biol. Chem.* **236**, 3252 (1961).
252. R. L. Wixom and J. H. Wikman, *Biochim. Biophys. Acta* **45**, 618 (1960).
253. R. L. Wixom, J. H. Wikman, and G. B. Howell, *J. Biol. Chem.* **236**, 3257 (1961).
254. J. W. Myers, *J. Biol. Chem.* **236**, 1414 (1961).
255. R. L. Wixom, J. B. Shatton, and M. Strassman, *J. Biol. Chem.* **235**, 128 (1960).
256. M. Kanamori and R. L. Wixom, *J. Biol. Chem.* **238**, 998 (1963).
257. T. Satyanarayana and A. N. Radhakrishnan, *Biochim. Biophys. Acta* **92**, 367 (1964).
258. K. Kiritani, S. Narise, and R. P. Wagner, *J. Biol. Chem.* **241**, 2042 (1966).
259. R. P. Wagner and A. Bergquist, *Proc. Natl. Acad. Sci. U.S.* **49**, 892 (1963).
260. H. E. Umbarger and B. Magasanik, *J. Biol. Chem.* **189**, 287 (1951).
261. L. I. Feldman and I. C. Gunsalus, *J. Biol. Chem.* **187**, 821 (1950).
262. H. E. Umbarger and B. Magasanik, *J. Am. Chem. Soc.* **74**, 4256 (1952).
263. D. Rudman and A. Meister, *J. Biol. Chem.* **200**, 591 (1953).
264. J. R. S. Fincham and A. B. Boulter, *Biochem. J.* **62**, 72 (1956).
265. R. L. Seecof and R. P. Wagner, *J. Biol. Chem.* **234**, 2689 (1959)
266. R. L. Seecof and R. P. Wagner, *J. Biol. Chem.* **234**, 2694 (1959).

267. R. T. Taylor and W. T. Jenkins, *J. Biol. Chem.* **241**, 4391 (1966).
268. R. T. Taylor and W. T. Jenkins, *J. Biol. Chem.* **241**, 4396 (1966).
269. R. T. Taylor and W. T. Jenkins, *J. Biol. Chem.* **241**, 4406 (1966).
270. O. Reiss and K. Bloch, *J. Biol. Chem.* **216**, 703 (1955).
271. C. Jungwirth, P. Margolin, H. E. Umbarger, and S. R. Gross, *Biochem. Biophys. Res. Commun.* **5**, 435 (1961).
272. S. R. Gross, C. Jungwirth, and H. E. Umbarger, *Biochem. Biophys. Res. Commun.* **7**, 5 (1962).
273. J. M. Calvo, M. G. Kalyanpur, and C. M. Stevens, *Biochemistry* **1**, 1157 (1962).
274. M. Strassman and L. N. Ceci, *J. Biol. Chem.* **238**, 2445 (1963).
275. C. Jungwirth, S. R. Gross, P. Margolin, and H. E. Umbarger, *Biochemistry* **2**, 1 (1963).
276. R. E. Webster and S. R. Gross, *Biochemistry* **4**, 2309 (1965).
277. R. E. Webster, C. A. Nelson, and S. R. Gross, *Biochemistry* **4**, 2319 (1965).
278. S. R. Gross, R. O. Burns, and H. E. Umbarger, *Biochemistry* **2**, 1046 (1963).
279. R. O. Burns, H. E. Umbarger, and S. R. Gross, *Biochemistry* **2**, 1053 (1963).
280. J. M. Calvo, C. M. Stevens, M. G. Kalyanpur, and H. E. Umbarger, *Biochemistry* **3**, 2024 (1964).
281. P. Margolin, *Genetics* **48**, 441 (1963).
282. F. Jacob and J. Monod, *Cold Spring Harbor Symp. Quant. Biol.* **26**, 193 (1961).
283. R. O. Burns, J. Calvo, P. Margolin, and H. E. Umbarger, *J. Bacteriol.* **91**, 1570 (1966).
284. M. Barat, C. Anagnostopoulos, and A.-M. Schneider, *J. Bacteriol.* **90**, 357 (1965).
285. E. V. Glanville and M. Demerec, *Genetics* **45**, 1359 (1960).
286. J. Pittard, J. S. Loutit, and E. A. Adelberg, *J. Bacteriol.* **85**, 1394 (1963).
287. S. R. Gross, *Proc. Natl. Acad. Sci. U.S.* **54**, 1538 (1965).

CHAPTER 17

Selected Aspects of Sulfur Metabolism

Ernest Kun*

INTRODUCTION

    Descriptive enumeration of diverse experimental results in the field of biochemistry of sulfur is not the main purpose of the present review. The excellent survey of Black (*1*) offers a concise starting point for the research worker who wants to gain rapid information in various areas of more recent vintage. A brief essay on *S*-amino acids gives a very good account of recent developments (*2*). The chapter on sulfur metabolism,

* Recipient of the Research Career Award of USPHS.

found in the preceding edition of this series (3) may serve as further extension; it contains references partly of historical importance. Attempts to construct a review with encyclopedic tendencies is considered pointless by this writer for several reasons. In the first place, reference type of information is readily available through various media (e.g., *Chemical Abstracts, Current Contents*) and plays an irreplaceable role in the work of the specialist, who cannot afford incomplete literature review for obvious reasons. Furthermore, reviews are notorious in their deficiency of citation of more recent work because of techniques of publication. On the other hand, research workers in search of ideas, or colleagues from other fields who wish to see conceptual trends, find no comfort in computerized compilation of information.* Selection of a subject matter on the basis of a chemical group characterization has an additional drawback. As in the case of sulfur-containing compounds, various substances, logically belonging to distinct biochemical areas, are more appropriately dealt with in treatises concerned with biochemical functional concepts (e.g., amino acid metabolism, catalytic role or metabolism of sulfur-containing vitamins, function of coenzyme A).

Analysis of the biochemistry of substances related to each other merely on the basis of their sulfur content is meaningful only when such reorientation is capable of inviting specific search leading toward novel information. It is of interest to note that the presence of S atoms in molecules of biological importance coincides with properties which may be appropriately described in terms of biological regulation. This role varies from contribution to the structure of proteins, participation in or regulation of catalytic functions of enzymes, or even more complex regulation of cell division (4). In general, metabolism of S-containing amino acids does not contribute significantly as an energy source for major cellular functions. The special importance of the metabolism of S-containing substances should be in the clarification of pathways of these "micrometabolites" leading to mechanisms of their incorporation into specific structures of macromolecules,† where biological regulation is actually exerted. Very little is known about this

---

* This writer attempted to do this in connection with the writing of the present chapter, receiving over 1000 reference cards from the National Library of Medicine (MEDLARS) as the result of machine selection. This vast collection was to contain only two years' references. Unfortunately, only a small fraction of the references received turned out to be relevant.

† Dynamic aspects of sulfur metabolism in various cellular constituents are dealt with by Z. Pokorny, *Acta Univ. Carolinae Med. Monograph* **21**, 70 (1965).

subject, probably because experimental approaches are notoriously hampered by inherent technical difficulties in studying the biosynthesis of specific macromolecules and mechanisms leading to specific structural organization of macromolecules.* The complexity of these problems is only moderately diminished by the fact that the number of precursors of S-containing macromolecules appears to be limited mainly to cysteine and methionine. Since metabolic aspects of the biochemistry of methionine are fully discussed in another chapter of this series, this chapter will be concerned to a large extent with cysteine and some of its derivatives. Other areas will be considered merely in connection with a search for conceptually novel trends.

In view of the projected limitations of material and the considerable ignorance characterizing this field, an attempt will be made to outline, whenever possible, areas of research which may lead to presently missing information. Since the orientation of this writer is primarily directed toward the biochemistry of animal organisms, detailed accounts of the microbiology of sulfur-containing substances will be left largely untouched.

I. BIOCHEMISTRY OF CYSTEINE

A. General

Ingested cysteine, in the form of protein, would be expected to follow a biosynthetic and degradative path. The regulatory mechanism which determines the prevalence of these two pathways in animal organisms is obscure. Clarification of mechanisms controlling cysteine incorporation in various cellular proteins is of obvious importance because of the well-known participation of protein-SH groups in structural and catalytic properties of enzymes. Apart from the well-known protective effect of cysteine on thiol enzymes, this amino acid was found to inhibit choline acetyl transferase (5) and alkaline phosphatase (6) as well as histidine uptake and K^+ retention in brain slices (7). It cannot be excluded that these effects of cysteine may be related to possible cellular control devices. The radio-protective effect of cysteine is most likely due to the presence of free cysteine in tissues, not to its participation as a protein constituent (8). Cellular transport of cysteine was studied in kidney tissue, indicating that the reduced form, which is the predominant intracellular form, is most probably the transported species (9,10).

---

* The genetic code for cysteine has been recently established: P. Leder and M. W. Nirenberg, *Proc. Natl. Acad. Sci. U.S.* **52**, 1521 (1964).

There appears to be a cysteine transport system in the intestine (*10*). According to Williamson and Clark (*11*), in regenerative tissue approximately 7% of cysteine is incorporated into tissue protein as disulfide attached to a cysteine residue of a protein. An obvious alternative mechanism of cysteine incorporation into protein is by way of peptide-bond synthesis, not counting glutathione biosynthesis, a mechanism which has not been explored.

B. Oxidations

The pathway of cysteine degradation has been outlined previously (cf. *3*). The oxidative pathway is probably the main catabolic mechanism of cysteine degradation. The initial steps of oxidation of cysteine, leading to cysteine sulfinic acid, are the most uncertain ones, and no unambiguous enzymic route has thus far been identified. It would appear logical to expect cysteine to undergo successive oxidations by way of the disulfide and possibly be followed by mono- and disulfoxides eventually yielding sulfinic acid (cf. *3*, p. 240). That this type of paper chemistry is not easily proved (or disproved) is the experience of the experimentalist. At present, two distinctly different oxidative pathways have been identified, one which appears to be localized in rat liver mitochondria (*12*), yielding sulfate, presumably without prior accumulation of sulfinate, and a system found in the cytosol, which converts cysteine to sulfinate (*13–16,20,21*) provided pyridoxal catalysis is blocked by hydroxylamine. Chemical characterization of the cytoplasmic system has advanced considerably over the largely problematic mitochondrial system. It is well known that further metabolism of cysteine sulfinate readily occurs *in vitro* by transaminative and oxidative systems of mitochondria, as discovered by Singer and Kearney (*16a*; cf. *3*).

1. CYTOPLASMIC SYSTEM

Since both cytoplasmic and mitochondrial pathways are novel, they will be discussed in more detail. The original observation of Sörbo and Ewetz (*13,14*) consisted of radiochemical detection of ^{35}S-labeled cysteine sulfinate after aerobic incubation of liver homogenates with cysteine in the presence of hydroxylamine, which prevented further degradation of the sulfinic acid. NADPH or NADP and NAD as well as ferrous ion (ferrous ammonium sulfate) stimulated cysteine sulfinate formation. The heat-labile catalytic system was localized in the cytosol fraction of liver homogenates and tentatively identified as a "mixed function" oxidase (*17*). Similar conclusions were drawn independently by Wainer (*15*). Further work of the Swedish investigators (*13,14*) indicated that particulate subcellular fractions stimulated cysteine sulfinate formation by the cytoplasmic system, an observation which remained unexplained and could not be readily confirmed by more recent work (*16,20,21*), yet could be of significance in the elucidation of cysteine metabolism in cellular systems. The cysteine → cysteine sulfinate system of liver cytosol is specific for L-cysteine, thus differs from the chemically analogous cysteamine → hypotaurine system of Cavallini *et al.* (*18,19*). Metal complexing agents (CN⁻, EDTA, *o*-phenanthroline) and certain thiol reagents inhibited the cysteine sulfinate-forming cytoplasmic system which resisted purification by usual methods of protein fractionation, a finding shared by the research group at the University of California (*16,20,21*). The cytoplasmic cysteine sulfinate-forming system of liver is apparently tissue-specific, as reported by Wainer (*15*). More recent studies concerned with the mechanism of cysteine sulfinate formation from cysteine catalyzed by liver cytosol (prepared in 0.14 M KCl or 0.25 M sucrose, centrifuged at $1444,000 \times g$ for 1 hour) revealed that the system exhibited a cofactor requirement for maximal activity discovered during attempts at purification of the catalytic protein by molecular filtration techniques (*16*). When the liver cytosol is passed through Sephadex G-25 (equilibrated with 0.02 M potassium phosphate of pH 6.9), 60–75% of the enzymic activity (measured by the rate of cysteine sulfinate formation) is lost and only partially recovered by addition of Fe^{2+} and NADH or NADPH (which, it appears, are equivalent). Complete reactivation occurred by further addition of heat-inactivated (boiled) cytosol. Cofactor requirements for the liver cytosol system are shown in Table I.

TABLE I

REQUIREMENT FOR COFACTOR IN THE CYSTEINE SULFINATE (CSA) FORMING SYSTEM OF LIVER CYTOSOL[a]

Components	CSA formed (μmoles)[b]
G-25 excluded (i.e., crude enzyme)	0.86
G-25 excluded + NADPH	1.48
G-25 excluded + NADPH + cofactor	2.67
G-25 excluded + NADH	1.68
G-25 excluded + NADH + cofactor	2.84

[a] From Lombardini *et al.* (*16*).
[b] Per 20 minutes at 37° per 22 mg protein.

The "cofactor" which could not be replaced by any known coenzyme (including ascorbate and tetrahydrofolate) is a relatively small
thermostable molecule with pronounced acid stability but unstable
toward alkaline pH. Further attempts to identify this apparently novel
natural product were considerably complicated by the inability to show
an absolute requirement for the coenzyme (see Table I). Nearly absolute
requirement for the cofactor was eventually demonstrated by gel
filtration on Sephadex G-50, a procedure which permitted the calculation
of the apparent molecular weight of the unknown coenzyme (20,21).
This was in the order of 750. No definite information concerning the
nature of the coenzyme or the catalytic protein itself is presently available. There are indications that the highly unstable enzyme has a molecular weight close to that of hemoglobin. The nature of the oxygenase
reaction itself was followed recently by ^{18}O incorporation experiments
(20). It appears that both O atoms of the sulfinic acid are derived from
molecular oxygen.

2. MITOCHONDRIAL SYSTEM

In contrast to the cytoplasmic system, the mitochondrial mechanism
of cysteine oxidation presumably does not involve cysteine sulfinate as
an intermediate. According to Wainer, mitochondrial oxidation of
cysteine is of "much greater quantitative significance" (12) than the
oxygenase type of oxidation catalyzed by the cytoplasmic system. This
claim cannot be unambiguously evaluated at present. Rates of SO_4^{2-}
formation of cysteine by mitochondria depend on the amount of cysteine
present, and the only variable cofactor requirement appears to be
glutathione (10–40 μmoles cysteine and 5 μmoles GSH). Sulfate production is inhibited under conditions where cysteine is oxidized to
cystine (e.g., in the absence of GSH), indicating that the reduced form
is the precursor of SO_4^{2-}. The mitochondrial system responsible for
this obviously complex series of reactions has not been identified.
Freezing of mitochondria drastically reduces the capacity to produce
SO_4^{2-} from cysteine, while it increases the rate of cystine formation
from cysteine. Potassium phosphate inhibits SO_4^{2-} production from
cysteine by fresh mitochondria, while GSH or GSSG restores activity
of the inhibited systems.

Uncoupling agents have similar effects to phosphate. Oxidizable
substrates seem to inhibit SO_4^{2-} formation from cysteine. Cysteine,
labeled in the carbon skeleton, gives rise to pyruvate, glutamic acid,
alanine, and acetic acid when incubated with mitochondria. Addition
of glutamate does not alter SO_4^{2-} production, but decreases acetic
acid and CO_2 production (from cysteine) with a corresponding increase

in alanine. β-Mercaptopyruvate and H_2S, likely intermediates of cysteine degradation, yield lower rates of SO_4^{2-} production than equivalent amounts of cysteine alone. This was interpreted as an argument against their participation as obligatory intermediates between cysteine and SO_4^{2-}. No increase in SO_4^{2-} production resulted from addition of known nucleotide cofactors. Hydroxylamine, KCN, or iodoacetate exerted inhibition of SO_4^{2-} formation.

Analysis of this complex mitochondrial system yields relatively few definite clues which would presently explain mitochondrial cysteine $\rightarrow SO_4^{2-}$ conversion in enzymic terms. Certain predictions related to cellular metabolism can be made on the basis of the interesting experiments of Wainer (12). In cellular systems of liver, extramitochondrial GSH, which is predominantly cytoplasmic (cf. 22), actually provides conditions for optimal operation of the presumed cysteine $\rightarrow SO_4^{2-}$ pathway of mitochondria. Disruption of coupling between electron transfer and energy conservation (or mitochondrial membrane functions) by uncouplers reduces SO_4^{2-} formation in favor of disulfide production, suggesting the operation of control devices not unfamiliar to research workers concerned with mitochondrial multienzyme systems. Optimal SO_4^{2-} production by fresh mitochondria apparently occurs within 30 minutes of in vitro incubation, and aging of mitochondria coincides with the decline of this pathway; therefore it appears unlikely that lipid peroxidation, a possible in vitro artifact, is indirectly responsible for oxidation of —SH to SO_4^{2-}. The potentially physiological importance of a distinct cysteine $\rightarrow SO_4^{2-}$ system of liver mitochondria should inspire further detailed studies necessary to settle uncertainties.* Conversion of cysteine sulfur to SO_4^{2-} by a pathway reconstructed from presently known enzymic components (e.g., transamination to thiol pyruvate followed by desulfuration and oxidation of H_2S to SO_4^{2-}) is unlikely in mitochrondia, since the thiol pyruvate desulfurase is predominantly a cytoplasmic enzyme (see later).

---

* According to A. Koj, J. Fredno, and Z. Janik, Biochem. J. 103, 791 (1967), liver mitochondria are capable of oxidizing thiosulfate to SO_4^{2-} in the presence of GSH. Inner labeled $S.^{35}SO_3^{2-}$ was converted to $^{35}SO_4^{2-}$ faster than outer labeled thiosulfate. It was proposed that GSH converts $S.SO_3^{2-}$ to SO_3^{2-}, which may be directly oxidized to SO_4^{2-}. If one assumes that a mixed disulfide formed from cystine + GSH can be decomposed by a reaction analogous to that catalyzed by cystathionase (Fig. 5), a product of the type R-SSH could yield sulfide, which, by way of rhodanese, may form $S.SO_3^{2-}$ provided some source of SO_3^{2-} exists in the mitochondria. Direct oxidation of HS to SO_4^{2-} appears to take place only to a very small extent (cf. above reference). No doubt the most obvious source of SO_3^{2-} should be cysteine sulfinic acid [cf. T. P. Singer and E. B. Kearney, Arch. Biochem. Biophys. 61, 397 (1956)] unless an equivalent sulfinite is formed from protein-bound cysteine residue, or some other complex —SH component (e.g., CoA, lipoic acid), a supposition which has no experimental basis at present.

C. S-Derivatives of Cysteine

The existence of metabolic pathways which do not degrade cysteine by disrupting the C—S bond of this molecule is suggested by the isolation of various S-derivatives of cysteine from biological material. For example, S-(2-carboxy-n-propyl)-L-cysteine and S-(2-carboxyethyl)-L-cysteine were identified in urine of children (23), S-(1,2-dicarboxyethyl)-L-cysteine in urine of adults and in pig kidney (24), S-methyl cysteine and its sulfoxide in human urine (25), and S-methyl glutathione in bovine brain (26). S-Sulfoglutathione has been isolated from the small intenstine of the rat (27), and the corresponding cysteine derivative, which was found to be synthesized from thio sulfate and serine by *Aspergillus nidulans* (28), was also identified in animal material (29). It would appear that the existence of S-derivatives of cysteine (and glutathione) point to a hitherto unrecognized series of enzymic pathways. Physiological significance of multienzyme systems catalyzing the metabolism of "micrometabolites" cannot be excluded, particularly in the case of cysteine which, as stated in the Introduction, plays a regulatory, not energy-yielding, metabolic role.

D. Transaminative Metabolism of Cysteine

The discovery of transamination of cysteine to yield thiol pyruvate, and its subsequent desulfuration, described by Meister *et al.* (30), gave impetus to more detailed investigation of this pathway. One may assume that transamination of cysteine with α-ketoglutarate is catalyzed by glutamic-aspartic transaminase. Actually, this reaction is not easily demonstrated since cysteine in substrate quantities inhibits this enzyme. The exact kinetic conditions required for transamination of cysteine with a keto acid are as yet unknown. It would be interesting to test S-alkyl cysteines or cystine as substrate of the transaminase, because blocking of the thiol group should eliminate its inhibitory effect on pyridoxal catalysis. The detailed mechanism of transamination of cysteine with keto acids obviously needs to be elucidated, particularly in cellular multienzyme systems where unfavorable kinetic conditions may represent a regulatory site.

E. Enzymic Reactions of Thiol Pyruvate

More specific information has been gathered on the metabolic fate of thiol pyruvate. The chemical structure of thiol pyruvate and its contribution to acidic properties have been investigated (*31*). This thioketo acid exists as an equilibrium mixture of thiolketo and enethiolol:

$$HS-CH_2-\overset{\overset{O}{\|}}{C}-\overset{\overset{O}{\|}}{C}-OH \rightleftharpoons HS-\overset{\overset{H}{|}}{C}=\overset{\overset{O}{|}}{\underset{H}{C}}-\overset{\overset{O}{\|}}{C}-OH$$

On the basis of accepted inductive constant of S and the known dissociation constant of pyruvic acid, the value of the first dissociation constant of thiol pyruvic acid was calculated to be $\log K_a = -2.14$, while the experimentally obtained value shows a midpoint of titration of the carboxyl group at pH 3.1. It is evident that thiol pyruvate is a much weaker carboxylic acid than anticipated from its keto-thiol structure. Furthermore, it was found that the acidic nature of the thiol group of thiol pyruvate is 0.8 of a $\log K_a$ unit stronger than the calculated value. These anomalous ionic properties of thiol pyruvate are readily predictable from electronic effects of the enethiolol (enol) form, which was shown by IR and UV spectra to be the predominant species. In the solid form, thiol pyruvate exists almost entirely as the enol. It is useful to bear in mind that the usual chemical properties of thiol pyruvate may eventually explain some of its metabolic (or regulatory) roles, although at present neither of these is properly defined.

Four enzymic reactions of thiol pyruvate have been so far studied in detail. Desulfuration, eventually leading to elemental sulfur (*30,32–34*), and trans-sulfuration to CN^- or SO_3^{2-} (*35*) are catalyzed by the same enzyme, reduction to thiolacetate is catalyzed by lactic dehydrogenase (*36*), and conversion to mercaptoethanol (*37*) by yeast enzyme systems occurs apparently as an alternative pathway, since yeast contains no desulfurating enzyme.

Thiol pyruvate cleaving enzyme was isolated from rat liver cytosol and the nature of the enzymic reaction studied more extensively. Enzymic activity was detected in all tissues (including human) thus far analyzed. As shown in Table II (cf. *33*), the highest specific activity was observed in rat kidney, which should serve as an alternative source for its isolation. Intracellular distribution of the desulfurase (Table III) indicates a predominantly cytoplasmic localization. Isolation of the rat liver enzyme yielded two types of preparations. Relatively simple salt fractionation of an extract of rat liver acetone powder yielded a preparation which, after about 10-fold purification (above the first extract of acetone powder), exhibited constant specific activity and migrated in the analytical ultracentrifuge as a protein fraction with an apparent average particle size of approximately 40,000.

TABLE II

DESULFURASE ACTIVITY OF TISSUES

Tissue	S.A.[a]
Mouse red corpuscules	1.2
Mouse plasma	0.0
Rat heart	5.5
Rat brain	2.1
Rat kidney	21.0
Rat muscle	4.0
Rat liver	10.0
Human red corpuscles	1.0
Human plasma[b]	0.1

[a] S.A. = Specific activity: μmoles pyruvate formed in 10 minutes at 30° (pH 7.45) per mg protein.

[b] A trace of hemolysis cannot be excluded.

Direct spectrographic analysis, as well as colorimetric tests, revealed Cu as the major metallic constituent that increased in concentration proportionally with purification (32,33). The absorption spectrum of the purified protein fraction is shown in Fig. 1; cf. (33). Mercaptoethanol (0.3 M) increased enzymic activity 6-fold without changing the apparent substrate constant of thiol pyruvate. Higher concentrations of mercaptoethanol were inhibitory. The smallest substrate constant for thiol pyruvate was obtained at pH 7.5, therefore this was considered to be the pH optimum of the enzyme, even though increasing alkalinity raised the apparent V_{max} (as well as the substrate constant), an effect

TABLE III

DESULFURASE ACTIVITY OF CENTRIFUGAL
FRACTIONS OF RAT LIVER[a]

Cell fraction	S.A.	Total activity
Rat liver homogenate	10.7	4000
Nuclear fraction	11.6	2350
Mitochondria	3.1	290
Supernatant[b]	16.7	1600

[a] Cf. (33).

[b] Contains soluble proteins and microsomes.

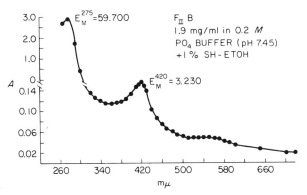

FIG. 1. Absorption spectrum of the purified enzyme fraction of MW 40,000 (F_{II}). Cf. (*33*).

attributed to the increased ionization of the SH group. It follows that the ionized thiol is the probable substrate for the desulfurase. Further search for the molecular nature of the trans-sulfurase (*34*) revealed that the originally obtained cytoplasmic function of average molecular weight 40,000 was a mixture of a family of copper-containing proteins (Table IV), confirming the suspicion that apparent homogeneity, defined

TABLE IV

DISTRIBUTION OF PROTEIN AND COPPER IN
CHROMATOGRAPHIC FRACTIONS OF F_{II}[a, b, c]

Fraction	Protein (mg/ml)	Copper, % (w/w)
II (dialyzed)	21.0	0.041
I	1.62	0.011
II	<0.01	—
III (active)	0.72	0.076
IV	0.33	0.060
V	0.17	0.220

[a] Cf. (*34*).

[b] Ammonium sulfate fraction (F_{II}) was dialyzed overnight and a sample containing 157 mg protein applied to a small column of DEAE cellulose (cf. *34*). All of the copper in the original sample could be accounted for in the protein fractions eluted (total of 0.063 mg copper/157 mg protein). While approximately half the initial copper was found in the active fraction, fraction V was found to retain the greatest concentration on a weight basis.

[c] F_{II} corresponds to the Cu protein fraction of average molecular weight 40,000.

by a symmetrical ultracentrifugal peak, does not necessarily mean that a single enzymically active protein has actually been isolated. The previously obtained cytoplasmic protein fraction (32,33) was resolved by chromatrography (34) on DEAE-cellulose columns (in the presence of mercaptoethanol) into a series of enzymically inactive Cu proteins and in one species with extremely high but rapidly vanishing enzymic activity. This protein was isolated and its molecular weight (10,000) determined, both by ultracentrifugation and by amino acid composition (Table V). Cysteine and thiol analyses, as well as the Cu content of this

TABLE V

AMINO ACID COMPOSITION OF THE CHROMATOGRAPHICALLY RESOLVED ENZYME[a]

Amino acid	μmoles amino acid/ 1.8 mg	μmoles amino acid/mole enzyme	No. of amino acid residues/ μmole enzyme	Concentration (%)
Lysine	1.130	6.30	6	8.72
Histidine	0.405	2.25	2	3.10
Ammonia	1.740	9.70	10	1.70
Arginine	0.663	3.70	4	6.69
Aspartic acid	1.410	7.85	8	10.64
Threonine	0.872	4.85	5	4.94
Serine	0.934	5.20	5	5.25
Glutamic acid	1.700	9.46	9	13.33
Proline	0.840	4.67	5	5.75
Glycine	1.090	6.07	6	4.50
Alanine	1.100	6.13	6	5.34
Valine	0.995	5.53	6	7.02
Methionine	0.171	0.95	1	1.49
Isoleucine	0.663	3.68	4	5.34
Leucine	1.300	7.23	7	9.17
Tyrosine	0.375	1.97	2	3.62
Phenylalanine	0.688	3.72	4	6.60
(Cysteine)	(0.208)	(1.11)	(1)	(1.50)

9.21 mg amino acids recovered from 10 mg (1 μmole) enzyme

[a] Cf. (34).

fraction, indicated that this protein is the most probable molecular species of the desulfurase.

Enzymic activity decays even during chromatographic isolation and is critically dependent on the protein content of the enzyme solution, as shown in Table VI. Since the rate of decay of enzymic activity is

TABLE VI

EFFECT OF STORAGE FOR 20 HOURS AT 4°
ON ENZYME ACTIVITY[a, b]

Protein concentration (mg/ml)	S.A.
0.585	900
0.292	660
0.146	330
0.073	40
0.036	0

[a] Cf. (34).
[b] Initial specific activity = 1400, measured within 5 hours of chromatographic elution. The decay of enzymic activity is a function of the concentration of enzyme protein. The dilutions were made in 0.05 M potassium phosphate (pH 7.4) containing 0.05% β-mercaptoethanol.

most rapid immediately after chromatographic isolation, the specific activity of approximately 1000–1500 (compared to 10 in homogenates) actually expresses only about one-third to one-fifth of the extrapolated specific activity which the enzyme possesses at the moment of separation from similar Cu proteins. Characteristics of the extreme instability of the chromatographically resolved species of molecular weight 10,000 indicate that this protein is most likely a subunit constituent of a larger molecular species (MW 40,000) which exists and is stabilized as a component of a family of Cu proteins of similar molecular properties isolated in the previously obtained cytoplasmic protein fraction (32,33). The absorption spectrum of the desulfurase monomer is shown in Fig. 2 (cf. 34) for comparison with the spectrum of the Cu protein fraction of average molecular weight 40,000 (see Fig. 1). The yellow band at 410 mμ is most likely the absorbance of the specific Cu complex present in the active site, as shown by spectral analysis of enzyme-substrate complexes, stabilized by the inhibitor o-thiolbenzoate which blocks desulfuration (i.e., enzyme-substrate decomposition). This is shown in Fig. 3. Inhibition of the enzyme by specific Cu complexing agents (38) strongly argues in favor of the metalloenzyme nature of the desulfurase. Inhibition by certain Cu complexing agents is counteracted by phosphate ion, indicating that the position of the metal in the enzyme protein is such that it can interact with different ligands, a behavior to be expected from catalytically active metal proteins.

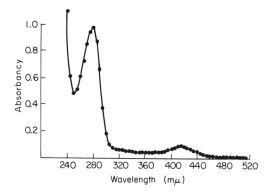

FIG. 2. Absorption spectrum of the purified enzyme in 0.2 M phosphate buffer of pH 7.4, containing 0.1 % β-mercaptoethanol. The dialysate served as a blank. The protein concentration was 1.1 mg/ml (1 cm light path). The spectrum of the Cu protein contaminant with no enzymic activity differed only by a broader peak between 250 and 280 mμ. No absorption peaks were detectable at wavelengths beyond 420 mμ, even in higher concentrations of protein solutions. Cf. (34).

The most probable mechanism for describing enzymatic desulfuration and trans-sulfuration of thiol pyruvate is shown in Fig. 4. Reaction (b) has been shown by Sörbo (39); reaction (d) is the most likely pathway of H$_2$S formation in this thiol system. The "persulfide"-type enzyme sulfur intermediate is probably difficult to prove unambiguously. Presently available evidence indicates an "enzyme-substrate" complex formation which kinetically precedes product formation. This

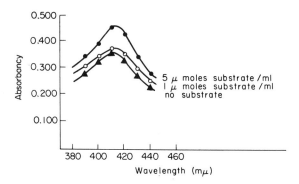

FIG. 3. The effect of o-thiolbenzoate on the absorption spectrum of the enzyme in the presence of substrate; 5 μmoles/ml o-thiolbenzoate were present in a solution of enzyme—4 mg/ml in 0.1 M potassium phosphate buffer, pH 7.4, in 0.01 % β-mercaptoethanol at 27° (lower curve). The effect of the addition of 1 and 5 μmoles β-mercaptopyruvate/ml is shown by the two upper curves. Cf. (34).

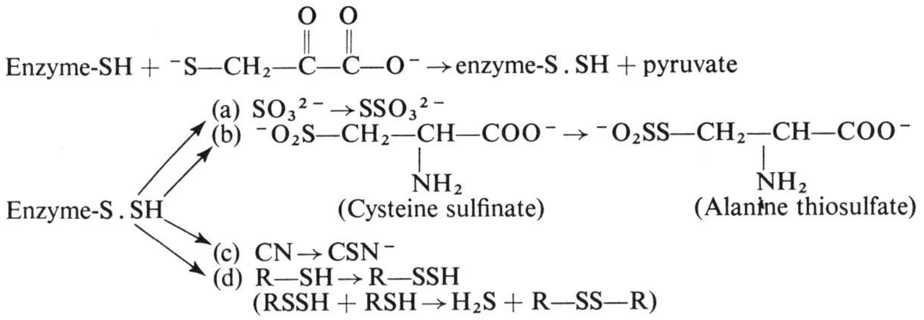

$$\text{Enzyme-SH} + {}^-\text{S--CH}_2\overset{\overset{\displaystyle O}{\|}}{\text{--C}}\overset{\overset{\displaystyle O}{\|}}{\text{--C}}\text{--O}^- \rightarrow \text{enzyme-S . SH} + \text{pyruvate}$$

(a) $SO_3^{2-} \rightarrow SSO_3^{2-}$

(b) $^-O_2S\text{--CH}_2\text{--CH--COO}^- \rightarrow {}^-O_2SS\text{--CH}_2\text{--CH--COO}^-$

Enzyme-S . SH

                                $\overset{|}{N}H_2$ $\overset{|}{N}H_2$

                      (Cysteine sulfinate)   (Alanine thiosulfate)

(c) $CN \rightarrow CSN^-$

(d) $R\text{--SH} \rightarrow R\text{--SSH}$

    (RSSH + RSH $\rightarrow H_2S$ + R--SS--R)

FIG. 4. Mechanism for enzymic desulfuration and trans-sulfuration of thiol pyruvate.

is based on absorbance changes at 415 mμ when stoichiometric amounts of enzyme and thiol pyruvate react in the presence of a sufficient amount of inhibitor (Fig. 3) to block product formation. These observations are suggestive, as in the case of many spectrophotometrically detectable intermediates, but require further avenues of experimental testing. Direct chemical characterization of persulfides, such as thiol cysteine or alanine hydrogen disulfide readily formed from thiols and elemental S, capable of donating S to acceptors like CN^- or iodoacetate (40), strengthens the proposed enzymic reaction sequence. Persulfides of the type mentioned are more stable at alkaline pH (approximately pH 9.0), but decompose to elemental sulfur and thiols at pH 7.5 (41).

The biochemical significance of the thiol pyruvate enzyme is not known, but may be presumed to be a regulatory one. Rapidly dividing cancer cells, but not embryonic cells, contain only traces of this enzyme (42), an observation which raises more challenging questions. It is of importance that thio lactate is not a substrate of the desulfurase; thus regulation of desulfuration can be dependent on the activity of lactate dehydrogenase (36).

In a recent paper Van Den Hamer et al. (43) reisolated the protein fraction containing the thiol pyruvate cleaving enzyme from rat liver by chromatography of an ammonium sulfate fraction in phosphate buffer. Treatment of this approximately 6-fold enriched enzyme fraction with diethyl dithiocarbamate removed most of the Cu^{2+} and did not alter enzymic catalysis. The results of these authors are readily predictable from the properties of the highly purified (at least 300- to 500-fold) but unstable 10,000 molecular weight species of the desulfurase (34), since it is clear that the protein fraction composed of 40,000 molecular weight proteins contains less than 1% monomeric enzyme, thus the remaining protein-bound Cu in their preparation (cf. 43, p. 2516) readily accounts for the thiol pyruvate enzyme. The observation of these authors (43) that the copper complexing agent, diethyl dithiocarbamate, does not inhibit enzymic activity in phosphate buffer, is in agreement with earlier experiments which demonstrated that phosphate counteracts the inhibitory effects of Cu complexing agents on the activity of the thiol pyruvate enzyme (cf. *38*, Table I, p. 187). On the basis of these experiments it cannot be concluded that Van Den Hamer *et al.* (*43*) performed appropriate experiments which are challenging the Cu-protein nature of this enzyme.

F. Cleavage of Cystine by Cystathionase and "Desulfhydrase"

It is now well established that cystathionase can cleave cystine according to the reaction shown in Fig. 5 (cf. *40*).

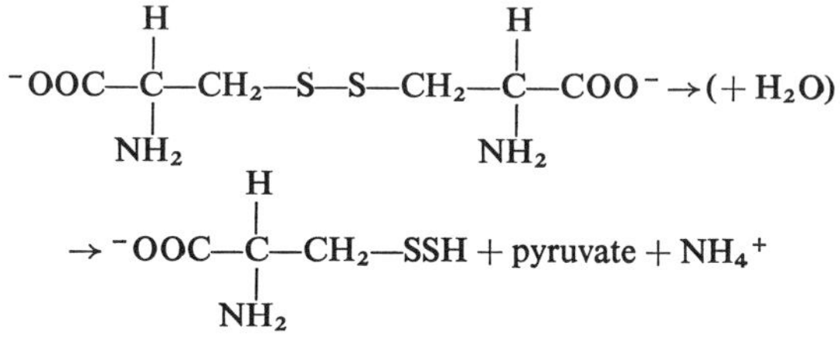

$$^-OOC—\overset{\overset{\text{H}}{|}}{\underset{\underset{\text{NH}_2}{|}}{C}}—CH_2—S—S—CH_2—\overset{\overset{\text{H}}{|}}{\underset{\underset{\text{NH}_2}{|}}{C}}—COO^- \rightarrow (+H_2O)$$

$$\rightarrow {}^-OOC—\overset{\overset{\text{H}}{|}}{\underset{\underset{\text{NH}_2}{|}}{C}}—CH_2—SSH + pyruvate + NH_4{}^+$$

Thiol cysteine

FIG. 5. Cleavage of cystine by cystathionase.

Cavallini *et al.* (*44*), in attempting to clarify the question of "cysteine desulfhydrase," indicated that a trace of cystine is necessary for the reactivity of cysteine with cystathionase. Thiol cysteine (alanine hydrogen disulfide or thiol sulfate) was the postulated intermediate, a mechanism elegantly demonstrated by Flavin (*40*).

G. Cysteine-S-sulfonate

A cysteine derivative first dealt with by Sörbo (*45*) has been shown to be decomposed by liver extracts to thiosulfate, pyruvate, and ammonia. The nature of the catalytic system has not yet been clarified. Cysteine-*S*-sulfonate readily transaminates with a keto acid to yield thiosulfate, pyruvate, and a corresponding amino acid (*46*).

A general scheme demonstrating the alternative pathways of cysteine may be proposed as shown in Fig. 6. Whether or not this distributive type of multienzymic system truly represents a ubiquitous multiplicity of potential pathways is not clear at present. It seems more probable that restrictive mechanisms may facilitate or inhibit certain systems, thus create preferential reaction sequences possessing pronounced tissue specificity. Elucidation of tissue-specific regulatory pathways is a prerequisite to biochemical interpretation of differentiated cellular functions. The search for participation of thiol-containing metabolites as part of metabolic regulatory devices in specific tissues is a profitable, yet difficult, task.

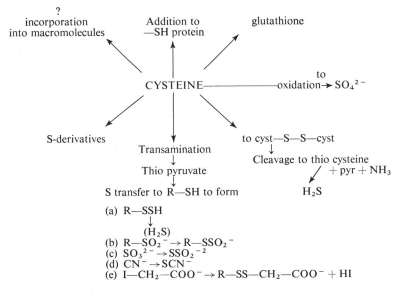

FIG. 6. Alternative pathways of cysteine.

II. FATE OF S DERIVED FROM CYSTEINE

Oxidation of sulfide to SO_3^- and SO_4^{2-} is catalyzed by a variety of oxidizing systems which will not be discussed here. No definite information exists which would point to a distinct pathway for the oxidation of various oxidation stages of S (e.g., by xanthine oxidase or cytochrome oxidase).

It appears doubtful that HS^- is actually oxidized significantly by cellular enzyme systems under physiological conditions (see footnote, Section I, B, 2).

More recent developments concerned with the mechanism of the rhodanese-catalyzed reaction clarified this system to a considerable extent, indicating participation of a persulfide intermediate (47,48). The double-displacement reaction, shown in Fig. 7 describes this mechanism. This mechanism replaces the one earlier proposed by Sörbo (*48a*) and discussed in the previous edition of this treatise (cf. *3*).

$$\text{Rhodanese} + 2\ SSO_3{}^- \rightleftharpoons \text{rhodanese—}S_2 + 2\ SO_3{}^-$$

$$\text{Rhodanese—}S_2 + 2\ CN^- \rightleftharpoons \text{rhodanese} + 2\ SCN^-$$

FIG. 7. Double-displacement mechanism of rhodanese-catalyzed reaction.

III. GLUTATHIONE

A. General

Although one of the first peptides recognized to have a fundamental regulatory role in cellular systems, glutathione remains a biochemical enigma. The apparently coenzymic function of GSH was recently recognized in liver systems for the synthesis of dimethyl selenide from selenite (*49*). It is hard to decide whether or not this curious reaction reflects the function of a physiologically significant metabolic reaction, a question left open at this time. The presently unexplained regulatory role of pyridoxine in glutathione metabolism is indicated by an increased GSH content of rat liver and erythrocytes which follows deficiency of this vitamin (*50*). Another undefined role of GSSG is its augmenting effect on formiminoglutamic acid excretion in vitamin B_{12}-deficient rats (cf. *50a*).

B. Oxidation of GSH

Glutathione in tissues, due to efficient enzymic reducing systems, is predominantly present as the thiol. This observation implies that oxidative mechanisms for GSH in cellular systems are either not operative or absent. It is common knowledge that GSH, when added to various tissue preparations, is rapidly oxidized; it is generally assumed to be oxidized by fairly trivial mechanisms (e.g., via reduced cytochromes, cytochrome oxidase, or catalase), but it is not known whether these enzymes actually oxidize GSH *in vivo*. Glutathione peroxidase (*51*) catalyzing the reduction of H_2O_2 by GSH was considered to function in cellular systems as a peroxide-removing device (*52*). The same enzyme was identified as a mitochondrial-contracting factor (*53*), suggesting a highly complex metabolic regulatory system which goes far beyond orthodox concepts of enzymology. More recent work dealing with the problem of aerobic oxidation of GSH by subcellular cell fractions (*54a*) revealed that cytosol, microsomes, and mitochondria mutually reinforce the rate of GSH oxidation, while each fraction alone causes little oxidation. Some low-molecular-weight compounds in the cytosol increased GSH oxidation. Products of lipid peroxidation (malonaldehyde) were diminished by GSH. The explanation is offered that GSH counteracts lipid peroxidation of membrane lipids (*54b*). It is somewhat mysterious that H_2O_2 itself does not induce lipid peroxidation in mitochondria (cf. *55*), suggesting that an organic peroxide may be the true oxidizing agent of unsaturated fatty acids of membrane structures. As a teleological argument: the "protective" role of GSH *in vitro* may have practical explanatory merit for artificial systems, yet leaves the biochemical significance of "lipid peroxidation" in an atmosphere of obscurity. It remains unclear whether or not this protective mechanism is an *in vitro* artifact or reflects a true cellular role of GSH in subcellular systems other than erythrocytes, where, by way of glutathione peroxidase, H_2O_2 is effectively detoxified by GSH (*56,57*). It is well-established that the life span of red blood cells of subjects possessing inherited GSH deficiency of blood is markedly reduced and the red cells are highly susceptible to oxidative breakdown of hemoglobin and other cellular constituents (*58*). It is, of course, another matter to consider other cellular systems (e.g., liver cell) containing a full complement of respiratory enzymes in which rates of peroxide formation are likely to be negligible. In simpler systems like red blood cells, competition of glutathione peroxidase (by way of H_2O_2-forming systems) with NADPH-linked G-S-S-G reductase, could effectively control intracellular NADPH levels, the latter generated by the hexose monophosphate pathway (*59*).

Oxidation of GSH may occur by somewhat unusual enzyme systems, such as the glutathione-organic nitrate reductase of liver (*60*) and GSH insulin transhydrogenase found in liver and pancreas (*61,62*). The mechanism of glyceryl trinitrate-GSH-oxidoreduction has been elucidated to a considerable extent (*60*) utilizing a partially purified liver enzyme (*63*) which apparently accounts for the formation of inorganic nitrite from organic nitro compounds in animal organisms. In elegant experiments Needleman and Hunter readily followed nitrate reduction by GSH in an assay coupled to G-SS-G reductase, according to the scheme shown in Fig. 8.

No glycerol is formed from trinitrates, but the enzyme reacts slowly with dinitrates, and to some extent with mononitrate. Mannitol hexanitrate and erythrityl tetranitrate are reduced at a considerably greater rate than glyceryl trinitrate.

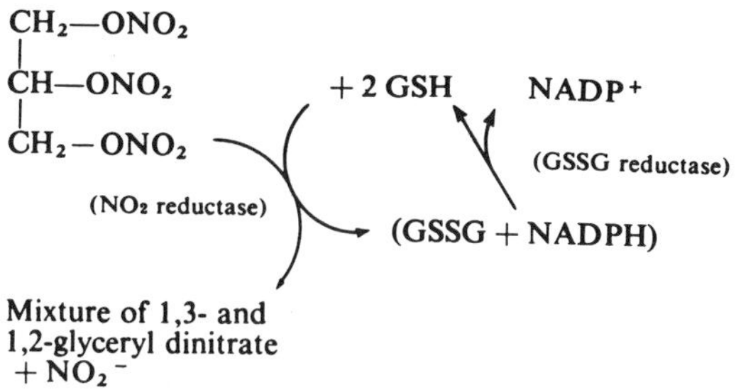

$$\begin{array}{l} \text{CH}_2\text{—ONO}_2 \\ | \\ \text{CH—ONO}_2 \\ | \\ \text{CH}_2\text{—ONO}_2 \end{array}$$

+ 2 GSH   NADP+

(NO₂ reductase)   (GSSG reductase)

(GSSG + NADPH)

Mixture of 1,3- and
1,2-glyceryl dinitrate
+ NO₂⁻

FIG. 8. Nitrate reduction by GSH.

The enzymology of the flavoprotein G-SS-G reductase has been clarified to a great extent utilizing highly purified yeast enzyme (64,65) and liver enzyme (65a). An analogous reaction is catalyzed by an apparently specific CoA-S-S-G disulfide reductase of liver and kidney (66,67). Ezymic reduction of penicillamine by GSH and excretion of the mixed disulfide has also been observed (68,69).

C. Mercapturic Acid Formation

This has been discussed in the previous edition of this series (cf. 3). Some progress was made in the elucidation of a more detailed reaction mechanism of this rather unusual metabolic pathway of glutathione. It was shown that iodomethane administered *in vivo* yields S-methyl-cysteine as well as methylmercaptoacetic acid and N-(methylmercapto-acetyl)-glycine (70). The last two substances were also identified as metabolites of S-methyl cysteine and S-methylglutathione (71), thus possible connections exist between the seemingly artificial " detoxification path " and probably physiological metabolic route of S-derivatives of cysteine (see Section I, C).

It is fairly certain that the cysteine moiety of mercapturic acids is derived from glutathione itself, since liver glutathione falls rapidly after administration of mercapturic acid precursors (72). At least four enzymic components have so far been shown to participate in mercapturic acid conjugation: S-aryltransferase (73), glutathione S-aryltransferase (74), GSH-epoxytransferase (75), and a GSH-transferase apparently specific for certain unsaturated compounds (76). It is about time that the elegant organic chemical work yielding identification of these complicated substances is followed by more definitive enzymological studies capable of elucidating the nature of this presently obscure "artificial" metabolic pathway of mercapturic acid formation. S-Substituted glutathione derivatives are hydrolyzed by the microsomal (kidney) glutathionase enzyme system (*77*). This enzyme may also be part of the mercapturic acid-forming system, since glutathione-S-derivatives presumably undergo hydrolytic cleavage to corresponding S-alkylcysteines, which are the actual precursors of mercapturic acids.

IV. METHIONINE

Like cysteine, this amino acid has its degradative pathways and also enters into peptide structures of proteins. Incorporation of methionine into various tissue proteins has been studied extensively (*78–80*). Conversion of the α-hydroxy derivative of methionine to methionine occurs via oxidative and transaminative reactions, as studied in rat tissues. A requirement for glutamine in this pathway, which can be partly replaced by asparagine, has also been demonstrated (*81*). An enzyme system of rat kidney oxidizes the S-methyl group of methionine (or other S-methyl compounds) without prior transmethylation, probably by a peroxidative mechanism involving catalase (*82*). An interesting series of enzymic reactions was disclosed during studies concerned with mechanisms of reduction of methionine sulfoxide (*83*) involving pyridine nucleotide and flavoprotein enzymes, some of which reduce various low-molecular-weight disulfides (e.g., insulin) but do not react with macromolecular disulfides (e.g., ribonuclease or serum albumin).

N-Formyl methionine deserves special attention, since this methionine derivative was found to be a unique initiator of protein synthesis, at least in bacterial systems, as sRNA-N-formyl-methionine (*84,85*). It appears that formyl methionine increases the rate of formation of the first peptide bond, while the ability to direct the location of methionine residue in the growing peptide chain is associated with sRNA (*86*). The true initiator of peptide synthesis is probably a specific met-sRNA species. Methionine analogs (norleucine or ethionine) can substitute for methionine in the acylation of sRNA, and they are readily formylated after attachment to sRNA (*87*).

S-Adenosylmethionine has been extensively studied both from the viewpoint of its large-scale isolation from yeast (*88,89*) and as a methyl donor in transmethylation reactions of protoporphyrin (*90*), histamine (*91*), and in phosphatidylcholine formation in rat liver (*92*). An enzyme, apparently uniquely localized in the pituitary, splits S-adenosyl-methionine to methanol and S-adenosylhomocysteine (*93*). This curious reaction is presumably responsible for traces of methanol found in biological material. More recent work indicates that S-adenosyl-methionine decarboxylating enzyme and spermidine synthetase may be part of the same enzyme complex (*94*).

V. CYSTATHIONINE

Cystathionine is a well-established intermediate in the reaction sequence of "trans-sulfuration" between cysteine and methionine, a subject dealt with in another section of this series. Distribution of the γ-cleavage enzyme, yielding cysteine, α-ketobutyrate, and NH_3 as well as cystathionine synthetase in various animal tissues, was determined (*95*).

VI. ETHIONINE

Ethionine has attracted much attention because of its interesting mechanism of toxicity, which apparently consists of ATP depletion (*96*). Complex metabolic effects, influencing RNA and lipid metabolism, are probably consequences of this primary toxic mechanism, although paradoxical effects (e.g., stimulation of RNA synthesis when fed to rats; cf. *97*) do not fit this picture.* S-Adenosylethionine formation is probably the expected intermediate in transethylation reactions (*98*). 5'-Ethylthioinosine appears to be the major excreted product of this amino acid in the rat (*99*).

VII. THIOL DISULFIDES IN BIOLOGICAL MACROMOLECULES

It has been pointed out in this review that the biological control exerted by thiol disulfide interconversion of cysteine residues in proteins is a significant aspect of sulfur metabolism. The oxidation-reduction of

---

* H. G. Sie and W. H. Fishman, *Biochem. J.* **106**, 769 (1968), reported that injection of ethionine (4–7 days) caused a large increase in hepatic glucose-6-phosphate dehydrogenase activity, uninfluenced by inhibitors of protein or mRNA synthesis, a further paradoxical phenomenon.

—SH residues of cysteine (or protein-bound lipoic acid) in enzyme proteins—in spite of the apparent simplicity of the chemical mechanisms—has far-reaching consequences in electron transfer, oxidative phosphorylation protein conformation, enzymic catalysis, as well as in complex processes (e.g., cell division, biological membrane functions). Specific aspects of these highly significant processes, involving —SH—SS interconversions are usually dealt with in reviews not primarily concerned with metabolism of small molecules. Certain less emphasized, yet equally important, roles of —SH groups bound to specific macromolecules will be briefly mentioned. Sulfur-containing nucleotides: 4-thiouridylic acid (100) and 2-thiopyrimidine (101) were found to be minor constituents of sRNA. Biosynthesis of these S-containing nucleotides occurs from cysteine, apparently on a macromolecular level [i.e., S is introduced into the uracyl residue of RNA; cf. (100) and (100a)].

Thiol groups in certain proteins, as in reduced ribonuclease, are specifically oxidized by microsomal systems, an enzymic activity of unusual nature (102) since thiol disulfide conversion in proteins has hitherto not been considered a reaction requiring enzymic catalysis. The reverse process is very likely brought about by transhydrogenase reactions from GSH to protein disulfides (103). It is not certain whether or not enzymes exist which are specific for certain protein-bound —SS— groups utilizing GSH as reducing agent. A much more specific reaction of obvious biological importance was recently identified by Anfinsen and collaborators as a sequel to previous studies (102). An enzyme was isolated in pure form from bovine liver microsomes which specifically catalyzes disulfide interchange between proteins (104). The enzyme protein contains 3 half-cystine residues, only 1 of which seems to be essential for enzymic activity. The unique residue must be in the —SH form for effective catalysis. The enzyme actually catalyzes the rearrangement of random or "incorrect" pairs of half-cystine residues to the native disulfide bonds in specific protein substrates. The significance of this novel type of "protein-modifying" enzyme can hardly be overemphasized, since it offers a mechanism of control of biological properties of macromolecules (proteins) at a highly integrated level. It follows that the information content of certain macromolecules is apparently dependent not only on the genetic chemically controlled replication mechanisms but also on subsequent specific modification of tertiary structure. It is evident that a new type of metabolic control emerges as a consequence of these important discoveries; it is exercised through control of the activity of enzymes catalyzing specific disulfide exchanges of specific proteins.

ACKNOWLEDGMENTS

The contribution of Miss C. Fegté in typing and editing this review is acknowledged with appreciation. Reproduction of Figs. 1, 2, and 3 from original articles (*33,34*) was permitted by the Editors of *Biochimica et Biophysica Acta*. The research work of the author quoted in this paper was supported by grants of the USPH, National Science Foundation, and American Heart Association, Inc.

Critical comments by Dr. E. B. Kearney and T. P. Singer are acknowledged with gratitude.

REFERENCES

1. S. Black, *Ann. Rev. Biochem.* **32**, 399 (1963).
2. G. A. Maw, The Biochemistry of sulfur containing amino acids: Chem. Soc. Annual Rep. Vol. LXIII, 639, 1966 (London).
3. E. Kun, in *Metabolic Pathways* (D. M. Greenberg, ed.), 2nd ed., Vol. 2, p. 237. Academic Press, New York, 1961.
4. Sulfur in Proteins, *in* "Proceeding of a Symposium, Falmouth, Massachusetts (R. Benesch, P. D. Boyer, I. M. Klotz, W. R. Middlebrook, A. D. Szent-Györgyi, and D. R. Schwartz, eds.). Academic Press, New York, 1958.
5. D. Morris, C. Hebb, and G. Bull, *Nature* **209**, 914 (1966).
6. E. Pojnar and E. C. Cocking, *Biochem. J.* **91**, 29P (1964).
7. K. D. Neame, *Nature* **203**, 1067 (1964); *J. Neurochem.* **11**, 67 (1964).
8. V. Schaedel, E. R. Lochmann, and W. Laskowski, *Nature* **211**, 431 (1966).
9. J. C. Crawhall and S. Segal, *Biochim. Biophys. Acta* **121**, 215 (1966); *Biochem. J.* **99**, 19C (1966).
10. R. P. Spencer, K. R. Brody, and H. G. Mautner, *Nature* **207**, 418 (1965).
11. M. B. Williamson and G. H. Clark, *Arch. Biochem. Biophys.* **114**, 314 (1966).
12. A. Wainer, *Biochem. Biophys. Res. Commun.* **16**, 141 (1964): *Biochim. Biophys. Acta* **141**, 466 (1967).
13. B. Sörbo and L. Ewetz, *Biochem. Biophys. Res. Commun.* **18**, 359 (1965).
14. L. Ewetz and B. Sörbo, *Biochim. Biophys. Acta* **128**, 296 (1966).
15. A. Wainer, *Biochim. Biophys. Acta* **104**, 405 (1965).
16. J. B. Lombardini, T. P. Singer, and E. Kun, Pacific Slope Biochemical Conference Proc., Aug. 1966, Univ. of Oregon, Eugene, Oregon, p. 34.
16a. T. P. Singer and E. B. Kearney, *in* "*Amino Acid Metabolism*" (W. D. McElroy and B. Glass, eds.), p. 558. Johns Hopkins Press, Baltimore, Maryland, 1955.
17. H. S. Mason, *Advan. Enzymol.* **19**, 79 (1957).
18. D. Cavallini, R. Scandurra, and C. De Marco, *Biochem. J.* **96**, 781 (1965).
19. D. Cavallini, C. De Marco, and B. Mondovi, *J. Biol. Chem.* **239**, 25 (1958).
20. P. Turini, J. B. Lombardini, and T. P. Singer, Cofactors of cysteine oxidation. Pacific Slope Biochem. Conf. Proc., June. 1967, p. 41, Univ. Calif., Davis.
21. J. B. Lombardini, Ph.D. Thesis. Univ. Calif. Med. Center, San Francisco, 1968.
22. C. E. Mize, T. E. Thompson, and R. G. Langdon, *J. Biol. Chem.* **237**, 1596 (1962).
23. S. Ohmori, T. Shimomura, T. Azumi, and S. Mizuhara, *Biochem. Z.* **343**, 9 (1965),
24. T. Kuwaki and S. Mizuhara, *Biochim. Biophys. Acta* **115**, 491 (1966).
25. F. Tominaga, S. Kobayashi, I. Muta, H. Takei, and M. Ichinose, *J. Biochem* (Tokyo) **54**, 220 (1963); F. Tominaga, K. Oka, and H. Yoshida, *J. Biochem.* (Tokyo) **57**, 717 (1965).

26. A. Kamazawa, Y. Kakimoto, T. Nakajima, and I. Sano, *Biochim. Biophys. Acts* **111** 90 (1965).
27. H. C. Robinson and C. A. Pasternak, *Biochem. J.* **93**, 487 (1964).
28. T. Nakamura and R. Sato, *Nature* **198**, 1198 (1963).
29. J. C. Crawhall and S. Segal, *Nature* **208**, 1320 (1965).
30. A. Meister, P. E. Fraser, and S. V. Tice, *J. Biol. Chem.* **206**, 561 (1954).
31. W. D. Kumler and E. Kun, *Biochim. Biophys. Acta* **27**, 464 (1958).
32. E. Kun and D. W. Fanshier, *Biochim. Biophys. Acta* **27**, 659 (1958).
33. E. Kun and D. W. Fanshier, *Biochim. Biophys. Acta* **32**, 338 (1959).
34. D. W. Fanshier and E. Kun, *Biochim. Biophys. Acta* **58**, 266 (1962).
35. E. Kun and D. W. Fanshier, *Biochim. Biophys. Acta* **33**, 28 (1959).
36. E. Kun, *Biochim. Biophys. Acta* **25**, 137 (1957).
37. E. Kun and H. G. Williams-Ashman, *Experientia* **18**, 261 (1962).
38. E. Kun and D. W. Fanshier, *Biochim. Biophys. Acta* **48**, 187 (1961).
39. B. Sörbo, *Biochim. Biophys. Acta* **24**, 324 (1957).
40. M. Flavin, *J. Biol. Chem.* **237**, 768 (1962); F. Flavin and C. Slaughter, *J. Biol. Chem.* **239**, 2212 (1964).
41. J. W. Hylin and J. L. Wood, *J. Biol. Chem.* **234**, 2141 (1959).
42. E. Kun, C. Klausner, and D. W. Fanshier, *Experientia* **16/2**, 55 (1960).
43. C. J. A. Van Den Hamer, A. G. Morell, and I. H. Scheinberg, *J. Biol. Chem.* **242**, 2514 (1967).
44. D. Cavallini, D. De Marco, C. Modovi, and B. Mori, *Enzymologia* **22**, 161 (1960); A. Scioscia-Santoro, *Arch Biochem. Biophys.* **96**, 456 (1962).
45. B. Sörbo, *Acta Chem. Scand.* **12**, 1900 (1958).
46. M. Coletta, S. A. Benerecetti, and C. De Marco, *Italian J. Biochem.* **10**, 244 (1961).
47. J. R. Green and J. Westley, *J. Biol. Chem.* **236**, 3047 (1961).
48. J. Westley and T. Nakamoto, *J. Biol. Chem.* **237**, 547 (1962); R. Mintel and J. Westley, *J. Biol. Chem.* **241**, 3381 (1966).
48a. B. Sorbo, *Acta Chem. Scand.* **11**, 628 (1957).
49. H. E. Ganther, *Biochemistry* **5**, 1089 (1966).
50. J. M. Hsu, E. Buddenmeyer, and B. F. Chow, *Biochem. J.* **90**, 60 (1964).
50a. N. P. Sen and P. L. McGeer, *Can. J. Biochem. Physiol.* **44**, 286 (1966).
51. G. C. Mills and H. P. Randall, *J. Biol. Chem.* **232**, 589 (1958); *Arch. Biochem. Biophys.* **86**, 1 (1960).
52. G. Cohen and P. Hochstein: *Biochemistry* **2**, 1420 (1963).
53. D. Neubert, A. B. Wojtczak, and A. L. Lehninger, *Proc. Natl. Acad. Sci. U.S.* **48**, 1651 (1962).
54a. B. O. Christophersen, *Biochem. J.* **100**, 95 (1966).
54b. B. O. Christophersen, *Biochem. J.* **106**, 515 (1968).
55. F. E. Hunter, A. Scottjun, P. E. Hoffsten, J. M. Gebicki, J. Weinstein, and A. Schneider, *J. Biol. Chem.* **239**, 614 (1964).
56. P. Hochstein and G. Cohen, *Acta Biol. Med. Ger. Suppl.* **3**, 292 (1964).
57. A. S. Hill, A. Haut, G. E. Cartwright, and M. M. Wintrobe, *J. Clin. Invest.* **43**, 17 (1964): H. S. Jacob, S. H. Ingbar, and J. H. Jandl, *J. Clin. Invest.* **44**, 1187 (1965).
58. H. K. Prins, M. Oort, J. A. Loos, C. Zurcher, and T. Beckers, *Blood* **27**, 145 (1966).
59. H. S. Jacob and J. H. Jandl, *J. Biol. Chem.* **241**, 4243 (1966).
60. P. Needleman and F. E. Hunter, Jr., *Mol. Pharmacol.* **1**, 77 (1965).
61. P. T. Varandani and H. H. Tomizawa, *Biochim. Biophys. Acta* **113**, 498 (1966); **118**, 198 (1966).
62. H. M. Katzen and F. Tietze, *J. Biol. Chem.* **241**, 3561 (1966).

63. L. A. Heppel and R. J. Hilmoe, *J. Biol. Chem.* **183**, 129 (1950).
64. R. F. Colman and S. Black, *J. Biol. Chem.* **240**, 1796 (1965).
65. V. Massey and C. H. Williams, *J. Biol. Chem.* **240**, 4470 (1965).
65a. C. E. Mize and R. G. Langdon, *J. Biol. Chem.* **237**, 1589 (1962); C. E. Mize, T. E. Thompson, and R. G. Langdon, *J. Biol. Chem.* **237**, 1596 (1962).
66. S. H. Chang and D. R. Wilken, *J. Biol. Chem.* **240**, 3136 (1965); **241**, 4251 (1966).
67. R. N. Ondarza, *Biochim. Biophys. Acta* **107**, 112 (1965); J. Martiner, *Biochim. Biophys. Acta* **113**, 409 (1966).
68. J. C. Crawhall and C. J. Thompson, *Science* **147**, 1459 (1965).
69. J. E. McDonald and P. H. Hanneman, *New Engl. J. Med.* **273**, 578 (1965).
70. E. A. Barnsley and L. Young, *Biochem. J.* **95**, 77 (1965).
71. E. A. Barnsley, *Biochim. Biophys. Acta* **90**, 24 (1964).
72. M. M. Barnes, S. P. James, and P. B. Wood, *Biochem. J.* **71**, 680 (1959); T. Suya, I. Ohata, and M. Akagi, *J. Biochem. (Tokyo)* **59**, 209 (1966).
73. J. Booth, E. Boyland, and P. Sims, *Biochem. J.* **74**, 117 (1960).
74. M. K. Johnson, *Biochem. J.* **98**, 44 (1966).
75. E. Boyland and K. Williams, *Biochem. J.* **94**, 190 (1965).
76. E. Boyland and L. F. Chasseaud, *Biochem. J.* **98**, 13P (1966); **104**, 95 (1967).
77. T. Suya, H. Kamaoka, and M. Akagi, *J. Biochem. (Tokyo)* **60**, 133 (1966).
78. J. L. Sirlin, J. Jacob, and C. J. Tandler, *Biochem. J.* **89**, 447 (1963).
79. Z. Pokorny, J. Neuwirt, J. Borova, and K. Sule, *Acta Biol. Med. Ger. Suppl.* **3**, 300 (1964).
80. E. L. Gadsden, C. H. Edwards, A. J. Webb, and G. A. Edwards, *J. Nutr.* **87**, 139 (1965).
81. B. N. Langer, *Biochem. J.* **95**, 683 (1965).
82. E. J. Kuchinskas, *Arch. Biochem. Biophys.* **112**, 605 (1965).
83. S. Black, E. M. Harte, B. Hudson, and L. Wartofsky, *J. Biol. Chem.* **235**, 2910 (1960).
84. J. M. Adams and M. R. Capecchi, *Proc. Natl. Acad. Sci. U.S.* **55**, 147 (1966); M. R. Capecchi, *Proc. Natl. Acad. Sci. U.S.* **55**, 1517 (1966).
85. R. E. Webster, D. L. Engelhardt, and N. D. Zinder, *Proc. Natl. Acad. Sci. U.S.* **55**, 155 (1966).
86. B. F. C. Clark and K. A. Marcker, *Nature* **211**, 378 (1966); M. S. Bretscher and K. A. Marcker, *Nature* **211**, 380 (1966).
87. J. Trupin, H. Dickerman, M. W. Nirenberg, and H. Weissbach, *Biochem. Biophys. Res. Commun.* **24**, 50 (1966).
88. F. Schlenk, C. R. Zydek, D. J. Ehminger, and J. L. Dainko, *Enzymologia* **29**, 283 (1965).
89. S. K. Shapiro and D. J. Ehminger, *Anal. Biochem.* **15**, 323 (1966).
90. K. D. Gibson, A. Neuberger, and G. H. Tait, *Biochem. J.* **88**, 325 (1963).
91. S. H. Snyder and J. Axelrod, *Biochem. Biophys. Acta* **111**, 416 (1965).
92. E. F. Marshall, T. Chojnacki, and G. B. Ansell, *Biochem. J.* **95**, 30P (1965).
93. J. Axelrod and J. Daly, *Science* **150**, 892 (1965).
94. A. E. Pegg and H. G. Williams-Ashman, *Biochem. Biophys. Res. Commun.* **30**, 76 (1968).
95. S. H. Mudd, J. D. Finkelstein, F. Irreverre, and L. Laster, *J. Biol. Chem.* **240**, 4382 (1965).
96. S. Villa-Trevino, K. H. Shull, and E. Farber, *J. Biol. Chem.* **241**, 4670 (1966): K. H. Shull, J. McConomy, M. Vogt, A. Castillo, and E. Farber, *J. Biol. Chem.* **241**, 5060 (1966).

97. M. K. Turner and E. Reid, *Nature* **203**, 1174 (1965).
98. J. A. Stekol, *Advan. Enzymol.* **25**, 369 (1963).
99. Y. Natori and H. Tarver, *Biochim. Biophys. Acta* **107**, 136 (1965).
100. A. Peterkofsky and M. N. Lipsett, *Biochem. Biophys. Res. Commun.* **20**, 780 (1965); M. N. Lipsett, *J. Biol. Chem.* **240**, 3975 (1965).
100a. R. S. Hayward and S. B. Weiss, *Proc Natl. Acad. Sci. U.S.* **55**, 1161 (1966).
101. J. A. Carbon, L. Hung, and D. S. Jones, *Proc. Natl. Acad. Sci. U.S.* **53**, 979 (1965).
102. C. B. Anfinsen, C. Epstein, and R. Goldberger, *in* Information of Macromolecules, Symp. Proc., Rutgers Univ., N.J., Sept. 1962.
103. H. M. Katzen, F. Tietze, and D. Stetten, Jr., *J. Biol. Chem.* **238**, 1006 (1963).
104. S. Fuchs, F. DeLorenzo, and C. B. Anfinsen, *J. Biol. Chem.* **242**, 398 (1967).

CHAPTER 18

Metabolism of Porphyrins and Corrinoids

Bruce F. Burnham

I. INTRODUCTION

There are innumerable pigmented compounds of natural origin in the biosphere. This chapter deals with the metabolism of three groups of pigments that are chemically related in that they are all tetrapyrroles. The red pigment heme is representative of one group, chlorophyll, a green pigment, represents a second group, and the cobalamins, a third. Perhaps the bile pigments should be set aside as a separate group, but in the context of this chapter they are considered derivatives of heme. The noncyclic tetrapyrroles, phycoerythrins and phycocyanins, accessory pigments of the blue-green algae, have been regretfully omitted from the present writing. It is unfortunate that this is necessary since there are a number of recent developments that might profitably have been included. For a recent review of this field the reader is referred to the chapter by O'hEocha (1).

This chapter is a revision of the excellent chapter written by Granick and Mauzerall (2) for the Second Edition of this work. Although the scope has been broadened and much new material included, the author has relied heavily upon the work of these authors.

Since the previous writing, several important book and review articles dealing with tetrapyrroles have been published (3–13). The book by Lascelles (3) is an excellent treatment of the biological aspects of the tetrapyrroles, and the book by Falk (4) has become the standard work on the chemistry of porphyrins and metalloporphyrins. A comprehensive book on the chemistry of the bile pigments by With (14) will soon be available.

As early as 1880, Hoppe-Seyler (15) obtained information indicating that heme and chlorophyll were related to each other. Historically the cobalamins enter the picture very much later. However, not long after the isolation of vitamin B_{12} in 1948, chemical studies on its composition were undertaken demonstrating that it was a tetrapyrrole and, therefore, related to heme and chlorophyll.

During the 1920's and 1930's, a tremendous number of basic studies on the chemistry of heme and chlorophyll were carried out. The work by Willstatter and his colleagues (16) established that heme and chlorophyll were both tetrapyrroles. Fischer and his colleagues (17) refined this work and succeeded in defining the composition and the position of the various side chains attached to the tetrapyrroles, allowing a clear distinction to be made between the two compounds. The structure of heme was established in 1928 and the chlorophyll structure in 1934.

Several teams of workers contributed to defining the structure of vitamin B_{12}. Major contributions were made by Folkers, Lester Smith, Todd, and Dorothy Hodgkin and their co-workers (*18–24*). These investigators approached the problem along two lines. One involved classical chemical analysis, and the second made use of X-ray crystallography, never before used to probe the structure of such a complex molecule. The outstanding success obtained has stimulated further work on far more complex molecules.

The structure of heme was confirmed by total synthesis by Fischer and his colleagues in 1929. The structure of chlorophyll has more recently been confirmed by total synthesis by Woodward and his co-workers (*25*). The total synthesis of vitamin B_{12} is now under way in the laboratory of Woodward.

A. Scope of Chapter

This chapter on tetrapyrroles will consider at length and in detail the metabolic aspects of tetrapyrrole chemistry. Our knowledge concerning the cobalamins has increased by several orders of magnitude since publication of the previous edition of this chapter. It seems proper in light of these developments to include the cobalamins in the present discussion along with the iron and magnesium tetrapyrroles. A discussion on the structure and function of hemoproteins is outside the scope of the chapter. The volume of research in this area has been tremendous during recent years, and it seems somewhat futile to attempt to review the field meaningfully in a chapter of this length.

1. SUMMARY OF BIOSYNTHESIS

The classic work by Shemin and his colleagues led to the recognition of the early precursors of the porphyrin ring. δ-Aminolevulinic acid (ALA) arising from glycine and succinyl-CoA was shown to be the primary building block. Porphobilinogen (PBG), the dicarboxylic monopyrrole formed by the condensation of two ALA molecules, was shown to be the first pyrrolic intermediate.

The similarity in the structures of heme, chlorophyll, and the cobalamins suggested that these compounds might well be synthesized along similar lines. Some of the clearest evidence for a common biosynthetic pathway for heme and chlorophyll came from the work of Granick. Working with *Chlorella* mutants unable to synthesize chlorophyll, he found that protoporphyrin IX and Mg-protoporphyrin accumulated.

Insertion of iron into protoporphyrin IX yields heme, and incorporation of magnesium yields Mg-protoporphyrin. Only one step past the formation of Mg-protoporphyrin leading to chlorophyll has been demonstrated at the enzymic level. This is the esterification of the propionic acid side chain at position 6 by S-adenosylmethionine to yield Mg-protoporphyrin-monomethyl ester.

It is known that ALA and PBG are precursors of vitamin B_{12}. There is some evidence that uroporphyrinogen III is the branch point leading to the cobalamins, though this has not yet been unequivocally demonstrated.

2. SUMMARY OF REGULATION

As our knowledge of the path of biosynthesis of the tetrapyrroles has increased, so has our ability to investigate the factors that control the system. Abnormalities in the control of porphyrin metabolism have been recognized for a long time. That this is true is probably not due so much to the prevalence of abnormalities as it is to the fact that colorful results are obtained when abnormalities exist, e.g., red urine and bacterial culture filtrates.

As indicated, the three tetrapyrroles under consideration arise from common precursors via branches in the biosynthetic chain. Any time that branches occur in biosynthetic chains, the problem of control becomes both more complicated and interesting.

Consider the control of biosynthesis of threonine and lysine which arise from a branched pathway. These amino acids are produced in roughly equal amounts. If there were slight defects in one branch of the chain, say 1%—so that there was a slight overproduction of one amino acid and an underproduction of the other—it is likely that this would go essentially unnoticed.

The branched pathways in tetrapyrrole biosynthesis, however, present a more complicated problem. Take, for example, the most interesting and complicated case which is presented by some of the photosynthetic bacteria. The relative abundance of the three tetrapyrroles in *Rhodopseudomonas spheroides* is: bacteriochlorophyll, 25,000; total heme, 300; cobalamin, 13 (measured as vitamin B_{12}) μmoles/gm dry weight of organism. These figures make it clear that a very high degree of metabolic traffic regulation is required to allow such tremendous differences in the end products to occur. A 1% slippage in production between bacteriochlorophyll and cobalamin would obviously play havoc with the normal metabolism of the cell.

There is also a coordination problem in the biosynthesis of hemoproteins. The problem is one of allowing heme synthesis to proceed at a rate to match the requirement for hemoprotein formation.

Studies on the control of porphyrin biosynthesis have shown that enzyme repression and feedback inhibition are employed as regulatory mechanisms. Many factors such as oxygen tension and light intensity (in photosynthetic organs) play important roles.

3. SUMMARY OF FUNCTION OF TETRAPYRROLES

It is possible to incorporate almost any metal ion of reasonable size into the tetrapyrrole nucleus. The tetrapyrroles isolated from biological systems (excluding petroleum products), however, are found to bind only copper, zinc, cobalt, iron, and magnesium. Of this group only the last three are known to be biologically functional. In most cases when a zinc or a copper tetrapyrrole is isolated from a biological source, it can be shown to be an artifact; the porphyrin has simply chelated metal ions which were present as trace contaminants of the system. Turacin, the copper complex of uroporphyrin III which is found in the flight feathers of the Turaco bird, is an exception in that it is biologically formed, though it has no apparent function other than decoration.

Considering now the functional tetrapyrroles, the iron complexes (hemes) serve as prosthetic groups of the cytochromes and hydroperoxidases and the oxygen-carrying heme proteins. Basically, they function in oxidative metabolism, i.e., in the transport of electrons to activate oxygen. Water is the end product of the sequence of reactions. Free energy which can be utilized by cells is released in this process. The magnesium dihydroporphyrins, e.g., chlorophyll, on the other hand absorbs light (energy). This energy is utilized to decompose water, giving rise to a reductant AH and an oxidant BOH. The oxidant breaks down to yield oxygen and the reductant serves as a source of reducing power (energy) that can be utilized by the cell or, as in the case of photosynthetic bacteria, the absorbed energy is almost exclusively utilized to generate ATP. The cobalamins have been demonstrated to function coenzymically in a number of rather diverse reactions such as isomerizations, oxidation-reductions, and group transfer reactions.

4. SUMMARY OF TETRAPYRROLE DEGRADATION

In higher organisms the catabolism of iron porphyrins is an elaborate and complicated process. The protein part of hemoproteins apparently is hydrolyzed to the component amino acids which enter the amino acid pool and may be reutilized for protein synthesis. The iron atom of the heme moiety is removed and saved for subsequent use by the organism. The porphyrin is oxidized, the macrocyclic ring is opened up, and the noncyclic tetrapyrrole is excreted into the intestine as bile pigment. In the intestine the bile pigment is further metabolized by microorganisms. The microorganisms reduce the bile pigments to colorless products which are ultimately excreted. In carrying out these reactions, it seems, the microorganisms may be using the bile pigments as terminal electron acceptors in their highly anaerobic environment.

Although quantitatively more chlorophyll is degraded than heme, little or nothing is known about the reactions involved or even the nature of the end products. Similarly, little is known about the degradation of the cobalamins.

5. Summary of Abnormal Tetrapyrrole Metabolism

There are numerous examples of abnormal tetrapyrrole metabolism. Those treated in the present chapter all involve the overproduction of porphyrins and their precursors. The anemias and other reflections of abnormal tetrapyrrole metabolism are beyond the scope of the present work.

Recent studies on metabolic regulatory mechanisms, in general, and studies on the biosynthesis of porphyrins combine to give us some understanding of abnormal porphyrin biosynthesis. Most of the evidence now indicates that the major lesion causing overproduction of porphyrins and precursors is connected with the enzyme ALA synthetase. However, as yet, the nature of the lesion is not understood.

B. General Properties of Tetrapyrroles

The structures of heme, chlorophyll, and vitamin B_{12} are given in Fig. 1. (Common usage defines heme as ferroprotoporphyrin IX, hemin as ferriprotoporphyrin IX chloride, and hematin as ferriprotoporphyrin IX hydroxide. Even though hemin is the starting material, or the isolated product of many iron-protoporphyrin investigations, it clearly does not remain as ferri-protoporphyrin chloride throughout. Rather than becoming involved in considerations about the oxidation state, or the identity of the extra ligands, in the context of this chapter the term heme is defined as ironprotoporphyrin IX. Nothing is intended to be implied concerning the oxidation state of the iron. When the latter is important, it is indicated by ferri- or ferroprotoporphyrin IX). Examination of the structures and comparison with that given in Fig. 2 shows that heme and chlorophyll are modified derivatives of porphine and

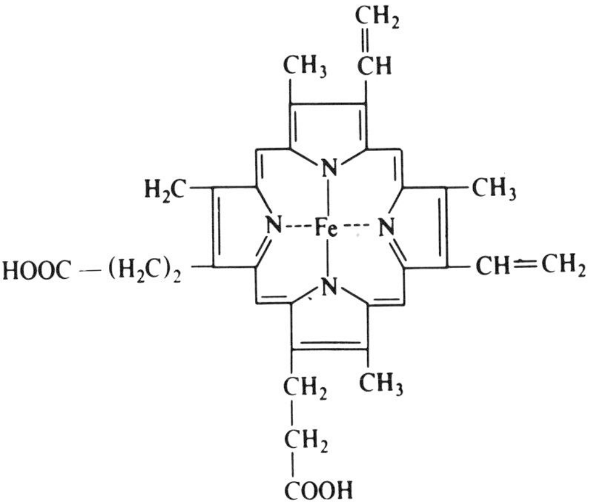

Heme

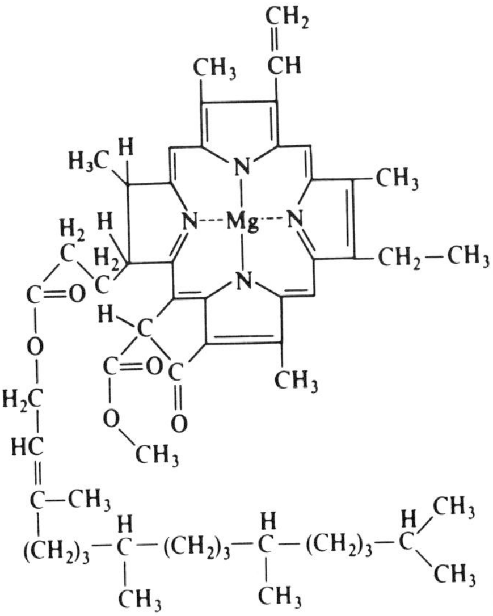

Chlorophyll a

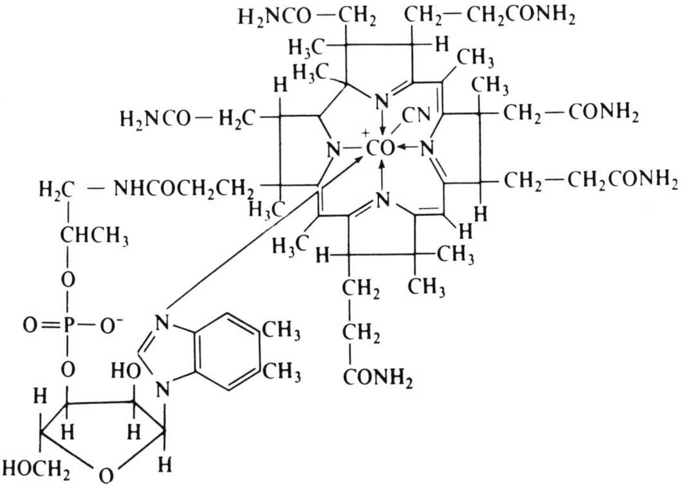

Vitamin B$_{12}$

FIG. 1. Structures of three representative metal tetrapyrroles.

vitamin B_{12}, a derivative of corrin. The similarity in properties between porphine, corrin, heme, chlorophyll, and cyanocobalamin decreases markedly in the order written. This decrease in similarity is not due so much to the metal atoms or to the various substituents at the β positions of the pyrroles (Fig. 3) as it is to the alterations in conjugation within the macrocyclic ring.

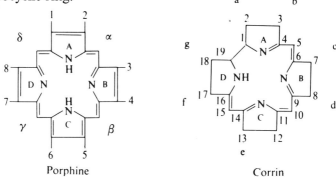

FIG. 2. Numbering system for porphine and corrin. Observe that two different conventions are followed. Some porphyrins of interest bear the following substituents:

uroporphyrin III 1,3,5,8 acetate; 2,4,6,7 propionate
coproporphyrin III 1,3,5,8 methyl; 2,4,6,7 propionate
protoporphyrin IX 1,3,5,8 methyl; 2,4 vinyl; 6,7 propionate
mesoporphyrin IX 1,3,5,8 methyl; 2,4 ethyl: 6,7 propionate
hematoporphyrin IX 1,3,5,8 methyl; 2,4 hydroxyethyl; 6,7 propionate
deuteroporphyrin IX 1,3,5,8 methyl; 2,4 hydrogen; 6,7 propionate.

1. CHEMICAL STRUCTURES

The tetrapyrroles can be conveniently broken down into several classes of compounds for separate consideration. Porphyrins by definition are metal-free cyclic tetrapyrroles. The rings are designated by the letters A, B, C, D, and they are joined by methine bridges α, β, δ, γ. The hydrogens in the β position of the pyrroles (Fig. 3) of porphine are substituted by alkyl groups in porphyrins.

FIG. 3. Numbering system for pyrroles. The pyrrole precursor of natural tetrapyrroles, porphobilinogen, is 2-aminomethyl, 3-acetate, 4-propionate.

The porphyrins are approximately planar molecules measuring about 10 Å on each side. The recent X-ray work from Hoard's laboratory (26–28), however, indicates that the porphine skeleton is somewhat more flexible than has been generally realized. This flexibility allows the pyrrole rings to ruffle out of the mean plane of the ring. Furthermore, the metal atom in some metalloporphyrin derivatives is considerably out of the plane of the porphyrin ring (28). This has also been noted by others (29). The macrocyclic tetrapyrrole structure as it occurs in porphyrins has 11 double bonds. The high stability of porphyrins is largely the result of resonance stabilization. Valence bond theory attributes the stability of the porphyrins to the large number of resonance hybrids of equal or nearly equal energy. Molecular orbital theory explains this stability as the result of delocalization of the electrons (double bonds) allowing them to move through the planar conjugated ring system, i.e., they are no longer restricted to the atomic orbital and now move in the molecular orbital. The more extended the conjugated system becomes, the smaller is the energy difference between the normal and excited states. Thus light absorption occurs at progressively longer wavelengths.

The stability of the porphyrin ring system is a thermodynamic stability, i.e., the conjugated planar cyclic arrangement has a greater heat of formation from the elements than has the hypothetical nonconjugated system. In addition, the porphyrin ring is chemically stable. It is stable to strong acids and bases and is substituted only with difficulty. All of the naturally occurring tetrapyrroles are fully substituted on the eight β positions of the ring, thus decreasing further the possibility of substitutions. Nonetheless, tetrapyrroles are highly susceptible to light-catalyzed reactions. They readily form very stable chelates with transition metals. However, magnesium, iron, and cobalt are the only metals found in the biologically functional tetrapyrroles.

Corrin, containing 6 double bonds, is the designation for the basic ring system found in vitamin B_{12} and its derivatives (30). A corrin differs from a porphine in that one of the methine bridge carbons (γ) is absent, and two of the pyrrole rings (A, D) are directly joined at the α positions. A corrin is also more reduced than a porphine. Corroles are defined as tetradehydrocorrins (i.e., minus 8 H), and contain 10 double bonds. The original designation of corroles as pentadehydrocorrins by the IUPAC-IUB Commission on Biochemical Nomenclature is incorrect (30).

Corroles (Fig. 5f) and porphines (Fig. 5d), having similar degrees of conjugation, etc., are, as would be expected, quite similar in their chemical and physical properties. Simple corroles have been chemically synthesized in the laboratories of Eschenmoser (31) and Johnson (32).

Cobyrinic acid (Fig. 4), a degradation product, and probably a biosynthetic precursor of the cobalamins, is a highly substituted corrin. It is more reduced than a corrole and, of particular importance, it has a methyl group attached at the α position of ring A, i.e., to carbon 1. This extra methyl group at that position effectively blocks cyclic conjugation. As a result, it is possible to reduce the central cobalt atom to the $+1$ oxidation state. When one attempts to reduce the cobalt in a cobalti-porphyrin past the $+2$ oxidation state, the porphyrin ring becomes reduced and the cobalt drops out. This is likely of importance in the biosynthesis of the cobalamin coenzymes (Section II, E, 3). Very recently the synthesis of 1-methyltetradehydrocorrins (1-methyl corrole) has been accomplished in the laboratory of Johnson (*33,34*).

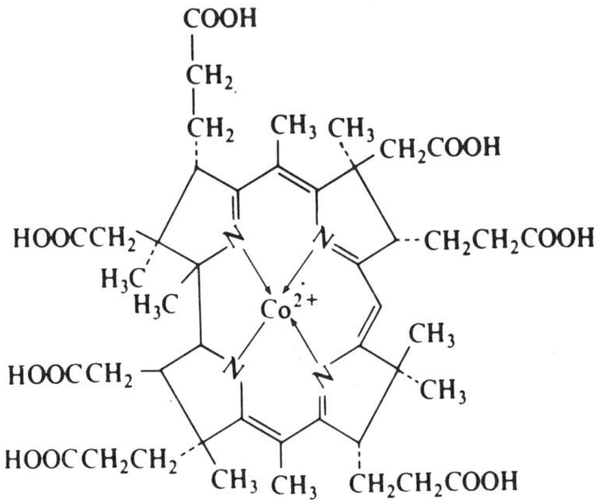

FIG. 4. Structure of cobyrinic acid.

2. SPECTRAL PROPERTIES AND NOMENCLATURE

There is a basic similarity in the absorption spectrum of all tetrapyr-roles with similar degrees of conjugation. Details of the spectral prop-erties are discussed in the section dealing with the examples of different types of tetrapyrroles.

a. Hexahydroporphyrins. It is possible to add 6 hydrogen atoms to the porphine nucleus. These hydrogens attach at the methine bridge and pyrrolinine nitrogen positions. Such a reduced porphyrin is known as a porphyrinogen. This structure is shown in Fig. 5a. These compounds have no conjugation, are quite unstable, and are colorless. Light absorption does occur in the 200 mμ region where monopyrroles characteristically absorb. It is worthwhile noting here that, as a tetra-pyrrole moves along the biosynthetic assembly line, it remains in the porphyrinogen form most of the way.

b. Tetrahydroporphyrins. A tetrahydroporphyrin (Fig. 5b) is essentially two isolated pyrroles joined to a dipyrrilmethene. The absorption maximum at 500 mμ with a molar extinction coefficient of 6×10^4 is similar to that of dipyrrilmethenes (*35*). The tetrahydroporphyrins are unstable in oxygen and in light, in these cases being converted to the porphyrins.

c. Dihydroporphyrins. There are two types of dihydroporphyrins, namely, the phlorins and the chlorins. Structures and spectra are shown in Fig. 5c and Fig. 5e. In the case of the phlorins, reduction has occurred at a bridge carbon position, and thus the cyclic conjugation is interrupted. Phlorins have absorption bands around 440 mμ and 740 mμ, with molar extinction coefficients of 6×10^4 and 2×10^4, respectively. Woodward (*25*) used these partially reduced porphyrins as intermediates in the chemical synthesis of chlorophyll. Mauzerall (*36*) has shown that they are the photoreduction products of porphyrins.

In the case of the chlorins, reduction has occurred at the β positions of one of the pyrrole rings. Cyclic conjugation is retained, and these compounds have much more resemblance to the porphyrins than when the cyclic conjugation is destroyed. Chlorins have major absorption bands around 400 mμ and 680 mμ with molar extinction coefficients of 2×10^5 and 5×10^4, respectively.

Among the porphyrin derivatives of biological interest a number of partially reduced compounds are found. The chlorophylls and corrins are examples of such partially reduced porphyrin derivatives. Chlorophyll a, for instance, has 2 hydrogen atoms added at the β positions of ring D and thus is a chlorin; bacteriochlorophyll is reduced similarly in both rings D and B and thus is a dihydrochlorin.

d. Effects of Substitutions at the β-Positions of Porphyrins and Corrins. Within limits, the effect of substituents at the β position of tetrapyrroles is minor compared to those modifications already discussed. These effects can be categorized as follows: effect on spectral properties; effect on redox properties of metallo derivatives; effect on ionization of ring nitrogens. As these effects are correlated, it is convenient to cite examples and to look at the effects on all of the above: an aliphatic side chain has little interaction with the conjugated ring sytem and, other than changing the solubility properties has little effect; electrophilic side chains, e.g., vinyl or aldehyde groups, shift the light absorption to the red, the pK of the pyrrolinine nitrogen is decreased, and $E°$ of the metalloporphyrin becomes more positive. If a carbonyl or other such electrophilic group is insulated from the ring by CH_2, the effect is markedly diminished.

The symmetry of substitution is of great importance in determining the net effect of substituents in the β position. This is best illustrated by

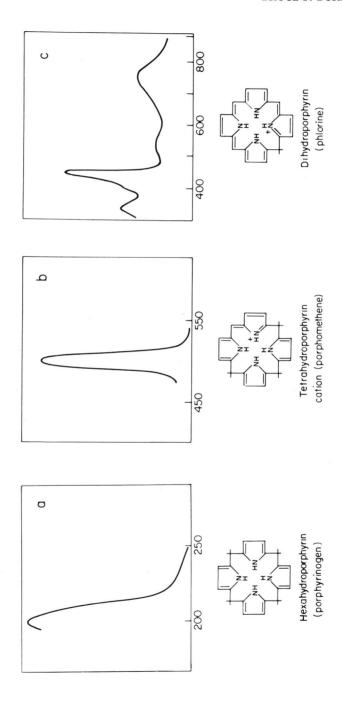

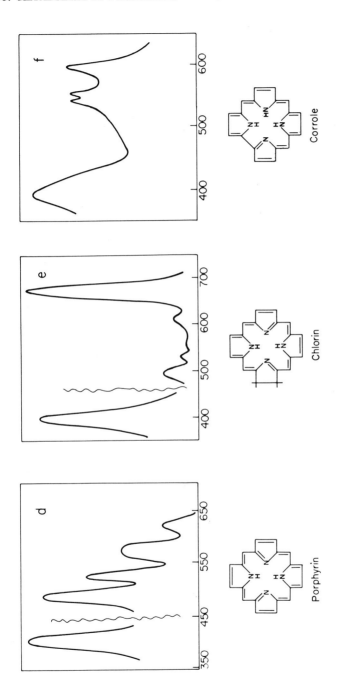

FIG. 5. Basic structures and generalized spectra of porphyrins, reduced porphyrins, and corrole. The wavelength, in millimicrons, is shown on the ordinates. Approximate molar extinction coefficients are given in the text.

changes in the spectral characteristics of the various porphyrin derivatives. There are four types of spectra (Fig. 6) and they will be discussed in turn.

Etio type (Fig. 6a). Etio-type porphyrins have four bands in the visible region of the spectrum, and the strength of the bands decreases going toward the red, i.e., band $IV > III > II > I$. These porphyrins have substituents such as methyl, ethyl, acetic acid, or propionic acid groups in six or more of the β positions. The common porphyrins such as uro-, copro-, deutero-, meso-, and hematoporphyrins are all of the etio type.

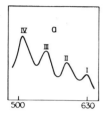 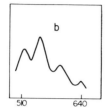 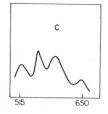 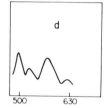

Fig. 6. Generalized spectra in the visible of: (a) etio-type, (b) rhodo-type, (c) oxorhodo-type, (d) phylo-type porphyrins. The wavelength, in millimicrons, is shown on the abcissas. The approximate molar extinction coefficient for the major band in each spectrum is 1×10^4.

Rhodo type (Fig. 6b). A single electrophilic group such as a carbonyl group or an acrylic acid side chain both shifts the absorption bands to longer wavelengths and causes band III to become the most intense. Thus the band order is $III > IV > II > I$. If two rhodo groups exist in the same molecule, the symmetry of substitution again becomes very important. If the groups are on adjacent pyrroles, they cancel each other and the spectrum is of the etio type. Two such adjacent groups, however, have an additive effect on the shift of the absorption bands to the red. Chlorocruoroporphyrin, derived from chlorocruorin, the oxygen transport pigment of certain polychaete worms, exhibits a rhodo-type spectrum.

Oxorhodo type (Fig. 6c). If electrophilic substituents are located on opposite pyrroles, then the shift of light absorption to the red is enhanced and the oxorhodo spectrum appears. The band order in oxorhodo porphyrins is $III > II > IV > I$. The porphyrin derived from heme a exhibits an oxorhodo-type spectrum.

Phylo type (Fig. 6d). Porphyrins with four or more unsubstituted β positions may have either etio-type spectra or phylo-type spectra depending on the symmetry of substitution. The etio-type spectrum occurs when the porphyrin is symmetric along the x and y axes. But, when the x and y axes are asymmetric, the phylo-type spectrum results. Thus a symmetrical tetraethylporphyrin will be of the etio type, but 1,4-diethyl-2,3-dimethyl porphine will have a phylo-type spectrum. The bands in phylo-type spectra do not decrease in order but go IV > II > III > I. A phylo-type spectrum is also observed when there is a single substitution at a methine bridge carbon, as is found in the chlorophyll derivative, γ-phylloporphyrin, and in the chlorobium chlorophylls.

The very intense absorption band found in the 350–450 mμ region called the Soret band is found in all tetrapyrroles having a complete cyclic conjugation. Substitutions such as discussed above will cause a shift in the absorbing wavelength and an alteration of the extinction, but the effect is relatively small compared to the dramatic changes seen in the visible region of the spectrum. The absorption of the Soret band is so strong ($E_m = 1$–5×10^5) that it is way off scale on the spectrophotometer when a tetrapyrrole solution is sufficiently concentrated to give

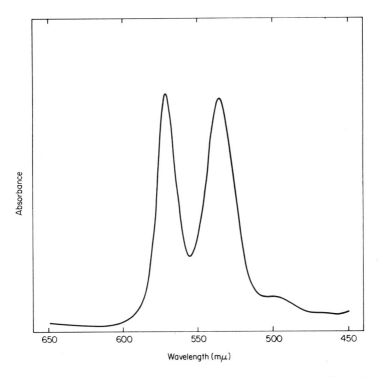

FIG. 7. Detailed visible absorption spectrum of Zn-meso-porphyrin dimethyl ester in chloroform. The molar extinction coefficient for the two bands is about 1.9×10^4.

good bands in the visible region of the spectrum ($E_m = 10^4$). Conversely, when a tetrapyrrole solution is sufficiently dilute to allow spectrophotometric measurement of the Soret band, the visible bands are generally little more than undulations of the base line.

The three types of tetrapyrroles with established biological function are all metal-substituted tetrapyrroles. The hemes contain iron, the chlorophylls contain magnesium, and the cobalamins contain cobalt. The incorporation of a metal atom into the center of a porphyrin nucleus has a marked effect on the properties of both the metal atom and the tetrapyrrole. The effect of the metal is most noticeable on the spectral properties of the tetrapyrrole (Fig. 7). Typical visible absorption spectra of free porphyrin and metalloporphyrin are given in Figs. 7 and 8. Incorporation of a metal into a porphyrin results in a compound having a two-banded visible spectrum (Fig. 7). The bands of metalloporphyrins are not designated by Roman numerals, as are the metal-free porphyrins; instead, Greek letters are used. The band absorbing

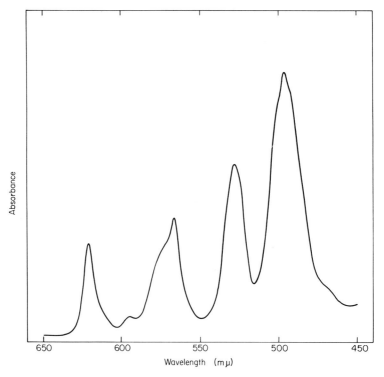

FIG. 8. Detailed visible absorption spectrum of coproporphyrin tetramethyl ester in chloroform. The molar extinction coefficient of band IV is 1.4×10^4.

at the longest wavelength is the α band. The relative intensity of the two visible bands is a function of the metal incorporated. In general, if the metal is very tightly bound, e.g., copper, the β band is smaller than the α band. Those metals that are bound with intermediate strength, e.g., zinc and magnesium, give rise to bands of approximately equal intensity. Metals that are weakly bound have spectra with the β band greater than the α band. There are exceptions to this generalization, tin porphyrins being an example.

It should be noted that there is some similarity between the spectra of metalloporphyrins (Fig. 7) and of porphyrin di-cations (Fig. 9). When

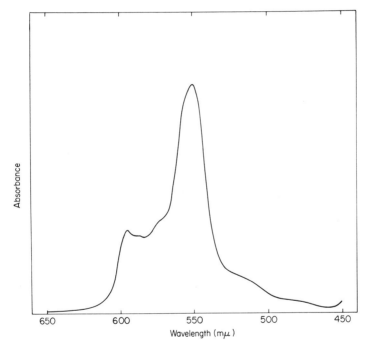

FIG. 9. Detailed visible absorption spectrum of the di-cation of coproporphyrin in 0.1 N HCl. The molar extinction coefficient of band II is 1.68×10^4.

the basic pyrrolinine nitrogens are protonated in acid solution, the di-cation has a two-banded spectrum. Band I is about 600 mμ and is much less intense than band II at about 550 mμ. Recent evidence has been obtained demonstrating that the "sitting atop" metal-porphyrin complexes described by Fleischer and Wang (*36a*) are in fact salts of porphyrin di-cations (*36b*). These complexes have spectra similar to, but distinctly different from, the di-cation formed by addition of acids such as HCl. The difference in the spectra is ascribed to the close association of the gegen-ion (alcohol-solvated metal ion) with the di-cation, formed from protons released by solvated metal in the anhydrous solvent required for demonstration of these complexes. Porphyrin di-anions have spectra very similar to the di-cation spectra. The similarity of these two spectra and the similarity to metalloporphyrin spectra is explained on grounds of symmetry. The acid and base ionizations of porphyrins are extensively reviewed by Falk (4) and by Dempsey et al. (37).

The metal-free cyclic tetrapyrroles all exhibit a very intense red fluorescence when irradiated with ultraviolet light. This property is of particular value as a quantitative tool for measuring porphyrins in biological systems. The fluorescence of a tetrapyrrole, however, is very sensitive to its molecular environment, and one must use great caution when doing quantitative work to ensure that proper standards are employed. This, of course, is true for any fluorescent molecule.

It is probably true that spectral studies in the visible region still predominate in the literature. However, of late, as instrumentation and theory (38–41) have advanced, more work is being carried out in other areas. These areas include infrared (42–44), nuclear magnetic resonance (45–50), electron spin resonance (51–55), and Mössbauer spectroscopic studies (56–59). Some specific studies will be cited in later sections. Extensive discussion of these areas is outside the scope of the present work, and the reader is referred to the above references for recent advances.

II. BIOSYNTHESIS OF TETRAPYRROLES

A. Demonstration of Common Pathways

Figure 10 summarizes the path of biosynthesis common to the three functional tetrapyrroles. Granick (60,61) provided the first clear experimental evidence that chlorophyll and heme share a common biosynthetic pathway. In order to isolate compounds that were intermediates along the path of chlorophyll synthesis, Granick (62) prepared mutants of Chlorella vulgaris by X-irradiation. The mutants that were unable to synthesize chlorophyll were examined for other tetrapyrrole compounds. One mutant accumulated relatively large amounts of protoporphyrin IX. Another accumulated Mg-protoporphyrin IX and Mg-2-vinyl-pheoporphyrin a_5. In more recent work, Granick has obtained yet another mutant that accumulates protoporphyrin monomethyl ester and

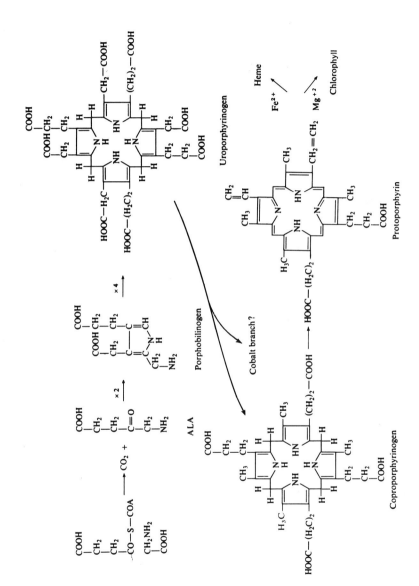

FIG. 10. Summary of the path of porphyrin biosynthesis from glycine and succinyl-CoA to protoporphyrin IX.

Mg-protoporphyrin monomethyl ester (63). The finding of protoporphyrin IX in a mutant unable to make chlorophyll indicated that heme and chlorophyll share a common biosynthetic pathway through protoporphyrin. It should be noted, however, that the accumulation of a compound by a mutant organism does not in itself prove that it is a biosynthetic precursor of some end product whose ultimate synthesis is blocked. An example of this problem is provided by the work of Granick and Bogorad (64,65), again working with *Chlorella* mutants. They found a mutant that accumulated hematoporphyrin. Subsequent work has shown that hematoporphyrin is not a biosynthetic precursor of protoporphyrin (66,67). Thus the hematoporphyrin produced by the *Chlorella* mutant must be considered a metabolic artifact whose significance is uncertain.

Evidence that the cobalamins also share a common pathway with heme and chlorophyll past the PBG stage has been more difficult to obtain. Experiments with mutants have led to the isolation and identification of numerous tetrapyrroles (68–72). All of these compounds, however, are well along on the way to vitamin B_{12}, they all contain cobalt, they are all corrins, and no information about common intermediates past porphobilinogen has resulted from this approach. There is, in fact, some evidence that corrins do not share a common pathway with the porphyrin derivatives. This evidence has come from experiments with *Propionibacterium shermanii*, where it has been shown that porphyrin and corrin synthesis can be independently altered by growth conditions and by inhibitors (73,74). More recent evidence, obtained in two laboratories, however, indicates that corrins and porphyrins do share a common pathway (75,77). In both cases *Clostridium tetanomorphum* was the experimental microorganism. The *Clostridia* are known to be unable to make heme enzymes, and the only tetrapyrroles known to be synthesized are the cobalamins. ALA, a precursor of vitamin B_{12} in *Actinomycetes* (78,79) and in *Clostridia* (75), was supplied to nongrowing cell suspensions of the organism. After treatment of the anaerobic incubation mixture with I_2 to oxidize porphyrinogens to porphyrins, it was possible to isolate uroporphyrin. Porra (77) claims this to be the type III isomer.

If porphyrins and corrins do share a common pathway, the branch must occur at the uroporphyrinogen level. Since the *Clostridia* cannot actually have a branch at this point, the occurrence of uroporphyrin (-ogen) is presumptive evidence that it is an intermediate in the biosynthesis of corrins. ALA dehydrase, however, is fairly active in *C. tetanomorphum*, and it is possible that the ALA is being converted enzymically to PBG, which then undergoes nonenzymic condensation to yield uroporphyrinogen. If the condensation had been nonenzymic, however, both type I and III isomers would be expected. The question of uroporphyrinogen as a vitamin B_{12} intermediate remains partially open and awaits proof by isotopic methods. Recently, a report has appeared claiming the isolation of a heme protein from *Clostridium pasteuranium* (*80*).

B. Biosynthesis of Common Precursors

The key necessary to unlock the mystery of common precursors is, of course, the identification of the precursors. Our knowledge of the early stages of porphyrin synthesis comes largely from the classic work of Shemin (*81–85*). In the first of these studies, ^{15}N amino acids were fed to a human subject and the incorporation of the ^{15}N atoms into newly synthesized heme was measured (*81*). While ^{15}N from most sources was incorporated to some extent, ^{15}N from glycine was incorporated much more directly, i.e., with minimal dilution, indicating that it was a direct precursor.

Experiments with ^{14}C-labeled glycine showed that the α carbon of glycine ends up in two different places in the porphyrin, namely, in the α position (Fig. 3) of each pyrrole and in the methine bridge positions (*83–87*). In further work, Shemin and Wittenberg (*88*) supplied duck erythrocytes with acetate labeled with ^{14}C in the methyl group or in the carboxyl group. After incubation, heme was isolated and degraded in a stepwise manner and the radioactivity of individual carbon atoms from each pair of pyrrole rings (A + B) and (C + D) was determined. The results showed that acetate labeled in the carboxyl position labeled the carboxyl groups of the heme. Little activity was found in the rest of the molecule. Methyl-labeled acetate, on the other hand, promoted a rather heavy labeling in the pyrrole rings and in the side chains with the exception of the carboxyl positions. The assumption was made that acetate first had to enter the citric acid cycle to form an asymmetric four-carbon acid which then reacted with glycine to give the porphyrin precursor molecule. The asymmetry was required by the ^{14}C-labeling pattern in the heme degradation products. It is now known that the asymmetric compound is succinyl-CoA (*89,90*).

Knowing these things, it was possible to predict the structure of the condensation product (*91*), namely α-amino-β-ketoadipic acid, which, being a β-keto acid, readily decarboxylates to yield ALA (*92*). When synthetic ALA was tested in the duck erythrocyte system it was found to be efficiently converted to heme (*91,93,94*). Thus ALA was established as the first precursor directly on the way to porphyrin.

1. BIOSYNTHESIS OF ALA

Several publications appeared in 1958 demonstrating for the first time the synthesis of ALA by biological systems. Both avian erythrocytes (90,95,96) and photosynthetic bacteria (97,98) were employed in these studies.

The enzyme ALA synthetase catalyzes the condensation of glycine and succinyl-CoA. Pyridoxal-P is apparently the only cofactor required (98,99). Nutritional studies by several groups of workers implicated pyridoxal-P as a cofactor in heme synthesis prior to the actual demonstration of its participation (100–102). Succinyl-CoA can be chemically prepared and added to the incubation mixture (90,98), or it can be generated in the reaction mixture from succinate, CoA, and ATP by succinic thiokinase (103,104). When it is generated in the reaction mixture during the course of the incubation, some of the troubles encountered in systems with deacylases can be at least partially avoided (105).

ALA synthetase is quite unstable (98,103) and, as a general rule, difficult to detect even in organisms and tissues where it almost certainly must exist. The level of ALA synthetase in normal guinea pig liver was barely detectable (105). The liver of guinea pigs with experimental porphyria, however, had sufficient enzyme for measurement (105). Marver et al. (106–108) using a new assay technique modified from that of Granick and Urata, have recently been able to measure ALA synthetase in normal liver tissue.

The most active ALA synthetase is found in the *Athiorhodacae* (nonsulfur purple photosynthetic bacteria) (90,98,109). The enzyme can not be detected in the *Thiorhodacae* (purple sulfur photosynthetic bacteria, such as *Chromatium*) which makes approximately as much total tetrapyrrole as the *Athiorhodacae*, nor can it be detected in iron-deficient microorganisms actively excreting porphyrins, such as *Arthrobacter globiformis* (110). Lascelles (111) has recently succeeded in demonstrating ALA synthetase in *Spirillum itersonii*. This is the only microorganism outside of the genus *Athiorhodacae* in which the enzyme has been found.

Why is this enzyme so difficult to work with? A complete answer to this question is not yet available, but enough facts have emerged to allow both a partial answer and some speculation. Some of these factors are first listed and then discussed. (a) The enzyme is quite unstable both in crude extracts and in partially purified form. (b) The enzyme is inhibited by a number of naturally occurring substances in the cell as well as by the usual protein reagents, e.g., PCMB (90). Heme and inorganic iron at low concentrations inhibit the enzyme (*103*), and a factor was described by Kikuchi *et al.* in aerobically grown *Rhodopseudomonas spheroides* that had inhibitory activity. (c) Some systems have a very active deacylase (*105*) which competes with the ALA synthetase for succinyl-CoA and, since the deacylase is more active, succinyl-CoA becomes depleted. (d) The cofactor requirement seems to vary somewhat from organism to organism. This is not to suggest that there is any question about the requirement for pyridoxal-P, or for CoA in systems generating succinyl-CoA from succinate, CoA, and ATP, but there apparently is a difference in how tightly bound the cofactors are (*98,108*). Thus it is necessary to add exogenous cofactors in some cases. The role of iron in the reaction is not yet clear. It is an inhibitor of the enzyme from *Athiorhodacae*, and there is evidence for its participation as a cofactor of the enzyme from avian erythrocytes (*112,113*).

The observed instability of the enzyme can perhaps be considered from two points of view, namely, the thermodynamic stability of the enzyme as a protein and the metabolic instability. The apparent spontaneous denaturation of the enzyme as a function of time, even when stored at $-15°$, seems to indicate thermodynamic instability. This is apparently an inherent property of the enzyme dictated by its structure. Not too much can be said about this aspect of the problem until the enzyme is available in pure form so that appropriate studies can be performed. The observation by Lascelles (*111*) that ALA synthetase from *Spirillum itersonii* can be somewhat stabilized by 10% ethanol is a step in the right direction.

The so-called metabolic instability of the enzyme is a more complex problem. For one thing, the enzyme has a rather high turnover rate. Granick (*114*) measured the half-life of the enzyme in tissue cultures of chick embryo liver and found it to be about 4–6 hours. The half-life of ALA synthetase in rat liver is about 70 minutes (*115*). Some of the experiments of Lascelles (*109,116*) with *Rhodopseudomonas spheroides*, while not actually measuring the half-life, show very clearly that under adverse conditions (presumably when protein synthesis is prevented) the level of the enzyme drops rapidly. The activity of the enzyme in nongrowing suspensions of *Rhodopseudomonas spheroides* or in unfractionated sonicates of the organism decreases rapidly (*103*), suggesting that perhaps the loss in enzyme activity is in itself an active process.

It is now generally accepted that ALA synthetase is the key enzyme in the regulation of porphyrin biosynthesis, and it may well be that the rapid response of the level of the enzyme is the most important regulatory mechanism (*103,114,117*). The role of ALA synthetase in the regulation of tetrapyrrole biosynthesis is fully discussed in Section V.

Several methods for the assay of ALA synthetase have been developed (105,106,114,118,119). The methods are, in general, those for determining aminoketones. Since aminoketones, such as aminoacetone, are normal constituents in biological material, care must be taken to distinguish these compounds from ALA. The reaction of ALA with alkaline picrate to give a brownish complex of unknown structure is perhaps the easiest to carry out, but it is also the least specific (119).

Mauzerall and Granick (118) developed a method dependent on the conversion of ALA to a pyrrole which is then measured in a colorimetric reaction. Recent modifications of this method have been developed in two laboratories (106,114) that increases both the sensitivity and the specificity of the determination. In the first step of both modifications, the ALA is reacted with acetylacetone and converted to the pyrrole. One method (106) then calls for the separation of the pyrroles (from both ALA and aminoacetone) by column chromatography. The other method (114) requires a direct determination of pyrroles with Ehrlich reagent on one aliquot of the pyrrole mixture, and a second determination of ALA pyrrole following extraction of the neutralized reaction mixture with ether to remove aminoacetone pyrrole.

At the time Shemin and Russell announced their discovery of the role of ALA in porphyrin synthesis, they also suggested other metabolic fates for this compound (91,120). In a series of reactions termed the succinate-glycine cycle, they suggested that ALA might lose NH_3 via transamination, yielding γ-ketoglutaraldehyde. γ-Ketoglutaraldehyde, upon losing the terminal CHO group, would regenerate succinate and yield a one-carbon fragment capable of undergoing further reactions, e.g., utilization in the biosynthesis of purines. Some experimental evidence has been obtained to support the postulated cycle (121,122). The quantitative significance of this metabolic pathway for ALA has not been evaluated. At least two factors, however, suggest that the succinate-glycine cycle may be of minor metabolic importance. In the first place, with very few exceptions, the capacity of most tissues and organisms for synthesizing ALA is very low. It is so low, in fact, that infinitely more hours are spent by biochemists wondering where the organism or tissue gets enough ALA to make all the porphyrin, heme, or chlorophyll that it does, as the case may be, than wondering where all the ALA can be going. The second objection centers on semi-theoretical grounds. Most of the studies on the regulation of porphyrin synthesis indicate that the primary control point is at the enzyme ALA synthetase (103,109,114). It seems well established that, as a general rule, control points come after, not before, major branches in biosynthetic pathways (see Section V). Since the succinate-glycine cycle amounts to a branch at the level of ALA, it seems doubtful that the cycle represents a significant metabolic fate for ALA. The evidence that porphyrin biosynthesis may be, at least in part, controlled at the level of utilization of ALA through ALA dehydrase is discussed in Section V.

Enzymes catalyzing the transamination of ALA and α-keto acids have been demonstrated both in mammalian tissues (123) and in microorganisms (124,125). The enzyme has been most extensively studied in *Rhodopseudomonas spheroides*, and it has been partially purified from this source (126,127). Though the reaction can be observed in both directions, i.e., formation and disappearance of ALA, the equilibrium seems to favor ALA formation. The enzyme has a rather broad specificity for the amino donor. Alanine, γ-amino-*n*-butyric acid, γ-amino-*n*-valeric acid, and ε-aminocaproic acid are all approximately equivalent as amino donors. A number of α-ketocarboxylic acid and ketoaldehyde analogs of ketoglutaraldehyde inhibit the formation of ALA. Data were not included to indicate whether this inhibition involved formation of ALA only, or of transamination in general. Since all of the compounds found to be inhibitory were capable of accepting amino groups via transamination, these compounds could have been acting as alternative substrates. The importance of the transaminase reaction in the formation and alternative metabolism of ALA has not been evaluated.

2. Biosynthesis of PBG

The monopyrrole precursor of all tetrapyrroles is porphobilinogen. This material was first isolated by Westal (128) from the urine of a porphyric human and the structure was worked out by Cookson and Rimington (129). Porphobilinogen proved to be identical with the chromogen which had been recognized for years in the urine of patients suffering from porphyria.

PBG is formed by the condensation of 2 molecules of ALA with concomitant loss of 2 molecules of water in a Knorr-type condensation reaction. The enzyme that catalyzes this reaction is widespread in nature, and has been demonstrated in plants (130), animals (131,132), and bacteria (103,131). The distribution and relative ease of measuring this enzyme offers a clue on which to base future searches of ALA synthetase. It seems reasonable to assume that, if an organism or tissue has an active ALA dehydrase, then it must have or have had an active ALA synthetase.

ALA dehydrase has been partially purified from *Rhodopseudomonas spheroides* (103,133), ox liver (131,133), reticulocytes (132), mouse liver and spleen (134). The enzyme from all sources is quite sensitive to thiol reagents (*103,131*). This seems to be a relatively straightforward inhibition in that it can be reversed or prevented by cysteine or glutathione. The SH group, or groups, on the enzyme must be moderately exposed since this reversible loss in activity occurs upon exposure to air during purification. The ALA dehydrase of wheat leaves is least susceptible to loss of activity in the absence of SH-protective compounds such as mercaptoethanol (*135*). The enzyme purified from mouse liver or spleen is remarkably stable toward heat, having a half-life of about 80 minutes at 75° (*134*). No cofactors, in the accepted sense, have been demonstrated for ALA dehydrase. Copper was implicated as a cofactor some time ago (*137*), but subsequent studies have shown that fully active enzyme can be prepared in the absence of copper (*138*). Burnham and Lascelles (*103*) reported that partially purified ALA dehydrase from *R. spheroides* required potassium ion for activity. More recently Shemin (*139*) has reported that this requirement for K^+ ion is not absolute and that the effect of the K^+ ion is to decrease the K_m of the enzyme. At very much higher concentrations of ALA the enzyme is active, and a plot of reaction velocity against substrate concentration yields a sigmoid curve in the absence of K^+ ion. Apparently the cation promotes the association of subunits of the enzyme. Some evidence has been obtained during gel filtration studies with ALA dehydrase from cow liver (*133*) that the enzyme from this source also undergoes dissociation into subunits. No mention was made as to the possible involvement of K^+ ions.

It is interesting that, in spite of the evidence indicating that no cofactor is involved with ALA dehydrase, EDTA is a potent inhibitor of the enzyme from mammalian sources (*131–136*) except the mouse (*134*). The enzyme from avian erythrocytes is only partially inhibited even by high concentrations of EDTA (*132*), and the enzyme from *R. spheroides* is insensitive to EDTA (*103*). In studies on ALA dehydrase from mouse tissues, Coleman (*134*) observed a stimulatory effect by EDTA, and inhibition or activation by Fe^{2+} or Hg^{2+} depending on their concentration relative to GSH. These effects, however, were less pronounced in experiments using the most highly purified enzyme, indicating that the action of these substances may have been on contaminating protein and only secondarily on ALA dehydrase. Experiments on partially purified ALA dehydrase from wheat leaves (*135*) revealed an inhibition by EDTA that could be overcome by addition of stoichiometric amounts of Mn^{2+}. The enzyme from this source retains activity during purification without having to be preincubated with cysteine or GSH prior to assay. Nevertheless, the enzyme is activated by GSH and by any one of several divalent metal ions, e.g., Mg^{2+}, Mn^{2+}, Fe^{2+}, Co^{2+}, Ni^{2+}. Interestingly, the activation by GSH and metal ion is synergistic. Enzyme from this source is the only one, aside from ALA dehydrase from *R. spheroides*, where the K^+ ion requirement has been studied. Wheat leaf ALA dehydrase does not require K^+ ion.

The original observation by Burnham and Lascelles (*103*) that ALA dehydrase from *R. spheroides* is inhibited by low concentrations of hemin has been extended and confirmed using enzyme from other sources (*134,136*). This evidence, coupled with the evidence that the enzyme is composed of subunits, showing characteristics now generally attributed to allosteric enzymes, opens the question of a second control point for porphyrin synthesis. This possibility is discussed in the section on regulation.

The enzymology of ALA dehydrase is particularly interesting, since the enzyme catalyzes one of the rather rare reactions where two identical substrate molecules participate. Granick and Mauzerall (*132*) observed that the affinity of the enzyme for the first molecule of ALA was at least ten times greater than for the second. A very interesting study on the mechanism has been reported by Nandi and Shemin (*140*). They found the enzyme from *R. spheroides* to be competitively inhibited by levulinic acid, not by α-keto acids. Furthermore, they reported that in the absence of ALA or levulinic acid the enzyme is not inhibited by $NaBH_4$. Addition of $NaBH_4$ to the enzyme in the presence of ALA or levulinate caused inhibition, suggesting formation of a Schiff-base enzyme-substrate intermediate. This was further substantiated when ALA-^{14}C and enzyme were treated with $NaBH_4$. Following reduction, the enzyme became labeled and the ^{14}C remained attached to peptide fractions after hydrolysis of the enzyme.

3. BIOSYNTHESIS OF UROPORPHYRINOGEN III

Chemical polymerization of the unsymmetrical pyrrole PBG to form a cyclic tetrapyrrole gives rise to all four possible isomers of uroporphyrin (*129,141*). These isomers differ only in the symmetry of the acetic and propionic acid side chains in the β positions (Fig. 11). The symmetry of substitutents in the β positions of all biologically active metal tetrapyrroles is related to that of uroporphyrin III. However, two porphyrins encountered with some frequency in nature are of the type I series. These porphyrins, uroporphyrin I and coproporphyrin I, are the products of a deranged metabolism and they are not functional, nor do they seem to be metabolized further.

Ever since PBG was recognized as the precursor of uroporphyrin III, the mechanism by which the condensation occurs has intrigued many chemists. Examination of the structure of PBG—CH_2—NH_3^+ shows

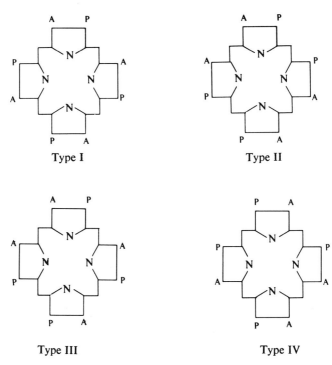

Fig. 11. Skeletal structure of uroporphyrin showing the four isomer types. A = acetate, P = propionate. All natural functional tetrapyrroles are related to uroporphyrin type III.

that, following loss of ammonia, the PBG—CH_2^+ can attack the free α position of another PBG—CH_2—NH_3^+ yielding a dipyrrylmethane with one free α position and CH_2—NH_3^+ in the distal α position. After this, the process repeats until there are four linked pyrroles. The free CH_2—NH_3^+ has a high probability of being in close proximity to the free α position on the first pyrrole of the tetramer (142). Ring closure occurs in preference to further lengthening of the pyrrole chain. Cyclization to yield a strainless ring is favored thermodynamically since it results in more free molecules being formed, and because of the proximity of the two reactive ends of the tetramer. This straightforward series of reactions yields the symmetrical uroporphyrin I type isomer. Uroporphyrin III does not have this symmetry, but instead has the acetate and propionate groups in the β positions on ring D in reverse order. One can mentally "flip" the D pyrrole to achieve the correct order of acetate and propionate substituents. In doing this, however, the aminomethyl group of ring D ends up on the "wrong" side. Numerous mechanisms have been proposed to explain this reaction

(*143–147*). The studies of Mauzerall (*142*) on the chemical condensation of PBG show that the reaction involves a free H_2CXR intermediate such as formaldehyde. This was demonstrated by adding ^{14}C-labeled formaldehyde to the reaction mixture and observing formation of labeled uroporphyrin. The enzymic condensation, on the other hand, does not involve free formaldehyde.

The nonenzymic condensation of PBG in acid gives rise to a mixture of uroporphyrin isomers with the composition: one-half type III, one-fourth type IV, one-eighth type II, and one-eighth type I. Mauzerall (*142*) in extensive studies on the condensation of PBG and stability of porphyrinogens has shown the above mixture of isomers to be in equilibrium in acid. The composition of this equilibrium mixture is what one would predict on statistical grounds for the totally random cyclization of a tetrapyrrole with two different R groups on each of the β positions. It is clear then that nature has simply perfected a mechanism that selects the porphyrin most favored on thermodynamic grounds.

Two enzymes are involved in the condensation reaction (*146–150*). The two enzymes can easily be distinguished by their susceptibility to heat inactivation. Uroporphyrinogen I synthetase, sometimes referred to as PBG deaminase, is quite heat-stable (*146*). It catalyzes the removal of the amino group from PBG and the head-to-tail condensation of four of the pyrroles yielding uroporphyrinogen I. The enzyme from spinach leaves and avian erythrocytes is competitively inhibited by opsopyrroledicarboxylic acid (*151,152*) (see Fig. 12c) and by iso-PBG (*153*) (see Fig. 12b). Hydroxylamine and high concentrations of ammonia are inhibitors of a special class (*67,154*). These compounds do not affect the rate of disappearance of PBG, but they do inhibit the formation of the cyclic tetrapyrrole. As a result, open-chain polypyrroles accumulate in the reaction mixture (*155*). That these polypyrroles may actually be intermediates in the formation of the cyclic tetrapyrrole is indicated by the fact that prolonged incubation of the reaction in the presence of inhibitor yields the expected cyclic tetrapyrrole, possibly by chemical (nonenzymic) condensation. The linear di-, tri-, and tetra-pyrroles are unstable, and they have not been fully characterized.

Bogorad (*11*) reports that uroporphyrinogen I is not formed by the condensation of two dipyrroles by PBG deaminase. Thus, when the synthetic dipyrrole, equivalent to one half of uroporphyrinogen I, is incubated with the enzyme, uroporphyrinogen I is not formed. The same dipyrrole plus PBG, when incubated with the enzyme, gives rise to uroporphyrinogen in excess of that accounted for by the PBG (*11*).

The second enzyme involved in the conversion of PBG to uroporphyrinogen III is called uroporphyrinogen III cosynthetase. This enzyme is quite unstable to heat (*146*), being rapidly inactivated at 60°. Uroporphyrinogen III cosynthetase, when free of uroporphyrinogen I synthetase, does not react with PBG (*11*). It has also been shown that this enzyme does not react with uroporphyrinogen I (*149,156*). The substrate for the cosynthetase thus appears to be one of the linear polypyrroles. This can be seen as an increased rate of consumption of PBG when the cosynthetase is added to a reaction mixture containing PBG and uroporphyrinogen I synthetase (*11*).

The various studies on the enzyme-catalyzed conversion of PBG to uroporphyrinogen III have yielded more negative than positive evidence with regard to the mechanism of the reaction. Though this is frustrating, it is helpful to a degree to be able to eliminate postulated mechanisms which are proved to be incorrect. Following are some of the facts that must be considered in explaining the mechanism. Opsopyrroledicarboxylic acid (Fig. 12c) is neither a cofactor for the reaction (*151,152*),

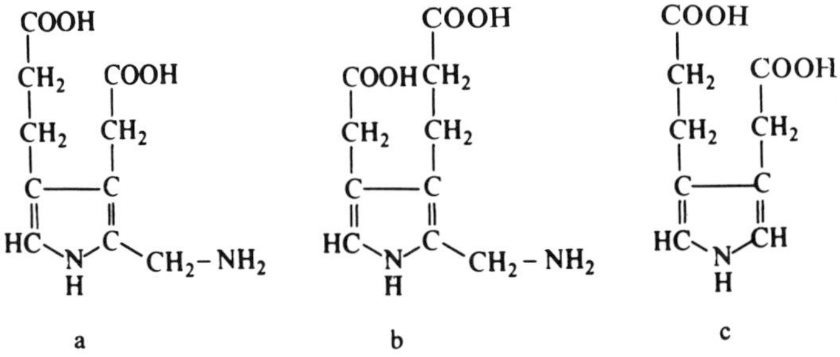

FIG. 12. Structures of important monopyrroles. (a) Porphobilinogen, (b) isoporphobilinogen, (c) opsopyrroledicarboxylic acid.

nor is it a product of the reaction, as postulated by some mechanisms (*147,151*). It is, in fact, an inhibitor of the reactions. Free formaldehyde is not involved in the reaction (*157,158*), thus excluding several mechanisms (*129,93,159*). No PBG molecules are wasted, i.e., the PBG : uroporphyrinogen III stoichiometry is 4 : 1 (*132,150,158,160*).

Of the existing postulated mechanisms that have not yet been ruled out on experimental grounds, two seem particularly attractive. One of them involves the migration of a —CH$_2^+$ group from one α position to the other α position of the same pyrrole, or to another pyrrole (*161*). This postulated sequence of reactions is shown in Fig. 13. Mathewson and Corwin (*143*) have postulated a mechanism based on work with Stuart models and existing data. The key intermediate in this mechanism is a pyrrolenine form of a tetrapyrrylmethane (Fig. 14). The retention of

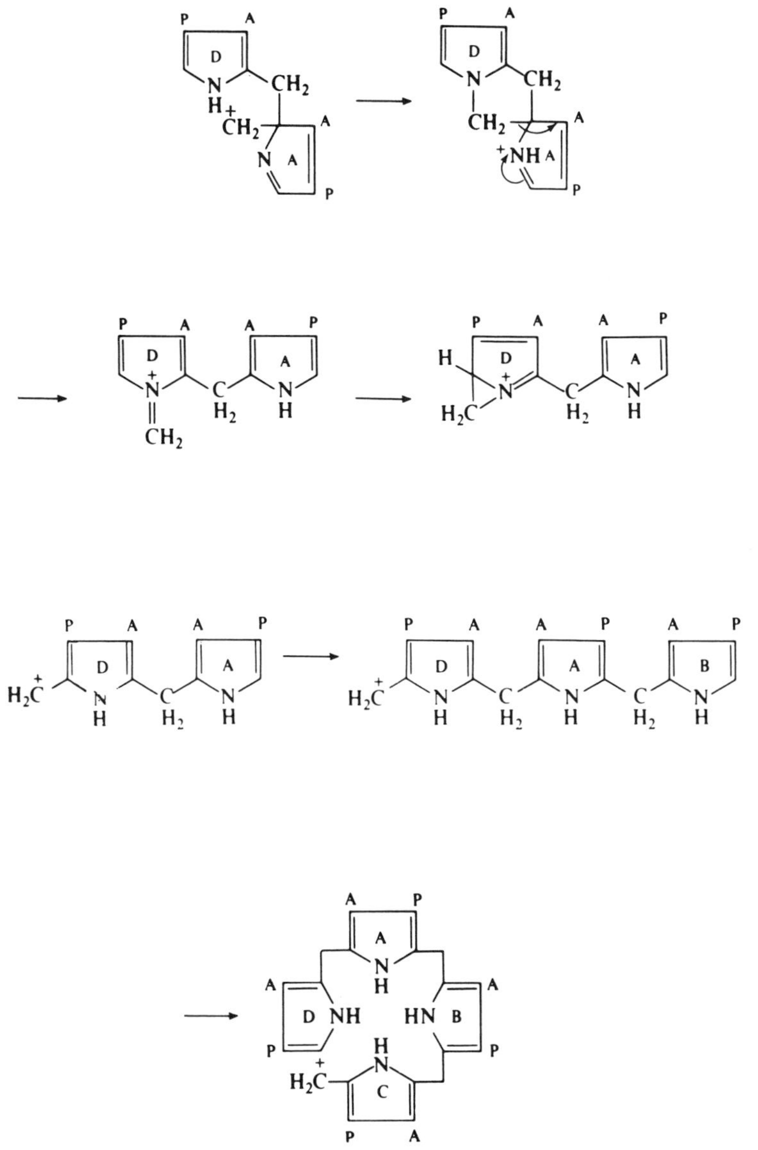

FIG. 13. Hypothetical mechanism for the condensation of PBG allowing formation of uroporphyrinogen III suggested by Bullock et al. (161).

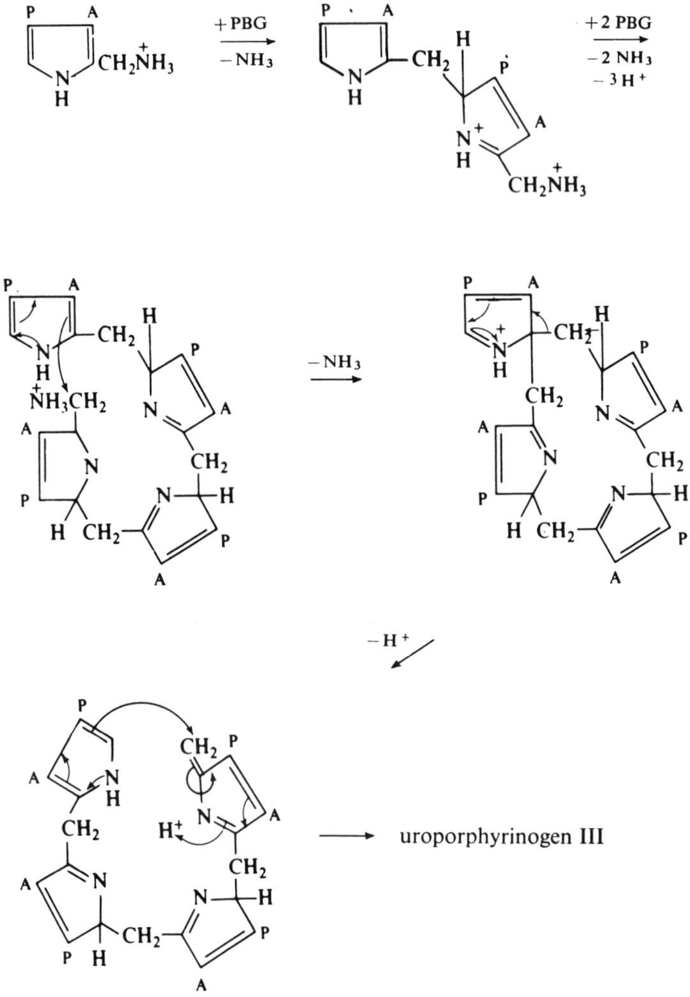

FIG. 14. Hypothetical mechanism for the condensation of PBG allowing formation of uroporphyrinogen III suggested by Mathewson and Corwin (143).

the α hydrogens following condensation of the PBG's allows sufficient flexibility for the tetramer to fold so that the terminal aminomethyl group of ring D lies over the inside α position of ring A. Following ring closure, the intermediate (Fig. 14) can reopen. The terminal pyrrole (ring D) can then rotate, and ring closure at the free α position of this pyrrole would yield a tetrapyrrole of the type III isomer series. One feature of this mechanism is that the formation of the corrin ring of vitamin B_{12} can be explained by a minor variation in the proposed scheme. This mechanism for formation of the corrin ring does not require uroporphyrinogen III as an intermediate.

4. BIOSYNTHESIS OF COPROPORPHYRINOGEN

Uroporphyrinogen is converted to coproporphyrinogen by the enzymic decarboxylation of the 4 acetic acid side chains (35,162,163). The reaction proceeds in a stepwise manner and porphyrins with 7, 6, and 5 carboxyl groups can be demonstrated, though they do not tend to accumulate. The first decarboxylation proceeds more readily than the subsequent 3 decarboxylations. One enzyme apparently is responsible for the removal of all of the carboxyl groups. The enzyme is active only on the fully reduced porphyrinogen (162). This fact was a puzzle for several years. Since uroporphyrin was the first visible product of the condensation of PBG, it was assumed to be the intermediate for subsequent reactions. Addition of labeled uroporphyrin to heme-forming systems, however, invariably failed to yield labeled heme. Nevé et al. (164) clarified this point when they demonstrated that addition of labeled uroporphyrinogen to a heme-forming system resulted in a highly labeled heme. Uroporphyrinogen decarboxylase is active with all uroporphyrinogen isomers. However, the type III isomer is decarboxylated by the enzyme from human red cells (165) at 7.5 times the rate of the type I isomer when the two uroporphyrinogen isomers are present in equal concentrations, and at only about twice the rate by the enzyme from rabbit reticulocytes (35). The decarboxylase has been partially purified from several sources (35,163,165).

The enzyme from R. spheroides seems to have a cofactor requirement (163), as does the enzyme from chicken erythrocytes (166). The enzyme from human erythrocytes does not require any cofactor (165).

Work with uroporphyrins has been hampered by the fact that the physical and chemical properties of the isomers are so nearly identical that there is no adequate method to separate them. Several methods have been described for the chromatographic partial separation of the different isomers, but success seems more art than science (167–170).

Unequivocal identification of uroporphyrin isomer mixtures is best obtained by decarboxylation in acid under N_2 (*171*) followed by chromatography of the coproporphyrin isomers. In a solvent containing 10 parts (v/v) 2,6-lutidine and 7 parts 0.7 M aqueous ammonia, the coproporphyrin isomers separate into 3 spots, i.e., II, III + IV, and I (*172*).

Pure uroporphyrin ester isomers cannot be distinguished from mixtures of the isomers by IR, X-ray powder patterns, or by melting-point determinations (*156*).

5. BIOSYNTHESIS OF PROTOPORPHYRIN

The conversion of coproporphyrinogen III to protoporphyrin IX is complex. It involves the decarboxylation and oxidation of the propionate groups at positions 2 and 4, yielding two vinyl groups, and the removal of 6 hydrogens from the protoporphyrinogen IX to yield the porphyrin. It is not entirely clear how many enzymes are involved in this process. An enzyme preparation purified 58-fold, which appeared as a single band upon Sephadex chromatography and upon electrophoresis had the capacity to carry out the oxidative decarboxylation (*173*). Apparently a single enzyme is responsible for the oxidative decarboxylation of both propionate residues. Some question, however, remains concerning the conversion of protoporphyrinogen to protoporphyrin. This reaction might be spontaneous, it might be catalyzed by the same enzyme that carries out the oxidative decarboxylation, or it might be catalyzed by a second enzyme. The experiments of Sano and Granick (*174*) indicate that the reaction is enzyme-catalyzed.

The first enzyme, coproporphyrinogen-(acceptor) oxidoreductase (decarboxylating), more often called coproporphyrinogen oxidase or coproporphyrinogen oxidative decarboxylase, has been demonstrated in *Euglena gracilis* (*175*), avian erythrocytes (*176*), and in a variety of mammalian tissues (*173,174,177*). It is a mitochondrial enzyme and can be solubilized by thioglycollate (*174*). The enzyme is specific for coproporphyrinogen III and has no activity toward the I isomer (*173,174,177*). There is no demonstrated cofactor requirement (*173,177*).

Several mechanisms have been suggested to explain the oxidative decarboxylation. A mechanism analogous to the oxidation of fatty acids can be visualized (*154*). The fact that Bogorad and Granick (*64*) found a *Chlorella* mutant that accumulated hematoporphyrin IX and monohydroxyethyl-monovinyl deuteroporphyrin IX seemed to lend experimental evidence to this hypothesis. Accumulation of hematoporphyrin, however, appears to be an artifact having nothing to do with protoporphyrin synthesis (*174*), and it represents a good example of the danger that comes from assuming that a compound that accumulates in a mutant is necessarily the precursor of something. This has been alluded to earlier (Section II, A). The evidence that hematopophyrin(-ogen) is not an intermediate between coproporphyrinogen and protoporphyrinogen is as follows: *Chlorella* that can make protoporphyrin IX from PBG after being made permeable by freezing and thawing do not convert hematoporphyrin or hematoporphyrinogen to protoporphyrin (*154*). Furthermore, hydroxylamine and semicarbazide do not inhibit coproporphyrinogen oxidase, as would be expected if a carbonyl group were formed during the reaction (*174,175*). Another possible reaction mechanism would involve the formation of deuteroporphyrin 2,4-diacrylic acid as an intermediate. This compound, and its hexahydro derivative, has been tested (*173,174*) and is not converted to protoporphyrin. Only *trans*-2,4-diacrylic deuteroporphyrin has been tested and, until the *cis* isomer is available, the possibility that it may be an intermediate must be considered open.

A mechanism involving hydride-ion removal from the β carbon of propionate with simultaneous decarboxylation has been proposed (*174,178*). The reaction must be concerted (simultaneous oxidation and decarboxylation) since protons from the solvent water are not incorporated into the vinyl groups (*179*).

One of the most attractive hypotheses, and one with experimental evidence to support it, involves 2,4-bis(β-hydroxypropionic acid) deuteroporphyrinogen IX as an intermediate (*180*). One of the features of this mechanism is that it accounts for the participation of molecular oxygen. According to this mechanism, the β carbon of the propionate is hydroxylated, then simultaneously decarboxylated and dehydrated. This ingenious mechanism (Fig. 15), suggested by Sano (*180*), is supported both by model experiments (nonenzyme, acid-catalyzed) and by

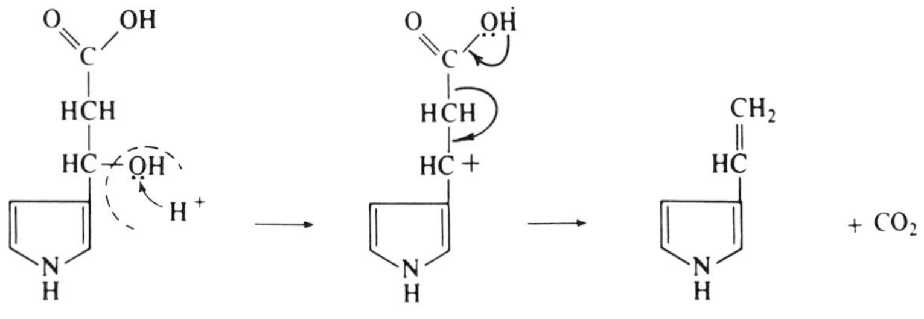

FIG. 15. Proposed mechanism of acid-catalyzed dehydration of 2,4-bis(β-hydroxypropionate) deuteroporphyrinogen IX, a suggested intermediate in the conversion of coproporphyrinogen III to protoporphyrin IX (*180*). Only the A ring is shown.

enzyme experiments. Mitochondrial preparations convert 2,4-bis(β-hydroxypropionate) deuteroporphyrinogen IX to protoporphyrinogen IX under both aerobic and anaerobic conditions. Coproporphyrinogen III, on the other hand, is converted to protoporphyrinogen only in the presence of molecular oxygen (*174,180*).

During the course of the conversion of coproporphyrinogen to protoporphyrin, Sano and Granick detected small amounts of a 3-carboxyl porphyrin. When the reaction is followed as a function of time, the 3-carboxyl porphyrin increases, then decreases, as would be expected of a reaction intermediate (*174*). The fact that the 3-carboxylmonovinyl intermediate accumulates even slightly indicates that it is a two-step reaction and that the affinity of either the original enzyme (or the second enzyme, if there are two) for the intermediate is equal to, or lower than, that for coproporphyrinogen III.

Porra and Falk (*177,181*) report that following short-term incubation of coproporphyrinogen with their mitochondrial enzyme, the recovery of total soluble porphyrin was low, ca. 50–70%. The "lost" porphyrin was found to be bound to protein, presumably the enzyme. Sano and Granick (*174*) emphasize the ease with which sulfhydryl reagents add to the vinyl groups of protoporphyrinogen. Porra and Falk (*177*), however, claim that their protein-bound porphyrin is not an artifact, and suggest that it may be an intermediate formed during the reaction.

Everyone that has worked with this enzyme from aerobic organisms agrees that only molecular oxygen will serve as oxidizing agent. The enzyme is inhibited by α,α-dipyridyl and *o*-phenanthroline (*173,174*). This inhibition can be reversed by dialysis. Batlle *et al.* (*173*), using a purified enzyme preparation, noted that the addition of these two chelating agents causes the formation of a red complex. These chelators are well known to form red complexes with ferrous iron. This behavior resembles that observed by Hayaishi (*182*), working with *meta*-pyrocatechuic acid oxidase, a ferrous iron enzyme. Other evidence for the participation of iron in this reaction is provided by studies on microorganisms. It has been repeatedly observed that, when many microorganisms are cultured under iron-deficient conditions, porphyrins accumulate in the growth medium (*183,184*). Under these conditions the predominant porphyrin that accumulates is invariably coproporphyrin III. When the microorganisms are given a limited amount of iron, they make some protoporphyrin in addition to the coproporphyrin (*185*). If they are provided with still more iron, they make protoporphyrin and iron protoporphyrin but no coproporphyrin (*185*).

It should be noted, however, that there is no evidence that coproporphyrinogen oxidative decarboxylase is similar in microorganisms and higher organisms. In fact, it is both curious and interesting that the enzyme has not been detected in cell-free extracts of microorganisms (186). This is surprising in the case of the *Athiorhodacae* since they generally have very active porphyrin-synthesizing enzymes. The inability to detect the enzyme in *Athiorhodacae* is unfortunate for two reasons. It would be interesting to discover the electron acceptor(s) used by the organisms when growing anaerobically. Of course, it may be that not knowing what electron acceptor to use explains why the reaction cannot be demonstrated in these organisms. Second, these organisms would be particularly suitable for investigating the participation of iron in the oxidative decarboxylase reaction since it is possible to critically regulate their iron nutrition.

It seems odd to the present writer that apparently the ten hydrogens removed from coproporphyrinogen are wasted in that no suitable electron acceptor other than oxygen has been found. That seems like a lot of reducing power to waste. If they were "run through" the oxidative phosphorylation sequence, 15 moles of ATP would be formed. Recall that 8 moles of ATP are required per porphyrin. Perhaps it can be argued that quantitatively the amount of energy available to the cell from this source would be insignificant compared to the overall energy metabolism of the cell, and thus it can be ignored.

One final point should be emphasized before leaving this section. Work with coproporphyrinogen oxidative decarboxylase is considerably complicated by the fact that the substrate, coproporphyrinogen, is rather unstable in air, and that vigorous aeration is required for activity. Thus autoxidation presents a problem of considerable magnitude.

C. The Iron Branch

Iron can be incorporated into porphyrins chemically without the participation of enzymes. The conditions for rapid and quantitive incorporation, however, are far from physiological. Laboratory preparations generally involve refluxing the porphyrin with a ferrous salt in acetic acid (4). Iron is also effectively incorporated in hot pyridine (187).

Under semiphysiological conditions, i.e., at neutral pH, at room temperature, and in the presence of air, iron incorporation tends to be slow, and the yield of iron porphyrin low. Two factors are probably responsible for the inefficient incorporation under these conditions. At neutral pH, solubility becomes a problem with porphyrins in general, and this is particularly true with protoporphyrin. This is not to say that the porphyrins actually precipitate out of solution, but rather that they aggregate to form dimers, trimers, and n-mers under these conditions (188). A solution may look clear to the eye even when the porphyrin molecules are markedly aggregated. Aggregated porphyrin solutions characteristically have broadened and rounded absorption bands. Such solutions do not obey Beer's law, and the addition of solubilizing agents such as detergents causes a marked increase at absorption maxima and a sharpening of the absorption bands. At neutral pH and in the presence of air, ferrous iron is rapidly oxidized to ferric iron, which tends to precipitate as the hydroxide. When air is excluded, if the porphyrin is solubilized with detergent and the metal is not complexed with buffer anions, metal incorporation proceeds at a reasonable rate. Model systems employing such refinements have been studied (37,189, 190). One such study has shown that sodium dithionite enhanced iron incorporation by several orders of magnitude (191). The dithionite apparently did not participate solely as a reductant, but rather caused formation of a black precipitate (iron sulfide?) which was somehow involved in the incorporation reaction.

Hexahydroporphyrins (porphyrinogens) cannot chelate metal ions. However, nonenzymic incorporation of iron into partially reduced porphyrins has been demonstrated (190,192). Evidence has been presented, however, that iron is not incorporated into partially reduced porphyrins in biological systems (193,194).

The enzyme that catalyzes the incorporation of iron into porphyrins has been given several names, but ferrochelatase seems to be most preferable. Ferrochelatase has been demonstrated in a wide variety of tissues and organisms (194–197). Considering the variation in enzyme source, the variation in preparation and incubation methods, and the different assay techniques employed, perhaps it is not too surprising that differences are found when the results from different laboratories are compared.

Some features of the system, however, appear to be reasonably constant. For instance, all systems are sensitive to air. This problem is circumvented by adding reducing agents such as GSH, ascorbate, or ergothionine (see the section on assay methods below for a warning about the use of these compounds), by flushing the system with nitrogen, or by running the assay in evacuated Thunberg tubes (194,196,198). There is some evidence that the enzyme has exposed SH groups required for activity (196,199), and possibly these groups are oxidized to the disulfide form by air. Other evidence suggests that the air is oxidizing the ferrous iron, and that reducing or anaerobic conditions are required to maintain the iron in the ferrous state (194,195,199). Studies on porphyrin specificity have consistently found that deuteroporphyrin (193,195,200) reacts at the fastest rate. Less consistency has been observed with other active porphyrins. Only porphyrins with two free carboxyl groups are active. Studies on metal specificity have yielded more complicated results in that the porphyrin substrate apparently effects metal specificity. For instance, Johnson and Jones (201) using ferrochelatase from avian erythrocytes found the following relative rates of metal incorporation: with protoporphyrin as substrate—Fe, 100; Co, 43; Zn, 300; with deuteroporphyrin as substrate—Fe, 266; Co, 270: Zn, 324. These rates are given in terms relative to the rate of incorporation of iron into protoporphyrin which is set at 100.

The experiments of Porra and Jones (194,195) on ferrochelatase from pig liver indicate that there may be more than one enzyme capable of catalyzing incorporation of metal into porphyrin. There is supporting evidence for this from studies using microorganisms (201). If there is more than one enzyme, this might account for some of the differences observed in solubility of the enzyme, and in the specificity toward the porphyrin and metal ion. Alternatively, the differences observed in specificity may reflect the nonspecific solubilizing effect of some proteins on porphyrins and iron.

Little or nothing is known about how or when iron is incorporated into tetrapyrroles that ultimately become cytochromes. Porphyrin c and porphyrin a have been tested as substrates of ferrochelatase, and neither of these porphyrins are active in the system studied (195). (Porphyrin c must be considered a model compound in these studies, because it seems clear that the two cysteines attached to the porphyrin could never be activated and incorporated into proteins.) It may be that specific ferrochelatases exist for each nonproto-type heme. More likely, protoheme is a precursor of other hemes like heme a. *Staphylococcus aureus* var 511 requires heme for growth except under certain conditions (3,202). When grown without heme the cells are devoid of cytochromes and other heme enzymes. When grown in the presence of heme, cytochromes and other heme enzymes are formed. Heme a has been identified under these conditions, indicating that these microorganisms can convert protoheme to heme a (203,204). Direct evidence for this conversion has recently been obtained (205) by incubating cells with ^{14}C-labeled protoheme and later isolating heme a with nearly the same specific activity.

Several reports have appeared recently describing an enzyme that specifically catalyzes the incorporation of zinc into protoporphyrin (206–208). This is a peculiar thing, because there is no evidence that zinc porphyrins have any biological function. In fact, zinc is incorporated into many porphyrins at neutral pH with such facility that it is sometimes a problem in getting porphyrins from natural sources free of zinc. The systems that catalyze the incorporation of zinc are rather specific both for zinc and for protoporphyrin (not in terms of absolute rates, but relative to noncatalyzed rates), and the general properties described for the system are consistent with those used in defining an enzyme. It seems possible that the proteins in these incubation systems are acting primarily as solubilizing agents. Since other porphyrins have somewhat less tendency than protoporphyrin to aggregate, they do not " benefit " so much from adsorption on the protein (enzyme), and this would explain the specificity. It is well known that the solubilization action of several detergents promotes metal-ion incorporation (2). It is also possible that apparent enzymic incorporation of zinc into protoporphyrin could result from the action of a partially denatured ferrochelatase or, perhaps, magnesium chelatase.

Several methods have been employed for measuring ferrochelatase activity. A recent re-evaluation of these methods has been made and the results show that major problems attend all methods (199). A key observation reveals that hemes are very unstable under aerobic conditions in the presence of thiols. Most ferrochelatase incubation mixtures include GSH, cysteine, or mercaptoethanol, and exposure of the system to air at the termination of the experiment is sufficient to permit degradation. Methods that have been used are as follows:

1. Iron incorporation measured with ^{59}Fe. Radioactive iron is included in the reaction system and, after terminating the reaction, carrier heme is added. The total heme is purified and specific radioactivity determined. This method is tedious and time-consuming, but gives the most unequivocal results (200,209). Precautions are required to prevent heme loss (199).

2. Direct measurement of the absorption spectrum of the sample. Either difference spectrum or absolute absorption changes can be followed. The major problem is high blank or background readings in crude systems (193,197). Use of a Thunberg cuvette with an inert atmosphere, however, circumvents most of the problems arising from heme degradation.

3. Indirect measurement of absorption spectrum. The reaction is stopped by the addition of strong acid, and the protein and metal porphyrin are precipitated. Free porphyrins remain in solution. Absorption measurements are made as either difference spectra or absolute absorption (196).

The latter two methods suffer somewhat from their lack of specificity in that disappearance of a reactant does not prove the formation of a given product (*194,195,210*).

4. Measurement of oxidized vs. reduced difference spectrum. The reaction mixture is divided into two equal samples and the oxidized vs. reduced pyridine hemochromogen difference spectrum is measured (*194,195*). This is a relatively sensitive analytical method, and has the advantage of allowing direct measurement of the product. Thiols interfere with hemochromogen formation, and the original procedure has been modified (*199*) to eliminate this problem.

None of the experiments with ferrochelatase from any of the sources tested has indicated that a coenzyme is involved in the reaction. Certain indirect information obtained from experiments on intact microorganisms, however, does indicate that a coenzyme might be involved (*211–214*). Since it is possible to explain why a cofactor might go unrecognized in experiments with cell-free systems, it is worthwhile to consider the positive evidence implicating such a cofactor.

The evidence to be considered is based to a large extent upon the nutritional requirements of certain microorganisms (*3,184,215*). A number of microorganisms are recognized to require heme for growth. In some cases this requirement is absolute (see Group 1 below). In other cases the heme requirement can be satisfied by heme precursors such as uro-, or coproporphyrinogen III or by protoporphyrin IX (Group 2). In certain other cases the heme requirement can be satisfied by one of several chelate compounds (Group 3).

The chelate compounds of particular interest are of natural origin and they have certain common features. The important common features are that these compounds are secondary hydroxamic acids and, as such, bind ferric iron specifically and tenaciously. The class name given to these compounds is siderochrome (*216*). Most siderochromes fall into one of two groups depending upon their biological activity. The sideromycins are potent antibiotics, having about 10 times the potency of penicillin, on a weight basis, against many bacteria (*216*). The sideromycins are chemically and biologically unstable, and the chemical structure has not been established for any of the several compounds in this class. The second major group of siderochromes are called sideramines. The sideramines have the biological potential of specifically antagonizing the antibiotic activity of the sideromycins (*212,213,216,217,218*). They are also potent growth factors for some microorganisms. The chemical structures of a number of sideramines has been established (*212,218,219–221*), and some have been synthesized. Investigations by Burnham and Neilands (*215*) and by Zahner *et al.* (*212,216,217*) indicate that siderochromes are widespread in nature. Zahner *et al.* have even found evidence for sideramines in liver (*217*), thought it is possible that the low level found could have been made by the intestinal flora and absorbed from the gut. Recently Morrison *et al.* (*222*) have found that some common synthetic chelate compounds, e.g., 8-hydroxyquinoline, can replace the heme or sideramine growth factor requirement of some bacteria.

Microorganisms that require heme for growth can be put into one of three groups. For purposes of the present discussion, they are arbitrarily called Groups 1, 2 and 3, and their nutritional requirements (with respect to heme) are shown in the accompanying tabulation.

	Heme	Protoporphyrin	Sideramine
Group 1	+	−	−
Group 2	+	+	−
Group 3	+	−	+

A hypothesis suggested in 1960 (*211*) to account for these observations, and presented twice since in slightly modified form (*212,214*) is shown in Fig. 16.

The absolute requirement for heme shown by microorganisms in Group 1 is explainable in terms of a genetic block at the level of synthesis of ferrochelatase. A heme-requiring strain of *Haemophilis aegyptus* studied by White and Granick (*223*) is a good example of such an organism. They were unable to demonstrate ferrochelatase in extracts of this microorganism under conditions where it was possible to demonstrate this enzyme in non-heme-requiring strains. Transformation experiments using DNA isolated from *Haemophilis influenzae*, a microorganism whose heme requirement can be replaced by protoporphyrin IX (Group 2), produced *H. aegyptus* transformants capable of growing on protoporphyrin IX. The transformants had demonstrable ferrochelatase. *Staphylococcus aureus* var 511, appears to be another bacterium that exemplifies Group 1 microorganisms (*3,224,225*).

The requirement for heme that can be replaced by protoporphyrin IX, as shown by microorganisms in Group 2, is explainable in terms of a genetic block along the path of porphyrin biosynthesis. *Haemophilus influenzae* (*223,226,227*) is an example of such a microorganism, and it has been shown that the heme- or protoporphyrin IX-requiring strains do, in fact, lack some of the enzymes for porphyrin synthesis (*223,227*). Specifically, ALA dehydrase, PBG deaminase (uroporphyrinogen I synthetase), and uroporphyrinogen decarboxylase were found missing

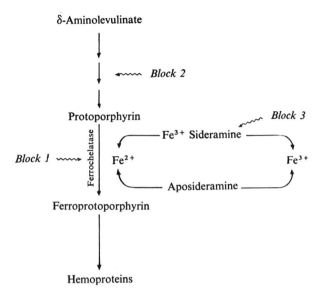

FIG. 16. Hypothesis to account for the participation of sideramines in the biosynthesis of heme (*211*). Known genetic blocks are indicated for Groups 1 and 2 microorganisms and a suggested block is indicated for Group 3 organisms.

or inactive under conditions when these enzymes could be detected in non-heme-requiring strains. However, when the heme-requiring strains of *H. influenzae* are supplied with protoporphyrin IX they can grow, indicating that they have the ability to incorporate iron. Ferrochelatase activity can be demonstrated in cell-free preparations of these microorganisms (*223*).

The microorganisms in Group 3 require heme for growth, and this requirement can be replaced by one of several sideramines (*212,215,217*). Generally the sideramines are at least 100 times as potent as heme on a weight basis. *Arthrobacter* JG-9 and *Microbacterium lacticum* 8181 are examples of Group 3 bacteria.

What are some explanations for the fact that protoporphyrin IX cannot replace heme? For one thing, this indicates that the failure to make heme is not due to a block in porphyrin biosynthesis and, in fact, *Arthrobacter* JG-9 excretes porphyrins (*211,217*). As is the case in Group 1 organisms, ferrochelatase might be inactive or missing. If ferrochelatase were missing, and yet replaceable by sideramine, that in effect says that sideramine and ferrochelatase are one and the same thing. It seems clear that this is not the case. On the other hand, if ferrochelatase were present but inactive, sideramine might function to activate the ferrochelatase. The activation of ferrochelatase by sideramines seems to be a reasonable possibility. If the genetic block in Group 3 microorganisms prevented the synthesis of a cofactor (sideramine) for ferrochelatase, this could explain the observation that sideramines can replace the heme requirement.

The evidence that ferrichrome (a representative sideramine) participates in heme synthesis, and more specifically that it may function as an iron-donating coenzyme for ferrochelatase, is summarized as follows:

1. When *Arthrobacter* JG-9 is cultured with a limiting amount of heme or ferrichrome, the bacteria contain very little of the heme enzyme catalase. Addition of either ferrichrome or heme results in a rapid and substantial increase in catalase activity (*213,217*).

2. When deficient cells of the same organism are supplied simultaneously with ferrichrome and ferrimycine A (a typical sideromycin which is a competitive antagonist against ferrichrome growth factor activity), the increase in catalase is markedly inhibited (*213*).

3. When deficient cells are supplied with heme and ferrimycin A simultaneously, the increase in catalase proceeds at the same rate and to the same extent as when heme is given alone (*213*).

4. Inorganic iron ($^{59}FeCl_3$) is permeable to *Arthrobacter* JG-9 cells grown in the presence or absence of ferrichrome. Iron is incorporated into heme only in the presence of ferrichrome (*228*).

5. Heme represses the uptake of ^{59}Fe-ferrichrome by *Arthrobacter* JG-9 (*228*).

These observations are compatible with the possibility that sideramines may function in the synthesis of heme by acting as a cofactor for ferrochelatase.

The very valid question then arises as to why there is no evidence for such a coenzyme, if that is the role of sideramines, in cell-free experiments with ferrochelatase. A possible explanation is as follows: The sideramines bind ferric iron very tightly, and ferrous iron very weakly (*229*). It is well established that, in order to incorporate into protoporphyrin, iron must be in the ferrous oxidation state. The proposition is made that sideramines, within the cell, transport ferric iron in a soluble form to the active site on ferrochelatase and, furthermore, that there is a reducing functional group (perhaps SH) at the active site of ferrochelatase that serves to reduce the iron in the sideramine, thus releasing it at the proper site for porphyrin to accept it when the porphyrin lands on the active site. This reaction scheme requires that some mechanism be available to re-reduce the active site reducing group on the enzyme. Perhaps NADPH could serve this function. In cell-free experiments on ferrochelatase employing the usual assay mixture this course of events could not be followed. As mentioned earlier, reducing agents such as cysteine and GSH are generally included in ferrochelatase assays. These reducing agents are able to reduce the iron in the sideramine, thus releasing it and making it essentially equivalent to added ferrous iron. Since ferrous iron alone is suitable for incorporation, the effect, if any, of sideramine would not be observed.

One set of ferrochelatase experiments employing an oxidizable substrate (succinate) in place of reducing agents has been reported (213). Synthesis of ^{59}Fe-heme was demonstrated in cell-free extracts of R. spheroides in the presence of succinate without addition of reducing agents when the only iron added to the system was ferrichrome-bound ^{59}Fe. This does not prove that ferrichrome was required, as other iron sources served equally well; however, it does show that the iron in sideramines can be used for heme synthesis. Final analysis of the validity of the hypothesis that sideramines are cofactors for ferrochelatase awaits the purification of ferrochelatase so that binding studies, etc., can be performed.

D. The Magnesium Branch

It is clear that heme and chlorophyll are synthesized using the same chemical intermediates up to protoporphyrin IX (2,3,5,11). Even though the early chemical intermediates are identical, it is not yet clear if a common pool of these intermediates exists for eventual use for heme or chlorophyll formation (230,231). There is some suggested evidence indicating that parallel pathways may exist, as in the case of amino acid biosynthesis. This will be more fully discussed in Section V on the regulation of tetrapyrrole synthesis.

1. EARLY STEPS LEADING TO CHLOROPHYLL

The first reaction leading directly to chlorophyll apparently is the incorporation of magnesium into protoporphyrin IX. The assignment of sequence is based on the observation of Granick (60,61), who found protoporphyrin IX and magnesium protoporphyrin IX in Chlorella mutants and upon the specificity of (−)-S-adenosyl-L-methionine-magnesium protoporphyrin methyltransferase (232–234). It should be pointed out that Granick (63) also found a Chlorella mutant that accumulated magnesium protoporphyrin monomethyl ester and protoporphyrin monomethyl ester, so it may be that there is some variation in the sequence in different systems, though it is more likely that Mg "fell out" during isolation.

Of all the metals incorporated into tetrapyrroles, quantitatively, magnesium heads the list by several orders of magnitude. In spite of the apparent ease with which photosynthetic organisms carry out the process "in the field," no one has succeeded in demonstrating the reaction in a cell-free system. Until recently, in fact, it was necessary to use activated forms of magnesium, e.g., Grignard reagent (61), magnesium alcohoxides (235), magnesium viologen (236), etc.—all under strictly anhydrous conditions. The requirement for anhydrous conditions precluded studies in model systems approaching physiological conditions. It could not be ascertained whether water exerted its effect by decomposing the activated magnesium complex, or if water was an inherent poison in the system.

Baum *et al.* (187,237) found that it was possible to incorporate magnesium under relatively mild (but far from physiological) conditions using pyridine as a catalyst and the perchlorate salt of Mg. This system functioned even under nonanhydrous conditions; up to 8% water was tolerated (238). In light of this development perhaps investigators, discouraged after years of failure, will reexamine magnesium incorporation.

The biosynthetic step following magnesium incorporation involves the esterification of one of the propionic acid side chains catalyzed by (—)-*S*-adenosyl-L-methionine-magnesium protoporphyrin methyltransferase. It is always assumed that the esterification occurs at position 6, since chlorophyll is esterified at position 6, though this has never been experimentally proved. Unfortunately, this is the only step leading to chlorophyll past formation of protoporphyrin that has been successfully studied at the enzyme level. Tait and Gibson (233) demonstrated that the methyl group that esterified the propionic acid comes from *S*-adenosylmethionine (SAM). In *R. spheroides* the enzyme is found in the chromatophore fraction (232), and it catalyzes the esterification of magnesium protoporphyrin IX about 15 times more rapidly than free protoporphyrin IX. Attempts to solubilize this enzyme were unsuccessful. The methyltransferase enzyme has also been demonstrated in plastides of *Zea maise* (234). Mg-protoporphyrin is the most active substrate compared to Zn-protoporphyrin or metal-free protoporphyrin, adding further strength to the supposition that the sequence goes:

$$\text{Protoporphyrin} + \text{Mg}^{2+} \rightarrow \text{Mg-protoporphyrin} \xrightarrow{\text{SAM}}$$
$$\text{Mg-protoporphyrin monomethyl ester}$$

Mg-protoporphyrin monomethyl ester has been detected in etiolated barley leaves incubated with ALA in the presence of α,α-dipyridil (239). Both *R. spheroides* (230,240) and *R. capsulata* (241), when cultured under conditions of moderate iron deficiency, synthesize and excrete relatively large quantities of Mg-protoporphyrin monomethyl ester. Severe iron deficiency strongly inhibits bacteriochlorophyll synthesis (*3,183*) and has long been recognized to cause chlorosis in higher plants. Numerous studies have been conducted to clarify the role of iron in the biosynthesis of chlorophyll and bacteriochlorophyll (*3,242–245*). No conclusive evidence has been obtained. The studies with photosynthetic bacteria show that low amounts of iron allow the microorganisms to synthesize protoporphyrin, but much higher levels are required to promote bacteriochlorophyll formation (*185,241*). Apparently iron is necessary for the conversion of coproporphyrinogen to protoporphyrin and, in addition, it plays an important role at some stage between Mg-protoporphyrin monomethyl ester and the final product.

The role of iron in the biosynthesis of chlorophyll, however, is less clear. Iron-deficient *Euglena gracilis*, though defective in the ability to form chlorophyll, produced protoporphyrin from PBG at rates equal to or greater than iron-sufficient controls, suggesting a role for iron past the formation of protoporphyrin (*244,246*). On the other hand, iron-deficient cowpea leaves incorporate ^{14}C-ALA into chlorophyll a at the same rate as iron-sufficient leaves (*243*). The incorporation of labeled Krebs cycle intermediates, however, was markedly reduced in iron-deficient cowpea leaves. This was interpreted as implicating iron in the synthesis of ALA (*242,243*). On the other hand, experiments with iron-deficient *Nicotiana tobacum* show that normal leaf discs incorporate labeled ALA into chlorophyll at faster rates than chlorotic leaf discs (*245*). The addition of low amounts (0.5 ppm) of iron to the incubation mixture significantly increased the incorporation of labeled ALA into chlorophyll in both chlorotic and normal leaf discs. When incorporation of labeled ALA into coproporphyrin and protoporphyrin was measured in similar experiments, addition of iron to normal leaf discs decreased incorporation into both porphyrins. Addition of iron to chlorotic leaf discs, however, increased the incorporation of labeled ALA into protoporphyrin while decreasing the incorporation into coproporphyrin (*245*).

Ethionine is recognized to inhibit bacteriochlorophyll formation in *R. spheroides*, and in the presence of ethionine these microorganisms excrete large amounts of porphyrin into the culture medium (*247*). The concentration of ethionine required for inhibition of bacteriochlorophyll is well below that required for interference with protein synthesis (*232*). Presumably the ethionine blocks the esterification of Mg-protoporphyrin, since the inhibition is competitive with *S*-adenosylmethionine (*232*). It is interesting to note that, even though the inhibition is at the esterification step, the porphyrin that accumulates is mainly coproporphyrin III (247), and not the immediate precursor to the blocked reaction, namely, Mg-protoporphyrin. Ethionine is known to cause a marked decrease in the level of ATP in the liver, and this decrease is larger than can be accounted for (248) owing to formation of S-adenosylethionine. It seems possible that a similar thing might also occur in R. spheroides, and might account for the observed pattern of response.

2. BIOSYNTHESIS OF CHLOROPHYLL FROM MAGNESIUM PROTOPORPHYRIN MONOMETHYL ESTER

Our picture of the biosynthetic sequence existing between Mg-protoporphyrin monomethyl ester, chlorophyll, and bacteriochlorophyll has been pieced together largely from evidence from two sources. They are the identification of pigments produced by mutants (230,231), and identification of pigments produced by normal organisms under controlled conditions, e.g., etiolated leaves (239), 8-hydroxyquinoline treated R. spheroides (249,250).

a. Complications of the System. Before examining the deduced and seemingly logical biosynthetic sequence, it is perhaps germane to consider some of the difficulties that have tended to retard developments in this area of biosynthesis.

One problem is the fact that, as we progress along the proposed sequence from Mg-protoporphyrin, the suspected substrates become progressively less soluble in water. Organic solvents and detergents can be utilized to some advantage, but the effect of these substances on the enzymes is unknown. Also, as we progress along the proposed sequence, the substrates become less stable. Corwin and Wei (251) have shown the following order of stability for Mg-tetrapyrroles when equilibrated with magnesium phenate in phenol solution. Mg-porphyrins are more stable than Mg-phaeoporphyrins (with the isocyclic ring E), which are in turn more stable than Mg-chlorins (porphyrins with the D ring reduced). They explain the stability series on the basis of progressive warpage of the essentially planar porphyrin ring, and the decreased resonance possibilities in the chlorin. The decreasing planarity and conjugation of the tetrapyrrole not only affects the stability of the Mg-complex, but also decreases the stability of the tetrapyrrole *per se.*

In addition to problems, with the tetrapyrrole part of the reaction system, the biological part of the system apparently contributes its share of troubles. For one thing, the enzymes are probably all insoluble. Certainly the first, (−)-S-adenosylmethionine-magnesium protoporphyrin methyltransferase (232), and the last, chlorophyllase (252), are insoluble. The questionable role of chlorophyllase as a biosynthetic enzyme is discussed in Section II, D, 2, b.

As the water solubility of the tetrapyrrole decreases, and the structural organization increases (assuming that particle-bound enzymes indicate structural organization), it becomes tempting to think of organelles as microfactories synthesizing the necessary components at,'or very near to, the ultimate functional site. The biosynthesis of chlorophyll and bacteriochlorophyll have been shown to be in lockstep with protein synthesis and the development of organelles (chloroplasts and chromatophore material (*116,253*)

Perhaps one should be heartened, however, by chlorophyllase and by the Zn-incorporating enzyme (*206–208*), both of which act on rather insoluble tetrapyrroles in the presence of organic solvents. Zinc-protoporphyrin chelatase functions in 28% ether (*206*) and chlorophyllase functions in 40% acetone (*254*).

b. Sequence of Reactions. To assist the reader in keeping straight the relationship of chlorophyll a to some of the derivatives mentioned in this section, a summary is presented in Fig. 17.

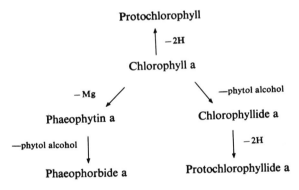

FIG. 17. Scheme showing relationship of chlorophyll a to various derivatives.

Jones has carried out extensive experiments with the photosynthetic bacterium *R. spheroides* (*240,249,250,255*). When wild-type organisms were grown in the presence of 8-hydroxyquinoline, bacteriochlorophyll synthesis was reduced, and a number of tetrapyrrole compounds accumulated in the growth medium (*240,250*). Jones has identified one of these compounds, as 2,4-divinylphaeoporphyrin a_5 monomethyl ester (*249*). Note that many of the tetrapyrroles accumulated under these and similar conditions lack magnesium. It is generally accepted that they are synthesized with Mg in, and that the Mg is spontaneously lost from these rather unstable molecules. The identification of this pigment, with the two vinyl groups intact and with the cyclopentanone ring (E), indicates that ring closure (E) occurs prior to reduction of the vinyl group at position 4. Bogorad (*11,154*) postulates that the carbon α to the number 6 position on the porphyrin is oxidized to the carbonyl prior to ring closure. The carbon β to the number 6 position, now between two carbonyls, would be highly reactive and would attack at the γ bridge carbon to form the cyclopentanone ring. Oxidation to the carbonyl might proceed via dehydrogenation, hydration, and dehydrogenation analogous to the β-oxidation of fatty acids. Alternatively, the reaction might proceed via direct hydroxylation followed by reduction, in a sequence analogous to that proposed by Sano (*180*), for the conversion of coproporphyrinogen to protoporphyrinogen.

The next pigment in the sequence is 2-vinylphaeoporphyrin a_5 monomethyl ester (protochlorophyllide). This pigment has been isolated and identified from several sources. Granick (*63*) has shown this to be formed from ALA by etiolated leaves and to accumulate in a *Chlorella* mutant that cannot form chlorophyll in the dark (*62*). The conversion of 2,4-divinylphaeoporphyrin a_5 monomethyl ester to the 2-vinyl-phaeoporphyrin a monomethyl ester involves only the reduction of the vinyl group at position 4.

Protochlorophyllide (2-vinylphaeoporphyrin a_5 monomethyl ester) is converted to chlorophyllide a by reduction at positions 7 and 8 (the β positions in the D ring). This is a light-catalyzed reduction in many plants and algae (*256,257*), and an enzymic reduction in others (*258*).

The photochemical reduction has been extensively studied by Smith and his colleagues at the Carnegie Institute (*256*). They have studied the effect of temperature, light intensity, and other effects of wavelength of illumination on the photoconversion reaction.

Koski *et al.* (*259*) showed that the photoreceptor for the conversion of protochlorophyllide is this pigment itself. The action spectra showed absorption maxima for protochlorophyllide *in vivo* at 445 mμ and 650 mμ.

The first evidence that the conversion of protochlorophyllide to chlorophyllide could be studied in cell-free systems came from the work of Krasnovsky and Kosobutskaya (*260*). These workers found that they could observe the conversion of protochlorophyllide in hemogenates prepared in the dark from etiolated leaves. Smith *et al.* (*256,261*) made similar observations, and found that they could purify a particulate fraction from the homogenate by differential centrifugation and ammonium sulfate fractionation. The active fraction is a protochloro-phyllide-protein complex, and was given the name protochlorophyll-holochrome by Smith. Protochlorophyll-holochrome has been extensively purified by Smith (256) and by Boardman (262). Molecular weight estimations of 6×10^5 and 9×10^5 have been made, and apparently there is one protochlorophyllide molecule per protein of the above size (256,257).

The size and shape of protochlorophyllide-holochrome is similar to two other materials extracted from chloroplasts of green leaves, namely, carboxydismutase (263) and the so-called fraction 1 protein (264). The fraction 1 protein seems to be an aggregate of several enzymes (phosphoribulokinase, phosphoribuloisomerase, and carboxydismutase).

Boardman has purified protochlorophyllide-holochrome and protein fraction 1, but cannot distinguish between them in the ultracentrifuge (257,262). It may be that they are two particles of similar size, or it could be that "true" protochlorophyllide-holochrome is a smaller molecule that makes up a fraction of the apparent enzyme complex fraction 1. Enzymic activities of protochlorophyllide-holochrome have not been reported. It would be particularly interesting to know if chlorophyllase activity is associated with the holochrome.

For relative simplicity in the above discussion, holochrome has been referred to as if the tetrapyrrole moiety were solely protochlorophyllide. This is not the case (257,262,265,266); a variable amount of protochlorophyll is present in addition to protochlorophyllide. The protochlorophyllide : protochlorophyll ratio varies considerably with different plants and is affected by such things as age. The major path for chlorophyll a synthesis seems to be:

$$\text{Protochlorophyllide} \xrightarrow{\text{2 H}} \text{chlorophyllide a} \xrightarrow{\text{phytol}} \text{chlorophyll a}$$

while a small amount goes via:

$$\text{Protochlorophyllide} \xrightarrow{\text{phytol}} \text{protochlorophyll} \xrightarrow{\text{2 H}} \text{chlorophyll a}$$

Thus there is clearly a branch at this point in the biosynthesis of chlorophyll a. The reason for this branch has not been explained. Perhaps it has something to do with control of chlorophyll synthesis.

The function of chlorophyllase as a biosynthetic enzyme has not been unequivocally established. Both thermodynamics and analogy argue against a biosynthetic role. The participation of water in the reaction suggests at first that hydrolysis rather than esterification would be favored. It is difficult to recall any other case where a biosynthetic reaction occurs via simple reversal of a hydrolytic reaction. The possibility has been considered in the past, e.g., protein synthesis, but it seems clear that in order to reverse a hydrolytic reaction it is always necessary to activate one of the reactants.

The chlorophyllase reaction, however, may be unique. It seems within the realm of possibility that the reaction is "pulled" in the direction of synthesis because of the extreme water-insolubility of the product, chlorophyll. Since the reaction presumably occurs in the immediate vicinity of lipid material, into which the chlorophyll could migrate, this would in effect remove the chlorophyll molecule from the equilibrium mixture, thus favoring synthesis over hydrolysis. There is some circumstantial evidence that chlorophyllase functions biosynthetically in that, upon induction of greening in etiolated leaves, the level of chlorophyllase activity increases markedly and rapidly (252,254,267,268).

The enzyme is remarkable in that it functions in high concentrations of methanol, ethanol, and acetone. In aqueous acetone, chlorophyll is hydrolyzed to chlorophyllide and phytol. In alcoholic solution, the enzyme catalyzes transesterification. The enzyme is quite nonspecific with regard to the alcohol moiety. More specificity is demonstrated concerning the tetrapyrrole. It is active with the usual chlorophylls, bacteriochlorophyll, chlorobium-chlorophyll, and upon the pheophytins derived from these chlorophylls, and it is not active with protochlorophyll (267,269). This indicates that reduction of the D ring to yield the chlorin is a necessary requisite. As already mentioned, however, varying amounts of protochlorophyll are observed along with chlorophyllide (262,265). This leaves an unanswered question; namely, how did the protochlorophyllide become phytolated? The enzyme has been solubilized and partially purified (252,270).

In summary, most evidence suggests that the principal biosynthetic pathway between protoporphyrin and chlorophyll a is as follows:

$\rightarrow$ Protoporphyrin IX $\xrightarrow{\text{Mg}}$ Mg-protoporphyrin IX $\xrightarrow{\text{SAM}}$

$\rightarrow$ Mg-protoporphyrin monomethyl ester $\rightarrow$ Mg-2,4-divinylphaeoporphyrin a_5 monomethyl ester $\rightarrow$

$\rightarrow$ Mg-2-vinylphaeoporphyrin a_5 monomethyl ester (protochlorophyllide a) $\xrightarrow{+2\,\text{H}}$

$\rightarrow$ chlorophyllide a $\xrightarrow[\text{alcohol}]{+\,\text{phytol}}$ chlorophyll a

c. Biosynthesis of Chlorophyll b. Chlorophyll b differs from chlorophyll a in that the methyl group in the number 3 position of chlorophyll a is replaced by CHO. Most higher plants contain chlorophyll a and chlorophyll b in a ratio of about 2.5 : 1 (3). The close similarity in structure suggests that one of these chlorophylls may be the precursor of the other. Alternatively, they could arise by totally independent biosynthetic pathways, or more likely via a branch in the pathways. This is a point that has not as yet been resolved.

Two types of experiments have been carried out to shed light on this question. Some conclusions can be drawn from studies on the distribution of the two pigments. Most higher plants contain both pigments. Some, however, are known to contain only chlorophyll a (5,271). No plant has yet been found to contain chlorophyll b in the absence of chlorophyll a. If chlorophyll b were a precursor of chlorophyll a, one would expect to find an occasional mutant that could not make chlorophyll a and, thus, would contain only chlorophyll b. Since the reverse is true, this suggests that chlorophyll a is the precursor of chlorophyll b. The same evidence also bears on the possibility of the two chlorophylls arising as the result of a branch in the biosynthetic pathway, for instance, at Mg-2,4-divinylphaeoporphyrin a_5. If a branch occurred at that point, it would again seem likely that a mutant might be found having a block in the conversion of Mg-2,4-divinylphaeoporphyrin a_5 to chlorophyll a yet retaining the ability to synthesize chlorophyll b.

The other major source of evidence has been provided through the use of ^{14}C-labeling. In long-term labeling experiments, if chlorophyll a and chlorophyll b had identical specific activities, this would be compatible with chlorophyll a being a precursor of chlorophyll b. On the other hand, different specific activities would indicate that synthesis occurred via different biosynthetic routes. Since the results of this type of experiment have not been clear-cut (272,273,274,275,276), the question remains partly open. From all the evidence, however, it does appear that the propositions that chlorophyll b is a precursor of chlorophyll a, and that the two chlorophylls are formed independently from different precursors, are eliminated. This topic has been recently reviewed by Smith and French (277).

d. Biosynthesis of Bacteriochlorophyll. The nonsulfur, purple photosynthetic bacteria *Athiorhodacae* and the sulfur, purple photosynthetic bacteria *Thiorhodacae* contain as their primary photosynthetic pigment bacteriochlorophyll.

There is strong circumstantial evidence that both chlorophyll a and bacteriochlorophyll are synthesized by very similar, or even the same, pathways. This evidence can be summarized briefly.

Mg-2,4-divinylphaeoporphyrin a_5 monomethyl ester and 2-vinylphaeoporphyrin a (phaeophorbide) have been identified in the culture medium of *R. spheroides* (249,250). The same pigments have been found in organisms synthesizing chlorophyll a (62,63). This is somewhat analogous to finding protoporphyrin IX in organisms that make heme and chlorophyll.

Our information of the intermediates arising after the phaeophorbide a stage comes mainly from the work of Jones, who has studied wild-type

R. spheroides in the presence of 8-hydroxyquinoline, and Lascelles (*230*), who has studied *R. spheroides* mutants. Jones has identified 2-divinyl-2-hydroxyethyl-phaeophorbide a in his systems (*255*), and Lascelles has identified the magnesium derivative of the same tetrapyrrole in her system (*230*). It is not difficult to visualize 2-divinyl-2-hydroxethyl-phaeophorbide as an intermediate between chlorophyllide a and bacteriochlorophyll. Jones (*255*) has suggested the reaction sequence shown in Fig. 18. Bogorad (*11*) reminds us to accept information such as the finding of this apparent intermediate with an open mind, remembering the *Chlorella* mutant that makes hematoporphyrin.

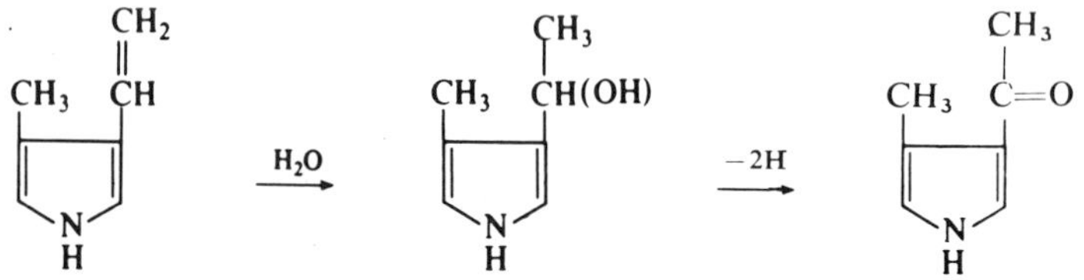

FIG. 18. Proposed reaction sequence during biosynthesis of bacteriochlorophyll taking into account the apparent intermediate 2-divinyl-2-hydroxyethylphaeophorbide a.

Chlorophyllase is active with bacteriochlorophyll (*11*) but there is no evidence yet at which stage the phytol group is incorporated during bacteriochlorophyll synthesis.

Comparatively little is known about the biosynthesis of chlorophylls other than the three already discussed. Richards and Rapaport (*278*) have studied the biosynthesis of *Chlorobium* chlorophylls-660 from *Chlorobium thiosulfatophilium*-660 in an effort to learn at which stage the meso-alkyl group is attached to the tetrapyrrole. Recall that a meso-alkyl-substituted porphyrin will exhibit a phyllo-type spectrum (Section I, B, 2, *d*). Interestingly, they found that the *C. thiosulfatophilium* makes mostly coproporphyrin III and uroporphyrin III under their incubation conditions; however, traces of other porphyrins were also observed. They did find a tetracarboxylic porphyrin with a phyllo-type spectrum in trace amount, but were unable to conclude whether it was an artifact or an intermediate.

E. The Cobalt Branch

Studies on the biosynthesis of the cobalamins can be grouped into two rather broad categories. In the first category are the studies employing radioisotopes. These have been of the classical type where a suspected intermediate in labeled form has been supplied to a microorganism and the extent and positions of labeling in the product determined. The second category includes experiments where a "factor" has been isolated from a microorganism (wild or mutant), its structure determined, and the further metabolism of the compound studied. These two approaches have yielded much information of value, but they have also left rather large gaps in our understanding of the overall picture. The largest gap exists in the early middle phase of the biosynthetic sequence, and includes formation of the corrin ring, substitutions on the corrin ring, and the incorporation of cobalt.

The development of the field of cobalamin biochemistry has come about during an era when the problems of nomenclature are well recognized. As a result, more order and logic prevail in the area of corrin nomenclature than in some of the older fields of biochemistry (30). Nevertheless, the "older" corrin literature (dating back 15 years) abounds with "factors" and "compounds" whose structures and relationships to the corrin were not recognized at the time. Some of the more important of these compounds are listed in Table I along with their scientific names to assist one in reading the original papers.

It is, of course, not strictly correct to refer to the biosynthesis of vitamin B_{12} (cyanocobalamin), since vitamin B_{12} is not a biosynthetic product, but rather a derivative of the natural corrinoid. Nevertheless, the term vitamin B_{12} is used in this context in this chapter, partly because of common usage of the term, and partly because many studies on the biosynthesis of the corrinoids have actually measured the biosynthetic product as vitamin B_{12}, since the vitamin form is more stable and convenient to assay.

1. TRACER STUDIES

Isotope studies have done much to define the origin of the corrin ring and its substituents. They have not, however, yielded much information regarding the actual sequence of reactions leading to the complete corrinoids.

Shortly after the structure of vitamin B_{12} became known, revealing its relationship to heme and chlorophyll, Shemin tested ^{14}C-labeled ALA as a vitamin B_{12} precursor. He found that an actinomycete used the added ALA to make vitamin B_{12} (78,79). Schwartz et al. also used an actinomycete and succeeded in demonstrating that PBG is a precursor of vitamin B_{12} (280), and L-threonine-^{15}N has been shown to give rise to the aminopropanol moiety of vitamin B_{12} (281).

Bray and Shemin (282,283) have shown that 6 of the 7 so-called "extra" methyl groups of vitamin B_{12} are derived from methionine.

TABLE I

CORRIN NOMENCLATURE[a]

Systematic name	Original name	Semisystematic name	References
α-(5,6-Dimethylbenziminazolyl) cobamide cyanide OR α-(5,6-dimethylbenziminazolyl) cyanocobamide	Vitamin B_{12} (Factor II)	Cyanocobalamin	(12,6,30)
α-(5,6-Dimethylbenziminazolyl) hydroxocobamide	Vitamin B_{12a}	Hydroxocobalamin	(12)
α-(5,6-Dimethylbenziminazolyl) aquocobamide chloride (or sulfate, etc.)	Vitamin B_{12b}	Aquocobalamin	(12)
α-(2-Methyladen-7-yl) cobamide cyanide	Factor A (B_{12m}, B_{12d})	—	(6,30,279)
Cobinamide dicyanide	Factor B (Etiocobalamin B_{12p}, Factor I)	—	(6)
Cobinamide phosphate dicyanide	Factor B phosphate (Factor C?)	—	(6)
Cobyrinic acid $a\ b\ c\ d\ e\ g$-hexamide dicyanide (cobyric acid dicyanide)	Factor V_{1a}	Cobyric acid	(6)
Cobamide dicyanide	(Etiocobalamin phosphoribose)		
α-(5,6-Dimethylbenziminazolyl) cobamide nitrite OR α-(5,6-dimethylbenziminazolyl) nitritocobamide	Vitamin B_{12c}	Nitritocobalamin OR nitrocobalamin	(12,30)
α-(5-Hydroxybenziminazolyl) cobamide cyanide	Factor III (Factor I)	—	(6,30,279)
α-Aden-7-ylcobamide cyanide	ψ Vitamin B_{12} ($\psi\ B_{12b}$, Factor IV, B_{12t})	—	(6,279)
α-(5,6-Dimethylbenziminazolyl) cobamide thiocyanate OR α-(5,6-dimethylbenziminazolyl) thiocyanatocobamide	—	Thiocyanatocobalamin	(12)
P^1-Guanosine-5′ OR guanylcobamide cyanide	(Factor C)	—	(6)
P^2-Cobinamide pyrophosphate	(Factor y_1)	—	(6)
α-(5,6-Dimethylbenzimidazolyl)-Co-5′-deoxyadenosylcobamide	Coenzyme B_{12}	5′-Deoxyadenosyl-cobalamin	(30,300)
α-(5,6-Dimethylbenzimidazolyl)-Co-methylcobamide	Methylcobalamin	—	(30)
α-(Benzimidazol)-Co-5′ deoxyadenosylcobamide	—	Benzimidozolylcobamide	(300)
α-(Adenyl)-Co-5′-deoxyadenosylcobamide	Coenzyme pseudo vitamin B_{12}	Adenyl cobamide	(30,300)

[a] Names in parentheses are less commonly used.

There are actually 8 free methyl groups on the molecule, but one of the two methyl groups at C-12 arises following the decarboxylation of the acetic acid residue at that position. The remaining methyl group not coming from methionine and not arising via decarboxylation is the one in the α position on pyrrole ring A. This methyl group apparently is derived from the δ-carbon of ALA, and corresponds to the "missing" δ bridge carbon.

2. ISOLATION TECHNIQUES

The isolation and identification of incomplete corrinoids, and the arrangement of these compounds in order of increasing complexity, have provided some understanding of the sequence of reactions leading to the completed molecule. This approach has, however, consistently revealed relatively complete corrinoid compounds. This is so because of the assay system required for their discovery and isolation. Although the corrinoid compounds are strikingly colored (purple, red, orange, yellow), they do not possess the sharp absorption bands characteristic of the porphyrin derivatives. That favorite tool of the porphyrin chemist, the hand spectroscope, is completely useless. Even recording spectrophotometers are unsatisfactory unless the concentration of corrin is fairly high, and it is relatively free of other light-absorbing material. None of the cobalt-containing compounds fluoresce.

Many of the compounds in question have been discovered and purified using microbiological assays. The different assay organisms, of course, have different specificities; however, for the most part they are able to utilize only those corrins that are near the terminal end of the biosynthetic sequence.

Recently it has become possible and practical to isolate molecules without microbiological activity. This is now practical because enough is known about the physical-chemical properties of these molecules that their isolation can be planned on theoretical grounds. Two factors are necessary for this type of work. The first factor is faith, and the second is a large batch of starting material. One simply carries out an isolation using techniques that would be expected to work if the compound being sought exists. Toohey's (284,285) remarkable discovery of the first cobalt-free corrinoid is exemplary of this approach.

During the search for vitamin B_{12} intermediates, an almost innumerable number of compounds have been isolated. Many have been fully characterized. It is not uncommon, however, to find that a single microorganism produces more or less equivalent amounts of several corrin derivatives. This presents something of a problem when it comes to

explaining on a rational basis the sequence of reactions under consideration. Rather than attempting a detailed discussion on these investigations,
the author chooses to present a general summary of the overall picture.

Figure 19 is an abbreviated summary of the steps between cobyrinic
acid and vitamin B_{12} (8).

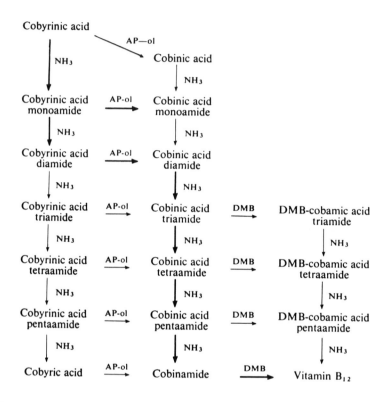

FIG. 19. Summary of apparent path of biosynthesis between cobyrinic acid and vitamin
B_{12} in *Propionibacterium shermanii*, adapted from Wagner (8). Heavy arrow = major
pathway; light arrow = minor pathway; AP-ol = 1-aminopropan-2-ol; NH_3 = amidation
of one carboxyl; DMB = 5,6-dimethylbenzimidazole.

The first stage of biosynthesis can be considered to be complete with
cobyrinic acid (7,8,286–289). It appears that cobyrinic acid (see Fig. 4)
can participate in one of two biosynthetic pathways, though this point
has recently been disputed (290). The carboxyl groups at a, b, c, d, e, and
g can be amidated stepwise. The final product of this series is cobyric
acid (cobyric acid hexaamide). Alternatively, the carboxyl group at
position f can be amidated with 1-aminopropan-2-ol, to yield the product

cobinic acid. The cobinic acid can then undergo stepwise amidation of the remaining carboxyl groups a, b, c, d, e, and g to yield cobinamide. Once cobyrinic acid starts down one of the above two pathways, however, it is not committed to continue through to the products mentioned, but can cross from one path to the other. To further complicate matters, once the f carboxyl is amidated with 1-aminopropan-2-ol and, if at least 3 of the other carboxyls are also amidated, the nucleotide moiety may be attached. The rest of the amides are subsequently attached stepwise.

Finally, apparently at any one of several stages during the biosynthesis, the molecule can be converted to the coenzyme form, i.e., 5'-deoxyadenosine, or a methyl group can be attached at the 6th coordination position of the cobalt.

Bernhauer *et al.* (7) believe that the 5'-deoxyadenosyl group is introduced at an early stage, probably at cobinic acid monoamide or diamide.

Since the 5'-deoxyadenosyl moiety can be attached as early as cobinic acid monoamide, and as late as aquocobalamin, as well as at intermediate stages, it would seem that either several enzymes with different specificities exist capable of catalyzing the condensation of ATP and corrins or that there is one enzyme with a very wide range of specificities. This apparent lack of specificity, demonstrated by the capacity to cope with a variety of corrins, is not confined to the addition of the 5'-deoxyadenosyl moiety to the 6th ligand position of the cobalt.

A recent report by Bartosinski *et al.* (*290*) disputes the conclusion that cobyrinic acid can be metabolized via either the cobinic or the cobyrinic pathway as shown in Fig. 19. They report experiments showing that *Propionibacterium shermanii* does not make detectable amounts of the cobinic acid amides. They could, however, detect small amounts of cobinamide. From these experiments they conclude that *P. shermanii* carries out the stepwise amidation of cobrinic acid yielding cobyric acid which they found to constitute 85% of the carboxylic acid-containing corrinoids, and which in turn constitute up to 25% of the total corrins in *P. shermanii*.

a. Studies on Intermediates. Experiments with cell-free extracts of *P. shermanii* have shown that cobyrinic acid, cobyrinic acid monoamide, and cobinic acid cannot be converted to the 5'-deoxyadenosyl forms (*8,286*). The investigations of Friedrich (*288*) indicate that *P. shermanii* can amidate cobyrinic acid triamide, tetraamide, and pentaamide to form cobyric acid. Di Marco *et al.* (*289,291*) have isolated a number of mutants of *Nocardia rugosa* that can incorporate the aminopropanol moiety and others that can incorporate the amide groups. *P. shermanii* cultures cannot decarboxylate cobyrinic a, b, c, d, e, g-hexaamide, f-N-threonine (*292*). This indicates that threonine is decarboxylated prior to being attached to the corrin and that aminopropanol is the substrate for the incorporation (*293*). There is, however, no evidence that threonine can be directly decarboxylated (*294*). As a result of his studies on the microbial metabolism of aminoketones, Turner (*295*) has suggested that 1-aminopropan-2-ol may arise through the following reactions:

$$\text{Threonine} \xrightarrow{-2H} \text{2-aminoacetoacetate} \xrightarrow{-CO_2} \text{aminoacetone} \xrightarrow{+2H} \text{1-aminopropan-2-ol}$$

Threonine dehydrogenase has been demonstrated in several microorganisms (*296,297*) as well as in mammalian tissues (*297,298*). 2-Aminoacetoacetate, being a β-keto acid, decarboxylates spontaneously. 1-Aminopropan-2-ol dehydrogenase has been demonstrated in *E. coli* (*295*). In addition to the above-mentioned origin of aminoacetone, it can be formed by the condensation of glycine and acetyl-CoA (*298*).

Microorganisms capable of synthesizing corrinoids are sometimes observed to be surprisingly unselective in the nucleotides that they use in forming the finished product (*12*). Thus the cobamide-requiring *E. coli* mutant 113–3, when supplied with cobinamide and a variety of bases, e.g., adenine, benzimidazole, and 5,6-dimethylbenzimidazole, utilizes the added base to make a B_{12} derivative. Alternatively, a variety of bases can be added to B_{12}-synthesizing microorganisms and the bases are incorporated into the corresponding B_{12} analog. Thus *C. tetanomorphum*, which normally synthesizes the adenine analog of the coenzyme, can be induced to make the benzimidazole analog or the 5,6-dimethylbenzimidazole analog, by growing the organisms with the appropriate base (*299,300*).

The accumulating weight of evidence indicates that the nucleotide is added to cobinamide via the following sequence of reactions (*68,290, 301–303*).

$$\text{Cobinamide} + \text{ATP} \rightarrow \text{Cobinamide-P} + \text{ADP}$$

$$\text{Cobinamide-P} + \text{GTP} \rightarrow \text{Cobinamide-PPRG} + \text{PP}$$

$$\text{Cobinamide-PPRG} + \alpha\text{-Ribazole*} \rightarrow \text{Vitamin } B_{12} + \text{GMP}$$

The evidence supporting this sequence has accumulated to a large extent upon identification of cobinamide-P (*68*), cobinamide-PPRG (*68,290*), and 1-α-D-ribofuranosyl-5,6-dimethylbenzimidazole (*304*) in suspensions of vitamin B_{12}-synthesizing bacteria, particularly *Nocardia rugosa* and *P. shermanii*. This evidence has been supported and extended

---

* α-Ribazole = 1-α-D-ribofuranosyl-5,6-dimethylbenzimidazole.

with labeling experiments. Friedman and Harris (*304–306*) have also provided supporting evidence in their studies on an enzyme from *P. shermanii* that catalyzes the *N*-transglycosidation of nicotinamide mononucleotide and 5,6-dimethylbenzimidazole. The initial product of the reaction, α-ribosole-PO_4, loses the $-PO_4$ by the action of a phosphatase in *P. shermanii*.

The studies of Bernhauer and his co-workers (*307–309*) led them to postulate the following reaction sequence:

$$\text{Cobinamide-P} \longrightarrow \text{cobinamide-phosphoribose-1-phosphate} \xrightarrow{\text{DMB}} \text{vitamin } B_{12} + P_i$$

The evidence supporting this hypothesis arises in part from studies on the comparative effectiveness of 5,6-dimethylbenzimidazole, α-ribazole, and α-ribozole-3'-PO_4 as precursors of vitamin B_{12}, where 5,6-dimethylbenzimidazole was found to be the preferred substrate. Bernhauer also interpreted the fact that *P. shermanii* suspensions have the ability to convert purine and benzimidazole cobamides into the vitamin B_{12} form when supplemented with 5,6-dimethylbenzimidazole as an example of base exchange, and not nucleoside exchange. More recent experiments utilizing ^{14}C residing exclusively in the ribosyl moiety have shown that the exchange involves benzimidazole riboside, and not just the benzimidazole (*310*).

3. BIOSYNTHESIS OF B_{12} COENZYMES

For a number of years following its discovery, vitamin B_{12} was isolated as cyanocobalamin. Several other forms of the vitamin were also isolated, one of the more prominent being called vitamin B_{12b}. In vitamin B_{12b} (aquocobalamin) the ligand in the 6th coordination position on the cobalt is water. None of the vitamin forms could be implicated in a coenzymic role. In 1958 Barker *et al.* (*311*) discovered that a light-sensitive cofactor was required for the conversion of glutamate to β-methylaspartate catalyzed by glutamate mutase. This light-sensitive cofactor was relatively quickly recognized as a corrin derivative, and the structure of the compound was shown to be adenyl-cobamide-5'-deoxyadenosine. Other forms of the coenzyme were soon isolated and identified. They differed from the above by having different nucleotides in the 5th coordination position (*300,312,313*). The striking feature of the B_{12} coenzymes is the cobalt-carbon bond. This bond is rather labile, and it is now recognized that this lability accounts for the fact that previous work with corrinoids had yielded a complex with either CN or H_2O as the 6th ligand. The CN apparently originated as an impurity from charcoal, an absorbant commonly used in isolating the vitamin.

Brady and Barker (*314,315*) were the first to demonstrate the enzymic conversion of cyanocobalamin and aquocobalamin to the coenzyme form in cell-free extracts of *P. shermanii*. This work was followed by that of Weissbach et al. (*316–319*) in a series of extensive studies using cell-free extracts of *C. tetanomorphum*. The enzyme system requires ATP, NADH, and either GSH or mercaptoethanol. Experiments with ^{14}C-labeled ATP showed that this compound is a substrate of the reaction (*317*). During this reaction the ATP is cleaved in a unique way, releasing inorganic tripolyphosphate (PPP) (*319*). This was demonstrated with the use of ^{32}P-labeled ATP (*319*).

Smith and co-workers (*320–322*) and others (*323–327*) have converted cyanocobalamin and aquocobalamin to the coenzyme by chemical means. Some understanding of the mechanism of the enzymic reaction is afforded by studies on the purely chemical conversion of the two cobalamins to the coenzyme.

A prerequisite of the chemical conversion is that the cobalt, normally in the +3 oxidation state in the corrins, must be reduced to the +1 oxidation state (*8,12,328,329*). Vitamin B_{12} with cobalt in the +2 oxidation state is known and is generally designated vitamin B_{12r}, while the +1 oxidation state is generally designated vitamin B_{12s}. When the B_{12} is reduced to B_{12s}, it will react with many alkylating agents, $(CH_3)_2SO_4$, or 2′,3′-isopropylidene-5′-tosyladenosine) to yield the corresponding alkylated cobalamin. The requirement for CO^{1+}, therefore, seems to explain the cofactor requirements in the enzymic reaction (*319,330*).

Two alternatives are available for the chemical synthesis of alkyl cobalamins. One employs the reduction of cobalamin with thiols to form a sulfide complex of bivalent cobalt as the reactive intermediate (*331*). In the dark and in presence of alkylating agent, alkyl cobalamins are formed. The other synthesis for the direct alkylation of the cobalt involves the utilization of alkyl Grignard compounds (*332*). In this instance, cobalt remains in the +3 oxidation state and the entering alkyl group participates as an anion. Using this technique, alkyl-cobalt porphyrins have been prepared (*333*). Other very interesting types of model alkyl-cobalt compounds have been prepared from cobalt-dimethylglyoxime complexes (*334*). The alkyl-cobalt porphyrins and alkylcobaloximes, since they can be prepared in quantity, should prove very useful in model enzyme studies.

Many suggestions have been made concerning the sequence of reactions that occur in the gaps already mentioned. For instance, Hodgkin has suggested that the "extra" methyl groups may be introduced prior to the ring closure (*335,336*). There is, however, no direct evidence on this point. The suggestion has also been made that the cobalt atom is introduced prior to ring closure, and that the ring actually closes around the cobalt (*8,337*). This conclusion was prompted largely by the observation that no tetrapyrroles, clearly precursors of vitamin B_{12}, had been found without the cobalt already in place. This situation is now changed, and several such compounds have been isolated (*284,285*). Before discussing these compounds, however, mention should be made of other evidence indicating that cobalt is not incorporated until the macrocyclic ring is formed. The first such evidence was obtained by Porra and Ross (*338*) with cell-free extracts of yeast and *C. tetanomorphum* in studies upon cobalt incorporation into porphyrins. They found that *C. tetanomorphum* apparently has a cobaltochelatase that incorporates Co^{2+} selectively over Fe^{2+}. As was pointed out in an earlier section, ferrochelatase will incorporate metals other than iron, cobalt being one of the metals that is actively incorporated. *C. tetanomorphum* was chosen since it is accepted that this microorganism does not make heme compounds, and thus presumably does not contain ferrochelatase. The converse experiment was carried out with yeast, since these organisms do make heme compounds but do not make corrins. In the latter case, they found that iron was inserted more efficiently than cobalt (*338*). It is clear that none of the porphyrins used in this study could be converted to the normal cobalamins, and they only represent model substrates. Additional evidence indicating that the macro-ring forms prior to incorporation of cobalt comes from the work of Burnham (*75,76*) and of Porra (*77*). Both of these workers found that cell suspensions and cell-free extracts of *Clostridia* convert ALA to uroporphyrinogen. They suggest that this represents the first macrocyclic ring precursor of B_{12}.

The recent work of Toohey (*284,285*) demonstrates quite conclusively that cobalt-free corrins can be made by microorganisms. He has found that it is possible to isolate cobalt-free corrinoids from *Chromatium* strain D and from *Rhodospirillum rubrum*. When reacted with cobalt under appropriate conditions, these corrins incorporate cobalt and have essentially identical spectral characteristics with cobinamide.

The cobalt-free corrin was extracted from *Chromatium* D or *R. rubrum* by basically the same procedure as is 5′-deoxyadenosylcobinamide. The purified material is orange-red in color and has an orange-red fluorescence similar to that of many porphyrins. The absorption spectrum is shown in Fig. 20. For comparison, the absorption spectrum of cyanocobalamin is given in Fig. 21. The general shapes of the two spectra are obviously similar, though the cobalt-containing molecule absorption bands are shifted about 30 mμ toward longer wavelengths. This material separates into several fractions during electrophoresis. The major

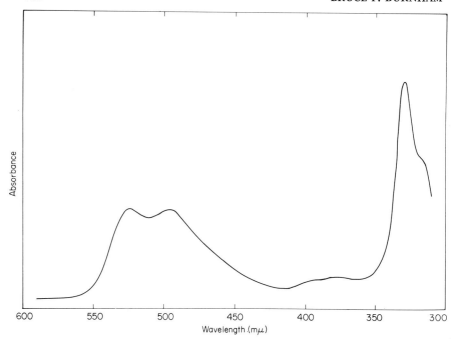

Fig. 20. Absorption spectrum of cobalt-free corrin isolated from *Chromatium* strain D.

pigment is neutral, and other components have +1, −1, −2, and −3 charges. The purified material has been examined by X-ray fluorescence and emission spectrography, and was found to contain no metal atoms. The chemical evidence indicates that the major pigment is equivalent to cobalt-free cobamide, i.e., there is no heterocyclic base. The neutral behavior of the material on electrophoresis can be explained by a −1 charge on the ionized phosphate ester, balanced by a +1 charge somewhere on the ring system, presumably on one of the ring nitrogens. The negative compounds probably differ in that they are not fully amidated, thus leaving free carboxyl groups. The positive compounds presumably do not contain the phosphate ester.

The compound as described thus far does not seem to be out of the ordinary. It seems possible to account for the charge properties logically. Other characteristics, however, are not so easily understood. The major compound which is electrophoretically neutral is the one most studied, and all of the following comments deal with this material. If the only difference between the several cobalt-free corrins is in the side chains, it is likely that they all behave the same way.

The color of the neutral compound is pH-dependent (*284*). The spectral changes, however, are not entirely the result of straightforward ioniza-

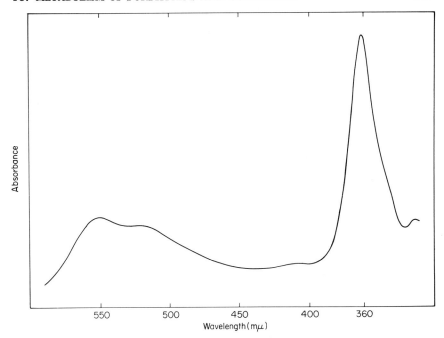

FIG. 21. Absorption spectrum of cyanocobalamin (vitamin B_{12}).

tions. If the solution is made 0.1 N with sodium hydroxide, the color changes immediately to yellow (designated yellow compound II by Toohey). The absorption spectrum of this material is shown in Fig. 22, Curve 3. When the basic solution is neutralized, the yellow color becomes more intense (yellow compound I) and the absorption shifts (Fig. 22, curve 1). The original red color (Fig. 22, curve 2) does not reappear. It is possible to convert the red compound directly to the yellow compound I, with an absorption maximum at 485 mμ by adjusting the pH slowly with dilute base, and either allowing the reaction mixture to stand for several hours, or by heating at 100° for 2 minutes. The conversion of yellow compound I to yellow compound II is reversible upon changing pH, but the conversion of red compound to yellow compound I is not reversible.

When the neutral red compound is titrated with acid, no spectral changes are observed. If the acidic solution (1.8 M HCl) is allowed to stand at room temperature in the dark for several hours, a slow loss of red color is observed, and a yellow product appears having the same spectral characteristics as yellow compound I (*339*). Toohey (*284*) also noted that the red compound is moderately unstable in light, again being converted to yellow compound I.

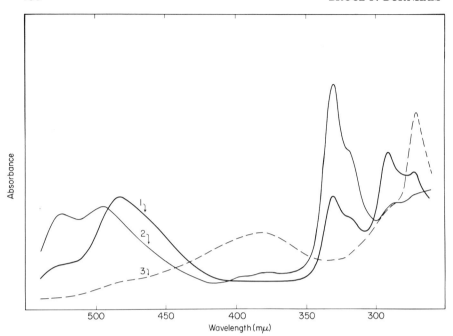

FIG. 22. Absorption spectra of the three forms of cobalt-free corrin. (1) Yellow compound I, in 0.07 *N* NaOH, then neutralized, (2) Red compound in neutral solution. (3) Yellow compound II, in 0.07 *N* NaOH.

If the neutral red corrin is heated in the presence of cobalt in dilute acid, under conditions where cobalt is rapidly incorporated by porphyrins, there is no incorporation of cobalt into the corrin. Similarly, Toohey (*284*) found that Fe^{3+} and Mg^{2+} were not incorporated under these conditions. Burnham (*339*) obtained similar results after converting the red corrin to yellow compound I. Cobalt can only be incorporated under basic conditions (*284*).

Since the structure of the product of the cobalt incorporation reaction is well established (cobinamide), one would suppose at first that it should be possible to predict the structure of the reactant. Certainly this is true for metalloporphyrins. Unfortunately, in the present case it is not so easy. Certain of the chemical properties of the cobalt-free corrin are difficult to rationalize with what might be termed expected properties.

For instance, if the compound had one or more pyrrolinine rings, the addition of acid would be expected to protonate the nitrogens, as is the case with porphyrins, and the spectral characteristics of the protonated species should be different from the unprotonated species (*4*). If there is a single pyrrolinine nitrogen, protonated at neutral pH, as seems to be the case judging from the $+1$ charge, then titration with base should reversibly convert it to a neutral molecule. Above pH 10, the molecule does in fact have a net negative charge, but this is not reversible (284). The spectral characteristics of the yellow compound I bear considerable resemblance to those of a dipyrrilmethene (35). The spectral characteristics of the red corrin suggest that it is more extensively conjugated than yellow compound I.

The exact structure of the red and yellow corrins thus appears to remain open. Chemical studies on the cobalt-free corrins are badly hampered by the paucity of material for investigation. The problem of obtaining more material is complicated by the fact that there is no good assay for the corrin. The red cobalt-free corrin behaves as a competitive antagonist of vitamin B_{12} in the microbiological assay with *E. coli* 113-3 and *Lactobacillus leichmanii* ATCC 4797, while the yellow compound has no effect (340). Perhaps a suitable assay can be developed taking advantage of this property.

One of the many interesting questions that comes to mind when considering the discovery of cobalt-free corrins is—why does *Chromatium* make this material? Toohey (284,285) reports that he cannot detect any vitamin B_{12}, nor any B_{12} coenzyme, in this organism. Perhaps, then, cobalt-free corrin is a vitamin B_{12} precursor, and it accumulates because it cannot be further metabolized. This situation would be analogous to Granick's discovery that protoporphyrin IX accumulates in a *Chlorella* mutant unable to make chlorophyll. On the other hand, *R. rubrum* (285) and *R. spheroides* (339) make cobalt-free corrin in amounts roughly equivalent to that produced by *Chromatium*, yet both of these organisms do make vitamin B_{12} (341) This argues against the cobalt-free corrin. being a metabolic intermediate in the biosynthesis of B_{12}, since intermediates do not generally accumulate. Recall, however, that uroporphyrin and coproporphyrin do sometimes accumulate. These are stable side products of the true intermediates, the corresponding porphyrinogens. The cobalt-free corrins could be in an analogous situation. Nevertheless, it is tempting to consider the cobalt-free corrins as possible precursors of the cobalamins, and this question is under investigation in the author's laboratory. The results obtained so far show that ^{14}C-labeled cobalt-free corrin isolated from *Chromatium*, when supplied to growing *Clostridium perfringens*, can be converted to ^{14}C-labeled vitamin B_{12} (339).

The possibility that cobalt-free corrins might have a metabolic function other than that of B_{12} precursor is very intriguing, and perhaps should be borne in mind. The following points can be raised in argument against the cobalt-free corrins   cobalamin precursors. It is very common to see branches in biosynthetic pathways. These branches, however, are always divergent and yield different end products. How many examples can one bring to mind of convergent biosynthetic pathways, where a single end product is made by two or more reaction sequences? Looking at Fig. 19, it appears that cobinamide can be synthesized via several different routes. If we now suggest that similar pathways can be followed by corrins lacking cobalt, presuming that the cobalt is incorporated near the end of the sequence, the pattern is unbelievably complex.

Alternatively, it might be suggested that some of the incomplete corrins, now suggested to be biosynthetic intermediates, might actually be functional end products in the organism which makes them. It seems quite likely that more reactions will be discovered that involve corrin coenzymes, and possibly some of these apparent intermediates will be found to be functional in these reactions.

III.   DECOMPOSITION OF IRON PORPHYRINS TO BILE PIGMENTS

A. General Pathway of Heme Breakdown

Heme breakdown to the ultimate mixture of excretory products takes place in two distinct stages. The first stage involves enzymic modification of the tetrapyrroles, primarily in the reticuloendothelial system of the liver. The second stage of breakdown is mediated by microorganisms in the gut.

All available evidence at present indicates that the bile pigments arise only from iron prophyrins (*13,14,342–344*). Metal-free porphyrins and other pyrrolic compounds are apparently excreted as such, without being converted to bile pigments. It is obvious that all iron porphyrins cannot give rise to the " usual " bile pigments. Examples of such compounds are the hemes of cytochrome c and cytochrome a.

Biliverdin (see Fig. 23) and all subsequent tetrapyrroles are open-chain compounds. The author prefers to avoid calling any of these compounds either straight-chain or noncyclic pyrroles, since both of these terms imply a stereochemical configuration that is either incorrect or unknown. "Open-chain " is meant to imply nothing about the configuration of the molecule. The rationale behind this choice will become apparent in the discussion on hydrogen bonding (Section III, D).

The convention regarding the presentation of structural formulas in the field of bile pigments has heretofore been notable for its absence.

Virtually everybody uses the linear presentation of pyrroles. However, some writers place ring B at the left while others place ring A at the left. The point can be made, of course, that it is the same molecule regardless of the position of the viewer relative to the molecule, but the author feels that the nonexpert must take too long to orient himself so as to be able to look at a bile pigment and relate what is seen to the parent compound, heme. To minimize this problem, the bile pigments are presented in this chapter in the cyclic form (Fig. 23). With this convention it is possible to see the relationships between the bile pigments and heme and porphyrins at a glance.

The bile pigments are capable of tautomerization, i.e., they can exist in either lactim or lactam form. Gray and co-workers (345) have shown that equilibrium favors the lactam. All bile pigments in this section are shown as the lactam.

The early studies of Shemin, London, Gray, and others (81,346) on the biosynthesis of heme provided valuable information on the metabolism of erythrocyte in addition to leading to an understanding of the biosynthetic precursors. In these experiments, labeled precursors were administered and the incorporation of the label was followed. After administration of glycine-^{15}N, for instance, the ^{15}N content of the heme in circulating erythrocytes rose steadily for a short time. Rather surprisingly, the ^{15}N in the erythrocytes stayed relatively constant for a period of about 120 days and then declined rather abruptly (83,347). A reciprocal labeling situation was observed when labeled bile pigments were studied. This indicated that there was no turnover of heme in the circulating erythrocytes, and suggested that an erythrocyte is formed with its full complement of hemoglobin and that it "lives" in the circulating blood essentially unchanged about 120 days, when it is removed and replaced by another erythrocyte.

More recently, isotopic experiments have shown that a measurable fraction of the administered isotope shows up in bile pigment much earlier than 120 days (346–348).

Heme is oxidized at the α bridge position (349) through one or more postulated intermediates yielding the bilitriene, biliverdin (17,342,350, 351). In model nonenzymic systems employing coupled oxidation of pyridine hemochrome, ring cleavage has been shown to occur also at other bridge positions (352). Biliverdin still retains much of the original conjugation, it is still basically planar and cyclic, and thus it can still bind iron. Iron is apparently lost at the next step, yielding iron-free biliverdin (Fig. 23). Biliverdin is enzymically reduced to bilirubin, and the propionic acid residues on the bilirubin are each esterified with glucuronic acid (*353–355*). The water-solubility of the bilirubin glucuronide, generally called conjugated or "direct" bilirubin, is greatly enhanced because of the large, polar side chains. Conjugated bilirubin is excreted as such into the intestine via the gall bladder. The conjugated bilirubin is hydrolyzed by the microorganisms of the gut. These microorganisms then further metabolize the bilirubin, mainly through reduction, to several colorless "end products." The two principal end products, stercobilinogen and urobilinogen, are unstable and they are air-oxidized to stercobilin and urobilin, both orange-red compounds (Fig. 23).

Several alternative possibilities have been omitted from this summary. They will not be discussed, but they are listed for completeness. Bilirubin monoglucuronide is apparently sometimes formed (*353,356,357*), though there is evidence suggesting that material behaving chromatographically as monoglucuronide may actually be a dimer of bilirubin diglucuronide and unconjugated bilirubin (*356,358*). Bilirubin sulfate has also been identified as one form of conjugated bilirubin (*359,360*). Free unconjugated bilirubin is released under some conditions, and in the plasma this is bound by serum albumin (*361,362*). Some of the bile pigments are reabsorbed from the gut, and they are then reexcreted into the bile (*363,364*). Only a small amount is excreted via the urine (*14*).

It is possible to distinguish conjugated bilirubin from the unconjugated bilirubin with the van den Bergh reaction (reaction of bilirubin with diazotized sulfanilic acid to yield colored azobilirubin) (*14,365*). The conjugated bilirubin reacts rapidly to yield the colored azobilirubin (hence the term "direct") while color development from the unconjugated form is slow. In the presence of ethanol, the reaction with unconjugated bilirubin is rapid.

The assumption is made, in many discussions on bile pigment formation, that the bile pigments arise principally from hemoglobin breakdown. This assumption is made largely as the result of quantitative considerations of the various hemoproteins in the body. About 90% of the total heme is present as hemoglobin, about 10% is present as myoglobin, and about 1% is present as other hemoproteins. At first glance, the assumption that hemoglobin is the major source of bile pigments seems logical, and it is probably basically true. However, it is now generally accepted that the nonhemoglobin hemoproteins have a considerably higher turnover rate than does hemoglobin (*366–369*). Therefore, even though there is a preponderance of hemoglobin, the fact that the other hemoproteins are being metabolized at a faster rate means that they contribute a greater percentage of the total bile pigment than might be anticipated.

B. Conversion of Heme to Biliverdin

The initial stages in the degradation of heme have been studied by a number of workers. Several first compounds have been claimed, but at present it seems best to be safe and call all of these compounds early products. The disagreements in the literature probably arise for several reasons. Many different "systems" have been examined, both model and biological. The compounds are themselves not notably stable, and it seems likely that the various workers are, in fact, looking at different reactions, and that an unstable initial product may be converted to another more stable compound during the chemical workup of the product.

Extensive studies have been carried out by Lemberg and co-workers (350,370–373) on the conversion of hemoglobin and heme to bile pigments. Two areas of this work will be discussed. Both methods involve the coupled oxidation of hemochromogen(s) in the presence of reducing agents. Since different results were obtained in the two methods, they are considered separately.

Hemoglobin in a solution containing a reducing agent such as ascorbic acid, when bubbled with oxygen, is partly converted to a green iron-tetrapyrrole-protein complex, choliglobin (350). In the early stages of the reaction, choliglobin is water-soluble, but it precipitates on prolonged incubation. The precipitate is soluble in alkali temporarily; however, extensive incubation yields an insoluble material. It is not possible to remove the green tetrapyrrole from the precipitated protein, indicating that the tetrapyrrole is covalently bound to the protein. The structure of the tetrapyrrole is not known. Biliverdin is formed in low yield (10%) during extensive incubation (350).

The other studies of the Lemberg school were on the coupled oxidation of pyridine hemochrome with hydrazine. The product of this reaction is an iron-containing green tetrapyrrole, verdohemochrome (350). Chemical and spectral evidence shows that verdohemochrome and choliglobin are not the same compound. Recently, Levin (374) has reexamined the conversion of pyridine hemochrome to verdohemochrome. The product of the reaction was not homogeneous. Up to ten pigments could be separated by thin-layer chromatography. Analysis of one fraction of material purified by chromatography on silica gel indicated a compound with a molecular weight of 818 with a formula $C_{43}H_{42}FeN_6O_8$. The N:Fe ratio suggested that verdohemochrome prepared by a coupled oxidation of pyridine hemochrome with ascorbic acid contains two pyridines. This is in agreement with the findings of Lemberg (350). It contrasts, however, to the results of Nakajima et al. (375,376), who have examined the conversion of pyridine hemochrome to verdohemochrome, and identified verdohemochrome as a monopyridine mono-O_2 complex. Levin confirmed that verdohemochrome prepared in the model system is a dipyridine complex with the use of tritiated pyridine (374).

Nakajima et al. (375,376) have studied the conversion of pyridine hemochrome and hemoglobin-haptoglobin to verdohemochrome. They claim that the reaction is catalyzed by the enzyme α-methenyl oxygenase, and they present evidence to demonstrate the enzymic nature of the reaction. Since the time of Nakajima's studies, a report by Murphy et al. employing the same reaction conditions has shown the conversion of pyridine hemochrome to verdohemochrome to be nonenzymic. Murphy et al. (377) showed that extracts of liver and red algae retained activity following boiling, or deproteinization, with trichloracetic acid. They further succeeded in demonstrating that a low molecular weight-reducing compound was responsible for the conversion of pyridine hemochrome to verdohemochrome. This, as-yet-unidentified, reducing compound was partly purified and was found to be twice as active as ascorbate solutions of the same reducing power, and many times more active than cysteine. This material appears to be the same as that which Nakajima et al. (375) refer to as cofactor for α-methenyl oxygenase. If the reaction is truly nonenzymic, as appears to be the case, it is remarkable that cleavage of the ring occurs specifically at the α bridge position, as Nakajima and Gray (378) have shown.

C. Conversion of Biliverdin to Bilirubin

Biliverdin can be chemically reduced to bilirubin at physiological pH by a number of nonspecific reducing agents such as ascorbate, cysteine, glutathione, and mercaptoethanol (350). Lemberg and Wyndham carried out the first study of the enzymic reaction in 1936 (379). They observed that the reduction was stimulated by the addition of several intermediary metabolites, and that biliverdin acted as a rather nonspecific terminal electron acceptor.

The problem has recently been reexamined by Singleton and Laster (380) using the more refined techniques now available, and they succeeded in purifying biliverdin reductase up to 15-fold from guinea pig liver by conventional techniques. The partially purified enzyme, which was rather unstable, catalyzed the reduction of biliverdin by NADH. NADPH also served as a reducing agent in this system. Neither enzyme alone, nor NADPH alone, reduced biliverdin. The stoichiometry of the biliverdin-bilirubin conversion was essentially unity, but about 2 moles of NADH were required per mole of substrate or product. The enzyme did not exhibit NADH oxidase activity, and the unexpected stoichiometry remains unexplained. The spectral properties of the bilirubin formed in this system were virtually identical with those of authentic bilirubin, indicating that reduction of one of the vinyl side chains had not occurred (which would explain the extra NADH). Several crystalline dehydrogenases, alcohol, lactate, and malate dehydrogenases did not convert biliverdin to bilirubin (*380*).

D. Conversion of Bilirubin to Bilirubin Diglucuronide

Prior to secretion into the bile, the rather water-insoluble bilirubin is esterified with two glucuronic acid residues, yielding bilirubin diglucuronide or, as it is commonly called, conjugated bilirubin (*353,354,381*). This enzymic esterification is carried out in the liver via the following sequence of reactions:

(1) UTP + glucose-1-P $\xrightarrow{\text{UDPG–pyrophosphorylase}}$ UDP-glucose + PP

(2) UDP-glucose + 2 NAD $\xrightarrow{\text{UDP–glucose dehydrogenase}}$ UDP-glucuronide + 2 NADH

(3) 2 UDP-glucuronide + bilirubin $\xrightarrow{\text{UDP–glucuronyl transferase}}$ 2 UDP + bilirubin diglucuronide

UDP-glucuronyl transferase has been studied by a number of workers in recent years (*382–386*). The enzyme is found in the microsomal fraction of liver homogenates. The relative insolubility of bilirubin, its chemical instability, particularly in the light, and its toxicity cause bilirubin to be something less than the ideal substrate for enzyme studies. As a consequence, a number of investigators have chosen to study UDP-glucuronyl transferase with the aid of aglycones other than bilirubin. The use of aglycones such as *p*-nitrophenol has had the unfortunate effect of confusing the issue. The question has arisen as to whether the UDP-glucuronyl transferase, studied with the aid of nonbilirubin aglycones, is in fact the specific enzyme responsible for formation of bilirubin diglucuronide; or is it a nonspecific detoxifying enzyme?

Recent work by Tomlinson and Yaffe (*387*) indicates that there are at least two UDP-glucuronyl transferases, one with specificity for bilirubin, and another, less specific, capable of utilizing aglycones such as *p*-nitrophenol.

It now seems clear that more than one UDP-glucuronyl transferase is present in the microsomes (*387–389*) and, in spite of the difficulties encountered using bilirubin as substrate, further understanding of this reaction will only be obtained with bilirubin.

E. Further Metabolism of Bilirubin Diglucuronide

In the intestine, bilirubin diglucuronide is hydrolyzed and extensively reduced to a colorless bilane by the intestinal flora (*14,390,391*). *In vitro* experiments indicate that the bacteria are able to utilize the diglucuronide more effectively than bilirubin (*392*). This may reflect the comparative solubility of the two forms of bilirubin, but it suggests that it is taken up by the bacteria in the intestine as the diglucuronide prior to hydrolysis. If this is the case, it is in apparent contrast to the ability of the intestinal mucosa to absorb the two compounds where free bilirubin is reabsorbed more readily than is bilirubin diglucuronide (*364,393*).

Three different bilanes make up the bulk of the final product of heme catabolism. They are: *d*-urobilinogen, *i*-urobilinogen, and *l*-stercobilinogen. In the normal individual, *l*-stercobilinogen seems to be the most common bilane. However, different ratios of the three bilanes are sometimes observed in a normal individual. In some diseases, and particularly following antibiotic therapy, the relative amounts of the three bilanes is commonly altered. There is, however, no apparent correlation between any pathological condition and a particular bilane excretion pattern.

A very interesting correlation, however, does exist between antibiotic therapy and the pattern of bilane excretion (*394*). Following extensive antibiotic therapy, sufficient to essentially sterilize the gut, bilirubin is not further metabolized and is excreted as such. As the intestinal flora becomes reestablished, *d*-urobilinogen appears as the primary bilane. After a time, *i*-urobilinogen formation is observed in addition to the *d*-urobilinogen, and at a still later stage *l*-stercobilinogen appears on the scene (*394,395*). This evidence suggests that the three bilanes are not alternative end products, but that there is a metabolic sequence involved and that *d*-urobilinogen is a precursor of *i*-urobilinogen, which in turn is a precursor of *l*-stercobilinogen, as shown in Fig. 23. There is additional evidence that this is the case. When the intestinal contents are examined at different levels (that is, at different positions from high in the intestine to low in the intestine), *d*-urobilinogen is found early, i.e., high, *i*-urobilinogen is found more or less in the middle, and *l*-stercobilinogen is found last, i.e., near the end. Furthermore, experiments with bacterial cultures have shown that *d*-urobilinogen is converted to *i*-urobilinogen and *l*-stercobilinogen, and that *i*-urobilinogen is converted to *l*-stercobilinogen (*391,392*). As the matter stands, one cannot say for certain that these conversions result from the action of a single microorganism or several types of microorganisms acting in sequence. Since the physiological state of a microorganism often reflects its surroundings, and since the nature of the surroundings changes considerably during the

FIG. 23. Apparent path of heme degradation showing structures of intermediates and products.

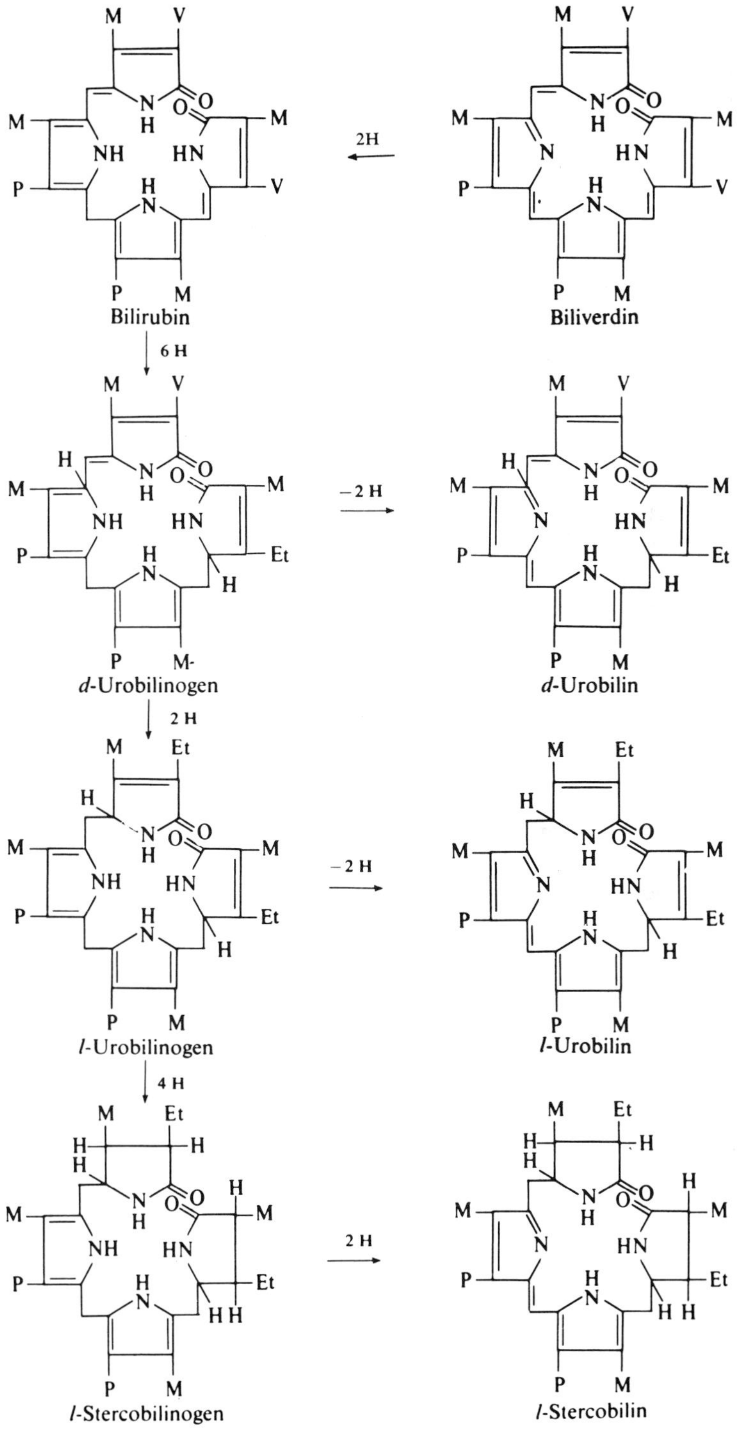

Bilirubin — 2H — Biliverdin

Bilirubin — 6 H →

d-Urobilinogen — −2 H → d-Urobilin

l-Urobilinogen — −2 H → l-Urobilin

l-Stercobilinogen — 2 H → l-Stercobilin

passage through the gut, it is possible that a single type of microorganism could carry out all three of the reactions, but that its ability to do so varies.

The chemistry involved in the conversion of d-urobilinogen to i-urobilinogen and to l-stercobilinogen is very interesting. As indicated by the names, d-urobilinogen is dextrorotary, i-urobilinogen is optically inactive, l-stercobilinogen is levorotary. The question whether i-urobilinogen is optically inactive because it is a racemic mixture, or whether it is inherently inactive, has been answered by the work of Watson et al. (395,396), who have succeeded in separating both d- and l-compounds from natural i-urobilinogen. While this accomplishment increases our knowledge concerning i-urobilin significantly, it should not be regarded as the total picture. The designation i-urobilin indicates only an optically inactive compound of the urobilin type. Nicholson (397) points out that there may be several distinct compounds in this class. Since d-urobilin (H_{40}) is not an isomer of i-urobilin (H_{42}), it is obvious that the choice of names leaves much to be desired. Reformation, however, is outside the scope of the present work, the most common names being retained.

The relatively enormous rotation due to the Cotton effect (398) (observed when rotation is measured at a wavelength close to the absorption band of an asymmetric chromophore) for d-urobilin and l-stercobilin has been explained in a beautiful set of experiments by Moscowitz and co-workers (399). They have demonstrated that this optical activity is due to an enforced molecular dissymmetry at the chromophore caused by internal hydrogen bond formation between the carbonyl oxygens on rings A and B and hydrogens on the nitrogens of rings B plus C and A plus D, respectively (Fig. 24) and that it is not due to the asymmetric carbon atoms in rings A and B per se. In order to achieve the internal hydrogen bonding, the tetrapyrrole assumes a tight distorted helical conformation. The molecule is not just asymmetric because it contains asymmetric carbon atoms, but it is dissymmetric because of the distorted helix. The direction of light rotation is thus dependent on whether the helix is right-handed or left-handed. Perhaps it is more correct to call the structure "helical" than "a helix." Nevertheless, there is a twist at the chromophoric center (the D,C:dipyrrylmethene part of the molecule) and the nitrogen in ring C may be "up" relative to D (right-handed) or it may be "down" relative to D (left-handed) (Fig. 25).

Helix conformation, i.e., right- or left-handed, is determined by the steric configuration of the two hydrogen atoms on the α positions of rings A and B. In order to get a stable helix, these two hydrogens must have opposite configuration, i.e., one is "up" and the other is "down"

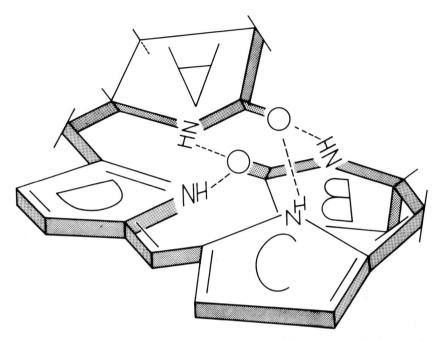

FIG. 24. Helical structure of urobilin showing internal hydrogen bonding between the oxygen on rings A and B and the pyrrole hydrogens on rings B, C and A, D, respectively.

(RR or SS, using the designation of Cahn, Ingold, and Prelog (*400*)) (Fig. 26). How these hydrogen atoms play this role is not obvious until one makes these molecules with molecular models.

The absolute configurations at the two α carbon atoms in rings A and B of *d*-urobilin and *l*-stercobilin have not yet been determined. However, the results obtained in the experiments of Moscowitz *et al.* (*399*) lead to the inescapable conclusion that the hydrogens must be RR or SS, and if *d*-urobilin is RR, then *l*-stercobilin must be SS, or vice versa (Fig. 26). Theoretically, *i*-urobilin could be optically inactive because it is RS or SR but, as mentioned above, *i*-urobilin has been resolved to *d*- and *l*-forms indicating that it is a racemic mixture of RR and SS forms. It should be noted that synthetic *i*-urobilin (also called urobilin IXα and mesobilirubinogen) prepared by chemical reduction of bilirubin would be expected to be a mixture of RR, RS, SR, and SS forms. While the RR and SS forms are sterioisomers, the RS and SR forms are diasterio-isomers and thus should be chemically distinct and separable.

The biological implications derived from these findings are very interesting, assuming that *d*-urobilinogen is, in fact, a precursor of

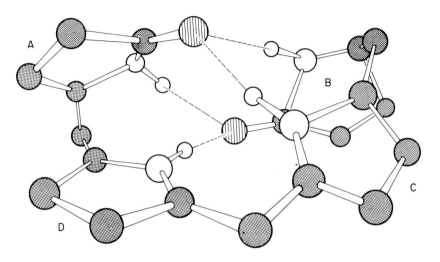

FIG. 25. Three-dimensional representation of urobilin demonstrating the distortion from planarity of the chromophoric group with the D ring nitrogen "down" and the C ring nitrogen "up."

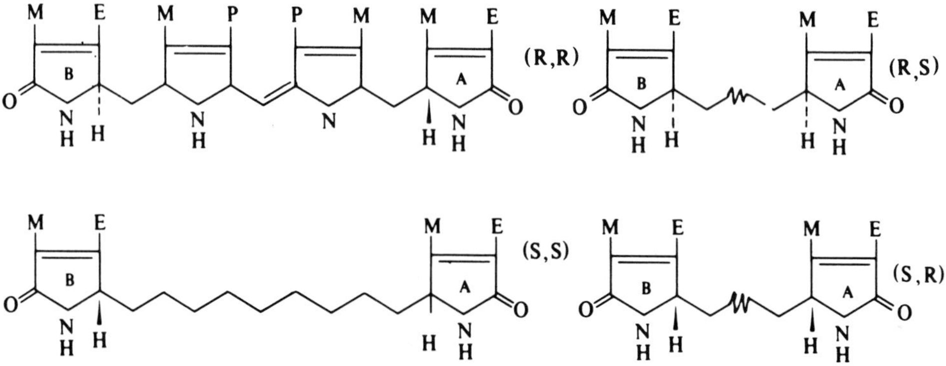

FIG. 26. Stereochemical relationships of urobilin isomer possibilities, adapted from Watson *et al.* (*395*).

l-stercobilinogen. It immediately becomes clear, for instance, that the *l*-stercobilinogen is not simply *d*-urobilinogen with 4 more hydrogens. It means that, during the conversion, the hydrogen on the α carbon of ring A (let's say it is R) must come off, and a hydrogen must be put on S. The same situation holds for the B ring. Experiments studying these reactions at the level of isolated enzymes are awaited with great interest.

F. Observations on the Pattern of Bile Pigment Excretion

As mentioned previously, about 10–20% of glycine-^{15}N incorporated into bile pigment appears very much earlier than that derived from senescent erythrocytes (*346–348,401*). This bile pigment, generally termed the early labeled peak, appears about 1–9 days after glycine-^{15}N administration to the normal individual. This early labeled fraction of stercobilin actually appears at a time when the isotope concentration in the hemoglobin of the circulating erythrocytes is still increasing (*346–348*). It is clear, therefore, that the early peak cannot be derived from hemoglobin of erythrocytes that have reached their physiological age limit, but that it must arise elsewhere.

Several theories have been put forward to explain the early peak (*348,402,403*). For instance, though the mature circulating erythrocyte does not turn over hemoglobin, it is possible that the nucleated, immature red cells could be doing so, and this might explain the early peak. Another explanation centers on the expulsion of the nucleus by the normoblast. Concomitant with the removal of the nucleus, it is more or less inevitable that some cytoplasm would be lost from the cell. This would include some hemoglobin, which upon being broken down could give rise to the early labeled peak of stercobilin (*402*). It has also been suggested that there might be a direct formation of bile pigment from porphyrin or its precursors, without ever going through the heme stage. It seems clear that this is not the case.

The early observations of London *et al.* (*346*) and Gray *et al.* (*347*) on the appearance of the early labeled peak have been followed by investigations in several laboratories. One of the refinements incorporated into the more recent work is the use of labeled bilirubin instead of stercobilin. Analysis of the bile pigment as stercobilin requires that feces be collected and stercobilin isolated. There is obviously a time lapse between first formation of labeled bile pigment and its actual appearance in the excreta and, furthermore, this lag is not constant. Under such conditions, short-term experiments are not possible. Labeled bilirubin in the plasma and in the bile (collected from fistulas) has provided much of the recent information on the early labeled fractions of bile pigment.

Short-term experiments using ^{14}C-labeled glycine and ALA show that the so-called early labeled peak consists of two components (*348,404,405*). There is a fast component, appearing within hours following administration of isotope, and a slow component appearing within the first few days (*406*). There is strong evidence that the fast component of the early labeled peak is not associated with hemoglobin synthesis in the newly forming erythrocytes, and that the slower component is associated with erythropoiesis (*404,405,407–410*). Several lines of evidence have been used to distinguish the erythropoietic and nonerythropoietic components of the early labeled peak.

It is possible to examine the relative amounts of the erythropoietic and nonerythropoietic components of the early labeled peak of bilirubin in the following way to demonstrate clearly that there are, in fact, two components of different origin (*401,411*). When glycine-^{14}C is used in labeling experiments, it is possible to see two components of early labeled bilirubin. ALA-^{14}C, on the other hand, causes extensive labeling of the fast component, and poor labeling of the slow component. Glycine-^{14}C is a good precursor of hemoglobin heme, while ALA-^{14}C is a much less effective precursor (*412*). Erythropoiesis can be stimulated in an experimental animal in a number of ways, e.g., by bleeding, administration of erythropoietin, or by making the animals hypoxic. Erythropoiesis can be inhibited in several ways, e.g., with repeated transfusions or by whole-body irradiation. Erythropoiesis can be quantitatively measured by following incorporation of glycine-^{14}C into the circulating hemoglobin heme in the erythrocytes. With these techniques, it is possible to demonstrate up to a 100-fold difference in the rate of erythropoiesis. The slower component of the early labeled peak varies by a large factor, approximately equivalent to the difference in measured erythropoiesis, while the rapidly appearing component varies only by a factor of less than 5. On the other hand, when the animals are given ALA-^{14}C there is little labeling of bilirubin in the slower component (erythropoietic) in spite of large differences in the rate of erythropoiesis (*401,411,413*).

Since the available evidence indicates that the fast component of the early labeled peak is nonerythropoietic in nature, one is left with the question of its origin (*404,414*). The rapidity with which this component shows up in the plasma following isotope administration, and the rapid rate at which this component falls after reaching the peak, indicate that this is a heme that has a very high turnover rate.

The approximate half-lives of many hemoproteins have been measured (*366–368,415–417*). None of them is short enough to explain the origin of the nonerythropoietic component of the early labeled bilirubin. An important advance in our understanding of this phenomenon has been made by Schmid *et al.* (*369*). These investigators administered phenobarbital to rats and observed a 6-fold enhancement of the nonerythropoietic component of the early labeled bilirubin from glycine-^{14}C compared to control rats. The only recognized heme proteins that were significantly increased by phenobarbital treatment were the microsomal cytochromes and, to a small extent, tryptophan pyrrolase. The increase in the microsomal cytochrome P-450 and the increase in incorporation of glycine-^{14}C into bilirubin were roughly of the same magnitude. This suggests that these cytochromes may be a major source of the nonerythropoietic component of the early labeled bile pigment. That nonhepatic, nonerythropoietic heme metabolism contributes significantly to early peak labeling of bile pigment is suggested by the observation that plasma bilirubin becomes rapidly and extensively labeled (*413*). Presumably nonerythropoietic bilirubin fromed in the reticuloendothelial system of the liver would be shunted directly into the biliary ductal system and would not be released into the plasma. Since labeled bilirubin does appear in the plasma, one supposes this must have originated in nonhepatic reticuloendothelium.

IV. LESIONS IN PORPHYRIN METABOLISM

Abnormalities in porphyrin metabolism are considered under two headings: naturally occurring and experimental. The regulation of porphyrin synthesis is discussed in this section, but detailed consideration of the mechanisms of regulation is defrerred to Section V. Focus in this section is on the pathological condition porphyria in its several manifest forms. Consideration of the various anemias is beyond the scope of the present work.

The porphyrias are generally now considered in two major categories: hepatic and erythropoietic. The subclassifications are as follows:

> Erythropoietic
>> Congenital erythropoietic porphyria
>> Congential erythropoietic protoporphyria
>
> Hepatic
>> Congenital
>>> Intermittent acute porphyria
>>> Cutaneous hepatic porphyria
>>
>> Acquired
>>> Toxic hepatic porphyria
>>> Chemical porphyria (experimental)

A. Erythropoietic Porphyria

1. CONGENTIAL ERYTHROPOIETIC PORPHYRIA

Erythropoietic porphyria is a very rare form of porphyria in man. In the past, however, because of some similarities in symptoms, some forms of hepatic porphyria have been mistakenly diagnosed as erythropoietic porphyria.

The disease is characterized by the excretion of large amounts of uroporphyrin and coproporphyrin in the urine (418,419). Quantitatively, uroporphyrin exceeds coproporphyrin and both are mostly the type I isomer. ALA and PBG in the urine are within normal limits (419). Both coproporphyrin and uroporphyrin are found in the feces, generally with coproporphyrin predominant.

This type of porphyria is seen both in man and in domestic animals. Pronounced photosensitivity is one of the symptoms of this type of porphyria, and this photosensitivity has caused some confusion with hepatic cutaneous porphyria. The neurological complications seen in conjunction with some of the other porphyrias are not found in erythropoietic porphyria. Hemolytic activity is generally increased. This is evidenced by increased excretion of fecal urobilinogen (420). Since hemoglobin is approximately normal, in spite of increased hemolysis, the enhanced destruction of red cells must be compensated by an increase in erythropoiesis.

The pattern of chemical findings, e.g., normal ALA, PBG, and hemoglobin heme, and elevated amounts of type I uroporphyrin and coproporphyrin, suggests that the biochemical lesion occurs at the level of uroporphyrinogen formation. In contrast to the normal situation where the uroporphyrin III : uroporphyrin I ratio is very large, the ratio is smaller in subjects with erythropoietic porphyria. In order to maintain the type III isomer at an adequate level, more total porphyrin is synthesized, resulting in an increased synthesis of the type I isomer.

The genetic data available from the relatively rare erythropoietic porphyria indicate an autosomal recessive mode of inheritance of the disease among humans. The genetics of erythropoietic porphyria have been more extensively investigated with domestic cattle. Although the bovine form of the disease (pink-tooth cattle) is relatively rare, it is becoming increasingly important owing to the increasing use of artificial insemination. A single bull is now able to sire 300,000 matings and, because of the recessive inheritable nature, an asymptomatic heterozygote bull can widely disseminate the characteristic throughout the cattle population. This is known to have happened at least once; a heterozygote bull belonging to a well-known artificial insemination company sired about 100,000 matings before he was recognized as an erythropoietic porphyria carrier (*421*).

Many of the symptoms of bovine erythropoietic porphyria are similar to those of the disease found in man. Nonpigmented areas of the animals are very sensitive to light. Following sunlight exposure, skin lesions occur, urinary levels of both uroporphyrin and coproporphyrin are elevated, and progressive macrocytic normochromic anemia is observed (*422*). Studies on the red cell life span of porphyric animals exposed to sunlight revealed two populations of red cells (*421*).

2. ERYTHROPOIETIC PROTOPORPHYRIA

This form of porphyria has relatively recently been discovered (*423*). It has remained unrecognized, in part, because it is not very common, and because the symptoms of the disease are rather dissimilar from those of other known porphyrias. There is no increase in urinary porphyrins, and both ALA and PBG levels in the urine are within normal limits (*423,424*). Although photosensitivity is observed, the exposed skin is not hyperpigmented, and it does not demonstrate the bullous lesions following exposure to light as commonly seen in other porphyrias complicated by photosensitivity. The most striking manifestation of protoporphyria is the elevated level of protoporphyrin in the erythrocytes and feces. The protoporphyrin in circulating red cells ranges from 5 to 30 times normal (*423–425*). The protoporphyrin is type IX isomer, i.e., normal (*423–426*). One might suspect that the increase in protoporphyrin could be due to a relative deficiency of ferrochelatase in the marrow, but both *in vivo* and *in vitro* experiments indicate that iron incorporation is normal (*423,426*), and it is generally accepted that protoporphyrin is simply overproduced.

There seems to be conflicting evidence indicating that protoporphyria is not entirely confined to the erythropoietic system (*424,427,428*). One study in particular, using glycine-^{15}N, strongly indicates that the protoporphyrin is not entirely erythropoietic in origin. In this study, protoporphyrin in the feces was much more heavily labeled than was the protoporphyrin found with the circulating erythrocytes (*427*). Though it is difficult to estimate pool sizes accurately, one can make order-of-magnitude estimates, and, after doing so it seems that the fecal protoporphyrin is about 10 times more heavily labeled than the erythrocyte protoporphyrin. The conclusion that the excess protoporphyrin derives only from the erythropoietic system may be premature.

B. Hepatic Porphyrias

Many difficulties attend the proper classification of the diseases where abnormalities in porphyrin metabolism are apparent. These problems are discussed at some length in an excellent recent review by Schmid (*13*). A good precentage of the classification of porphyrins is based on the convenience of grouping apparently similar pathological conditions on the basis of symptomatic evidence. It seems very likely that, as our knowledge of the fundamental biochemical defects that ultimately cause the symptoms increases, the classification will be considerably revised.

Following the present system, however, we have seen that certain of the metabolic defects giving rise to porphyric symptoms seem to be associated with erythropoiesis. With the possible exception of erythropoietic protoporphyria where the liver may be involved (a point yet to be proved), the liver is apparently not affected in erythropoietic porphyria. Several of the erythropoietic porphyria cases are reported where the liver is not normal, but this is generally attributed to causes other than the porphyria. On the other hand, the more common forms of porphyria seem to more or less specifically involve the liver, and the hematological picture is essentially normal. In such cases the porphyrin content of the bone marrow and circulating erythrocytes is normal and only the liver shows a high level of porphyrins and porphyrin precursors (*13,429*).

It is very common in hepatic types of porphyria that liver function is badly impaired, and liver disease (in addition to overproduction of porphyrin) is common. Hepatic forms of porphyria are apparently genetically transmitted, but some forms are apparently induced by drugs and chemicals in individuals with no family history of the disease.

1. INTERMITTENT ACUTE PORPHYRIA

This is the most common form of abnormal porphyrin metabolism. It is characterized, as the name implies, by acute attacks, lasting days or months, and periods of remission of varying duration during which symptoms all but disappear. The attacks are characterized by abdominal pain, variable neurological complications, and the excretion of large amounts of porphyrin precursors in the urine. The extreme photosensitivity characteristic of the erythropoietic porphyrias is generally conspicuous by its absence. Latent porphyrics generally show somewhat increased levels of ALA and PBG in the urine, but this is neither dramatic nor invariable (*430,431*). Attacks of the disease may be precipitated by drugs, most commonly barbiturates, sulfonamides, and estrogens

(*13,432*). There is as yet no explanation for the abdominal pains or neurological disorders associated with acute attacks.

During an attack, ALA, normally excreted less than 1 mg/day in the urine, raises to as much as 180 mg/day, while PBG also raises from less than 1 mg/day to several hundred milligrams per day (*118,430,431,433, 434*). The porphyrin content of porphyric urine, while higher than normal, does not increase to nearly the same extent as ALA and PBG. However, if porphyric urine is allowed to stand for any length of time, it turns to a deep, wine-red color. The color derives from two sources. In part, it is due to porphyrin which evidently forms by nonenzymic condensation of PBG to yield uroporphyrin. Other pigments are di-, tri-, and polypyrrolmethenes as well as porphobilin, a brown oxidation product of PBG (*435*). Fecal porphyrins are also generally somewhat elevated (*429,436*).

The metabolic defect of this type of porphyria seems to clearly be associated with the liver. In some cases, uroporphyrin and coproporphyrin have been identified in hepatic tissue; however, most significantly, the porphyric liver contains large amounts of PBG (*437,438*). Since ALA dehydrase activity is many times more active than ALA synthetase in the livers of all species measured, attention has been focused on ALA synthetase to explain the overproduction of ALA and PBG. Granick and Urata (*105*) were the first to show that the activity of ALA synthetase is increased in animals with experimental porphyria. The increased activity apparently represents an increased amount of enzyme (*117*). Similar results have been obtained in other studies on experimental porphyria (*108,114,115*). Experimental porphyrias are discussed in Section IV, B, 4.

Several studies have recently been carried out on humans suffering from hepatic porphyria (*107,107a,439*). These studies showed that ALA synthetase activity in porphyric hepatic tissue was elevated by a factor of at least 7 (*107,107a*).

A positive correlation between diet and severity of acute attacks has been demonstrated (*440*). High carbohydrate consumption relieves symptoms, whereas fasting will sometimes precipitate an attack (*440*). Similar findings have been reported in studies with experimental animals. Administration of glucose to rats in a state of chemically induced porphyria, depending on the time of administration, either prevents the induction of porphyria or relieves the already induced condition (*115,441,442*).

Acute intermittent hepatic porphyria is most severe in young adults (*443,444*). It seldom becomes manifest before puberty. That there is some connection between manifestation of the disease symptoms and the sex hormones is indicated by the following observations. In female porphyrics there is sometimes a correlation between periodic appearance of symptoms and the menstrual cycle (*13,445*). While androgens have been reported to suppress the symptoms (*13*), estrogens have been found to stimulate expression of symptoms in porphyric humans (*445,446*). Normal individuals apparently are not effected by estrogenic compounds, at least as far as porphyrin biosynthesis is concerned (*445*). It is interesting to note that, as oral contraceptives are gaining wide acceptance and use, a number of reports are appearing about the appearance of porphyric symptoms among women (*447,448*). It seems likely that these women are latent porphyrics and that the contraceptive estrogens precipitate onset of symptoms.

2. CUTANEOUS HEPATIC PORPHYRIA (SOUTH AFRICAN HEPATIC PORPHYRIA)

This disease as found in white South Africans is characterized by acute abdominal pain and neurological complications. During attack, the ALA and PBG concentration in the urine is greatly increased (*449,450*). These symptoms are similar to those seen in acute intermittent porphyria (*451*). However, several symptoms clearly distinguish the South African form of the disease. For instance, there is skin involvement which is manifested by photosensitivity (*452*) and an associated hyperpigmentation, the exposed areas of the skin being particularly sensitive to mechanical injury. Of the two effects the mechanical fragility of the skin is generally more prominant, often leading to scarring (*450,453*). As is the case with acute intermittent porphyria, severe attacks of symptoms may be precipitated by drugs and chemicals such as barbiturates and sulfonamides (*445,451,454*).

In addition to the increased excretion of ALA and PBG in the urine, there is excessive exretion of coproporphyrin and protoporphyrin in the feces (*449,451,455*). While the ALA and PBG in the urine are greatly in excess of normal only during acute attacks of the South African form of the disease (*449,450*), the excretion of elevated amount of coproporphyrin and protoporphyrin is generally continuous (*449,455*).

3. TOXIC HEPATIC PORPHYRIA (ACQUIRED)

Another form of hepatic porphyria is seen among the native (non-white) population of South Africa. This form of porphyria is particularly predominant among members of the Bantu race (*456,457*). In this disease there is an elevation in urinary porphyrin excretion (*450,455*);

however fecal porphyrins are generally within normal limits (*450,453, 458*). Photosensitivity demonstrated by lesions, scarring, and hyperpigmentation is common (*453,458*). A number of cases of porphyria demonstrating symptoms similar to those above have been reported throughout the world (*429,430,444*). As yet there is no evidence indicating that this form of the disease is inheritable, as is the case with other types of porphyria. There is, however, apparently a common factor in that virtually all patients suffereing from this disease have chronic liver disease of one sort or another (*13*). Often this is associated with high alcohol consumption (*430,433,444,458*). This is particularly prominent among the Bantu, who consume large quantities of a home-brewed beer having a notoriously high iron content (*458,459*). It should be noted, however, that this acquired type of porphyria does not inevitably follow iron overload or high alcohol consumption. Symptoms associated with porphyria do not as a rule show up until the individual is middle-aged, and until enormous quantities of iron have been consumed. As the human body characteristically hoards iron, these people generally have very pronounced siderosis at the time of attack. There is further evidence for a correlation between iron deposits in liver and expression of porphyric symptoms. Patients who have been bled repeatedly and extensively to deplete the reserve of stored iron have been observed to go into remission (*460,461*).

The iron-free form of one of the sideramines mentioned in Section II, C, known as desferrioxamine (Desferal, Ciba Pharmaceutical Co., Summit, N. J.), has in recent years been employed to "deironize" the liver of people with siderosis (*462*). This trihydroxamic acid chelates the ferric iron in the iron deposits, and is excreted in the urine as ferrioxamine. Only one case has been reported on the use of this drug in relieving hepatic porphyria (*460*). Desferrioxamine therapy was not maintained for very long, and no effect on porphyrin excretion was noted (*460*). Its use should have obvious advantages over bleeding in allowing conclusions to be drawn relating siderosis to hepatic porphyria since it is much more specific in its action than the loss of blood, which may well establish other imbalances.

4. EXPERIMENTAL OR CHEMICALLY INDUCED PORPHYRIA

It has been recognized for some time that drugs such as Sulfonal, Trional, and barbiturates sometimes precipitate acute attacks of porphyria (*432,445,463*). It seems probable that these drugs are precipitating an attack in an individual with latent porphyria rather than inducing the disease.

The first chemically induced porphyria in a nonporphyric subject was reported by Schmid and Schwartz (464). They made use of the drug allyisopropyl urea (Sedormid) by administering it in large doses to normal rabbits. The rabbits developed many of the symptoms of hepatic porphyria, most noticeably by excreting massive amounts of ALA and PBG. Two interesting observations that were made in these studies have not been thoroughly followed up, yet they seem of possible fundamental importance: (a) The activity of liver catalase in animals treated with Sedormid drops to about 10% of the normal values. (b) When the liver of a porphyric animal is removed and perfused with isotonic saline to remove the residual blood in the hepatic circulatory system, the color of the hepatic tissue is green. Schwartz has shown the green pigment to be a tetrapyrrole (465), but it has not been identified.

Since the discovery of the possibility of producing porphyria in experimental animals under controlled conditions, a large number of reports have appeared on both the biochemistry of the induced porphyria and on the nature of the inducing chemicals. Studies on the relationship between chemical structure and porphyria-inducing activity in the laboratory of Marks (467,468) reveal that many active inducers have in common a steric hindrance to hydrolysis of the amide or ester groups in the molecule. Apparently, this is important in that it retards detoxification of the chemicals.

The work from Granick's laboratory has been instrumental in providing an understanding of the nature of chemically induced porphyria. In a series of experiments with guinea pigs, Granick and Urata (105) demonstrated that the enhanced production of ALA and PBG was due to an increase in ALA synthetase in the mitochondira. The ALA synthetase in mitochondria from animals treated with 3,5-dicarb-ethoxyl-1,4-dihydrocollidine (DDC) was increased by a factor of more than 40. Aminoacetone synthetase, the enzyme that catalyzes the condensation of acetyl-CoA and glycine, is present in liver mitochondria, and Urata and Granick (298) provide convincing evidence that aminoacetone (and other α-aminoketones) and ALA are formed by different enzymes.

Subsequent work by Granick has been carried out using chick embryo liver tissue cultures as experimental material (114,117,466). These experiments are considered in detail in Section V.

V. REGULATION OF TETRAPYRROLE BIOSYNTHESIS

Metabolic processes are known to be controlled by a variety of factors. Even in the case of higher organisms where organs and hormone systems are involved, it is now sometimes possible to examine and explain regulatory phenomena at the molecular level. Such understanding may help in the treatment of the rare but striking porphyrias of man and other animals.

If we consider any series of reactions, anabolic or catabolic, and wish to examine the regulation of the rate at which one metabolite is converted to another several steps away, our attention must focus on the enzymes that catalyze each reaction. The rate at which compound A is converted to compound Z is a function of the activity and quantity of enzymes, the availability of substrate, and the relationship in space between different enzymes.

As our knowledge of the ultrastructure of the cell expands, we increasingly recognize that the study of cellular architecture contributes to our understanding of function. The cell no longer can be visualized as a bag of enzymes with a nucleus, mitochondria, and a few other organelles. Even the so-called soluble portion of the cytoplasm is increasingly recognized as being structurally organized. For a recent review see (469).

When enzyme molecules are anchored and not freely in solution in the cytoplasm, then the effective concentration can be almost anything. If we arbitrarily say that there are 1000 molecules of a given enzyme in a cell, knowing the volume of the cell, we can calculate the apparent concentration of enzyme. This is largely what one measures, or sees, when working with homogenates. However, if the 1000 molecules are concentrated within organelles in the cell, which have a combined volume one-tenth that of the cell, the functional concentration of enzyme at these sites is 10 times the average. This must be the situation with respect to certain of the enzymes of porphyrin biosynthesis such as ALA synthetase and coproporphyrinogen oxidase, which are localized in the mitochondira. Other enzymes required for the formation of tetrapyrroles such as ALA dehydrase are cytoplasmic. Compartmentalization, no doubt, plays a significant role in the regulation of tetrapyrrole biosynthesis (105), although little can be said at this point concerning the mechanism.

The phenomenon of feedback control is well established as an important regulatory mechanism in the metabolism of many compounds in addition to tetrapyrroles (470–473).

Negative-feedback systems have certain constant characteristics. The enzyme under control is the first one leading exclusively to a given end product, and the end product governs its own rate of synthesis by reacting with the first enzyme to affect its activity. In those cases where the enzymology of feedback control has been investigated, it has been shown that the inhibitor (end product) reacts at a site other than the catalytically active site (*474*). This is designated the allosteric site. When the inhibitor has combined with this site, presumably the tertiary structure of the enzyme is altered in such a way that the components of the catalytic site are no longer situated in such a way as to cooperate in catalyzing the reaction. An important property of the inhibition is the fact that it is readily reversible. The significance of feedback control lies in the provision of a delicate and rapid mechanism for stopping and starting a series of reactions, thus the requirement for reversibility. Another characteristic of feedback systems is that the point of control is generally one involving a high-energy substrate and, therefore, a reaction with an equilibrium far to the right. In the case of the tetrapyrroles, succinyl-CoA represents the high-energy intermediate. Control at the point of condensation of succinyl-CoA and glycine prevents utilization of the high-energy compounds beyond that required by the cell for the formation of end product.

Several interesting complications are known that appear to place the cell in a dilemma. There are a number of instances where several end products result following sub-branching of the main biosynthetic pathways. The tetrapyrroles present such a situation. ALA is the first compound leading directly to all three biologically functional tetrapyrroles. Protoporphyrin is at the branch point between heme and chlorophyll, and unroporphyrinogen may be at the branch point leading to cobalamins. If feedback control is exerted on ALA synthetase, then the question of which end product is the proper one to regulate ALA formation must be considered. In the case of *R. spheroides*, experimental evidence indicates that heme regulates its own synthesis (see Fig. 27) without affecting bacteriochlorophyll synthesis (*103*).

It is then necessary to explain how chlorophyll and cobalamin can regulate their own synthesis independently or, in some cases, in concert with action of heme. Possible answers to this problem are suggested by analogy with other situations where multiple end products arise from a common precursor. The mechanism by which the cell avoids the problem of allowing one amino acid to regulate the biosynthesis of another has been elegantly demonstrated by Stadtman *et al.* (*475*). The first reaction leading ultimately to the formation of lysine, methionine, and threonine is:

$$\text{Asparate} + \text{ATP} \rightarrow \text{aspartyl-PO}_4 + \text{ADP}$$

As is typical of reactions at branch points, the equilibrium lies far to the right because of the participation of ATP. It has been shown that at least two aspartokinases catalyze the first reaction. Presumably they have identical catalytic active sites since they catalyze the same reaction;

however, each apparently has a specific allosteric site sensitive to only one of the amino acid end products. Furthermore, there are additional control points, this time as sub-branches leading to the three individual amino acids. Thus, balanced distribution of the intermediates from the main branch can occur.

The converse of branching pathways is the converging of independent pathways. This is relevant to the regulation of tetrapyrrole biosynthesis for, apart from certain diseases, tetrapyrroles do not exist free in any significant concentration. Thus heme is always associated with a specific protein, chlorophyll with phospholipoprotein, and cobalamin with specific enzymes. It is apparent that the cell regulates the concerted synthesis of the tetrapyrrole and the specific protein. Convergence is seen in the case of chlorophyll and cobalamin even prior to consideration of the associated proteins. For instance, in a greening leaf, chlorophyllide formation must be coordinated with phytol formation. In cobalamin biosynthesis the supply of methyl groups from methionine, the supply of nucleotide, etc., must all be controlled in the converging system. Recently a report by Newell and Tucker (476,477) on the regulation of the biosynthesis of thiamine has appeared that may be helpful by analogy. The thiazole and pyrimidine moieties of thiamine are independently synthesized, and thiamine is formed when these two pathways converge. These investigators found that thiamine exerted a feedback control on both of the independent pathways, and also that this feedback occurred to the same extent on each pathway. They found, too, that thiazole controlled its own synthesis by a feedback mechanism.

As mentioned, not only the activity but the amount of enzyme determines the quantity of product formed. The amount of enzyme can be regulated by induction or repression of enzyme synthesis. The same key control points are involved in repression that are involved in feedback inhibition. However, the effector in repression is a protein, whereas in inhibition a low molecular weight compound is involved.

While there are obvious advantages in a regulatory system that has the ability to respond very rapidly to changes imposed upon the system from without, there is a disadvantage in relying exclusively on such a system, namely, unusable enzymes would continue to be synthesized. This would represent a waste of energy for the cell and could be disadvantageous. A system, therefore, exists that signals the cells to stop making the enzyme in question.

This system is known as enzyme repression. According to the model of Jacob and Monod (478), the end product of a biosynthetic sequence reacts with an aporepressor substance to form a repressor. The repressor in turn reacts with the operator gene to prevent the synthesis of messenger

RNA, thus preventing formation of the enzymes under control of the gene. In this case the only thing affected is the synthesis of new enzyme. Since a finite time is required for a cell to grow and dilute out active enzyme already present, control by repression is more sluggish than control by feedback inhibition. This sluggishness in response may be partly circumvented since the turnover rate of different enzymes varies considerably.

Recently modifications and alternative possibilities have been added to the model proposed by Jacob and Monod (479–482). As will be seen in the discussion on hemoglobin synthesis (Section V, C), certain modifications at the level of translation must occur since the evidence indicates that synthesis can be regulated after the mRNA for hemoglobin has been formed. Regulation at the translational level is also seen with other enzymes.

Both feedback inhibition and repression (see Fig. 27) seem to be important in regulating tetrapyrrole biosynthesis in microorganisms (3,103,483,484), but only represssion seems to operate in controlling heme synthesis in the liver of higher animals (108,114,115,485). That such a difference should exist seems logical. The environment of a microorganism, and particularly the available nutrients can fluctuate over rather wide ranges. As a result, it is to the advantage of the microorganism to be able to adapt quickly to these changes. The environment of a liver cell, on the other hand, is relatively constant. Wide fluctuations are incompatible with the life of the organism, and the machinery of the cell does not have to be geared to handle extremely rapid changes. It is, therefore, not too surprising that the ALA synthetase of liver has apparently lost its allosteric site.

A. Factors that Influence Tetrapyrrole Biosynthesis as Studied with Intact Organisms

1. OXYGEN

Oxygen is reported to have an effect on tetrapyrrole biosynthesis in almost every system where it has been studied. In spite of the prevalence of observations, however, not much is understood about the mechanism of the various responses. No doubt, part of the reason that it is difficult to explain the action of oxygen is the fact that its effects are multifarious. It enhances hemoprotein synthesis in some cases (486–488), yet in others it represses hemoprotein formation (489–492). Oxygen also has a profound effect upon the synthesis of bacteriochlorophyll (3), and it is also known to affect cobalamin synthesis (493).

Only one step in the biosynthesis of tetrapyrroles has been shown to directly require oxygen, and that is the oxidative decarboxylation of coproporphyrinogen III to protoporphyrin IX (*66,180*).

This reaction, however, has never been demonstrated at the subcellular level in experiments with bacteria. Since a number of microorganisms are known that have the ability to synthesize heme under anaerobic conditions, it appears that other mechanisms for oxidatively decarboxylating coproporphyrinogen III exist. Conversely, it seems likely that microorganisms unable to synthesize heme under anaerobic conditions rely on oxygen for formation of protoporphyrin IX.

The apparent paradox that oxygen both represses and induces tetrapyrrole synthesis can perhaps, in part, be explained on the basis of experiments by Falk and Porra (*494*) and by Hammel and Bessman (*495*). These investigators have studied the effect of oxygen upon overall heme synthesis, and at the level of individual enzyme reactions. They found that there is an optimum level of oxygen for porphyrin synthesis. The effect of oxygen appeared to be exerted at the level of conversion of PBG to uroporphyrinogen III. If the optimum oxygen level for porphyrin synthesis varied considerably for different organisms, this might explain the inductive and repressive effect of oxygen. Microorganisms such as *Pseudomonas fluorescens* (*489*), *Micrococcus denitrificans* (*490*), and *E. coli* (*492*), where oxygen has a repressive effect on cytochrome formation, may have an exceedingly low optimum oxygen requirement. On the other hand, microorganisms such as *Saccharomyces cerevisiae* (*496–498*), *Pasturella pestis* (*487*), and *Bacillus cereus* (*486*), which show increased cytochrome formation in the presence of oxygen, may require higher levels of oxygen for optimal hemoprotein synthesis. That this generalization is too simple to be accurate is demonstrated by *M. denitrificans*. Hemoprotein synthesis in this organism is maximal in the absence of oxygen (*490*); thus, one appears to be looking only at repression any time oxygen is present.

Oxygen is also recognized to have a pronounced effect upon photopigment synthesis in photosynthetic bacteria. As with higher photosynthetic organisms, light also effects photopigment synthesis (*3,499*). The effects of light and oxygen, however, can be studied and considered separately with some bacteria (see Fig. 27). *R. spheroides* can be cultured either aerobically or anaerobically. When the cells are grown aerobically under high oxygen pressure, the presence or absence of light is without effect. Cells grown under these conditions are essentially devoid of photopigments. Aerobic cells grown under low oxygen pressure, however, form photopigments to about the same extent as photosynthetic, anaerobic cultured cells. See Table II.

TABLE II

LEVELS OF TETRAPYRROLES FORMED BY *R.*
Spheroides UNDER DIFFERENT GROWTH CONDITIONS

Growth conditions	Bacterio-chlorophyll (*3*) (μmoles/mg dry wt. cells)	Total heme (*186*) (μmoles/mg protein)	Cyano-cobalamin (*341*) (μg/g dry wt. cells)
Aerobic, high O_2	0.2	0.32	20
Aerobic, low O_2	22	—	—
Anaerobic, in light	24	0.82	102

When the oxygen supply of aerobically grown, high oxygen cells (nonpigmented) is decreased, the cells commence making photopigments (*109*). Introduction of oxygen to cultures of cells actively synthesizing photopigment causes immediate cessation of this synthesis. Upon removal of oxygen, bacteriochlorophyll synthesis recommences without a lag at the same rate as prior to oxygen treatment. Associated with the synthesis of bacteriochlorophyll, the level of the enzyme ALA synthetase responds to oxygen (*109,500*). Upon lowering the available oxygen to nonpigmented cells, ALA synthetase is preferentially formed and the increase in ALA synthetase precedes increased synthesis of bacteriochlorophyll. Synthesis of this enzyme is represented upon introduction of oxygen (*116,500*).

Both *Chromatium* D and *Rhodospirillum molischianum* require light and anaerobic conditions for growth. In a pair of recent studies it has been shown that oxygen is not toxic to these cells, except in that it inhibits bacteriochlorophyll synthesis (*501,502*).

The synthesis of photopigment and its relationship to oxygen and light intensity appears to be under control of a "regulatory substance" (*502,503*). The evidence leading to this conclusion is discussed in Section V, A, 2 on light and photopigment production.

The extensive series of studies from Slominski's laboratory has contributed significantly to our understanding of the regulation of hemoprotein synthesis. However, again, it is difficult to explain the exact role of oxygen in the process. The situation is basically as follows. Anaerobically grown *Saccharomyces cerevisiae* do not contain mitochondria, nor do they appear to contain the complete complex of cytochromes (*497,498, 504,505*). Cytochromes a and b are the only spectroscopically observable cytochromes. When the anaerobic yeasts are transferred to aerobic conditions, mitchondria develop, and cytochromes a, a_5, b, and c appear (*496,506*). One of the most fascinating aspects of the adaptation to aerobic conditions has come to light following the key observation that the cytochrome c of *S. cerevisae* can be resolved into two components (*507*). These two cytochromes are distinct molecular species, and they have been named iso-1-cytochrome c and iso-2-cytochrome c by Slonimski. The relative proportions of these two cytochromes depend upon the growth conditions of the yeast. Anaerobically grown cells in the stationary phase contain about 20% iso-1-cytochrome c and 80% iso-2-cytochrome c. Aerobically grown cells in the stationary phase, on the other hand, contain about 92% iso-1-cytochrome c and 8% iso-2-cytochrome c (*508*). During adaptation from anaerobic to aerobic conditions, the total cytochrome c content increases about 40- to 50-fold (*508*).

A study of the kinetics of biosynthesis of the two cytochromes during adaptation revealed that initially the rate of iso-2-cytochrome c formation was fast, while the initial rate of iso-1-cytochrome c was slow. Later during adaptation the rate of iso-1-cytochrome c synthesis increased (*508*). Further studies employing amino acid analogs, amino acid-requiring mutants (*509*), and the protein synthesis inhibitor, cycloheximide (*510*), revealed additional differences in the formation of the two cytochromes. Most importantly, inhibition of protein synthesis, while markedly inhibiting formation of iso-1-cytochrome c had relatively little effect on formation of iso-2-cytochrome c (*510*). This was interpreted as indicating that iso-2-cytochrome c existed in "precursor" form in anaerobic cells. Presumably, the "precursor" form is the heme-free polypeptide, i.e., apo-iso-2-cytochrome c. The data further suggested that apo-iso-2-cytochrome c had the capacity to function as the repressor of iso-1-cytochrome c. This possibility is supported by data of the following type. Addition of cycloheximide at different times during adaptation allowed the determination of the apo-iso-2-cytochrome c concentration at the time of addition. It was found that initially apo-iso-2-cytochrome c decreased rapidly and approached zero after about 1 hour. Simultaneously iso-2-cytochrome c appeared in the cells. During the initial stages of adaptation, while the level of apo-iso-2-cytochrome c was high, the synthesis of iso-1-cytochrome c was slow. The rate of synthesis of iso-1-cytochrome c, however, continuously increased and, by the time the apo-iso-2-cytochrome c level dropped to zero, the rate of iso-1-cytochrome c synthesis became maximal and constant (*510*). Apparently, all that is required for the conversion of the active repressor, apo-iso-2-cytochrome c, to iso-2-cytochrome c, which is inactive as repressor, is the attachment of heme. The role of oxygen in this process is not obvious. It would appear, however, that the oxygen is not required for protoporphyrin formation since the anaerobic cells do synthesize heme.

In conjunction with this work the experiments of Somlo and Fukuhara (*511*) with *S. cerevisiae* showed that extraordinary care must be
taken to ensure that anaerobic conditions are truly anaerobic. These
investigators found that there was sufficient oxygen even in the highest-
quality commercial nitrogen to induce cytochrome formation. Treatment
of the nitrogen with alkaline pyrogallol allowed truly anaerobic conditions to be achieved. Bearing this in mind, it is difficult to know how
one should judge some experiments where anaerobic conditions are
reported.

2. LIGHT

The effect of light intensity upon chlorophyll synthesis has been
recognized for many, many years. Certainly anyone who has ever mowed
a lawn has observed that shade-grown grass is greener than grass
grown in the sun. Likewise, the requirement that some plants have for
light in order to synthesize chlorophyll is commonly observed outside
the laboratory.

The requirement or nonrequirement of light to synthesize chlorophyll
is relatively well understood. Angiosperms require light for the reduction
of protochlorophyllide a to form chlorophyllide a (Section II, D, 2, *b*).
Gymnosperms, on the other hand, are capable of carrying out this
reduction in the dark. Some algae require light for chlorophyll formation, while others do not.

An inverse relationship between light intensity and chlorophyll content
is observed in photosynthetic bacteria, algae, and higher plants. Studies
on this problem with *R. spheroides* indicates that there is a direct correlation between bacteriochlorophyll content and specific growth rate (*503*).
As mentioned in Section II, D, 2, *a*, pigment synthesis and protein synthesis are in lockstep association in this organism (*116,253*), and this may,
at least in part, explain the observed relationship between growth rate
and bacteriochlorophyll content. In one of the early comprehensive
studies on this problem, it was suggested that the effect of light intensity
might be mediated through some postulated redox-sensitive "regulatory substance" in the cell (*499*). Presumably, this "regulatory substance" is converted to the oxidized form at high light intensity or at
high oxygen pressures, and in the oxidized form this "regulatory substance" inhibits bacteriochlorophyll synthesis. A regulatory scheme
incorporating these observations is depicted in Fig. 27. The reduced
form of the "regulatory substance" apparently exists at low light
intensities and at low oxygen pressures. Support for this hypothesis has
come from recent experiments on the obligate phototroph *Rhodospiril-*

lum molischianum (*501*). Unlike the facultative phototrophs, e.g., *Rhodopseudomonas spheroides* and *Rhodospirillum rubrum*, bacteriochlorophyll synthesis by *R. molischianum* under semi-aerobic conditions is not independent of light intensity. In these experiments, at a given light intensity, oxygen decreased bacteriochlorophyll synthesis ("regulatory substance" oxidized). The inhibitory effect of oxygen, however, could be offset by decreasing the light intensity ("regulatory substance" reduced). The converse experiments gave supporting evidence.

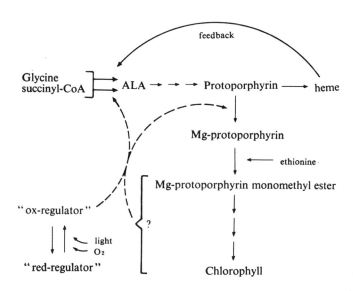

FIG. 27. Schematic representation of tetrapyrrole regulation mechanisms. Experimental support has been obtained primarily with photosynthetic bacteria, but the scheme is probably also applicable to green plants. The double arrows at ALA synthetase represent the possible existence of two such enzymes, one for the iron branch and sensitive to heme, the other for the magnesium branch and sensitive to some Mg-tetrapyrrole. The solid feedback arrow from heme to ALA synthetase included inhibition of enzyme activity and repression of enzyme formation, both of which have been experimentally demonstrated. The two dashed feedback arrows represent hypothetical controls. Circumstantial evidence supports these additional controls, though the details are not yet understood.

These experiments indicate that the inhibition of bacteriochlorophyll synthesis in *R. molischianum* by oxygen is not caused by the direct inhibition of any of the biosynthetic steps leading to bacteriochlorophyll. On the other hand, this evidence supports the suggestion that a redox "regulatory substance" controls bacteriochlorophyll synthesis.

3. IRON

It has been recognized for some time that iron nutrition plays a major role in the regulation of tetrapyrrole biosynthesis (3,184,512). The effect of iron deficiency is manifest in several ways. In mammals, iron deficiency is associated with lowered hemoglobin values, plants develop chlorosis, and microorganisms excrete porphyrins.

There is some evidence that iron may be involved at the level of ALA synthetase and the oxidative decarboxylation of coproporphyrinogen (66,173,185). The only function thus far unequivocally demonstrated is at the stage of formation of heme.

Since it is comparatively easy to regulate the iron nutrition of microorganisms, they have been used most extensively in studying the effect of iron on tetrapyrrole biosynthesis. It is commonly observed that, when microorganisms are cultured in an iron-deficient medium, they excrete porphyrins.

This phenomenon has been extensively studied by Lascelles using the photosynthetic bacterium *R. spheroides* (3,183). She observed that, when these microorganisms are cultured at very low levels of iron, large amounts of coproporphyrin III accumulate in the growth medium. Under these iron-deficient conditions, bacteriochlorophyll synthesis is inhibited. The addition of very small amounts of iron suppressed the accumulation of porphyrin. The effect of the added iron is clearly catalytic in that the ratio of added iron to porphyrin that fails to accumulate is approximately 100 (*183*). This ratio makes it abundantly clear that the iron is doing more than "siphoning" off porphyrin to form heme. In these studies, low concentrations of iron were found to suppress porphyrin accumulation when glycine and α-ketoglutarate were provided as porphyrin precursors. On the other hand, when ALA was employed as the porphyrin precursor, iron had no effect on porphyrin accumulation (*109,183*). This was the first direct evidence indicating that tetrapyrrole synthesis in these organisms was regulated at the step involving formation of ALA. The mechanism of this regulation is considered in Section V, B, and is summarized in Fig. 27.

4. DRUGS AND CHEMICALS

a. Chemicals that Influence Regulation of Porphyrin Biosynthesis. That certain drugs can aggravate, and sometimes precipitate, attacks of hepatic porphyria in man has been recognized for a number of years (*432*). The first intentionally induced experimental porphyria in animals was produced with Sedormid (allylisopropyl urea) by Schmid and

Schwartz in 1952 (*464*). Since that time experimental porphyria has been induced by a wide spectrum of chemicals in a number of experimental animals.

Studies on chemically induced porphyria have been conducted in an effort to understand normal regulation of porphyrin synthesis and its possible correlation to the breakdown in regulation seen in diseased conditions. In the course of these investigations, many questions concerning the regulation of porphyrin synthesis have been, at least in part, answered. Unfortunately, the mechanism by which chemicals cause the breakdown of regulation is not yet entirely clear.

Upon considering why the mechanism of action of porphyrinogenic compounds is not completely understood, several factors must be recalled. One factor, apparently not considered often enough, is the fact that *an* explanation may not be possible. Any chemical that causes an experimental animal to excrete porphyrins and porphyrin precursors can be classed as porphyrinogenic. It is quite possible, however, that this particular effect is only of secondary importance with regard to the action of the chemical. The so-called porphyria symptoms could arise as a result of other metabolic imbalances directly caused by the inducing chemical. If porphyrins were not red, and if some other class of metabolites were, the same disease with the same symptoms might well be discussed under another name in another chapter of this book! This is not to suggest all chemically induced porphyrias are different and/or secondary manifestations, but simply to stress that it is not proved one way or another in all cases.

In spite of the diverse nature of the chemicals that induce porphyria, there are some generally common characteristics. Likewise, certain generalizations may be stated about the symptoms brought about by the action of these chemicals.

The following chemicals are representative of some of the many that have been reported to induce porphyric type symptoms: (a) allylisopropylacetamide (AIA), (b) 3,5-dicarbethoxy-1,4-dihydrocollidine (DDC), (c) phenobarbital, (d) griseofulvin, (e) pregnandiol, (f) hexachlorobenzene, (g) orotic acid, (h) ethionine, (i) lead.

The author prefers to break this list of chemicals into two groups. In some cases there are reasons to suspect that the effect of the chemical is indirect. The chemicals that fall into the indirect group have comparatively little in common with each other or with the other group. The second group of chemicals appear to the author to act more directly, but not necessarily directly.

Representative compounds in the more direct group are AIA, DDC, phenobarbital, and steroids, and the following generalizations hold, within certain limits. The compounds are rather insoluble in water, many have ester or amide linkages that are sterically hindered with respect to hydrolysis, and they are detoxified by the body with some difficulty (467,468,513). A series of reports from Labbe's laboratory has established that many of the chemicals with porphyrinogenic activity are inhibitors of oxidative metabolism (514–517). These investigators have found that NADH oxidase is inhibited by low concentrations of most of these chemicals. Succinate oxidation, however, is not impaired. Tissue culture studies revealed that, during inhibition of oxidative metabolism, the cells produced and accumulated excess lactate (515). In the indirect group, ethionine (518) and orotic acid (519), while not tremendously soluble in water, are, nevertheless, more soluble than most of the compounds in the first group. They have in common the ability to cause a depression in the level of hepatic ATP (248,519–521). As discussed later, several factors suggest that decreased ATP levels may have something to do with manifestation of porphyric symptoms. Lead seems to fall in a class by itself (522–524). It is recognized as a potent inhibitor of numerous enzymes, including certain of the porphyrin biosynthetic ones. In spite of the fact that one of the lead-sensitive enzymes is ALA dehydrase, one of the symptoms of lead poisoning is increased urinary PBG. Also, ALA synthetase is suggested to be lead-sensitive, yet increased urinary ALA is one of the first symptoms of lead poisoning. The explanation for this is left as an exercise for the reader.

   b. *Metabolic Changes Caused by Porphyrinogenic Chemicals in Addition to Induction of Porphyria Symptoms.* The following generalizations apply with regard to many of the symptoms associated with the overproduction of porphyrins.

   Cell size is increased (525–527), RNA per cell is increased (425,526, 527), DNA per cell remains essentially constant (525–527), protein per cell is increased (115,525–527), and total heme per cell is increased (528,529). There are reports that ATP levels decrease (525,530), yet at least one report shows little change in the level of all nucleosides and nucleotides (531). In those instances where it has been measured, ALA synthetase is found to be markedly increased (105,115), catalase is decreased (115,532), succinyl-CoA synthetase increased (533), and ALA dehydrase increased (115,131,534). It is possible to explain many of these changes on the basis that heme exerts a strong controlling influence on its own synthesis. Granick has suggested that heme functions as a corepressor and acts in conjunction with a protein aporepressor (114).

$$\text{Heme} + \text{aporepressor} \rightleftharpoons \text{heme-repressor}$$

He postulates that porphyrinogenic chemicals interact with the repressor, causing the displacement of heme and forming aporepressor which can no longer exert a controlling affect. It would seem equally possible that the chemical might cause some change in the cell whereby "free" heme is decreased, pulling the equilibrium to the left.

Evidence for this sort of situation was found by Marver *et al.* (*535*) in studies on rat liver tryptophan pyrrolase and AIA. Tryptophan pyrrolase is the only heme enzyme where it is recognized that the heme is readily dissociable (*368,536–540*).

$$\text{Heme} + \text{apotryptophan pyrrolase} \rightleftharpoons \text{heme-tryptophan pyrrolase}$$

Tryptophan is known to promote formation of the active heme-enzyme, i.e., the equilibrium is forced to the right. This decreases the concentration of "free heme" and results in a shift in heme-repressor to reestablish equilibrium. With Granick's model in mind, it is possible to rationalize many of the findings in chemical porphyria studies. The modified model presented in Fig. 28 allows for both of these possibilities.

In trying to understand chemical porphyria it is useful to consider yet another aspect of the experiments. Almost all studies on chemical porphyria focus on the liver, either in intact animals or in tissue cultures. The reason for this is obvious, namely, the liver appears to be the only organ affected by these chemicals. This could mean that heme synthesis in the liver is regulated by a mechanism different from that in other tissues, or it could mean that the liver has some special property that renders it particularly susceptible to impairment of regulation. The second alternative appears the more attractive of the two. The liver is the principal organ for detoxification. If, when performing this function, metabolic imbalances become established, this might well alter the normal metabolism of the tissue, and, furthermore, the effect of the imbalance would be largely localized in the liver. Imbalance might arise in a number of ways. As mentioned, steric hindrance with respect to hydrolysis is one of the more common features of porphyrinogenic chemicals. Thus, it might be expected that clearance would be delayed, the drugs might concentrate in the liver tissue and interfere with normal operations. Any one of several situations could ensue. Heme might be displaced from association with repressor by the inducers (*114,513*), increased heme utilization for microsomal cytochromes could deplete "free heme" and promote dissociation of the active repressor, interference with electron transport might upset the heme-repressor balance, e.g., by inducing a compensatory increase in cytochrome synthesis (*541–545,369*), or an increased common metabolite (lactate) could react at an allosteric site on the repressor displacing the heme. Any one or a combination of these events could trigger the onset of porphyric symptoms. The evidence so far does not suggest the answer to these questions as simple.

c. Parameters that Affect Response of Organism to Porphyrinogenic Chemicals. In considering the symptoms of chemical porphyria, and the chemicals that induce the condition, it is of some value to review some of the factors that influence the condition.

A relationship between induction of porphyric symptoms (measured either by examining the urine for ALA and PBG, or by quantitating liver ALA synthetase), and the diet of the experimental animals has been noted *(441,442)*. Pursuit of this observation has revealed that carbohydrate in the diet strongly affects the induction of chemical porphyria. Probably the first thing that comes to mind upon discovering that carbohydrate (glucose) represses the induction of an enzyme is catabolite repression. Close examination of the compounds and metabolic pathways involved in the present case, however, quickly eliminates catabolite repression as it is commonly defined. It has been observed that administration of ATP to experimental animals can sometimes relieve or prevent chemical porphyria and, as already noted, decreased hepatic ATP has on occasion been observed during chemical porphyria *(530)*.

Could it not be that the glucose effect and the ATP effect are related? It seems quite possible that these two compounds could disturb the balance of a common metabolite which in turn might affect the equilibrium condition of heme and aporepressor. It is interesting that Granick *(114)* did not observe a glucose effect in his extensive studies employing embryonic chick liver tissue cultures. Ascorbate, NADPH, and in another study UDP-glucuronic acid retarded porphyrin formation. No other normal metabolite, except heme, showed this effect.

Hydrocortisone is recognized to have some role in the induction of chemical porphyria. Adrenalectomized rats do not respond to AIA by developing porphyria *(546)*. Hydrocortisone itself does not induce porphyria but, when administered along with AIA to adrenalectomized rats, induction is observed. Thus hydrocortisone is described as permissive in its action *(546)*.

The recent work of Granick and Kappas *(466)* on the induction of porphyria with 5β-H (A:B *cis*), C-19, and C-21 steroids further substantiates that steroids may somehow be involved in the regulation of porphyrin biosynthesis. Corroborating evidence is available in that, as already noted, there is some relationship between acute intermittent hepatic porphyria and sex steroids. The oxidative metabolism of steroids preparatory to elimination apparently involves the same enzyme system as employed for the detoxification of drugs and chemicals. Induction of the enzyme system with phenobarbital, for instance, enhances the clearance of steroids administered at a later date and vice versa *(544)*.

As the relationship between steroid metabolism and the control of porphyrin biosynthesis gains recognition, we can look forward to reports on steroids in the urine following induction of chemical porphyria.

B. Feedback Inhibition and Repression in the Regulation of Tetrapyrrole Biosynthesis

The studies of Lascelles on the relationship between iron nutrition and porphyrin accumulation by *R. spheroides* clearly established that tetrapyrrole biosynthesis was regulated at the level of ALA synthetase (*109,183*). The early work showed that heme repressed formation of ALA synthetase by this organism, but left open the question of feedback inhibition. Studies on the inhibition of ALA synthetase activity were performed by Burnham and Lascelles (*103*). These investigators partially purified three porphyrin biosynthetic enzymes from light-grown *R. spheroides*. Succinyl-CoA synthetase was utilized in these studies for generation of succinyl-CoA from succinate, CoA, and ATP, and was included in the ALA synthetase incubation mixture (*104*). Succinyl-CoA synthetase was tested independently for inhibition by all of the compounds tested with ALA synthetase and was found not be be inhibited.

Partially purified (20-fold) ALA synthetase was extremely sensitive to heme. Marked inhibition was observed with heme concentrations as low as 0.1 μM. Interestingly, even at the highest concentration tested, 100 μM, the inhibition was never complete. Other metalloporphyrins as well as metal-free dicarboxylic porphyrins were also somewhat inhibitory. It was necessary to use higher concentrations of these compounds to achieve inhibition equal to that caused by heme. For instance, to achieve 50% inhibition, protoporphyrin was required at 10 times the concentration of iron protoporphyrin. The inhibition by heme was noncompetitive with any of the substrates or cofactors, but it was reversed upon dilution. Since some question remains about the concentration of nonprotein-bound heme within a cell, inhibition experiments were also carried out with several common hemoproteins. The heme moiety of both hemoglobin and myoglobin is known to exist in a state of equilibrium with the protein moiety (*547–549*). The heme of cytochrome c, being covalently bound to the protein, is, of course, never free in solution, and the heme of catalase, though not covalently bound to the enzyme, nevertheless is apparently not in equilibrium with free heme. Both hemoglobin and myoglobin caused marked inhibition of ALA synthetase, while globin, cytochrome c, and catalase did not affect enzyme activity. Lascelles has more recently found that ALA synthetase from *Spirrilum itersonii* is also inhibited by low concentrations of heme (*111*).

In conjunction with the study of the inhibition of ALA synthetase by heme, ALA dehydrase was also partially purified and studied (*103*). This enzyme is also inhibited by heme, the inhibition being strongly influenced by the incubation protocol. When the enzyme was preincubated with cysteine, addition of heme at the start of the incubation caused a 23% inhibition of activity. On the other hand, if heme were added to the enzyme solution prior to preincubation with cysteine, ALA dehydrase was completely inhibited. ALA dehydrase is much less sensitive to heme than ALA synthetase, each being inhibited about 50% by 40 μM and 1 μM heme, respectively (*103*). The significance of the inhibition of ALA dehydrase by heme as a control mechanism was investigated with the use of intact *R. spheroides* in cell suspensions. These investigators found that exogenous heme suppressed the synthesis of porphyrins, but not bacteriochlorophyll, when glycine and α-ketoglutarate were the porphyrin precursors (Fig. 27). When ALA was the porphyrin precursor, exogenous heme did not suppress porphyrin synthesis. These results indicate that little control is exerted at the level of ALA dehydrase. Nevertheless, heme does inhibit ALA dehydrase, and this has also been observed by other workers using ALA dehydrase from different sources (*134,136*). It has been suggested that a second control point exists at this step in the biosynthesis of porphyrins. Since all available evidence indicates that ALA synthetase is present in very low amounts, and, thus, is the rate-limiting enzyme in porphyrin synthesis, it is difficult to understand why a second control point might exist. Taking, for example, the mature non-nucleated erythrocyte, ALA synthetase with its high turnover rate has long since ceased to exist. Unless ALA is supplied from some exogenous source, there should be no need on the part of the cell to regulate the further metabolism of ALA, yet the ALA dehydrase from this source is inhibited by heme, and this is one of the cases where the second control point has been postulated (*136*).

In spite of these arguments questioning the logic of a second control point, ALA dehydrase has other characteristics that are compatible with the existence of such a control. Shemin (*139*) has reported that, at low K^+ ion concentrations, ALA dehydrase from *R. spheroides* shows a sigmoid response to increasing substrate concentration. Furthermore, he claims that the enzyme consists of subunits and that K^+ ions promote association of the enzyme to an equilibrium mixture of monomer, dimer, and trimer. Both sigmoid curves and subunits are characteristics of allosteric enzymes (*550,551*).

Theoretical support for a control point at ALA dehydrase comes from Shemin's postulated succinate-glycine cycle (*91,120–122*). If this cycle is of metabolic significance, this represents a branch at ALA, and a control point is then required after the formation of ALA. Thus, the apparent allosteric nature of the enzyme, and the postulated succinate-glycine cycle, are mutually supporting.

In summary, partially purified ALA dehydrase behaves as if it were a rate-controlling enzyme of porphyrin biosynthesis. Experimental evidence with intact organisms [R. spheroides (103), embryonic chick liver tissue cultures (114,564,582), etc.], however, does not indicate that control is exerted through ALA dehydrase.

Indirect evidence for the inhibition of ALA synthetase by heme has come from the work of Karibian and London (552) in studies employing rabbit reticulocyte preparations. These investigators studied the formation of heme-^{14}C using glycine-^{14}C and ALA-^{14}C as heme precursors. When heme was included in the incubation mixture, the conversion of glycine-^{14}C into heme-^{14}C was inhibited, while heme had no effect upon the conversion of ALA-^{14}C to heme-^{14}C. This indicates that heme is regulating its own synthesis by regulating the rate at which ALA is formed. These experiments represent the only nonmicrobial system where evidence has been obtained indicating heme might have a feedback inhibition on ALA synthetase. The evidence is circumstantial, however, since ALA synthetase activity was not measured directly. Taddeini has studied ALA synthetase from rabbit reticulocytes, and has been unable to detect an effect of heme on the activity of the enzyme (553).

A regulatory mechanism involving repression of ALA synthetase has been directly demonstrated in R. spheroides (109) and in primary cultures of embryonic chick liver (114). Indirect support has come from several sources.

Formation of ALA synthetase by growing cultures of R. spheroides is repressed by low concentrations (0.01 mM) of exogenous heme (109). Neither metal-free porphyrins nor other metalloporphyrins represented formation of this enzyme.

Granick found that the induction of ALA synthetase by porphyrinogenic chemicals was repressed by exogenous heme. In a comprehensive and elegant study, Granick (114) used chick embryo liver tissue cultures to examine the mechanism of production of chemical porphyria. The results of this study have permitted us to obtain a rather broad understanding of the normal regulation of heme biosynthesis. Recalling the events in sequence, Granick and Urata (105) demonstrated that, following administration of porphyria-inducing drugs, the level of ALA synthetase in mitochondria from guinea pig liver increased markedly. At that time it was not possible to say with complete assurance whether the inducer accelerated the rate of de novo synthesis of enzyme, activated a preexisting zymogen, decreased the rate of destruction of the enzyme, or perhaps a combination of these possibilities. The chick embryo liver studies have answered many of these questions (*114*). Numerous key points were revealed. (a) Induction with allylisopropylacetamide (AIA) caused an increase in ALA synthetase, and the increase was due to an increased rate of formation of the enzyme. (b) The inducing action of the chemical was reversible. (c) The rate of formation of ALA synthetase was apparently controlled by a feedback repression, heme being implicated as the corepressor.

The technique consisted, briefly, in growing liver cells from 16- to 17-day old chick embryos in primary culture for 24 hours on coverslips on which they form a monolayer of colonies. On the second day the culture medium was replaced with fresh medium containing test chemicals. After approximately 20 hours' incubation, the colonies were examined by fluorescence microscopy, and in some cases by quantitative extraction of the porphyrins.

When inducer was removed from the cells, the rate of decrease in enzyme activity was the same as when acetoxycycloheximide was added to prevent protein synthesis. Had inducer acted by stabilizing ALA synthetase, slower degradation following addition of cycloheximide would be expected, since the inducer remained in the system.

That inducer did not convert inactive enzyme to active enzyme through an allosteric interaction was shown by the fact that washed mitochondria prepared from induced tissue retained an elevated enzyme activity. Since induction was reversible, the assumption was made that the inducer would wash out of isolated mitochondria and restore the original low level of acitivity.

Another possible action of inducing chemicals might be interaction with ALA synthetase to prevent an allosteric inhibitor such as heme from inhibiting the enzyme. To test this, inducing chemicals were added to mitochondria prepared from normal liver. No change in ALA synthetase activity was observed. The possible displacement of an allosteric inhibitor such as heme was tested by measuring ALA synthetase activity in the presence of heme and several other metal tetrapyrroles, and again no inhibition of ALA synthetase activity was observed.

The only time when inducing chemicals had an affect on ALA synthetase was when they were added to tissue preparations incubated under conditions permitting protein synthesis. Inhibition of protein synthesis at various levels with actinomycin D, puromycin, or acetoxycycloheximide invariably prevented further increase in ALA synthetase activity. In fact, not only was the increase in ALA synthetase prevented, but activity declined after a short time, indicating the enzyme had a rather high turnover rate. From these experiments the half-life of ALA synthetase in chick embryo liver was determined to be 4–6 hours.

The results summarized so far indicate that formation of ALA synthetase is repressed in normal liver, and that inducing drugs and chemicals act in some way to prevent this repression. In order to examine how repression is brought about and how inducing chemicals might interfere with the repression, Granick made some hypotheses that could be tested. One practical mechanism might involve a negative-feedback system whereby the repressor would respond to heme. He suggested that the active repressor might consist of a protein aporepressor and a heme corepressor (Fig. 28). Derepression then would occur when the heme

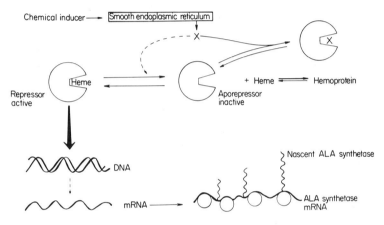

FIG. 28. Suggested control mechanism for porphyrin biosynthesis via repression of ALA synthetase modified from Granick (*114*). The hypothetical metabolite "X" arising as a result of the action of inducing chemicals on the smooth endoplasmic reticulum may displace heme from the repressor driving the equilibrium to the right, or it may react with free aporepressor, in either case forming an inactive complex with the repressor. Derepression can also occur following increased hemoprotein synthesis, pulling the equilibrium to the right. If that part of the scheme including chemical inducers is omitted, this control scheme assumes general applicability outside of the liver, for instance, in reticulocytes and in microorganisms.

level was low, or when the heme was displaced from the repressor by an inducing chemical. To test this, the chick embryo liver was treated with inducing chemical in the presence and absence of heme and other metalloporphyrins. After 18 hours' incubation, porphyrin fluorescence in the metalloporphyrin-treated cultures was less than half that observed in controls incubated with inducer alone. Thus, apparently heme can prevent the induction of ALA synthetase. According to the hypothesis of Granick, the corepressor heme and the inducing chemical compete for the same site on the aporepressor (*114*). To test this, two chemical inducers, AIA and DDC, were added singly and together, each at a concentration that would cause maximum fluorescence when administered alone. The rate of appearance of fluorescence when administered together was no greater than when administered singly. At low concentrations when they were tested together, they produced a greater fluorescence than when given alone, indicating that under these conditions their effect was additive, and supporting the single-site-of-action hypothesis.

This hypothesis, that heme is corepressor and that inducing chemicals compete with heme and prevent its combination with aporepressor, is attractive on many grounds including experimental ones. One point, however, causes concern, and that is the wide variety of compounds that are capable of inducing chemical porphyria. Granick divides these chemicals into four classes: barbiturates, collidines, sex steroids, and miscellaneous. The most active inducing compounds are the barbiturates, collidines, and the steroid metabolites (*114,466*). Granick has suggested a model showing how the barbiturates and collidines might fit on a heme template (*114*). It is much more difficult, however, to visualize an analogous situation with the steroids and miscellaneous compounds. Rather than attempt to rationalize how all, or even most, porphyrinogenic chemicals might compete with heme for a common binding site on the aporepressor, the present author suggests a modification of Granick's postulated scheme for regulation of porphyrin synthesis.

Since the site of action of porphyrinogenic chemicals is the liver, it seems likely that the primary action of the chemical might be upon something exclusively found in this tissue, namely, upon the detoxification enzyme system. As a result of the interaction of the inducer with the detoxification system, a metabolic imbalance may result, causing some metabolite "X" to be formed in excess. Metabolite "X" is then proposed to react allosterically with the repressor, either by displacing the corepressor heme, or by trapping aporepressor with the resultant formation of inactive repressor. Metabolite "X" might be almost anything. The 5β-H(A:B *cis*), C-19, or C-21 steroids that Granick and Kappas (*466*) found to possess high porphyrinogenic activity, being natural metabolites, are certainly possibilities. The advantage of this proposal lies in the fact that a specific receptor site on the aporepressor is not required to recognize a wide spectrum of inducing chemicals. The aporepressor would only need to recognize heme and metabolite "X." Since the primary event is postulated to occur between inducing chemical and an entire enzyme system, i.e., the detoxification system of the smooth endoplasmic reticulum, a number of primary sites might be expected which would account for the ability of so many different chemicals to act as inducers.

Granick (*114,466,513*) believes that the porphyrinogenic effect is associated with the detoxification system—the significant difference between his proposed mechanism and the present modification being that his mechanism involves interaction of inducer and repressor while the present proposal suggests that metabolite " X " interacts with the repressor. The modified model is presented in Fig. 28.

An adjunct to this hypothesis is that ALA synthetase in all tissues may be under the control of the same repressor. The reason that breakdown of regulation is observed only in the liver during chemical porphyria is due to the fact that the inducing chemical does not interact with a detoxification system with the resultant formation of metabolite " X " in these other tissues. The mechanism by which imbalance causing accumulation of metabolite " X " might occur remains open. In keeping with Labbe's observations (*516,517*), it is possible that inhibition of electron transport might be important.

In addition to the chemicals that are capable of inducing chemical porphyria, Granick studied a number of compounds to see which ones could block induction of porphyria. As already noted, inhibitors of nucleic acid and protein synthesis have the capacity to prevent induction of ALA synthetase. Only a few compounds were discovered that inhibited induction of porphyrin fluorescence without an apparent effect on the cell growth. Ascorbic acid, UDP-glucuronic acid, and NADPH decreased porphyrin fluorescence. Interestingly, these compounds have all been implicated in drug detoxification. DeMatteis (*554*) has observed that the level of ascorbic acid excreted in the urine during drug-induced experimental porphyria increases significantly. Glucose, which has been shown to have a marked effect on induction of chemical porphyria in intact rats, guinea pigs, and man, did not have any action-preventing induction of porphyrin fluorescence in chick embryo liver tissue.

Studies on the induction of ALA synthetase in rat liver have been recently reported. The assay system employed by Marver *et al.* (*106,108, 115*) has been modified from the original one of Granick and Urata (*105*). The principal changes are the use of unfractionated liver homogenates instead of mitochondria, inclusion of EDTA in the incubation mixture, and reliance on endogenous citric acid cycle intermediates to provide succinyl-CoA. With this assay system, ALA formation is proportional to enzyme added, the rate of ALA formation is linear for a longer time than with mitochondria, and the yield of ALA is enhanced over that formed by mitochondrial suspensions.

The results of induction experiments with rats are consistent with the findings in the chick embryo liver studies in that inhibitors of mRNA and protein biosynthesis block induction (*115*). The half-life of the rat liver enzyme was calculated to be about 70 minutes. Glucose was about as effective in preventing the induction of ALA synthetase as was actinomycin D.

The same laboratory has also investigated the effect of heme on the induction of ALA synthetase in rats (*555*). Administration of heme to rats previously treated with AIA caused an initial depression of ALA synthetase followed by pronounced oscillations in the level of the enzyme, persisting for several days. These results support the hypothesis that heme participates in the regulation of its own biosynthesis. The oscillations in the level of the enzyme, which increase in amplitude for a time, are a rather curious phenomenon. One would *a priori* expect a more precise control of the level of enzyme than is observed. These results contribute to our understanding of an earlier report upon the effect of ALA on the level of ALA dehydrase in mouse liver. Onisawa and Labbe (*556*) found that, following administration of ALA, the level of ALA dehydrase increased, then decreased, and then continued to oscillate for several hours. Coupled with the oscillations in hepatic ALA synthetase, this has the appearance of coordinate repression. Lascelles also observed coordinate repression of ALA dehydrase in *R. spheroides* (*109*).

C. Coordination of Tetrapyrrole and Protein Synthesis

Coordination of tetrapyrrole and protein synthesis has been most extensively studied employing hemoproteins. Hemoglobin (Hb) has received the lion's share of attention, though some valuable experiments have been conducted with other hemoproteins.

Two types of experimental systems have been employed in the Hb studies, each providing unique vantage points. Studies on Hb formation by chick embryos and blood islands of deembryonated chick blastoderms have contributed to our understanding of the sequence of events during Hb synthesis (*557,560,564,565,568*). Studies employing reticulocytes, nucleated erythrocytes, and cell-free protein-synthesizing systems have contributed to our understanding of the control and coordination of the synthesis of the heme and protein moieties of Hb.

Wilt studied globin and Hb synthesis in chick blastoderms (*558–560*). For these studies he prepared antibodies to highly purified adult chicken Hb. Adult chicken Hb is composed of two major fractions which are separable by ion-exchange chromatography or electrophoresis. Both fractions react with the antibody.

When extracts of blastoderms of different ages were reacted with the antibody, a very interesting picture emerged. The blastoderms of unincubated eggs contained two immunologically distinct components that cross-reacted with the antibody. One component was identical with one of the adult Hb's. Both components persist for 36–48 hours of development. At about 48 hours, the component not identical with either adult Hb disappeared, while at the same time a new component identical with the other Hb fraction of adult appeared.

Whereas the immunological test does not differentiate globin from Hb, staining with benzidine does allow this distinction. Wilt found that the benzidine reaction first became positive at 8–9 somites (36 hours), and concluded that globin was present in the blastoderm prior to heme synthesis. A double labeling experiment was also carried out in which blastoderms of different ages were incubated with tritium-labeled leucine and with ^{59}Fe. The labeled leucine was incorporated into antibody precipitable material at all stages while almost no iron was incorporated prior to the 7–8 somite stage. (This again confirms that the globin is synthesized somewhat ahead of heme.) Following the 7–8 somite stage, iron incorporation increased markedly.

It seems possible that the cross-reacting protein in the early blastoderms, not electrophoretically identical to adult Hb, might be free globin. Winterhalter (583) has shown that heme-free globin exists as an $\alpha\beta$ dimer. One would expect the $\alpha\beta$ dimer to migrate at a different rate than the $\alpha_2\beta_2$ tetramer.

Saha has studied the two Hb's (existing in unequal amounts) of adult and embryonic chickens (561,562). His results indicate that the relative proportions of the two fractions in the developing embryo change as the embryo develops. As the one component decreases, the other increases. His experiments on the resolved polypeptide chains from the two adult components, however, indicate that the polypeptide chains of the major component are distinctly different from those of the minor component. Thus, the change in relative proportion of the two components must be the result of the different rates of synthesis, and not the conversion of one component into another.

Levere and Granick have studied Hb formation in deembryonated chick blastoderm (563,564). When the blastoderms were incubated with glycine and succinate, they did not produce excess porphyrins. However, when incubated with low concentrations of ALA, porphyrin formation could be detected by fluorescence. They concluded that ALA synthetase was present in limiting amounts and thus prevented excess porphyrin formation. They then asked the question: What effect does the limitation of heme have on the production of globin? To answer this they supplied

ALA in nonlimiting amounts to increase the heme synthesized by the cells. (Heme added directly had no effect, probably because not enough gets into the proper site in the tissue.) After appropriate incubation of the blastoderms with ALA, Hb was extracted and measured. Tissue incubated in the presence of ALA formed an average of 2.7 times as much Hb as the control tissue. Actinomycin D did not have an effect on Hb synthesis, but puromycin prevented Hb formation in both ALA-treated tissues and controls. These results indicate that mRNA for globin synthesis was already present at the beginning of the experiment, and that there was apparently not a pool of preformed globin. They, therefore, concluded that heme may have an effect on globin formation at the ribosomal level.

Wainwright and Wainwright (565) have extended this line of investigation by placing emphasis on the effect of actinomycin D as a function of the age of blastoderm to determine when the synthesis of mRNA was critical for Hb formation. They found that Hb synthesis was inhibited by actinomycin D in the early stages of blastoderm development and that ALA reversed this inhibition. These results indicate that mRNA for globin is formed at a very early stage, and that mRNA for ALA synthetase is formed later. These investigators also found that Hb synthesis continued for some time following inhibition of mRNA synthesis, and they concluded that the mRNA for globin is metabolically stable.

It is interesting that, when ALA was added in addition to actinomycin D, the inhibition caused by the actinomycin D was overcome and Hb synthesis was equal to that in the control. On the other hand, when ALA was added to the control, protein synthesis was stimulated by a factor of about 3. Such results suggest that heme might play a regulatory role at the level of transcription as well as at the level of translation.

The fact that, during development of the embryo, one protein component cross-reacting with Hb-antibody disappears, while another one appears, suggests that a different mRNA is being read during this process. Apparently this mRNA is already present. In a situation like the present one, it is, of course, tempting to think in terms of analogy with established systems, and the cytochrome c regulatory system observed in yeast comes to mind. One wonders whether the fact that there is a major and minor Hb component, just as there is iso-1-cytochrome c (major) and iso-2-cytochrome c (minor), if the regulatory systems could be analogous? In both cases the minor component is apparently present as the apoprotein prior to hemoprotein formation. In both cases the major component seems to be preferentially formed after the initial stages of hemoprotein synthesis. As there have been no experiments where Hb development has been examined with the yeast analogy in mind, one is left to speculate about these possibilities. Messenger RNA coded for both types of Hb apparently is present from very early stages. Both types of globin might be synthesized initially; however, one of them might function as repressor to prevent overproduction of globin until heme is available. The cross-reacting protein that disappears (and perhaps acts as repressor) could possibly be an $\alpha\beta$ dimer of globin. Later during development of the embryo, mRNA for ALA synthetase is formed, and, once ALA is available, heme can be synthesized. The heme would then combine with the $\alpha\beta$ dimer, allowing tetramer formation. The $\alpha\beta$ dimer, which could have been functioning as repressor, would no longer exist to repress globin synthesis, and coordinated synthesis of Hb could then proceed.

Perhaps it is germane, to point out that the presence of two Hb's in chickens is not unique (recent work shows that there are additional Hb's in very small amounts); many birds have two Hb's, while some have three, and others only one (562,566–568).

Hammel and Bessman (569) studied the biosynthesis of Hb in pigeon red cell nuclei and demonstrated that the site of biosynthesis was actually within the nuclei. They found the optimum concentration of O_2 for Hb synthesis to be about 10% (495). At higher O_2 levels, Hb synthesis was inhibited. This inhibition was overcome by addition of heme to the incubation system. They conclude, therefore, that O_2 in this system affected only heme synthesis and not globin synthesis. They suggest that this effect may be at the uroporphyrinogen formation level, as previously observed by Falk and Porra (494).

Bruns and London studied the effect of heme on globin synthesis, using rabbit reticulocytes (570). They found that the incorporation of valine-^{14}C into Hb was increased in the presence of exogenous heme. A similar stimulation was produced by iron. However, the stimulatory effect of iron and that of heme were not additive, suggesting that iron functioned in the system by enhancing heme formation.

Grayzell et al. (571) looked at the effect of added exogenous heme on Hb synthesis in an attempt to elucidate the mechanism of its action in stimulating Hb formation. They used reticulocytes prepared from iron-deficient rabbits as experimental material, and found that in the presence of heme there was an increase in the proportion of ribosomes existing as polysomes. There was also an increase in the specific activity of both the soluble Hb and of the polypeptide chain attached to the polysomes. They concluded from this work that heme effects globin synthesis by affecting the size, stability, and functional activity of polysomes.

In a cell-free protein-synthesizing system from rabbit reticulocytes,

Gribble and Schwartz (572) found that protoporphyrin promoted the release of newly synthesized protein from the ribosomes.

In following up the work carried out on the reticulocyte systems, where heme has been shown to have an effect on the microsome-polysome equilibrium, it would be interesting to see if added labeled heme were actually bound to the polysomes.

In a study employing rabbit reticulocytes, Colombo and Baglioni (573,574) have reported on the relative proportions of completed α and β chains that are bound to polysomes and "free" in solution. They found that the completed, or nearly completed, α chains remain attached to polysomes while very few complete β chains remained on polysomes. They suggest two hypotheses that appear possible (ruling out a third): α chains are not "soluble" and remain on polysomes; and α chains are released from polysomes only upon combination with β chains.

The accumulation of α chains on the polysomes suggests that the rate-limiting step in the assembly of the Hb is the rate of synthesis and, thus, the availability of β chains. If the α chain remained attached to the polysome, there would be a pile up of ribosomes on the polysome. This would lead to a temporary halt in the reading of the mRNA and further synthesis of α chains. Thus, the rate of α chain synthesis is geared to the rate of synthesis of β chains.

An alternative model, supported by elegant experimental design, has been recently proposed by Schaeffer (575,576). Pulse-labeling experiments with rabbit reticulocytes, followed by separation of lysate into soluble and particulate fractions by centrifugation, showed 89% of the protein radioactivity in the soluble fraction and 11% in the ribosomal fraction (576). Electrophoresis of the soluble fraction showed that 45% of the activity of this fraction migrated as rabbit Hb. The remainder of the radioactivity was located in a disperse band between the line of sample application and the Hb bands. When the band of Hb was eluted from the electrophoretogram and separated into α and β chains, 94% of the activity in the Hb was found in the β chains. This indicates that newly synthesized labeled β chains had combined with endogenous unlabeled α chains to form the complete Hb molecule. Other workers have also found the β chain to be more highly labeled in pulse experiments than the α chain when the two were separated after purification of the Hb (573,579).

In an effort to learn about the composition of the disperse band of radioactivity on the electrophoretogram, the soluble fraction following pulse-labeling was incubated with isolated human α chains and isolated human β chains [prepared by the method of Bucci and Fronticelli (577)] in separate experiments. Electrophoresis of the mixture to which human β chains had been added revealed two major radioactive bands. One band corresponded to rabbit Hb, and the other to $\alpha_2^{Rab}\beta_2^A$ (A = human Hb-A), and most of the disperse band disappeared. Electrophoresis of the mixture to which human α chains had been added, did not show an Hb-hybrid ($\alpha_2^A\beta_2^{Rab}$). These experiments indicate that there are soluble rabbit α chains but no soluble β chains in the reticulocyte. These studies were extended by cell-free protein-synthesizing experiments (575). The addition of isolated human α chains or β chains prior to the synthesis of protein did not affect total protein formation. The presence of human β chains, however, did lead to a 45% reduction in labeling of the rabbit β chain in the soluble fraction, while the polysome fraction showed a proportionate increase in labeling. The rabbit α chain labeling under these conditions was the same as the control without human β chains. These results confirm those obtained with intact reticulocytes in demonstrating a pool of soluble rabbit α chains. The inhibition of release of rabbit β chains by human β chains suggest that the release of β chains from the polysomes is influenced by the presence of soluble rabbit α chains. Release of newly synthesized rabbit β chains was retarded, since the soluble α chains were removed by interaction with excess added human β chains. An attractive supposition is that the soluble α chain combines with the nascent β chain while the latter is still growing on the polysome, and promotes the release of the β chain.

This model is the exact opposite of that of Colombo and Baglioni (574). However, their data can be rationalized to fit Schaeffer's model, while Schaeffer's data cannot be rationalized to fit the model of Colombo and Baglioni. Recall that Colombo and Baglioni found very few complete β chains on the polysomes, while they did find completed α chains on the polysomes. If one makes the assumption that the complete α chains they found on the polysomes were not on polysomes reading α chain mRNA, but instead were on polysomes reading β chain mRNA by virtue of the fact that the α chain had reacted with the growing β chain (see Fig. 29), then the results are compatible with Schaeffer's model. Cline and Bock (480) have also suggested this interpretation of Colombo and Baglioni's data.

Winslow and Ingram (578) have measured the rate of synthesis of the α and β chains in human bone marrow preparations. Short pulse-labeling caused the β chain to have a higher specific activity than the α chain in agreement with the above. These investigators also succeeded in demonstrating that the rate of synthesis of both the α and β chains markedly decreased between position 90 and the carboxyl end of the molecule. They suggest that this represents a control point, and that it may result from the assumption of a tertiary conformation by the growing chain following heme incorporation. This author ventures the prediction that, when experiments of the type reported by Winslow and Ingram are repeated in a system supplemented with exogenous heme, the rate of synthesis of both the α chain and β chain will be found to be linear.

In spite of the fact that heme is recognized to stimulate the synthesis of globin, it has not yet been possible to demonstrate how this is done. Experimentally this is a difficult problem to investigate since the endogenous level of heme in most systems is generally high compared to the amount of Hb synthesized during short intervals. The very early chick embryos seem to provide the best experimental system where this problem could be avoided.

The author proposes a model assigning heme a role in promoting release of α chains from the α chain polysome. This is shown in Fig. 29.

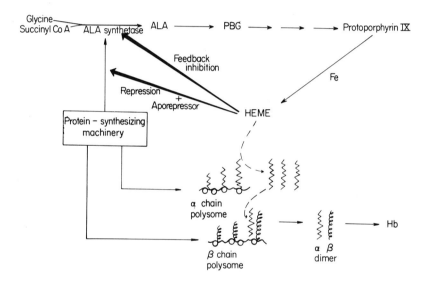

FIG. 29. Integrated control of heme and protein biosynthesis. The action of heme in controlling its own synthesis is discussed in the text and in Figs. 27 and 28. The possible way in which heme enhances Hb formation is indicated by the dashed line indicating interaction of heme with nascent α chains to promote their completion and release. The released α chains in turn interact with nascent β chains to promote their completion and release. The continued utilization of heme in this process keeps the free heme concentration low, maintaining ALA synthetase derepressed and the two systems coordinated.

The released α chain would then participate as already described in promoting the release of the β chain from the β chain polysome. The preliminary data of Vanderhoff et al. (579) appear compatible with such a mechanism. They found that the specific radioactivity ratio of $\alpha : \beta$

chains in purified Hb was 0.59 when no heme was included in the incubation. When heme was included, not only was total Hb increased, but the ratio of specific activities of $\alpha : \beta$ chains increased to 1.02. Since endogenous unlabeled α chains that exist at the start of the incubation combine with newly synthesized labeled β chains to form Hb, the specific activity of the β chains of the purified Hb would be greater than that of the α chain, as the experiments reveal. Now, if α chain release is enhanced by interaction with heme, and assuming that β chain formation is the rate-limiting step in Hb formation (since there are free α chains this must be true), the newly synthesized labeled α chains would dilute out the endogenous unlabeled α chains, and the specific activity ratio of the α and β chains should approach unity, as observed. The enhanced synthesis of Hb would, in turn, enhance heme formation by derepressing ALA synthetase (Fig. 28), providing a well-coordinated system that seems to fit with the experimental observations.

There are metabolic abnormalities known where excess β chains are known to exist, e.g., β-thalassemia (580,581), apparently indicating that the presence of complementary α chains is not an absolute requirement for release of β chains. That apoproteins do not invariably repress further synthesis of themselves seems to be clearly shown by the experiments of Chang and Lascelles with *Staph. aureus*. These investigators demonstrated that addition of heme to free cell extracts of this microorganism permitted formation of cytochrome b. Such results could only be obtained if the apoenzyme were already present in the cell extract.

Another interesting case of control of tetrapyrrole synthesis and coordination of heme and protein synthesis is afforded by studies on tryptophan pyrrolase (TPO). The level of this enzyme in liver is markedly increased by administration of cortisone or L-tryptophan to an experimental animal. Tryptophan increases the affinity of the enzyme for its cofactor, heme. Tryptophan pyrrolase with heme bound is more stable than the apoenzyme (540). Apparently the rate of degradation of the enzyme is decreased. The net result, therefore, is an increase in the actual amount of enzyme present (368,540). The effect of cortisone treatment is distinctly different, in that the rate of synthesis of tryptophan pyrrolase is increased (368).

The barrage of excellent papers published in 1966 by Marver *et al.* includes a study of this problem (535), which provides an interesting and well-integrated picture.

The primary action of tryptophan, namely, the enhancement of the binding of heme to the apoprotein has two effects. First, the enzyme is relatively more stable. The second, and for the moment, more important effect, however, is the decrease in the concentration of "free" heme caused by the shift in the equilibrium to the right. When the heme

$$\text{Heme} + \text{apo-TPO} + \text{tryptophan} \rightleftharpoons \text{heme-TPO-tryptophan}$$

concentration drops, ALA synthetase is derepressed and heme formation is stimulated (Fig. 28). This postulated sequence of events is supported by the observation that administration of L-tryptophan caused a rise in the level of both tryptophan pyrrolase and ALA synthetase (535). Additional support was provided by the corollary observations that the porphyria-inducing drug AIA stimulated synthesis of both ALA synthetase and tryptophan pyrrolase. In this case, apparently the primary event was derepression of ALA synthetase. As the level of this rate-limiting enzyme for heme synthesis increased, the formation of heme increased, and the higher concentration of heme then forced the heme-apotryptophan pyrrolase equilibrium to the right. Again, the enzyme became stabilized, degradation was retarded, and the total tryptophan pyrrolase level increased. These observations (535), for the most part, support the recent work of Granick (114) on the induction of ALA synthetase by porphyria-inducing drugs where heme was implicated in the repression of ALA synthetase.

VI. COBALAMIN COENZYMES

At the time the chapter on tetrapyrroles was prepared for the preceding edition of *Metabolic Pathways*, cobalamin coenzyme chemistry was in its infancy. Limited discussion of the metabolic role of the cobalamins, therefore, seems warranted.

Even before the discovery of the corrinoid coenzymes (311), circumstantial evidence was available linking vitamin B_{12} with certain areas of metabolism. Nutritional studies revealed a relationship between vitamin B_{12} and the synthesis of metabolically labile methyl groups, the biosynthesis of purines, pyrimidines, and their deoxyribosides. At present, not only is the relationship between vitamin B_{12} and these areas much more fully understood, but a sizable number of specific reactions are known that require the coenzyme form of the vitamin. There seems little doubt that this number will increase considerably in the future.

In retrospect, it is apparent that two factors delayed the discovery of the coenzyme form of the corrins. In the first place the concentration of the corrinoids in most organisms and tissues is low. Second, and of more importance, is the fact that the coenzyme forms of vitamin B_{12} are very sensitive to light, and no doubt, in early studies before this sensitivity was recognized, the coenzyme was inadvertently decomposed to the vitamin form by room light as the material was purified.

Reactions involving corrinoid coenzymes are:

(a) Glutamate $\xleftrightarrow{\text{glutamate mutase}}$ β-methylaspartate

(b) Methylmalonly-CoA $\xleftrightarrow{\text{methylmalonyl–CoA mutase}}$ succinyl-CoA

(c) Propane-1,2-diol $\xrightarrow{\text{diol dehydrase}}$ proprionaldehyde

(d) GTP $\xrightarrow[\text{reductase}]{\text{ribonucleoside tri–P}}$ deoxy-GTP

(e) Methyltetrahydrofolate-N^5 + homocysteine $\rightarrow$ $\xrightarrow[\text{FADH}_2]{\text{S–adenosylmethionine}}$ tetrahydrofolate + methionine

(f) $CO_2 \rightarrow \rightarrow \rightarrow \rightarrow$ acetate

(g) $CO_2 \rightarrow \rightarrow \rightarrow \rightarrow$ methane

In his recent Hopkins Memorial Lecture, Barker divides these reactions into two groups depending on the type of corrinoid required (584). In turn, this also separates the two groups according to the apparent mechanism of the reaction. Barker places those reactions requiring a vitamin B_{12} analog and apparently involving the formation of a cobalt-methyl corrinoid in one group (e, f, g, above). The second group (a, b, c, and d) includes the reactions in which the 5′-deoxyadenosyl form of the corrinoids participate.

The reactions in which methyl corrinoids participate are considerably more involved than those requiring 5′-deoxyadenosyl corrinoids, and the reader is referred to the following recent reviews for coverage of these reactions (8,584–586).

There is an obvious similarity in the glutamate mutase reaction in which a glycine residue moves from C-3 to C-4 (587), and the methylmalonyl-CoA mutase reaction where a carbonylthio ester moves from C-2 to C-3 (588). In both cases a hydrogen atom simultaneously moves in the opposite direction. The diol dehydrase reaction is analogous, though the carbon skeleton is not rearranged. Thus, during the conversion of propane-1,2-diol to propionaldehyde, the oxygen atom on C-2 moves to C-1 (589), while the hydrogen on C-1 moves to C-2 (590).

In the case of all three reactions it has been demonstrated that hydrogen transfer occurs without exhange with protons from the solvent (590–592). Direct evidence has been obtained in the laboratory of Abeles (593) that the corrinoid coenzyme participates in the nonexchangeable hydrogen-atom transfer and, furthermore, that the transfer is intermolecular. The design of the experiments that clarified this point was both simple and clever. Diol dehydrase will accept both propane-1,2-diol and ethylene glycol as substrates. These investigators took advantage of this fact and carried out the enzyme reaction using propane-1,2-diol-1-^{3}H and unlabeled ethylene glycol as simultaneous substrates. Analysis of the products of the reaction revealed that tritium was present in both the propionaldehyde and acetaldehyde (593). These results indicated that during the course of the reaction the enzyme or coenzyme removed a tritium atom from the propane-1,2-diol and transferred it to an incoming ethylene glycol with formation of acetaldehyde. Proof that the transferred tritium was carried on the corrin coenzyme was obtained in the same laboratory when tritium-labeled propane-1,2-diol was allowed to react with diol dehydrase from which the 5′-deoxyadenosyl corrinoid was subsequently isolated (594,595). The coenzyme was not only labeled, but the specific radioactivity was such that it could only be accounted for if two moles of tritium had been incorporated per mole of coenzyme (594). Actually, these results had to follow to account for the intermolecular transfer of tritium already noted. That is, if the coenzyme accepted and donated a single hydrogen atom employing a single hydrogen binding site, it follows that the hydrogen accepted would also be the one donated in an intramolecular reaction. Since this is not what happens, there must be two equivalent hydrogens at the transfer site. This focuses attention upon the 5′-methylene group linked to the 6th coordination position of the cobalt. It has been suggested on theoretical grounds that the corrinoid coenzymes may be biological Grignard-type compounds; the 5′-methylene group participating as a carbanion stabilized by the nearby cobaltic ion (596). Frey and Abeles have tested this by chemically synthesizing 5′-deoxyadenosyl cobalamin labeled with tritium in the 5′-methylene position (594,595). When the tritium-labeled synthetic coenzyme was allowed to react with enzyme and unlabeled substrate, the tritium was transferred to the product. The technique of preparing tritium-labeled corrinoid coenzyme and adding it to the appropriate enzyme and unlabeled substrate has been applied to both the glutamate mutase and methylmalonyl-CoA systems (584,597). Results analogous to those in the diol dehydrase experiments were obtained, further supporting the conclusion that the mechanisms of these reactions are similar.

REFERENCES

1. C. O'hEocha, in "Chemistry and Biochemistry of Plant Pigments" (T. W. Goodwin, ed.), p. 175. Academic Press, London, 1965.
2. S. Granick and D. Mauzerall, in "Metabolic Pathways" (D. M. Greenberg, ed.), Vol. II, p. 525, Academic Press, New York, 1961.

3. J. Lascelles, "Tetrapyrrole Biosynthesis and Its Regulation." Benjamin, New York, 1964.
4. J. E. Falk, "Porphyrins and Metalloporphyrins," Vol. 2. Elsevier, New York, 1964.
5. G. S. Marks, *Botan. Rev.* **32**, 56 (1966).
6. R. Bonnett, *Chem. Rev.* **63**, 573 (1963).
7. K. Bernhauer, O. Muller, and F. Wagner, *Angew. Chem. Intern. Ed. Engl.* **3**, 200 (1964).
8. F. Wagner, *Ann. Rev. Biochem.* **35**, 405 (1966).
9. R. Hill, *in* "Comprehensive Biochemistry" (M. Florkin and E. H. Stotz, eds.), Vol. 9, p. 73. Elsevier, New York, 1963.
10. S. Schwartz, M. H. Berg, I. Bossenmaier, and H. Dinsmore, *Methods Biochem. Anal.* **8**, 221 (1960).
11. L. Bogorad, *in* "The Chlorophylls" (L. P. Vernon and G. R. Seely, eds.), p. 481. Academic Press, New York, 1966.
12. E. L. Smith, "Vitamin B_{12}." John Wiley and Sons, New York, 1965.
13. R. Schmid, *in* "The Metabolic Basis of Inherited Disease" (J. B. Stanbury, J. B. Wyngaarden, and D. S. Fredrickson, eds.), pp. 813, 871. McGraw-Hill, New York, 1966.
14. T. K. With, "Chemistry of the Bile Pigments." Academic Press, New York, 1967.
15. F. Hoppe-Seyler, *Z. Physiol. Chem.* **4**, 193 (1880).
16. R. Willstatter, *Uber Pflanzenfarbstoffe, in* "Nobelstiftelsen, Stockholm, Les Prix Nobel en 1914–1918," p. 1. P. A. Norstedt, Stockholm, 1920.
17. H. Fischer and H. Orth, "Die Chemie des Pyrrols," 2 vols. Akademische Verlagsgescellschaft, Leipzig, 1937.
18. E. L. Rickes, N. G. Brink, F. R. Koniuszy, T. R. Wood, and K. Folkers, *Science* **107**, 396 (1948).
19. E. L. Smith and L. F. J. Parker, *Biochem. J.* **43**, viii (1948).
20. D. E. Wolf, W. H. Jones, J. Valiant, and K. Folkers, *J. Am. Chem. Soc.* **72**, 2820 (1950).
21. J. B. Armitage, J. R. Cannon, A. W. Johnson, L. F. J. Parker, E. L. Smith, W. H. Stafford, and A. R. Todd, *J. Chem. Soc.* 3849 (1953).
22. C. Brink, D. C. Hodgkin, J. Lindsey, J. Pickworth, J. H. Robertson, and J. G. White, *Nature* **174**, 1169 (1954).
23. D. C. Hodgkin, J. Pickworth, J. H. Robertson, R. J. Prosen, R. A. Sparks, and K. N. Trueblood, *Proc. Roy. Soc. (London)* **A251**, 306 (1959).
24. P. G. Lenhert and D. C. Hodgkin, *Nature* **192**, 937 (1961).
25. R. B. Woodward, W. A. Ayer, J. M. Beaton, F. Bickelhaupt, R. Bonnett, P. Buchschacher, G. L. Closs, H. Dutler, J. Hannah, F. P. Hauck, S. Ito, A. Langemana, E. LeFoff, W. Leimgruber, W. Levowski, J. Sauer, J. Valenta, and H. Volz, *J. Am. Chem. Soc.* **82**, 3800 (1960).
26. J. L. Hoard, M. J. Hamor, and T. A. Hamor, *J. Am. Chem. Soc.* **85**, 2334 (1963).
27. T. A. Hamor, W. S. Caughey, and J. L. Hoard, *J. Am. Chem. Soc.* **87**, 2305 (1965).
28. J. L. Hoard, M. J. Hamor, T. A. Hamor, and W. S. Caughey, *J. Am. Chem. Soc.* **87**, 2312 (1965).
29. D. F. Koenig, *Acta Cryst.* **18**, 663 (1965).
30. IUPAC-IUB Commission on Biochemical Nomenclature, *Biochim. Biophys. Acta* **117**, 285 (1966).
31. E. Bertele, H. Boos, J. D. Dunitz, F. Elsinger, A. Eschenmoser, I. Filner, H. P. Gribi, H. Gschwend, E. F. Meyer, M. Pesaro, and R. Scheffold, *Angew. Chem. Intern. Ed. Engl.* **3**, 490 (1964).

32. A. W. Johnson and I. T. Kay, *J. Chem. Soc.* 1620 (1965).
33. R. L. N. Harris, A. W. Johnson, and I. T. Kay, *Chem. Commun.*, 355 (1965).
34. A. W. Johnson, *Chem. Brit.* **3**, 253 (1967).
35. D. Mauzerall and S. Granick, *J. Biol. Chem.* **232**, 1141 (1958).
36. D. Mauzerall, *J. Am. Chem. Soc.* **84**, 2437 (1962).
36a. E. B. Fleischer and J. H. Wang, *J. Am. Chem. Soc.* **82**, 3498 (1960).
36b. B. F. Burnham and J. J. Zuckerman (manuscript in preparation, 1967).
37. B. Dempsey, M. B. Lowe, and J. N. Phillips, *in* "Haematin Enzymes" (J. E. Falk, R. Lemberg, and R. K. Morton, eds.). Pergamon Press, New York, 1961.
38. M. Gouterman, G. H. Wagniere, and L. C. Snyder, *J. Mol. Spectr.* **11**, 108 (1963).
39. K. Ohno, Y. Tanabe, and F. Sasaki, *Theoret. Chim. Acta* **1**, 378 (1963).
40. C. Weiss, H. Kobayashi, and M. Gouterman, *J. Mol. Spectr.* **16**, 415 (1965).
41. M. Gouterman, *J. Mol. Spectr.* **6**, 138 (1961).
42. L. J. Boucher and J. J. Katz, *J. Am. Chem. Soc.* **89**, 1340 (1967).
43. W. S. Caughey, J. O. Alben, W. Y. Fujimoto, and J. L. York, *J. Org. Chem.* **31**, 2631 (1966).
44. L. J. Boucher, H. H. Strain, and J. J. Katz, *J. Am. Chem. Soc.* **88**, 1341 (1966).
45. S. Sano, T. Shingu, J. M. French, and E. Thronger, *Biochem. J.* **97**, 250 (1965).
46. M. C. Dougherty, H. L. Crespi, H. H. Strain, and J. J. Katz, *J. Am. Chem. Soc.* **88**, 2854 (1966).
47. R. J. Abraham, P. A. Burbidge, A. H. Jackson, and D. B. Macdonald, *J. Chem. Soc.*, 620 (1966).
48. C. B. Storm and A. H. Corwin, *J. Org. Chem.* **29**, 3700 (1964).
49. J. H. Mathewson, W. R. Richards, and H. Rapoport, *J. Am. Chem. Soc.* **85**, 364 (1963).
50. H. A. O. Hill, J. M. Pratt, and R. J. P. Williams, *J. Chem. Soc.*, 2859 (1965).
51. J. M. Assour, *J. Chem. Phys.* **43**, 2477 (1965).
52. W. E. Blumberg and J. Peisach, *J. Biol. Chem.* **240**, 870 (1965).
53. T. C. Hollocher, *J. Biol. Chem.* **241**, 1958 (1966).
54. N. S. Hush and J. R. Rowland, *J. Am. Chem. Soc.* **89**, 2976 (1967).
55. D. Mauzerall and G. Leher, *Biochim. Biophys. Acta* **79**, 430 (1964).
56. L. M. Epstein, *J. Chem. Phys.* **36**, 2731 (1962).
57. U. Gonser and R. W. Grant, *Biophys. J.* **5**, 823 (1965).
58. W. S. Caughey, W. Y. Fujimoto, A. J. Bearden, and T. H. Moss, *Biochemistry* **5**, 1225 (1966).
59. G. Lang and W. Marshall, *J. Mol. Biol.* **18**, 385 (1966).
60. S. Granick, *J. Biol. Chem.* **172**, 717 (1948).
61. S. Granick, *J. Biol. Chem.* **175**, 333 (1948).
62. S. Granick, *J. Biol. Chem.* **183**, 713 (1950).
63. S. Granick, *J. Biol. Chem.* **236**, 1168 (1961).
64. L. Bogorad and S. Granick, *J. Biol. Chem.* **202**, 793 (1953).
65. S. Granick, L. Bogorad, and H. Jaffe, *J. Biol. Chem.* **202**, 801 (1953).
66. S. Sano and S. Granick, *J. Biol. Chem.* **236**, 1173 (1961).
67. L. Bogorad, *in* "Comparative Biochemistry of Photoreactive Systems" (M. B. Allen, ed.), Vol. 1, p. 227. Academic Press, New York, 1960.
68. R. Barchielli, G. Borretti, A. DiMarco, D. Julita, A. Migliacci, A. Minghetti, and C. Spalla, *Biochem. J.* **74**, 382 (1960).
69. G. Borretti, A. DiMarco, L. Fuoco, M. P. Marnati, A. Migliacci, and C. Spalla, *Biochim. Biophys. Acta* **37**, 379 (1960).
70. B. Bartosinski, *Bull. Acad. Polon. Sciences* **14**, 143 (1966).

71. A. Migliacci and A. Rusconi, *Biochim. Biophys. Acta* **50**, 370 (1961).
72. K. Bernhauer, E. Becher, G. Gross, and G. Wilharm, *Biochem. Z.* **332**, 562 (1960).
73. V. N. Bukin and G. V. Pronyakova, *J. Biochem. (Tokyo)* **47**, 781 (1960).
74. G. Mantrova, V. N. Bukin, and V. V. Pchelkina, *5th Intern. Congr. Biochem.*, *Moscow, 1961.*
75. B. F. Burnham, *Federation Proc.* **24**, 223 (1965).
76. B. F. Burnham and R. A. Plane, *Biochem. J.* **98**, 13c (1966).
77. R. J. Porra, *Biochim. Biophys. Acta* **107**, 176 (1965).
78. J. W. Corcoran and D. Shemin, *Biochim. Biophys. Acta* **25**, 661 (1957).
79. D. Shemin, J. W. Corcoran, C. Rosenblum, and I. M. Miller, *Science* **124**, 272 (1956).
80. A. J. D'Eustachio, *Bacteriol. Proc.*, 71 (1966).
81. D. Shemin and D. Rittenberg, *J. Biol. Chem.* **166**, 621 (1946).
82. D. Shemin and D. Rittenberg, *J. Biol. Chem.* **166**, 627 (1946).
83. J. Wittenberg and D. Shemin, *J. Biol. Chem.* **185**, 103 (1950).
84. N. S. Radin, D. Rittenberg, and D. Shemin, *J. Biol. Chem.* **184**, 745 (1950).
85. D. Shemin, *Harvey Lectures Ser.* **50**, 258 (1956).
86. J. Wittenberg and D. Shemin, *J. Biol. Chem.* **178**, 47 (1949).
87. H. M. Muir and A. Neuberger, *Biochem. J.* **47**, 97 (1950).
88. D. Shemin and J. Wittenberg, *J. Biol. Chem.* **192**, 315 (1951).
89. D. Shemin and S. Kumin, *J. Biol. Chem.* **198**, 827 (1952).
90. K. D. Gibson, W. G. Laver, and A. Neuberger, *Biochem. J.* **70**, 71 (1958).
91. D. Shemin and C. S. Russell, *J. Am. Chem. Soc.* **75**, 4873 (1953).
92. W. G. Laver, A. Neuberger, and J. J. Scott, *J. Chem. Soc.* 1474 (1959); 1483 (1959).
93. D. Shemin, C. S. Russell, and T. Abramsky, *J. Biol. Chem.* **215**, 613 (1955).
94. A. Neuberger and J. J. Scott, *Nature* **172**, 1093 (1953).
95. W. G. Laver, A. Neuberger, and S. Udenfriend, *Biochem. J.* **70**, 4 (1958).
96. E. G. Brown, *Biochem. J.* **70**, 313 (1958).
97. K. D. Gibson, *Biochim. Biophys. Acta* **28**, 451 (1958).
98. G. Kikuchi, A. Kumar, P. Talmadge, and D. Shemin, *J. Biol. Chem.* **233**, 1214 (1958).
99. S. Granick, *J. Biol. Chem.* **232**, 1101 (1958).
100. J. P. Shulman and D. A. Richert, *J. Biol. Chem.* **226**, 181 (1957).
101. J. Lascelles, *Biochem. J.* **66**, 65 (1957).
102. M. M. Wintrobe, *Harvey Lectures Ser.* **45**, 87 (1951).
103. B. F. Burnham and J. Lascelles, *Biochem. J.* **87**, 462 (1963).
104. B. F. Burnham, *Acta Chem. Scand.* **17**, 123 (1963).
105. S. Granick and G. Urata, *J. Biol. Chem.* **238**, 821 (1963).
106. H. S. Marver, D. P. Tschudy, M. G. Perlroth, A. Collins, and G. Hunter, Jr., *Anal. Biochem.* **14**, 53 (1966).
107. D. P. Tschudy, M. G. Perlroth, H. S. Marver, A. Collins, and G. Hunter, Jr., *Proc. Natl. Acad. Sci. U.S.* **53**, 841 (1965).
107a. M. G. Perlroth, D. P. Tschudy, H. S. Marver, C. W. Berard, R. F. Zeigel, M. Rechcigl, and A. Collins, *Am. J. Med.* **41**, 149 (1966).
108. H. S. Marver, D. P. Tschudy, M. G. Perlroth, and A. Collins, *J. Biol. Chem.* **241**, 2803 (1966).
109. J. Lascelles, *J. Gen. Microbiol.* **23**, 487 (1960).
110. B. F. Burnham and J. Lascelles, unpublished observations.
111. J. Lascelles, personal communication.
112. W. Vogel, D. A. Richert, B. Q. Pixley, and M. P. Shulman, *J. Biol. Chem.* **235**, 1769 (1960).
113. E. G. Brown, *Nature* **182**, 314 (1958).

114. S. Granick, *J. Biol. Chem.* **241**, 1359 (1966).
115. H. S. Marver, A. Collins, D. P. Tschudy, and M. Rechcigl, *J. Biol. Chem.* **241**, 4323 (1966).
116. M. J. Bull and J. Lascelles, *Biochem. J.* **87**, 15 (1963).
117. S. Granick, *J. Biol. Chem.* **238**, PC 2247 (1963).
118. D. Mauzerall and S. Granick, *J. Biol. Chem.* **219**, 435 (1956).
119. L. Shuster, *Biochem. J.* **64**, 101 (1956).
120. D. Shemin, *in* "Ciba Foundation Symposium on Porphyrin Biosynthesis and Metabolism" (G. E. W. Wolstenhome and E. C. P. Millar, eds.), p. 4. Little, Brown and Co., Boston, 1955.
121. A. M. Nemeth, C. S. Russell, and D. Shemin, *J. Biol. Chem.* **229**, 415 (1959).
122. N. I. Berlin, A. Neuberger, and J. J. Scott, *Biochem. J.* **64**, 80 (1956).
123. E. Kowalski, A. M. Dancewicz, and Z. Scot, *Intern. Congr. Biochem., 4th, Vienna, 1958, Abstr.* *1–6* (1959).
124. M. Bagdasarian, *Nature* **181**, 1399 (1958).
125. K. D. Gibson, A. Neuberger, and G. H. Tait, *Biochem. J.* **83**, 539 (1962).
126. A. Neuberger and J. M. Turner, *Biochim. Biophys. Acta* **67**, 342 (1963).
127. J. M. Turner, *Biochim. Biophys. Acta* **77**, 697 (1963).
128. R. G. Westall, *Nature* **170**, 614 (1952).
129. G. H. Cookson and C. Rimington, *Biochem. J.* **57**, 476 (1954).
130. S. Granick, *Science* **120**, 1105 (1954).
131. K. D. Gibson, A. Neuberger, and J. J. Scott, *Biochem. J.* **61**, 618 (1955).
132. S. Granick and D. Mauzerall, *J. Biol. Chem.* **232**, 1119 (1958).
133. A. M. del C. Batlle, A. M. Ferramola, and M. Grinstein, *Biochem. J.* **104**, 244 (1967).
134. D. L. Coleman, *J. Biol. Chem.* **241**, 5511 (1966).
135. D. L. Nandi and E. R. Waygood, *Can. J. Biochem.* **45**, 327 (1967).
136. P. Callissano, D. Bonsignore, and C. Cartasegna, *Biochem. J.* **101**, 550 (1966).
137. A. A. Iodice, *Federation Proc.* **17**, 248 (1958).
138. M. L. Wilson, A. A. Iodice, M. P. Shulman, and D. A. Richert, *Federation Proc.* **18**, 352 (1959).
139. D. Shemin and D. L. Nandi, *Federation Proc.* **26**, 745 (1967).
140. D. L. Nandi and D. Shemin, *Federation Proc.* **24**, 531 (1965).
141. D. Mauzerall, *J. Am. Chem. Soc.* **82**, 2601 (1960).
142. D. Mauzerall, *J. Am. Chem. Soc.* **82**, 2605 (1960).
143. J. H. Mathewson and A. H. Corwin, *J. Am. Chem. Soc.* **83**, 135 (1961).
144. E. Bullock, A. W. Johnson, E. Markham, and K. B. Shaw, *Nature* **185**, 607 (1960).
145. J. B. Wittenberg, *Nature* **184**, 876 (1959).
146. L. Bogorad and S. Granick, *Proc. Natl. Acad. Sci. U.S.* **39**, 1176 (1953).
147. A. H. Jackson and S. F. MacDonald, *Can. J. Chem.* **35**, 715 (1957).
148. L. Bogorad, *J. Biol. Chem.* **233**, 501 (1958).
149. L. Bogorad, *J. Biol. Chem.* **233**, 510 (1958).
150. L. Bogorad, *J. Biol. Chem.* **233**, 516 (1958).
151. A. T. Carpenter and J. J. Scott, *Biochem. J.* **71**, 325 (1959).
152. L. Bogorad, *Plant Physiol.* **32**, xli (1957).
153. A. T. Carpenter and J. J. Scott, *Biochim. Biophys. Acta* **52**, 195 (1961).
154. L. Bogorad, *in* "5th International Congress of Biochemistry," p. 101. Pergamon Press, Oxford, 1961.
155. L. Bogorad, *Ann. N.Y. Acad. Sci.* **104**, 676 (1963).
156. L. Bogorad and G. S. Marks, *Biochim. Biophys. Acta* **41**, 356 (1960).
157. L. Bogorad and G. S. Marks, *J. Biol. Chem.* **235**, 2127 (1960).

158. W. H. Lockwood and A. Benson, *Biochem. J.* **75**, 372 (1960).
159. A. Treibs and W. Ott, *Ann. Chem.* **615**, 137 (1958).
160. E. I. B. Dresel and J. E. Falk, *Biochem. J.* **63**, 80 (1956).
161. E. Bullock, A. W. Johnson, E. Markham, and K. B. Shaw, *J. Chem. Soc.*, 1430 (1958).
162. A. M. del C. Batlle and M. Grinstein, *Biochem. Biophys. Acta* **82**, 13 (1962).
163. D. S. Hoare and H. Heath, *Biochem. J.* **73**, 679 (1959).
164. R. A. Nevé, R. F. Labbe, and R. A. Aldrich, *J. Am. Chem. Soc.* **78**, 691 (1956).
165. P. Cornford, *Biochem. J.* **91**, 64 (1964).
166. G. Urata and H. Kimura, *J. Biochem. (Tokyo)* **47**, 150 (1960).
167. J. E. Falk and A. Benson, *Biochem. J.* **55**, 101 (1953).
168. T. C. Chu and E. J. Chu, *J. Biol. Chem.* **227**, 505 (1957).
169. T. C Chu and E. J. Chu, *J. Biol. Chem.* **234**, 2741, 2747, 2751 (1959).
170. P. Cornford and A. Benson, *J. Chromatog.* **10**, 141 (1963).
171. P. R. Edmonson and S. Schwartz, *J. Biol. Chem.* **205**, 605 (1953).
172. J. E. Falk, *J. Chromatog.* **5**, 277 (1961).
173. A. M. del C. Batlle, A. Benson, and C. Rimington, *Biochem. J.* **97**, 731 (1965).
174. S. Sano and S. Granick, *J. Biol. Chem.* **236**, 1173 (1961).
175. S. Granick and D. Mauzerall, *Federation Proc.* **17**, 233 (1958).
176. S. Granick and D. Mauzerall, *Ann. N.Y. Acad. Sci.* **75**, 115 (1958).
177. R. J. Porra and J. E. Falk, *Biochem. J.* **90**, 69 (1964).
178. S. Granick and S. Sano, *Federation Proc.* **20**, 376 (1961).
179. D. Mauzerall, *J. Pediatrics* **64**, 5 (1964).
180. S. Sano, *J. Biol. Chem.* **241**, 5276 (1966).
181. R. J. Porra and J. E. Falk, *Biochem. Biophys. Res. Communs.* **5**, 179 (1961).
182. O. Hayaishi, *Bacteriol. Rev.* **30**, 720 (1966).
183. J. Lascelles, *Biochem. J.* **62**, 78 (1956).
184. J. Lascelles, *in* "The Bacteria" (I. C. Gunsalus and R. Y. Stanier, eds.). Vol. 3, p. 335. Academic Press, New York, 1962.
185. J. Lascelles, *J. Gen. Microbiol.* **15**, 404 (1956).
186. R. J. Porra and J. Lascelles, *Biochem. J.* **94**, 120 (1965).
187. S. Baum, Ph.D. dissertation, Cornell Univ., Ithaca, N.Y., 1965.
188. D. Mauzerall, *Biochemistry* **4**, 1801 (1965).
189. M. B. Lowe and J. N. Phillips, *Nature* **190**, 262 (1961).
190. T. Heikel, W. H. Lockwood, and C. Rimington, *Nature* **182**, 313 (1958).
191. R. Tokunaga and S. Sano, *Biochem. Biophys. Res. Commun.* **25**, 489 (1966).
192. J. Orlando, Ph.D. dissertation, Univ. of California, Berkeley, California, 1958.
193. R. F. Labbe, N. Hubbard, and W. S. Caughey, *Biochemistry* **2**, 372 (1963).
194. R. J. Porra and O. T. G. Jones, *Biochem. J.* **87**, 181 (1963).
195. R. J. Porra and O. T. G. Jones, *Biochem. J.* **87**, 186 (1963).
196. R. F. Labbe and N. Hubbard, *Biochim. Biophys. Acta* **41**, 185 (1960).
197. H. Oyama, Y. Sugita, Y. Yoneyama, H. Yoshikaya, *Biochim. Biophys. Acta* **47**, 413 (1961).
198. A. Goldberg, *Brit. J. Haematol.* **5**, 150 (1959).
199. R. J. Porra, Y. S. Vitols, R. F. Labbe, and N. A. Newton, *Biochem. J.* **104**, 321 (1967).
200. Y. Yoneyama, H. Ohyama, Y. Sugita, and H. Yoshikawa, *Biochim. Biophys. Acta* **62**, 261 (1962).
201. A. Johnson and O. T. G. Jones, *Biochim. Biophys. Acta* **93**, 171 (1964).
202. J. E. Gardner and J. Lascelles, *J. Gen. Microbiol.* **29**, 157 (1962).
203. E. Thofern, *Ergeb. Mikrobiol.* **34**, 213 (1961).
204. M. Kiese, H. Kurz, and E. Thofern, *Biochem. Z.* **330**, 541 (1958).

205. P. Sinclair, D. C. White, and J. Barrett, *Biochim. Biophys. Acta* **143**, 427 (1967).
206. A. Neuberger and G. H. Tait, *Biochem. J.* **90**, 607 (1964).
207. H. N. Little and M. I. Kelsey, *Federation Proc.* **23**, 223 (1964).
208. W. E. C. Wacker, G. H. Tait, and A. Neuberger, *Biochemistry* **4**, 940 (1965).
209. G. Nishida and R. F. Labbe, *Biochim. Biophys. Acta* **31**, 519 (1959).
210. Y. Sugita, Y. Yoneyama, and H. Ohyama, *J. Biochem.* (*Tokyo*) **51**, 450 (1962).
211. B. F. Burnham, Ph.D. dissertation, Univ. of California, Berkeley, California, 1960.
212. H. Zahner, E. Bachmann, R. Hutter, and J. Nuesch, *Pathol. Microbiol* **25**, 708 (1962).
213. B. F. Burnham, *J. Gen. Microbiol.* **32**, 117 (1963).
214. J. B. Neilands, *Science* **156**, 1443 (1967).
215. B. F. Burnham and J. B. Neilands, *J. Biol. Chem.* **236**, 554 (1961).
216. H. Bickel, E. Gaumann, W. Keller-Schierlein, V. Prelog, E. Vischer, A. Wettstein, and H. Zahner, *Experientia* **16**, 129 (1960).
217. H. Zahner, R. Hutter, and E. Bachmann, *Arch. Mikrobiol.* **36**, 325 (1960).
218. F. Knusel, J. Nuesch, and H. J. Triechler, *Naturwiss.* **54**, 242 (1967).
219. T. Emery and J. B. Neilands, *J. Am. Chem. Soc.* **83**, 1626 (1961).
220. H. Bickel, B. Fechtig, G. E. Hall, W. Keller-Schierlein, V. Prelog, and E. Vischer, *Helv. Chim. Acta* **43**, 901 (1960).
221. H. Bickel, H. Keberle, and E. Vischer, *Helv. Chim. Acta* **46**, 1385 (1963).
222. N. E. Morrison, A. D. Antoine, and E. E. Dewbrey, *J. Bacteriol.* **89**, 1630 (1965).
223. D. C. White and S. Granick, *J. Bacteriol.* **85**, 842 (1963).
224. J. Jenson, *J. Bacteriol.* **73**, 324 (1957).
225. J. Jenson, *Biochem. Biophys. Res. Commun.* **8**, 271 (1962).
226. S. Granick and H. Gilder, *J. Gen. Physiol.* **30**, 1 (1946).
227. E. L. Biberstein, P. D. Mini, and M. G. Gills, *J. Bacteriol.* **86**, 814 (1963).
228. B. F. Burnham, *Arch. Biochem. Biophys.* **97**, 329 (1962).
229. G. Anderegg, F. L'Epplattenier, and G. Schwartzenbach, *Helv. Chim. Acta* **46**, 1409 (1963).
230. J. Lascelles, *Biochem. J.* **100**, 175 (1966).
231. J. Lascelles, *Biochem. J.* **100**, 184 (1966).
232. K. D. Gibson, A. Neuberger, and G. H. Tait, *Biochem. J.* **88**, 325 (1963).
233. G. H. Tait and K. D. Gibson, *Biochim. Biophys. Acta* **52**, 614 (1961).
234. R. J. Radmer and L. Bogorad, *Plant Physiol.* **42**, 463 (1967).
235. R. Willstatter and L. Forsin, *Ann. Chem.* **396**, 180 (1913).
236. P. E. Wei, A. H. Corwin, and R. Arellano, *J. Org. Chem.* **27**, 3344 (1962).
237. S. J. Baum, B. F. Burnham, and R. A. Plane, *Proc. Natl. Acad. Sci. U.S.* **52**, 1439 (1964).
238. S. J. Baum and R. A. Plane, *J. Am. Chem. Soc.* **88**, 910 (1966).
239. S. Granick, *Plant Physiol.* **34**, XVIII (1959).
240. O. T. G. Jones, *Biochem. J.* **86**, 429 (1963).
241. R. Cooper, *Biochem. J.* **89**, 100 (1963).
242. H. V. Marsh, Jr., H. J. Evans, and G. Matrone, *Plant Physiol.* **38**, 632 (1963).
243. H. V. Marsh, Jr., H. J. Evans, and G. Matrone, *Plant Physiol.* **38**, 638 (1963).
244. E. F. Karali and C. A. Price, *Nature* **198**, 708 (1964).
245. W. Hsu and G. Miller, *Biochim. Biophys. Acta* **111**, 393 (1965).
246. E. F. Cavel and C. A. Price, *Plant Physiol.* **40**, 1 (1965).
247. K. D. Gibson, A. Neuberger, and G. H. Tait, *Biochem. J.* **83**, 550 (1962).
248. K. H. Shull, J. McConomy, M. Voget, A. Castillo, and E. Farber, *J. Biol. Chem.* **241**, 5060 (1966).
249. O. T. G. Jones, *Biochem. J.* **89**, 182 (1963).

250. O. T. G. Jones, *Biochem. J.* **88**, 325 (1963).
251. A. H. Corwin and P. E. Wei, *J. Org. Chem.* **27**, 4285 (1962).
252. M. Holden, *Biochem. J.* **78**, 359 (1961).
253. M. Gassman and L. Bogorad, *Plant Physiol.* **43**, 774 (1967).
254. P. Boger, *Phytochemistry* **4**, 435 (1965).
255. O. T. G. Jones, *Biochem. J.* **91**, 572 (1964).
256. J. H. C. Smith, *in* "Comparative Biochemistry of Photoreactive Systems" (M. B. Allen, ed.). Academic Press, New York, 1960.
257. N. K. Boardman, *in* "The Chlorophylls" (L. P. Vernon and G. R. Seely, eds.). Academic Press, New York, 1966.
258. S. Granick, *Ann. Rev. Plant Physiol.* **2**, 115 (1951).
259. V. M. Koski, C. S. French, and J. H. C. Smith, *Arch. Biochem. Biophys.* **31**, 1 (1951).
260. A. A. Krasnovsky and L. M. Kosobutskaya, *Dokl. Akad. Nauk SSSR* **82**, 761 (1952).
261. J. H. C. Smith, *Carnegie Inst. Wash. Year Book* **57**, 287 (1958).
262. N. K. Boardman, *Biochim. Biophys. Acta* **62**, 63 (1962).
263. P. W. Trown, *Biochemistry* **4**, 908 (1965).
264. D. W. Kupke, *J. Biol. Chem.* **237**, 3287 (1962).
265. J. B. Wolff and L. Price, *Arch. Biochem. Biophys.* **72**, 293 (1957).
266. H. I. Virgin, *Physiol. Plantarium* **13**, 155 (1960).
267. E. G. Sudyina, *Photochem. Photobiol.* **2**, 181 (1963).
268. S. Shimazo and E. Tamaki, *Arch. Biochem. Biophys.* **102**, 152 (1963).
269. M. Holden, *Photochem. Photobiol.* **2**, 175 (1963).
270. A. O. Klein and W. Vishniac, *J. Biol. Chem.* **236**, 2544 (1961).
271. M.B. Allen, *in* "The Chlorophylls" (L. P. Vernon and G. R. Seely, eds.). Academic Press, New York, 1966.
272. J. M. Anderson, U. Blass, and M. Calvin, *in* "Comparative Biochemistry of Photoreactive Systems" (M. B. Allen, ed.). Academic Press, New York, 1960.
273. D. W. Kupke and T. E. Dorrier, *Plant Physiol.* **37**, lxiii (1962).
274. A. A. Shlyk, V. L. Kaler, L. I. Vlasenok, and V. I. Gaponenko, *Photochem. Photobiol.* **2**, 129 (1963).
275. R. J. Della Rosea, K. I. Altman, and K. Salmon, *J. Biol. Chem.* **202**, 771 (1953).
276. W. Brzeski and W. Rucker, *Nature* **185**, 922 (1960).
277. J. H. C. Smith and C. S. French, *Ann. Rev. Plant Physiol.* **14**, 181 (1963).
278. W. R. Richards and H. Rapoport, *Biochemistry* **5**, 1079 (1966).
279. T. W. Goodwin, *in* "The Biosynthesis of Vitamins and Related Compounds," p. 172. Academic Press, New York, 1963.
280. S. Schwartz, K. Ikeda, I. M. Miller, and C. J. Watson, *Science* **129**, 40 (1959).
281. A. I. Krasna, C. Rosenblum, and D. B. Sprinson, *J. Bioi. Chem.* **225**, 745 (1957).
282. R. Bray and D. Shemin, *Biochim. Biophys. Acta* **30**, 647 (1958).
283. R. C. Bray and D. Shemin, *J. Biol. Chem.* **238**, 1501 (1963).
284. J. I. Toohey, *Proc. Natl. Acad. Sci. U.S.* **54**, 934 (1965).
285. J. Toohey, *Federation Proc.* **25**, 1628 (1966).
286. K. Bernhauer, F. Wagner, H. Beisbarth, P. Rietz, and H. Vogelmann, *Biochem. Z.* **344**, 289 (1966).
287. W. Friedrich and W. Sandeck, *Biochem. Z.* **340**, 465 (1964).
288. W. Friedrich, *Biochem. Z.* **342**, 143 (1965).
289. A. Di Marco and C. Spalla, *Giorn. Microbiol.* **9**, 237 (1961).
290. B. Bartoskinski, B. Zagalsk, and J. Pawelkiewicz, *Biochim. Biophys. Acta* **136**, 581 (1967).

291. A. Di Marco, M. P. Marnati, A. Migliacci, A. Rusconi, and C. Spalla, *in* "2nd European Symposium on Vitamin B_{12} and Intrinsic Factor" (H. C. Heinrich, ed.), p. 69, Enke Verlag, Stuttgart, Germany, 1961.
292. K. Bernhauer and F. Wagner, *Biochem. Z.* **335**, 325 (1962).
293. K. Bernhauer and F. Wagner, *Z. Physiol. Chem.* **322**, 184 (1960).
294. B. M. Guirard and E. E. Snell, *in* "Comprehensive Biochemistry" (M. Florkin and E. H. Stotz, eds.), Vol. 15, p. 138. Elsevier, Amsterdam, 1964.
295. J. M. Turner, *Biochem. J.* **99**, 427 (1966).
296. A. Neuberger and G. H. Tait, *Biochem. J.* **84**, 317 (1962).
297. M. L. Green and W. H. Elliott, *Biochem. J.* **92**, 537 (1964).
298. G. Urata and S. Granick, *J. Biol. Chem.* **238**, 811 (1963).
299. H. Weissbach, J. Toohey, and H. A. Barker, *Proc. Natl. Acad. Sci. U.S.* **45**, 521 (1959).
300. H. A. Barker, R. D. Smyth, H. Weissbach, J. I. Toohey, J. N. Ladd, and B. E. Volcani, *J. Biol. Chem.* **235**, 480 (1960).
301. G. Boretti, *Biochem. J.* **89**, 3p (1963).
302. P. Barbieri, G. Borretti, A. DiMarco, A. Migliacci, and C. Spalla, *Biochim. Biophys. Acta* **57**, 599 (1962).
303. A. G. Lezius and H. A. Barker, *Biochemistry* **4**, 510 (1965).
304. H. C. Friedman and D. L. Harris, *Biochem. Biophys. Res. Commun.* **8**, 164 (1962).
305. H. C. Friedman and D. L. Harris, *J. Biol. Chem.* **240**, 406 (1965).
306. H. C. Friedman, *J. Biol. Chem.* **240**, 413 (1965).
307. H. Dellweg, E. Becher, and K. Bernhauer, *Biochem. Z.* **327**, 422 (1956).
308. H. Dellweg and K. Bernhauer, *Arch. Biochem. Biophys.* **69**, 74 (1957).
309. K. Bernhauer, E. Becher, and G. Wilharm, *Arch. Biochem. Biophys.* **83**, 248 (1959).
310. P. Renz, *Angew. Chem. Intern. Ed. Engl.* **4**, 527 (1965).
311. H. A. Barker, H. Weissbach, and R. D. Smyth, *Proc. Natl. Acad. Sci. U.S.* **44**, 1093 (1958).
312. H. A. Barker, R. D. Smyth, H. Weissbach, A. Munch-Petersen, J. I. Toohey, J. N. Ladd, B. E. Volcani, and R. M. Wilson, *J. Biol. Chem.* **235**, 181 (1960).
313. J. I. Toohey, D. Perlman, and H. A. Barker, *J. Biol. Chem.* **236**, 2119 (1961).
314. R. O. Brady and H. A. Barker, *Biochem. Biophys. Res. Commun.* **4**, 464 (1961).
315. R. O. Brady, E. G. Castanera, and H. A. Barker, *J. Biol. Chem.* **237**, 2325 (1962).
316. A. Peterkofsky, B. Redfield, and H. Weissbach, *Biochem. Biophys. Res. Commun.* **5**, 213 (1961).
317. H. Weissbach, B. Redfield, and A. Peterkofsky, *J. Biol. Chem.* **236**, PC40 (1961).
318. H. Weissbach, B. G. Redfield, and A. Peterkofsky, *J. Biol. Chem.* **237**, 3217 (1962).
319. A. Peterkofsky and H. Weissbach, *J. Biol. Chem.* **238**, 1491 (1963).
320. E. L. Smith, L. Mervyn, A. W. Johnson, and N. Shaw, *Nature* **194**, 1175 (1962).
321. E. L. Smith and L. Mervyn, *Biochem. J.* **86**, 2p (1963).
322. A. W. Johnson, L. Mervyn, N. Shaw, and E. L. Smith, *J. Chem. Soc.*, p. 4146 (1963).
323. K. Bernhauer, O. Muller, and G. Muller, *Biochem. Z.* **336**, 102 (1962).
324. O. Muller and G. Muller, *Biochem. Z.* **336**, 299 (1962).
325. O. Muller and G. Muller, *Biochem. Z.* **337**, 179 (1963).
326. K. Bernhauer, O. Muller, and F. Wagner, *Advan. Enzymol.* **26**, 233 (1964).
327. K. Bernhauer and E. Irion, *Biochem. Z.* **339**, 530 (1964).
328. G. H. Beaven and E. A. Johnson, *Nature* **176**, 1264 (1955).
329. S. L. Tackett, J. W. Collatt, and J. C. Abbott, *Biochemistry* **2**, 919 (1963).
330. H. Weissbach, N. Brot, and W. Lovenberg, *J. Biol. Chem.* **241**, 317 (1966).
331. D. H. Dolphin and A. W. Johnson, *Proc. Chem. Soc.*, p. 311 (1963).

332. F. Wagner and K. Bernhauer, *Ann. N.Y. Acad. Sci.* **112**, 580 (1962).
333. D. Dolphin and A. W. Johnson, *Chem. Commun.* 494 (1965).
334. G. N. Schrauzer and R. J. Windgassen, *J. Am. Chem. Soc.* **88**, 3738 (1966).
335. D. C. Hodgkin, *Federation Proc.* **23**, 592 (1964).
336. H. Booth, A. W. Johnson, F. Johnson, and R. A. Langdale-Smith, *J. Chem. Soc.*, 650 (1963).
337. R. Bonnett, J. R. Cannon, A. W. Johnson, I. Sutherland, A. R. Todd, and E. L. Smith, *Nature* **176**, 328 (1955).
338. R. J. Porra and B. D. Ross, *Biochem. J.* **94**, 557 (1965).
339. B. F. Burnham, unpublished observations.
340. D. Perlman and J. I. Toohey, *Nature* **212**, 300 (1966).
341. S. E. Cauthen, J. R. Pattison, and J. Lascelles, *Biochem. J.* **102**, 774 (1967).
342. C. H. Gray, "The Bile Pigments." Methuen, London, 1953.
343. R. Lester and R. Schmid, *New Eng. J. Med.* **270**, 779 (1964).
344. I. A. D. Bouchier and B. H. Billing (eds.), "Bilirubin Metabolism." Blackwell Scientific Publications, Oxford, 1967.
345. C. H. Gray, A. Lichtarowicz-Kulczycka, D. C. Nicholson, and Z. Petryka, *J. Chem. Soc.*, p. 2264 (1961).
346. I. M. London, R. West, D. Shemin, and D. Rittenburg, *J. Biol. Chem.* **184**, 351 (1950).
347. C. H. Gray, A. Neuberger, and P. H. A. Sneath, *Biochem. J.* **47**, 87 (1950).
348. L. G. Israels, T. Yamamoto, J. Skanderberg, and A. Zipursky, *Science* **139**, 1054 (1963).
349. C. H. Gray, D. C. Nicholson, and R. A. Nicolaus, *Nature* **181**, 183 (1958).
350. R. Lemberg and J. W. Legge, "Hematin Compounds and Bile Pigments." Interscience, New York, 1949.
351. C. H. Gray and D. C. Nicholson, *J. Chem. Soc.*, p. 3085 (1958).
352. Z. Petryka, D. C. Nicholson, and C. H. Gray, *Nature* **194**, 1047 (1962).
353. B. H. Billing, P. G. Cole, and G. H. Lathe, *Biochem. J.* **65**, 774 (1957).
354. R. Schmid, *Science* **124**, 76 (1956).
355. E. Talafant, *Nature* **178**, 312 (1956).
356. B. H. Billing and G. H. Lathe, *Am. J. Med.* **24**, 111 (1958).
357. A. P. Weber, L. Schalm, and J. Witmans, *Acta Med. Scand.* **173**, 19 (1963).
358. C. H. Gregory, *J. Lab. Clin. Med.* **61**, 917 (1963).
359. K. J. Isselbacher and E. A. McCarthy, *J. Clin. Invest.* **38**, 645 (1959).
360. B. A. Noir, A. T. de Walz, and R. Groszman, *Biochem. Biophys. Acta* **117**, 297 (1966).
361. G. Klatskin and L. Bungards, *J. Clin. Invest.* **35**, 537 (1956).
362. J. D. Ostrow and R. Schmid, *J. Clin. Invest.* **42**, 1286 (1963).
363. A. S. Gilbertson, I. Bossenmaier, and R. Cardinal, *Nature* **196**, 141 (1962).
364. R. Lester and R. Schmid, *New Engl. J. Med.* **269**, 178 (1963).
365. H. van den Bergh and P. Mueller, *Biochem. Z.* **77**, 90 (1916).
366. D. L. Drabkin, *Proc. Soc. Exptl. Biol. Med.* **76**, 527 (1951).
367. V. E. Price, W. R. Sterling, V. A. Tarantola, R. W. Hartley, Jr., and M. Rechcigl, Jr., *J. Biol. Chem.* **237**, 3468 (1962).
368. R. T. Schimke, E. W. Sweeney, and C. M. Berlin, *J. Biol. Chem.* **240**, 322 (1965).
369. R. Schmid, H. S. Marver, and L. Hammaker, *Biochem. Biophys. Res. Commun.* **24**, 319 (1966).
370. R. Lemberg, *Biochem. J.* **29**, 1322 (1935).
371. R. Lemberg, J. W. Legge, and W. H. Lockwood, *Nature* **142**, 148 (1938).
372. J. W. Legge and R. Lemberg, *Biochem. J.* **35**, 353 (1941).

373. R. Lemberg, W. H. Lockwood, and J. W. Legge, *Biochem. J.* **35**, 363 (1941).
374. E. Y. Levin, *Biochemistry* **5**, 2845 (1966).
375. H. Nakajima, T. Takemura, O. Nakajima, and K. Yamaoka, *J. Biol. Chem.* **238**, 3784 (1963).
376. H. Nakajima, *J. Biol. Chem.* **238**, 3797 (1963).
377. R. F. Murphy, C. O'hEocha, and P. O'Carra, *Biochem. J.* **104**, 6c (1967).
378. O. Nakajima and C. H. Gray, *Biochem. J.* **104**, 20 (1967).
379. R. Lemberg and R. A. Wyndham, *Biochem. J.* **30**, 1147 (1936).
380. J. W. Singleton and L. Laster, *J. Biol. Chem.* **240**, 4780 (1965).
381. R. Schmid, *J. Biol. Chem.* **229**, 881 (1957).
382. J. Axelrod, R. Schmid, and L. Hammaker, *Nature* **180**, 1426 (1957).
383. A. K. Brown, W. W. Zuelzer, and H. H. Burnett, *J. Clin. Invest.* **37**, 332 (1958).
384. K. J. Isselbacher, M. F. Chrabas, and R. C. Quinn, *J. Biol. Chem.* **237**, 3033 (1962).
385. G. J. Dutton, D. E. Langelaan, and P. E. Ross, *Biochem. J.* **93**, 4p (1964).
386. I. D. E. Storey, *Biochem. J.* **95**, 201 (1965).
387. G. A. Tomlinson and S. J. Yaffe, *Biochem. J.* **99**, 507 (1966).
388. G. J. Dutton and I. D. E. Storey, *Biochem. J.* **57**, 275 (1954).
389. I. D. E. Storey, *Biochem. J.* **95**, 209 (1965).
390. C. H. Gray, "Bile Pigments in Health and Disease" (I. N. Kugelmass, ed.). Charles C Thomas, Springfield, Ill., 1961.
391. C. J. Watson, *J. Clin. Pathol.* **16**, 1 (1963).
392. C. J. Watson, M. Campbell, and P. I. Lowry, *Proc. Soc. Exptl. Biol. Med.* **98**, 707 (1958).
393. R. Lester and R. Schmid, *J. Clin. Invest.* **42**, 736 (1963).
394. C. J. Watson, P. Lowry, S. Collins, A. Graham, and N. R. Ziegler, *Trans. Assoc. Am. Physicians* **67**, 242 (1954).
395. C. J. Watson, A. Moscowitz, D. Lightner, W. C. Krueger, and M. Weimer, *J. Biol. Chem.* **241**, 5037 (1966).
396. C. J. Watson, M. Weimer, W. Krueger, D. A. Lightner and A. Moscowitz, *Federation Proc.* **34**, 520 (1965).
397. D. C. Nicholson, *in* "Bilirubin Metabolism" (I. A. D. Bouchier and B. H. Billing, eds.). Blackwell Scientific Publications, Oxford, 1967.
398. C. H. Gray, P. M. Jones, W. Klyne, and D. C. Nicholson, *Nature* **184**, 41 (1959).
399. A. Moscowitz, W. C. Krueger, I. T. Kay, G. Skewes, and S. Bruckenstein, *Proc. Natl. Acad. Sci. U.S.* **52**, 1190 (1964).
400. R. S. Cahn, *J. Chem. Educ.* **41**, 116 (1964).
401. S. Schwartz, *in* "Bilirubin Metabolism" (I. A. D. Bouchier and B. H. Billing, eds.). Blackwell Scientific Publications, Oxford, 1967.
402. M. Bessis, J. Breton-Gorius, and J. P. Thiery, *Compt. Rend.* **252**, 2300 (1961).
403. S. Berendoshn, J. Lowman, D. Sundberg, and C. J. Watson, *Blood* **24**, 1 (1964).
404. S. Schwartz, G. Ibrahim, and C. J. Watson, *J. Lab. Clin. Med.* **64**, 1003 (1964).
405. T. Yamamoto, J. Skanderberg, A. Zipursky, and L. G. Israels, *J. Clin. Invest.* **44**, 31 (1965).
406. L. G. Israels, T. Yamanoto, J. Skanderberg, and A. Zipursky, *in Proc. 9th Congr. European Soc. Haemotol., Lisbon,* **2**, 891 (1963).
407. L. G. Israels, J. Skanderberg, H. Guyda, W. Zingg, and A. Zipursky, *Brit. J. Haematol.* **9**, 50 (1963).
408. S. H. Robinson and R. Schmid, *Medicine,* **43**, 667 (1964).
409. S. H. Robinson, C. A. Owen, E. V. Flock, and R. Schmid, *Blood* **26**, 823 (1965).

410. S. H. Robinson, M. Tsong, B. W. Brown, and R. Schmid, *J. Clin. Invest.* **45**, 1569 (1966).
411. S. H. Robinson, *in* "Bilirubin Metabolism" (I. A. D. Bouchier and B. H. Billing, eds.). Blackwell Scientific Publications, Oxford, 1967.
412. N. I. Berlin, A. Neuberger, and J. J. Scott, *Biochem. J.* **64**, 90 (1956).
413. L. G. Israels, M. Levitt, W. Novak, J. Foerster, and A. Zipursky, *in* "Bilirubin Metabolism" (I. A. D. Bouchier and B. H. Billing, eds.). Blackwell Scientific Publications, Oxford, 1967.
414. S. Schwartz and R. Cardinal, *Federation Proc.* **24**, 485 (1965).
415. H. Theorell, M. Beznak, R. Bonnichsen, K. G. Paul, and A. Akeson, *Acta. Chem. Scand.* **5**, 445 (1951).
416. A. Akeson, G. V. Ehrenstein, G. Hevesy, and H. Theorell, *Arch. Biochem. Biophys.* **91**, 310 (1960).
417. M. J. Fletcher and D. R. Sanadi, *Biochim. Biophys. Acta* **51**, 356 (1961).
418. S. Varadi, *Brit. J. Haemotol.* **4**, 270 (1958).
419. L. E. Heilmeyer, R. Clotten, L. Kerp, H. Merker, C. A. Porra, and H. P. Wetzel, *Deut. Med. Wochschr.* **88**, 2449 (1963).
420. R. Schmid, S. Schwartz, and R. D. Sandberg, *Blood* **10**, 416 (1955).
421. L. Johnson, Ph.D. dissertation, Univ. of Minnesota, Minneapolis, Minn., 1966.
422. W. M. Wass and H. H. Hoyt, *Am. J. Vet. Res.* **26**, 659 (1965).
423. I. A. Magnus, A. Jarrett, T. A. Pranherd, and C. Rimington, *Lancet* **2**, 448 (1961).
424. E. S. Peterka, R. M. Fusaro, W. J. Runge, M. O. Jaffe, and C. J. Watson, *J. Am. Med. Assoc.* **193**, 1036 (1965).
425. A. G. Redeker and M. Berke, *Arch. Dermatol.* **86**, 569 (1962).
426. S. Porter, *Blood* **22**, 532 (1963).
427. C. H. Gray, A. Kulezcha, D. C. Nicholson, I. A. Mangus, and C. Rimington, *Clin. Sci.* **26**, 7 (1964).
428. A. G. Redeker and H. G. Bryan, *Lancet* **1**, 1449 (1964).
429. A. Goldberg and C. Rimington, "Diseases of Porphyrin Metabolism." Charles C Thomas, Springfield, Ill., 1962.
430. J. Waldenstrom and B. Haeger-Aronsen, *Brit. Med. J.* **2**, 272 (1963).
431. T. K. With, *Z. Klin. Chem.* **1**, 134 (1963).
432. J. Waldenstrom, *Acta Med. Scand. Suppl.* 82 (1937).
433. L. Eales, *Ann. Rev. Med.* **12**, 251 (1961).
434. B. Ackner, J. E. Cooper, C. H. Gray, M. Kelly, and D. C. Nicholson, *Lancet* **1**, 1256 (1961).
435. J. Waldenstrom and B. Vahlquist, *Z. Physiol. Chem.* **260**, 189 (1939).
436. C. J. Watson, S. Schwartz, V. Hawkinson, *J. Biol. Chem.* **157**, 345 (1945).
437. R. Schmid, S. Schwartz, and C. J. Watson, *A.M.A. Arch. Internal Med.* **93**, 167 (1954).
438. S. G. Smith, *Arch. Pathol.* **70**, 361 (1960).
439. K. Nakao, O. Wada, T. Kelamura, M. Ueno, and G. Urata, *Nature* **210**, 838 (1966).
440. F. H. Welland, E. S. Hellman, E. M. Gaddis, A. Collins, G. W. Hunter, and D. P. Tschudy, *Metabolism* **13**, 232 (1964).
441. J. A. Rose, E. S. Hellman, and D. P. Tschudy, *Metabolism* **10**, 514 (1961).
442. D. P. Tschudy, F. H. Welland, A. Collins, and G. Hunter, Jr., *Metabolism* **13**, 396 (1964).
443. C. J. Watson, *Advan. Internal Med.* **6**, 235 (1954).
444. J. Waldenstrom, *Am. J. Med.* **22**, 758 (1957).
445. Proceedings of the International Conference on the Porphyrias, *S. African J. Lab. Clin. Med.* **9**, 143 (1963).

446. C. J. Watson, W. Runge, and I. Bossenmaier, *Metabolism* **11**, 1129 (1962).
447. L. Wetterberg, *Lancet* **2**, 1178 (1964).
448. C. Rimington and F. DeMatteis, *Lancet* **1**, 270 (1965).
449. G. Dean and H. D. Barnes, *S. African Med. J.* **33**, 246 (1959).
450. L. Eales, *S. African J. Lab. Clin. Med.* **9**, 151 (1963).
451. G. Dean, "The Porphyrias." Pitman Med. Publ. Co., London, 1963.
452. I. A. Magnus, *S. African J. Lab. Clin. Med.* **9**, 238 (1963).
453. L. Eales, *S. African J. Lab. Clin. Med.* **6**, 63 (1960).
454. L. Eales and G. C. Linder, *S. African Med. J.* **36**, 284 (1962).
455. H. D. Barnes, *S. African Med. J.* **32**, 680 (1958).
456. H. D. Barnes, *S. African Med. J.* **29**, 781 (1955).
457. H. D. Barnes, *S. African Med. J.* **33**, 274 (1959).
458. N. M. Lamont, M. Hathorn, and S. M. Joubert, *Quart. J. Med.* N.S. **30**, 373 (1961).
459. T. H. Bothwell, H. Seftel, P. Jacobs, J. D. Torrances, and N. Baumslag, *Am. J. Clin. Nutr.* **14**, 47 (1964).
460. S. J. Saunders, *S. African J. Lab. Clin. Med.* **9**, 277 (1963).
461. J. H. Epstein, and A. G. Redeker, *Arch. Dermatol.* **92**, 286 (1965).
462. C. Albahary, *Presse. Med.* **73**, 73 (1965).
463. A. G. Redeker, R. E. Slerting, and R. S. Bronow, *J. Am. Med. Assoc.* **188**, 466 (1964).
464. R. Schmid and S. Schwartz, *Proc. Soc. Exptl. Biol. Med.* **81**, 685 (1952).
465. S. Schwartz, personal communication.
466. S. Granick and A. Kappas, *Proc. Natl. Acad. Sci. U.S.* **57**, 1463 (1967).
467. G. S. Marks, E. G. Hunter, U. K. Terner, and D. Schneck, *Biochem. Pharmacol.* **14**, 1077 (1965).
468. G. H. Hirsch, J. D. Gillis, and G. S. Marks, *Biochem. Pharmacol.* **15**, 1006 (1966).
469. L. J. Reed and D. J. Cox, *Ann. Rev. Biochem.* **35**, 57 (1966).
470. D. E. Atkinson, *Ann. Rev. Biochem.* **35**, 85 (1966).
471. H. E. Umbarger, *Cold Spring Harbor Symp. Quant. Biol.* **26**, 301 (1961).
472. E. R. Stadtman, *Bacteriol. Rev.* **27**, 170 (1963).
473. H. S. Moyed and H. E. Umbarger, *Physiol. Rev.* **42**, 444 (1962).
474. J. C. Gerhart and A. B. Pardee, *J. Biol. Chem.* **237**, 891 (1962).
475. E. R. Stadtman, G. N. Cohen, G. LeBras, and H. deRobichon-Szulmajster, *J. Biol. Chem.* **236**, 2033 (1961).
476. P. C. Newell and R. G. Tucker, *Biochem. J.* **100**, 512 (1966).
477. P. C. Newell and R. G. Tucker, *Biochem. J.* **100**, 517 (1966).
478. F. Jacob and J. Monod, *J. Mol. Biol.* **3**, 318 (1961).
479. M. Gruber and R. N. Campagne, *Perspectives Biol. Med.* 125 (1966).
480. A. L. Cline and R. M. Bock, *Cold Spring Harbor Symp. Quant. Biol.* **31**, 321 (1966).
481. S. W. Englander and L. A. Page, *Biochem. Biophys. Res. Commun.* **19**, 565 (1965).
482. F. T. Kenney and W. L. Aebritton, *Proc. Natl. Acad. Sci. U.S.* **54**, 1693 (1965).
483. T. G. Lessie and W. R. Sistrom, *Biochim. Biophys. Acta* **86**, 250 (1964).
484. K. Goto, M. Higuchi, H. Sakai, and G. Kikuchi, *J. Biochem. Tokyo* **61**, 186 (1967).
485. R. C. Gallo, *J. Clin. Invest.* **46**, 124 (1967).
486. P. Schaeffer, *Biochim. Biophys. Acta* **9**, 261 (1952).
487. E. Englesburg, J. B. Levy, and A. Gibor, *J. Bacteriol.* **68**, 178 (1954).
488. R. K. Clayton, *J. Biol. Chem.* **235**, 405 (1960).
489. H. M. Lenhoff, D. J. D. Nicholas, and N. O. Kaplan, *J. Biol. Chem.* **220**, 983 (1956).
490. J. P. Chang, D.Phil. Thesis, Oxford Univ., 1963.
491. J. P. Chang and J. Lascelles, *Biochem. J.* **89**, 503 (1963).

492. J. W. T. Wimpenny, M. Ranlett, and C. T. Gray, *Biochim. Biophys. Acta* **73**, 170 (1963).
493. L. Mervyn and E. L. Smith, *Progr. Indust. Microbiol.* **5**, 153 (1964).
494. J. E. Falk and R. J. Porra, *Biochem. J.* **90**, 66 (1964).
495. C. L. Hammel and S. P. Bessman, *Arch. Biochem. Biophys.* **110**, 622 (1965).
496. P. P. Slonimski, "La Formation des Enzymes Respiratoires chez la Levure." Masson, Paris, 1953.
497. D. R. Biggs and A. W. Linnane, *Biochim. Biophys. Acta* **78**, 785 (1963).
498. A. W. Linnane, *in* "Oxidases and Related Redox Systems" (T. E. King, H. S. Mason, and M. Morrison, eds.), Vol. 2, p. 1102. John Wiley and Sons, New York, 1964.
499. G. Cohen-Bazire, W. R. Sistrom, and R. Y. Stanier, *J. Cell. Comp. Physiol.* **49**, 25 (1957).
500. J. Lascelles, *Biochem. J.* **72**, 508 (1959).
501. W. R. Sistrom, *J. Bacteriol.* **89**, 403 (1965).
502. R. E. Hurlbert, *J. Bacteriol.* **93**, 1346 (1967).
503. G. Cohen-Bazire and W. R. Sistrom, *in* "The Chlorophylls" (L. P. Vernon and G. R. Seely, eds.), p. 313. Academic Press, New York, 1966.
504. A. W. Linnane, E. Vitols, and P. G. Nowland, *J. Cell. Biol.* **13**, 345 (1962).
505. P. G. Wallace and A. W. Linnane, *Nature* **201**, 1191 (1964).
506. P. P. Slonimski, *Proc. Intern. Congr. Biochem., 3rd, Brussels, 1955*, 242 (1956).
507. P. Slonimski, R. Acher, G. Péré, A. Sels, and M. Somlo, "Dans Intern. Symp. Mechanisms of Regulation of Cellular Activities in Micro-organisms," Marseille. Gordan and Breach, New York, 1963.
508. A. A. Sels, H. Fukuhara, G. Péré, and P. P. Slonimski, *Biochim. Biophys. Acta* **95**, 486 (1965).
509. H. Fukuhara and A. Sels, *J. Mol. Biol.* **17**, 319 (1966).
510. H. Fukuhara, *J. Mol. Biol.* **17**, 334 (1966).
511. M. Somlo and H. Fukuhara, *Biochem. Biophys. Res. Commun.* **19**, 587 (1965).
512. A. M. Pappenheimer, Jr., *J. Biol. Chem.* **167**, 251 (1947).
513. S. Granick, *Ann. N.Y. Acad. Sci.* **123**, 188 (1965).
514. M. L. Cowger, R. F. Labbe, and B. Mackler, *Arch. Biochem. Biophys.* **96**, 583 (1962).
515. M. L. Cowger, R. F. Labbe, and M. Sewell, *Arch. Biochem. Biophys.* **101**, 96 (1963).
516. M. L. Cowger and R. F. Labbe, *Biochem. Pharmacol.* **16**, 2189 (1967).
517. R. F. Labbe, *Lancet* **1**, 1361 (1967).
518. A. Palma-Carlos, L. Palma-Carlos, M. Gajdos-Torok, and A. Gajdos, *Nature* **211**, 977 (1966).
519. A. Gajdos, M. Gajdos-Torok, A. Palma-Carlos, and L. Palma-Carlos, *Nature* **211**, 974 (1966).
520. L. H. von Euler, R. J. Rubin, and R. E. Handschumacher, *J. Biol. Chem.* **238**, 2464 (1963).
521. E. Farber, K. H. Shull, S. Villa-Trevino, B. Lombardi, and M. Thomas, *Nature* **203**, 34 (1964).
522. B. Haeger-Aronsen, *Scand. J. Clin. Lab. Invest.* **12** (*Suppl.* 47), 45 (1960)
523. M. Kreimer-Birnbaum and M. Grinstein, *Biochim. Biophys. Acta* **111**, 110 (1965).
524. J. R. Davis and S. L. Andelman, *Arch. Environ. Health* **15**, 53 (1967).
525. M. L. Cowger, R. F. Labbe, and M. Sewell, *Arch. Biochem. Biophys.* **101**, 96 (1963).
526. F. DeMatteis and E. D. Gray, *Biochem. J.* **94**, 1c (1965).
527. F. I. Lottsfeldt and R. F. Labbe, *Proc. Soc. Exptl. Biol. Med.* **119**, 226 (1965).
528. R. F. Labbe, Y. Hanawa, and F. I. Lottsfeldt, *Arch. Biochem. Biophys.* **92**, 373 (1961).
529. J. Onisawa and R. F. Labbe, *J. Biol. Chem.* **238**, 724 (1963).

530. A. Gajdos, M. Gajdos-Torok, A. Palma-Carlos, and L. Palma-Carlos, *Nature* **213**, 1022 (1967).
531. F. DeMatteis, T. F. Slater, and D. Y. Wang, *Biochem. Biophys. Acta* **68**, 100 (1963).
532. R. Schmid. J. F. Figen, and S. Schwartz, *J. Biol. Chem.* **217**, 263 (1955).
533. R. F. Labbe, T. Kurumada, and J. Onisawa, *Biochim. Biophys. Acta* **111**, 403 (1965).
534. D. P. Tschudy, J. Rose, E. Hellman, A. Collins, and M. Rechcigl, Jr., *Federation Proc.* **21**, 400 (1962).
535. H. S. Marver, D. P. Tschudy, M. G. Perlroth, and A. Collins, *Science* **154**, 501 (1966).
536. P. Feigelson, M. Feigelson, and O. Greengard, *Recent Progr. Hormone Res.* **18**, 491 (1962).
537. K. Koike and S. Okni, *J. Biochem. (Tokyo)* **56**, 308 (1964).
538. K. Tokuyama and W. E. Knox, *Biochim. Biophys. Acta* **81**, 201 (1964).
539. O. Greengard, M. A. Smith, and G. Acs, *J. Biol. Chem.* **238**, 1548 (1963).
540. R. T. Schimke, E. W. Sweeney, and C. M. Berlin, *J. Biol. Chem.* **240**, 4609 (1965).
541. S. Orrenius and L. Ernster, *Biochem. Biophys. Res. Commun.* **16**, 60 (1964).
542. T. Omura and R. Sato, *J. Biol. Chem.* **239**, 2379 (1964).
543. A. H. Conney and J. J. Burns, *Advan. Enzyme Regulation* **1**, 189 (1963).
544. A. H. Conney, K. Schneidman, M. Jacobson, and R. Kuntzman, *Ann. N.Y. Acad. Sci.* **123**, 98 (1965).
545. H. Remmer and H. J. Merker, *Ann. N.Y. Acad. Sci.* **123**, 79 (1965).
546. H. S. Marver, A. Collins, and D. P. Tschudy, *Biochem. J.* **99**, 31c (1966).
547. R. Banerjee, *Biochem. Biophys. Res. Commun.* **8**, 114 (1962).
548. A. Rossi-Fanelli and E. Antonini, *J. Bioi. Chem.* **235**, PC4 (1960).
549. H. F. Bunn and J. H. Jandl, *Proc. Natl. Acad. Sci. U.S.* **56**, 974 (1966).
550. J. Monod, J. P. Changeux, and F. Jacob, *J. Mol. Biol.* **6**, 306 (1963).
551. J. Monod, J. Wyman, and J. P. Changeux, *J. Mol. Biol.* **12**, 88 (1965).
552. D. Karibian and I. M. London, *Biochem. Biophys. Res. Commun.* **18**, 243 (1965).
553. L. Taddeini, personal communication, 1966.
554. F. DeMatteis, *Biochim. Biophys. Acta* **82**, 641 (1964).
555. A. D. Waxman, A. Collins, and D. P. Tschudy, *Biochem. Biophys. Res, Commun.* **24**, 675 (1966).
556. J. Onisawa and R. F. Labbe, *Biochim. Biophys. Acta* **56**, 618 (1962).
557. B. R. A. O'Brien, *J. Embryol. Exptl. Morphol.* **9**, 202 (1961).
558. F. Wilt, *Proc. Natl. Acad. Sci. U.S.* **48**, 1582 (1962).
559. F. Wilt, *Science* **147**, 1588 (1965).
560. F. H. Wilt, *J. Mol. Biol.* **12**, 331 (1965).
561. A. Saha, *Biochim. Biophys. Acta* **93**, 573 (1964).
562. A. Saha, R. Dutta, and J. Gosh, *Science* **125**, 447 (1957).
563. R. D. Levere and S. Granick, *Proc. Natl. Acad. Sci. U.S.* **54**, 134 (1965).
564. R. D. Levere and S. Granick, *J. Biol. Chem.* **242**, 1903 (1967).
565. S. D. Wainwright and L. K. Wainwright, *Can. J. Biochem.* **44**, 1543 (1966).
566. K. Hashimoto and F. H. Wilt, *Proc. Natl. Acad. Sci. U.S.* **56**, 1477 (1966).
567. V. D'Amelio, *Biochim. Biophys. Acta* **127**, 59 (1966).
568. J. A. Simons, *J. Exptl. Zool.* **162**, 219 (1966).
569. C. L. Hammel and S. P. Bessman, *J. Biol. Chem.* **239**, 2228 (1964).
570. G. P. Bruns and I. M. London, *Biochem. Biophys. Res. Commun.* **18**, 236 (1965).
571. A. I. Grayzel, P. Horchner, and I. M. London, *Proc. Natl. Acad. Sci. U.S.* **55**, 650 (1966).
572. T. J. Gribble and H. C. Schwartz, *Biochim. Biophys. Acta* **103**, 333 (1965).

573. C. Baglioni and B. Colombo, *Cold Spring Harbor Symp. Quant. Biol.* **29**, 347 (1964).
574. B. Colombo and C. Baglioni, *J. Mol. Biol.* **16**, 51 (1966).
575. J. R. Shaeffer, P. K. Trostle, and R. F. Evans, *Science* **158**, 488 (1967).
576. J. R. Shaeffer, *Biochem. Biophys. Res. Commun.* **28**, 647 (1967).
577. E. Bucci and C. Fronticelli, *J. Biol. Chem.* **240**, PC551 (1965).
578. R. M. Winslow and V. M. Ingram, *J. Biol. Chem.* **241**, 1144 (1966).
579. G. A. Vanderhoff, M. K. Williams, A. I. Grayzel, and I. M. London, *Federation Proc.* **26**, 673 (1967).
580. V. M. Ingram and A. D. W. Stretton, *Nature* **184**, 1903 (1959).
581. C. Baglioni, *in* "Molecular Genetics" (J. H. Taylor, ed.), Part I, p. 451. Academic Press, New York, 1963.
582. S. Granick and R. D. Levere, *Progr. Hematol.* **4**, 1 (1964).
583. K. H. Winterhalter and E. R. Helehns, *J. Biol. Chem.* **239**, 3699 (1964).
584. H. A. Barker, *Biochem. J.* **105**, 1 (1967).
585. Biochemistry Society Symposium on B_{12} Coenzymes, *Federation Proc.* **25**, 1623 (1966).
586. Vitamin B_{12} Coenzymes, *Ann. N.Y. Acad. Sci.* **112**, 550 (1964).
587. H. A. Barker, F. Suzuki, A. A. Iodice, and V. Rooze, *Ann. N.Y. Acad. Sci.* **112**, 644 (1964).
588. H. Eggerer, E. R. Stadtman, P. Overath, and F. Lynen, *Biochem. Z.* **333**, 1 (1960).
589. J. Retey, A. Umani-Ronchi, J. Seibl, and D. Arigoni, *Experientia* **22**, 502 (1966).
590. A. M. Brownstein and R. H. Abeles, *J. Biol. Chem.* **236**, 1199 (1961).
591. P. Overath, G. M. Kellerman, F. Lynen, H. P. Fritz, and H. J. Keller, *Biochem. Z.* **335**, 500 (1962).
592. A. A. Iodice and H. A. Barker, *J. Biol. Chem.* **238**, 2094 (1963).
593. R. H. Abeles and B. Zagalak, *J. Biol. Chem.* **241**, 1245 (1966).
594. P. A. Frey and R. H. Abeles, *J. Biol. Chem.* **241**, 2732 (1966).
595. R. H. Abeles and P. A. Frey, *Federation Proc.* **25**, 1639 (1966)
596. L. L. Ingraham, *Ann. N.Y. Acad. Sci.* **112**, 713 (1964).
597. J. Retey and D. Arigoni, *Experientia* **22**, 783 (1966).

Numbers in parentheses are reference numbers and are included to assist in locating references in which author's names are not mentioned in the text. Numbers in italic refer to pages on which the references are listed.

(93), 257(93), *309*, 334(91), 341(121), 348(164), 349(91, 164), *368, 369, 370,* 492(475), *534*

Cohen, P. P., 2(1), 7(1, 25, 27, 28, 29), 8(25), 9(25, 80, 83, 84), 28(1, 203), 29(203), 30(1), 31(203, 220, 221), 32(220), 33 (203, 220, 221, 244), 34(203, 220), 35 (203, 220, 221), 40(1), 51(2), 59(1), 60(1), 62(2, 468, 469, 472, 473, 476, 481, 483, 484, 485, 486, 488, 489, 490, 491, 492), 63(473, 495, 496, 497, 499, 500, 502), 64(499, 500, 502), 65(499), 66(468, 496, 499, 513), 67(203, 220, 476, 488, 489, 491), 68(468, 490, 514), 69 (514, 520, 521), 70(468, 481, 534, 540, 541), 72(2, 469, 553, 555), 75(221, 484, 491, 606, 608, 639), *79, 80, 81, 84, 85, 90, 91, 92, 93, 94,* 207(90), *231,* 303 (292), *315,* 327(69), *368*

Cohen-Bazire, G., 495(499), 496(503), 498 (499, 503), *535*

Cole, P. G., 472(353), 475(353), *531*

Cole, S. W., 153(307), *186*

Coleman, D. L., 427(134), 428, 429(134), 506(134), *526*

Coletta, M., 390(46), *399*

Collatt, J. W., 464(329), *530*

Collins, A., 425(108, 115), 426(106, 107, 107a, 108), 487(107, 107a, 108, 115, 440, 442), 494(108, 115), 502(115, 534), 503(535), 504(442, 546), 511(106, 108, 115), 512(115, 555), 519(535), 520(535), *525, 526, 533, 536*

Collins, S., 476(394), *532*

Colman, R. F., 31(227, 228), 35(267), 36 (267), 77(620, 621, 622), *85, 86, 94,* 394(64), *399*

Colombo, B., 516, 517, *537*

Connelly, J. L., 195(20a, 20b), *229*

Conney, A. H., 503(543, 544), 504(544), *536*

Connors, W. M., 153(305), *186*

Contractor, S. F., 175(448), *190*

Contsera, J. F., 299(271), *314*

Cooksey, K. E., 226(243, 244), *235*

Cookson, G. H., 427, 429(129), 432(129), *526*

Coon, M. J., 9(51), 10(51), *81,* 192(4, 6), 194, 195, 196(9, 10), 197(6, 8, 28), 199 (9, 37, 38, 46, 47), 200(37, 38, 46, 48), 202(9, 65, 66, 67), 203(9, 65), *229, 230*

Cooper, J. E., 487(434), *533*

Cooper, J. R., 145, 174(429), *185, 189,* 296(254), *314*

Cooper, R., 448(241), 449(241), *528*

Copenher, J. H., Jr., 28(198), *84*

Corcoran, B. J., 300(283), *314*

Corcoran, J. W., 422(78, 79), 457(78, 79), *525*

Cordes, E. H., 16(134), 22(134), *82*

Corman, L., 35(270), 77(626, 627), *86, 94*

Cornford, P., 435(165, 170), *527*

Coronado, A., 323, *367*

Corrigan, J. J., 61(463), *90*

Corrivaux, D., 255, *309*

Corsey, M. E., 358(239), *372*

Corwin, A. H., 420(48), 431(143), 432, 434, 448(236), 450, *524, 526, 528, 529*

Cossins, E. A., 112(99), *181*

Costilow, R. N., 206(89), 200(191), *231, 233*

Cotton, R. G. H., 280(187), 287(187), 288 (187), 289(187), *312*

Coursaget, J., 45(323), *87*

Coval, M. L., 300(287), 301(287), *314*

Cowger, M. L., 502(514, 515, 516, 525), 511(516), *535*

Cowgill, R. W., 244(27), 245(27), *308*

Cox, D. J., 491(469), *534*

Craig, J. M., 195(15), *229*

Crandall, D. I., 151, 152(304), *186*

Crane, F. L., 199(44), *230*

Crawford, I. P., 54(424), 55(424), 57(424), *89,* 290, 293(226, 229, 230, 234, 244), 294(226, 230, 234), 295(234, 244), *313*

Crawford, L. V., 100(24), *179*

Crawhall, J. C., 377(9), 382(29), 394(68), *398, 399, 400*

Creighton, T. E., 293(237, 238), 294(237), 295(238), *313*

Crespi, H. L., 420(46), *524*

Creveling, C. R., 299(273, 274), *314*

Cronin, J. R., 253(74), *309*

Crosbie, G. W., 113(101), *181*

Cross, D. C., 34(252), *85*

Cross, D. G., 29(209), 33(209), 35(265, 266), *84, 86*

Cruickshank, D. H., 9(81), *81*

Crumpler, H. C., 200(53), *230*

Curran, J., 30(217, 218), 33(217, 218), *84*

Curran, J. F., 31(226), 34(256), 35(256), *85*

Curti, B., 48(348, 349), *87*

A

Acetoacetate
  formation, 153
  oxidation by tricarboxylic acid cycle, 153
Acetohydroxy acid synthetase (s)
  (2-hydroxy-3-oxoacid lyase) 354, 355, 356
    357
  biosynthetic, properties of, 357
  cofactor for, 355
  feedback inhibiting by isoleucine, 366(T)
  functions of, 356, 357
  occurrence, 356, 357
  properties, differences in, 357
  reactions of, 355, 357
α-Acetohydroxybutyrate, formation, 354,
    355
α-Acetolactate
  synthesis of, 354
    from pyruvate, 355
N-Acetylglutamate
  biosynthesis of, 327–328
    enzymes catalyzing, 325, 326(T), 328
    proline biosynthesis from 318–319
    pathway for, 319
  role in liver carbamyl phosphate synthe-
    tase, 63, 64–66
N-Acetylglutamate γ-semialdehyde
  conversion to N^a-acetylornithine, 329–330
N-Acetylglutamate   γ-semialdehyde   dehydrogenase
  reaction of, 325
  substrate of, 326(T)
N-Acetylglutamate-γ-synthetase   (Acetyl-
    CoA: L-glutamate N-acetyltransfer-
    ase EC 2.3.1.1)
  feedback inhibition by arginine, 328
  reaction of, 325
  substrate of, 326(T)
N-Acetyl-γ-glutamokinase (ATP: N-acetyl-
    glutamate 5-phosphotransferase)
  activity, requirements for, 328
  as key regulatory enzyme in Chlamydo-
    monas, 328
  inhibitors of, 328, 329

reaction of, 325, 328
role in ornithine and 'arginine biosynthesis, 329
separation from N-acetylglutamate γ-
  semialdehyde dehydrogenase in E.
  coli extracts, 328
N-Acetyl-γ-glutamyl phosphate, 328–329
O-Acetyl-L-homoserine
  synthesis of cystathione in Neurospora
    and yeast from, 260, 261, 262
Acetylornithinase
  activators of, 331
  kinetic data, 331
  occurrence, 331
    as guide to taxonomy, 331
  reaction of, 325, 329
  relative specific activity in microorgan-
    isms, 332(T)
  substrate of, 326(T)
N^a-Acetylornithine
  conversion to ornithine, 330–331
  enzymes catalyzing, 330, 331
Acetylornithine δ-transaminase (N^a-acetyl
    L-ornithine: 2-oxoglutarate amino-
    transferase, EC 2.6.1.11)
  as site of action of growth inhibitors of
    wild-type S. typhimurium, 330
  binding of pyridoxal-P by, 330
  inhibitors of, 330
  isolation and purification, 329, 330
  kinetic data, 329
  reaction of, 325, 329
cis-Aconitase [citrate (isocitrate) hydro-
    lyase, EC 4. 2.1.3], 343
Actinomycin D
  effect on hemoglobin synthesis, 514
Actinomycin peptide, 4-hydroxyproline and
    4-ketoproline in, 324
"Active pyruvate" (α-hydroxy α-carboxy-
    ethyl-thiamine pyrophosphate)
  formation, 356
Acyl-CoA   dehydrogenase   [acyl-CoA:
    (acceptor) oxidoreductase, EC 1.3.99.
    3]
  reaction of, 199

Catechol
  formation of, 169, 179
  oxidation of, 170–173
Central nervous system, γ-aminobutyrate
    enzyme system in, 102–103
Cephalin, ethanolamine in, 248
Cephalopods, D-glutamic acid oxidase in
    liver of, 52
Chelating agents, effect on glutamate de-
    hydrogenase activity, 30–31
Chemicals, see also Drugs
  porphyrogenic, 490, 500–505,
  see also individual compounds
    common characteristics, 501–502
    effect on liver, 503
      mechanism of, 503
      on regulation of porphyrin synthesis,
        500–502
    induction of ALA synthetase by, 508–
      512
    inhibitors of, 508–512
    inhibition of oxidative metabolism by,
      502
    metabolic changes caused by, 502–503,
      504
    response to, factors affecting, 504–505
    sterical hindrance with respect to hydro-
      lysis in, 502, 503
Chickens, hemoglobins of, 512ff
Chlamydomonas reinhardti, ornithine bio-
    synthesis in, enzyme catalyzing, 331
Chlorella mutants, accumulation of tetra-
    pyrrole compounds in, 421–422
  hematoporphyrin synthesis by, 422, 456
Chlorins
  absorption spectra of, 413
  structure, 413
Chlorobium chlorophylls, 456
  absorption spectra, phylotype of, 417
Chlorobium thiosulfatophilium-660, por-
    phyrin biosynthesis in, 456
Chlorocruoroporphyrin, rhodo-type
    absorption spectrum of, 416
Chlorophyll, 404
  biological function, 407
  common biosynthetic pathways for heme
    and, 405–406, 420
    through protoporphyrin IX, 422, 447
  degradation of, 408
  distribution of, 455

historical aspects, 404, 405
  synthesis, chemical, phlorins as inter-
    mediates in, 413
Chlorophyll a
  biosynthesis of, 405, 447–454
    intermediates in, 420
    light requirement for, variations in, 452,
      498
    from magnesium protoporphyrin
      monomethyl ester, 450–454
    pathway of, 451, 454
      difficulties in establishment of,
        450–451
    pathway for, 451–454
    regulation of, 492
    relationship to protein synthesis and
      organelle development, 451
  structure of, 409
Chlorophyll b
  as possible precursor of chlorophyll a,
    455
  biosynthesis of, 454–455
  distribution, 455
  structural difference between chlorophyll a
    and, 454
Chlorophyllase, 450, 453–454
  bacteriochlorophyll as substrate of, 456
  reaction of, 454
  specificity of, 454
Chlorophyllide, formation of, 452, 498
Choliglobin, formation, 473
Choline
  demethylation of, 250–252
    enzymes catalyzing, 251–252
    factors regulating, 252
    pathway of metabolic, 251
  formation from serine, 248
    enzyme catalyzing, 248, 249, 250
    mechanism of, 249, 248
  permeability of mitochondria to, 252
Choline dehydrogenase (EC 1.1.99.1), 251–
    252
  association with mitochondrial electron
    transport system, 251
  composition, 251
  properties of, 251
Chorismate mutase P, 287
  association with phenylalanine synthesis,
    287
  prephenate dehydratase as part of, 287

in rat, effect of age, 239–240
thiol pyruvate-cleaving, 383–390
tritium-labeled, reactions of, 522
Epinephrine compounds
   biosynthesis of, 296–300
      enzymes catalyzing, 297, 298–300
      pathways of, 300
Erythrocytes
   detoxification of H_2O_2 by GSH in,
      393
   metabolism of, 471
Erythropoiesis
   rate of, differences in, 482
   stimulation of, 482
Escherichia coli
   aspartate metabolism in, 348
   biosynthesis of aromatic acids in, genetic
      regulation, 289
      of cystathione in, control of, 261
      of homoserine and threonine in,
         254
      of lysine in, 336
      of proline in, 318
      of putrescine in, 207–209
         pathways for, 207, 208–209
      of O-succinyl-L-homoserine in, 261
   enzymes of serine biosynthetic pathways
      in, 240–241, 242
   genes of arginine synthesis in, 331,
      332
      enzymes corresponding to, 331, 332
   isolation of D-phosphoglycerate dehydro-
      genase from, 243
   methionine methyl formation in, path-
      ways of, 264–265
   ornithine transcarbamylase of, 207
      physiological function, 207
      reaction of, 206
   L-threonine dehydrases of, 56–57
   tryptophanase of, 176–178
Ethionine
   effect on hepatic ATP level, 450
      on glucose 6-phosphate dehydrogenase,
         396
      on RNA and lipid metabolism, 396
   inhibition of bacteriochlorophyll synthe-
      sis by, 449–450
      mechanism of, 449
   toxicity, mechanism of, 396
5'-Ethylthioinosine, excretion in rat, 396

F

Factor B, 266
   methyl donor for, 266
Ferrichrome
   as iron-donating coenzyme for ferrochela-
      tase, evidence for, 446
Ferriprotoporphyrin IX chloride, *see*
   Hemin
Ferriprotoporphyrin IX hydroxide, *see*
   Hematin
Ferrochelatase, 440ff
   assay of, 442–443
   distribution, 440
   cofactor requirement, 443
   relative rates of metal incorporation into
      porphyrins by, 441
   reaction of, 440
   sensitivity to air, 440
   specificity, 441
Flavin adenine dinucleotide (FAD)
   as coenzyme of snake venom amino acid
      oxidase, 39
   as prosthetic group of mammalian D-
      amino acid oxidase, 47
   binding to apoenzyme, 49
Flavin mononucleotide as prosthetic group
   of amino acid oxidases, 44–45, 46
4-Fluorophenylalanine, conversion to tyrosine, 147
Folic acid, tyrosine metabolism in vitamin C
   deficiency and, 150
Formiminoglutamic acid, 133
   conversion to glutamic acid, 136
   formation of, 135
      enzyme catalyzing, 136
      histidine as precursor of, 133
Formiminoglutamate formiminotransferase
   purification of, 136
   reaction of, 137
Formylisoglutamine
   formation, nonenzymic, 137
      half-life of, 137
      mechanism of, 137, 138
Formylkynurenine, conversion to kynurenine, 160
N-Formyl methionine, initiation of protein
   synthesis by, 395
Fumarate
   formation of, 153